Ecological Engineering

Ecological Engineering

Edited by **Jeffery Clarke**

SYRAWOOD
PUBLISHING HOUSE
New York

Published by Syrawood Publishing House,
750 Third Avenue, 9th Floor,
New York, NY 10017, USA
www.syrawoodpublishinghouse.com

Ecological Engineering
Edited by Jeffery Clarke

International Standard Book Number: 978-1-68286-299-5 (Hardback)

Printed in China.

Contents

Preface

Ecological engineering is an emerging field focused on the design and monitoring of ecosystems through the principles of engineering. Its potential applications can be found in landscape architecture, urban horticulture, etc. This all inclusive book encapsulates some of the most significant topics in this field such as evaluation of water quality, waste management, environmental landscaping, sustainable development, etc. It brings together case studies of renowned experts from across the globe to help students in understanding the present status and future prospects of this discipline.

The information shared in this book is based on empirical researches made by veterans in this field of study. The elaborative information provided in this book will help the readers further their scope of knowledge leading to advancements in this field.

Finally, I would like to thank my fellow researchers who gave constructive feedback and my family members who supported me at every step of my research.

Editor

1

AN EVALUATION OF WATER QUALITY FROM MOJEN RIVER, BY NSFWQI INDEX

Allahbakhh Javid[1], Kamyar Yaghmaeian[2], Ebrahim Abbasi[1], Aliakbar Roudbari[3]

[1] School of Public Health, Shahroud University of Medical Sciences, Shahroud, Iran.

[2] School of Public Health, Tehran University of Medical Sciences, Tehran, Iran.

[3] Department of Environmental Health Engineering, School of Public Health, Shahroud University of Medical Sciences, Shahroud, Iran, e-mail: Roudbari@shmu.ac.ir

ABSTRACT

Water quality index (WQI) is valuable and unique rating to depict the overall water quality status in a single term that is helpful for the selection of appropriate treatment technique to meet the concerned issues. The aim of the study was to evaluating water quality from Mojen River by Water Quality Index based on National Science Foundation (NSFWQI). For this purpose, samples were collected from stations at up, middle and downstream of Mojen River in Semnan province (the biggest river in region) in a 2 years interval of 2013-2014 years. Nine parameters namely Turbidity, Biochemical Oxygen Demand, Dissolved Oxygen, Fecal Coliform, nitrate, pH, temperature, total solids and total phosphate were considered to compute the index. Our findings highlighted the deterioration of water quality in the river due to industrialization and human activities. According to NSFWQI, the best condition was recorded in the Darkhaniab (Upstream) and the worst condition concerned the Pole (Midstream).

Keywords: NSFWQI, Mojen, River, Pollution, Water Quality.

INTRODUCTION

Water, a prime natural resource and precious national asset, forms the chief constituent of ecosystem. Water sources may be mainly in a form of rivers, lakes, glaciers, rain water or ground water. The availability and quality of water either surface or ground, have been deteriorated due to some important factors like increasing population, industrialization and urbanization [1]. The first step for keeping river water quality and purification of polluted parts is obtaining information on the qualitative changes of river water in dimensions of time and place and also, determination of major sources and various water pollutants [2].

Water quality of any specific area or specific source can be assessed using physical, chemical and biological parameters. The values of these parameters are harmful for human health if they exceed the defined limits [3, 4, and 5]. Therefore, the suitability of water sources for human consumption has been described in terms of Water quality index (WQI), which is one of the most effective ways to describe the quality of water. WQI utilizes the water quality data and helps in the modification of the policies, which are formulated by various environmental monitoring agencies. It has been realized that the use of individual water quality variable in order to describe the water quality for common public is not easily understandable [6]. That's why; WQI has the capability to reduce the bulk of the information into a single value to express the data in a simplified and logical form [1]. Also, application of WQI specifies the process of variations and qualitative trends of water resources and also allows the classification of water quality (Brian O., Calculating NSF Water Quality Index, Wilkes University Center for Environmental Quality Geo Environmental Sciences and Engineering Department, Water Quality Index [7], National Sanitation Foundation

Water Quality Index (NSFWQI) [8], Canadian Council of Ministers of the Environment Water Quality Index (CCME WQI) [9], Oregon Water Quality Index (OWQI) [10] and Weighted Arithmetic Water Quality Index Method [11] are some of the important water quality indices used in water quality assessment.

Among various indexes which are applicable for water quality zoning, NSFWQI was selected because of its high precision, simplicity and availability of the required parameters [12]. It summarizes data in a single index value in an objective, rapid and reproducible manner [13].

According to previous studies, Mirmoshtaghi in 2011, studied the water quality of Sefidrood River by investigation of 20 samples at 5 sampling stations according to NSFWQI index and compared the results with OWQI index. The results showed that maximum and minimum values of NSF were 57 and 32, respectively. And the average value of NSFWQI along with Sefidrood River was obtained equals to 47.5, which is placed at bad region. Also, calculation of OWQI index showed the very bad quality of Sefidrood River during the study period [12].

In this study, classification of Mojen River's water quality in Semnan province (North east of Iran) is investigated by NSFWQI index.

MATERIALS AND METHODS

Study Area

Mojen River flows through a vast stretch with numerous perennial and non-perennial streams in the study area and forms a network of large and small rivers before joining the Gorgan Bay. The river is mainly used for fishing, propagation of wild life and irrigation. The Mojen at this stretch is classified as "Class B" as per the best designated use of water quality. Eighteen water samples from Mojen River were analyzed to assess the water quality of Mojen during the three seasons. The sampling location description of Mojen River are given in Table 1 and shown in Figure 1.

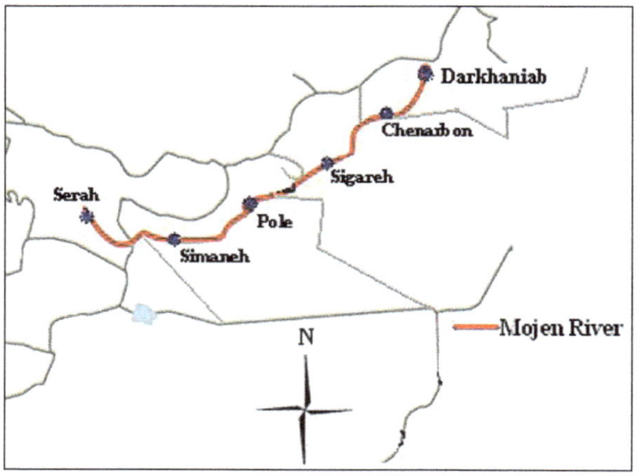

Figure 1. Studied area and sampling points

Sampling Procedure and Sample analysis

Samples were collected seasonally from stations at upstream (2 points), middle stream (2 points) and downstream (2 points) of in a 2 years interval of 2013–2014. The samples were analyzed using standard procedure [14, 15]. The pH of water samples was measured in the field. Samples were subjected to filtration prior to chemical analysis. The determination of TDS was done by a gravimetric process. The Winkler's method was followed for the analysis of DO and BOD. Nitrate was determined by colorimetric procedure. Fecal coliform population was analyzed by MPN/100ml method, by growing on M-FC medium at temperature 44.5° ± 1°C and counted after 48 hrs. Phosphate was determined by Ascorbic Acid Method. Turbidity was determined by Lenntech turbidimeter (LT 550).

Table 1. Sampling locations – Mojen River

Sampling No.	Location	Code	Description
1	Drakhaniab	M_1	The sampling location is near the bridge. The water is extensively used for irrigation.
2	Chenarbon	M_2	Sampling point is near the Chenarbon village. The main activity observed in the river is fishing.
3	Sigareh	M_3	Sampling point is at Sigareh village. The water is used for domestic purpose.
4	Pole	M_4	The sampling station is near the factory. The water is extensively used for irrigation and for factory process.
5	Simaneh	M_5	The sampling location is near the factory. The water is used for domestic usage like washing and cleaning.
6	Serah	M_6	The sampling location is at Serah village. The main activity observed in the river is irrigation and cleaning.

National Sanitation Foundation Water Quality Index (NSF WQI)

After measuring 9 of the above mentioned factors, each sub-index is obtained according to the conversion curves (appendix). The following equation (1) is applied for calculation of final index:

$$NSFWQI = \sum_{i=1}^{p} W_i I_i \qquad (1)$$

where: I_i – the sub-index for i^{th} water quality parameters obtained from conversion curves according to Table 2,
W_i – the weight (in terms of importance) associated with i^{th} water quality parameter,
p – the number of water quality parameters [16].

NSFWQI index is a reduction index, namely it is decreases with increasing water pollution. This index has a value between 0 to 100 and is classified according to Table 3.

Table 2. Weight factor of NSFWQI

Parameters	Weighting factor
Turbidity	0.08
BOD	0.11
DO	0.17
Fecal Coliform	0.16
Nitrate	0.1
pH	0.11
Temperature	0.1
TS	0.07
Total Phosphate	0.1

Table 3. Water quality classification according to NSFWQI

Water quality	Index
Excellent	91–100
Good	71–90
Medium	51–70
Unsuitable	26–50
Very unsuitable	0–25

RESULTS AND DISCUSSION

Water quality of Mojen River during different seasons

The water quality of Mojen River during different seasons are presented in Table 4 and highlighted as follow:

- **Winter Season:** pH of water samples varied between 7.03 at Darkhaniab to 7.21 at Pole. Total Solids (TS) varied between 1490 mg/l at Sigareh to 1430 mg/l at Chenarbon. Dissolved Oxygen varied between 6 mg/l to 6.4 mg/l. Likewise BOD values varied between 4.2 mg/l at Darkhaniab to 4.7 mg/l in Serah. Nitrate varied between 16.2 mg/l at Serah to 18.9 mg/l at Sigareh.
- **Spring Season:** pH of water samples varied between 7.14 at Sigareh to 7.3 at Chenarbon. Total Solids (TS) varied between 1340 mg/l at Serah to 1360 mg/l at Darkhaniab. Dissolved Oxygen varied between 6.1 mg/l to 6.3 mg/l. BOD values varied between 4.3 mg/l to 4.4 mg/l. Nitrate varied between 15.8 mg/l at Chenarbon to 17.2 mg/l at Sigareh.
- **Autumn Season:** pH of water samples varied between 7.01 at Drakhaniab to 7.31 at Serah. Total Solids (TS) varied between 1290 mg/l at Drakhaniab to 1320 mg/l at Pole. Dissolved Oxygen varied between 6.1 mg/l to 6.4 mg/l. BOD values varied between 4.2 mg/l to 4.6 mg/l. Nitrate varied between 15.2 mg/l at Drakhaniab to 17.8 mg/l at Sigareh.

Low concentration of BOD was recorded during winter due to a huge volume of fresh water that diluted the organic matter resulting in the decrease in the BOD values.

Value of National Sanitation Foundation Water Quality Index (NSF WQI)

NSF WQI of Mojen River is depicted in Table 5. The classification criteria standards based on NSF WQI are given in Table 6. The following observations were made based on the National Sanitation Foundation Water Quality Index of Mojen River at up, mid and downstream.

The obtained values of NSFWQI at stations (Table 5) shows that the water quality at all stations are Medium.

Figure 2 shows that downstream of Mojen River in Pole Sampling point had a much worse situation than other points, due to discharge of numerous factories swages in to the river. However, Mojen River condition is relatively more appropriate than other rivers due to less villages and residential area around the river. Among all the sampling points, the water quality at Drakhaniab (Upstream) is the less polluted as the NSF WQI is the highest.

Table 4. Water quality during different seasons – Mojen River

Location	Season	Turbidity [NTU]	BOD [mg/l]	DO [mg/l]	Fecal Coliform [MPN/100nl]	Nitrate [mg/l]	pH	Temperature [°C]	TS [mg/l]	Total Phosphate [mg/l]
Drakhaniab (Upstream)	Winter	5	4.2	6.4	2380	16.8	7.1	17.3	1450	0.22
	Spring	5	4.4	6.2	2350	15.9	7.21	18.9	1360	0.22
	Autumn	6	4.6	6.1	2340	15.1	7.01	19.6	1290	0.21
Chenarbon (Upstream)	Winter	5	4.4	6.2	2350	16.3	7.03	17.2	1430	0.22
	Spring	6	4.6	6.1	2380	15.8	7.3	19.3	1350	0.21
	Autumn	6	4.3	6.3	2350	15.7	7.11	19.5	1300	0.21
Sigareh (Midstream)	Winter	5	4.4	6.2	2360	18.9	7.12	17.2	1490	0.2
	Spring	5	4.4	6.2	2360	17.2	7.14	19.1	1340	0.2
	Autumn	5	4.2	6.4	2370	17.8	7.16	19.2	1305	0.2
Pole (Midstream)	Winter	6	4.3	6.3	2390	16.8	7.21	17.1	1470	0.21
	Spring	5	4.3	6.3	2390	16.9	7.19	19	1360	0.21
	Autumn	4	4.4	6.2	2360	17.2	7.31	19.2	1320	0.21
Simaneh (Downstream)	Winter	5	4.6	6.1	2350	17.5	7.09	17.3	1460	0.22
	Spring	6	4.4	6.2	2340	17.2	7.24	19.1	1350	0.22
	Autumn	4	4.6	6.1	2350	17.6	7.14	19.3	1307	0.22
Serah (Downstream)	Winter	5	4.7	6	2370	16.2	7.11	17.1	1440	0.23
	Spring	4	4.4	6.2	2380	16.5	7.22	19.2	1340	0.21
	Autumn	4	4.6	6.1	2370	17.5	7.31	19.3	1300	0.21

Table 5. NSFWQI values in Sampling points-Mojen River

Station No.	Sampling Point	Turbidity [NTU]	BOD [mg/l]	DO [mg/l]	Fecal Coliform [MPN/100nl]	Nitrate [mg/l]	pH	Temperature [°C]	TS [mg/l]	Total Phosphate [mg/l]	NSFWQI	Description	Class
1	Drakhaniab	5.3	4.4	6.23	2356	15.93	7.11	18.6	1366	0.22	62	Medium	C
2	Chenarbon	5.6	4.4	6.23	2360	15.93	7.15	18.67	1360	0.21	61	Medium	C
3	Sigareh	5	4.3	6.23	2363	17.97	7.14	18.5	1378	0.2	60	Medium	C
4	Pole	5	4.3	6.23	2380	16.97	7.24	18.43	1383	0.21	52	Medium	C
5	Simaneh	5	4.5	6.13	2346	17.41	7.15	18.57	1372	0.22	57	Medium	C
6	Serah	4.3	4.5	6.1	2373	16.73	7.22	18.53	1360	0.21	56	Medium	C

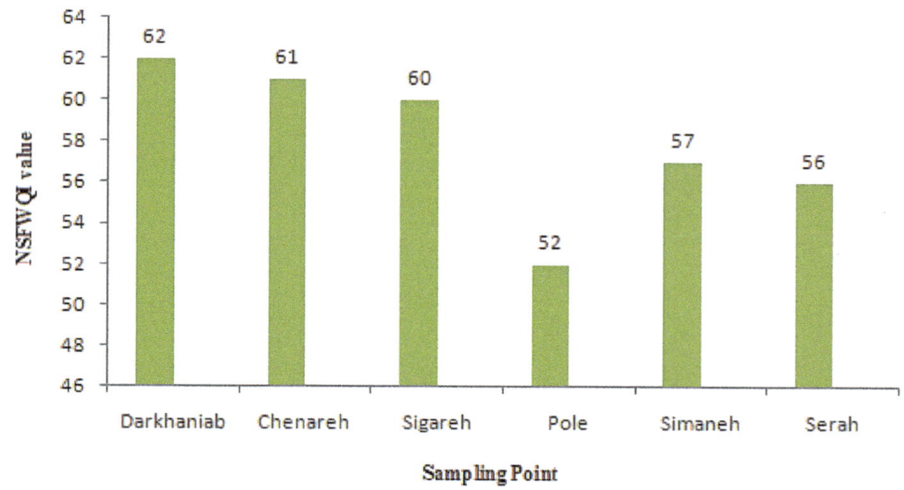

Figure 2. NSFWQI values in different stations- Mojen River

Table 6. Classification criteria standards based on NSF-WQI

NSFWQI	Descriptor	Category
91–100	Excellent	A
71–90	Good	B
51–70	Medium	C
26–50	Bad	D
0–25	Very Bad	E

Damo and Icka in 2012 studied the water quality of drinking water in the city of pogradec, Albania and the average value of NSFWQI was obtained equals to 87.81, which is placed at good region. The good quality can be attributed to the measured turbidity that exceeds the objective and to its large excursion. It reflects the intervention between natural effects and those of anthropogenic activities [17].

Mukherjee and Dora in 2012 studied water quality of Damodar River in Jharkhand and the average value of NSFWQI was obtained equals to 35, which is placed at Unsuitable region. The study shows that the river water is not potable at maximum sampling stations and are highly fecal contaminated. According to the desirable limit, maximum locations are also not even suitable for bathing [18].

CONCLUSION

We have studied the National Sanitation Foundation Water Quality Index of various sampling points of Mojen River i.e. Darkhaniab, Chenarbon, Sigareh, Pole, Simaneh and Serah in different seasons were located on the level of Medium conditions. According to NSFWQI, the best condition was related to the Darkhaniab station and the worst condition was related to the Pole station due to discharge of numerous factories swages in to the river. The obtained results showed that qualitative condition of Mojen River is medium and inappropriate management measures such as population load and excess urban activity in the basin of this river, industrial activities, excessive consumption of chemical fertilizers and pesticides, discharge of rural, urban and industrial wastewater and also solid wastes into the river which have a continuous increasing trend are the main source of river pollution.

So, human factor is the main cause of river pollution. Besides human factors, natural factors such as low rainfall, water consumption for agricultural and industrial purposes, development of agricultural lands at the expense of natural lands wastefulness and finally, all increased the physical and chemical pollution of the river and leads to natural disruption of its biological and bioavailability capacity.

Acknowledgements

The authors would like to thank theresearch affair of Shahroud University of Medical Sciences for their warm support and also to members of Central Laboratory of Shahroud University of Medical Sciences for their technical support.

REFERENCES

1. Babaei Semiromi F, Hassani A.H., Torabian A., Karbassi A.R., Hosseinzadeh Lotfi F. 2011. Water quality index development using fuzzy logic: A case study of the Karoon river of Iran. African Journal of Biotechnology 10(50), 1012–1013.
2. Banjaka D., Nikolic J. 2012. Hydrochemical characteristics and water quality of the Musnica River catchment, Bosnia and Herzegovina. Hydrological Sciences Journal 57(3), 562–575.
3. Das K.K., Panigrahi T., Panda R.B. 2012. Evaluation of water quality index (WQI) of drinking water of Balasore district, Odisha, India. Discovery life 1(3), 48–52.
4. Jena V., Dixit S., Gupta S. 2013. Assessment of water quality index of industrial area surface water samples. International Journal of ChemTech Research 5(1), 278–283.
5. Saberi Nasr A., Rezaei M., Dashti Barmaki M. 2013. Groundwater contamination analysis using Fuzzy Water Quality index (FWQI): Yazd province, Iran. Journal of Geope 3(1), 47–55.
6. Abdulwahid S.J. 2013 Water quality index of Delizhiyan Springs and Shawrawa River within Soran District, Erbil, Kurdistan Region of Iraq. Journal of Applied Environmental Biological Sciences 3(1), 40–48.
7. Jagadeeswari P.B., Ramesh K. 2012. Water quality index for assessment of water quality in south Chennai coastal aquifer, Tamil Nadu, India. International Journal of ChemTech Research 4(4), 1582–1588.
8. Chowdhury R.M., Muntasir S.Y., Hossain M.M. 2012. Study on ground water quality and its suitability or drinking purpose in Alathur block – Perambalur district. Archives of Applied Science Research 4(3), 1332–1338.

9. Lumb A., Halliwell D., Sharma T. 2006. Application of CCME water quality index to monitor water quality: a case of the Mackenzie River basin, Canada. Environmental Monitoring and Assessment 113, 411–429.
10. Lumb A., Halliwell D., Sharma T. 2002. Canadian water quality index to monitor the changes in water quality in the Mackenzie river–Great Bear. Proceedings of the 29th Annual Aquatic Toxicity Workshop, (Oct. 21–23), Whistler, B.C, Canada.
11. Rao C.S., Rao B.S., Hariharan A.V., Bharathi N.M. 2010. Determination of water quality index of some areas in Guntur district Andhra Pradesh. International Journal of Applied Biology and Pharmaceutical Technology 1(1), 79–86.
12. MirMoshtaghi M. 2012. Qualitative investigation of Sefid-rood river water and its zoning according to NSFWQI and OWQI. Journal of Wetlands 9, 25–32.
13. Giordani G., Zaldívar J.M., Viaroli P. 2009. Simple tools for assessing water quality and trophic status in transitional water ecosystems. Ecological Indicators 9(5), 982–991.
14. Terrado M., Barcelo D., Tauler R., Borrell E., Campos S.D. 2010. Surface-water-quality indices for the analysis of data generated by automated sampling networks. Trends in Analytical Chemistry 29(1), 40–52.
15. APHA, 2005. Standard Methods for the Examination of Water and Wastewater.
16. Ramakrishna C.R., Sadashivaiah C., Ranganna G. 2009. Assessment of water quality index for the ground water in Tamkur Taluk, Karnataka state, India. E-Journal of Chemistry 6, 523–530.
17. Damo R., Icka P. 2012. Evaluation of water quality index for drinking water. Polish Journal of Environmental Studies 22(4), 1045–1051.
18. Diptangshu Mukherjee S., Lata D., Tiwary R.K. 2012. Evaluation of Water Quality Index for Drinking Purposes in The Case of Damodar River. Bioremediation & Biodegradation 3(9), 56–61.

2

SOIL MULCHING WITH STRAW IN BROCCOLI CULTIVATION FOR EARLY HARVEST

Edyta Kosterna[1]

[1] Department of Vegetables Crop, Siedlce University of Natural Sciences and Humanities, B. Prusa 14, 08-110 Siedlce, Poland, e-mail: edyta.kosterna@uph.edu.pl

ABSTRACT

All treatments which protect soil from degradation and use of plant protection methods, other than chemicals are of great importance in the cultivation. This effect is attributed, among others, to organic mulches. By limiting the growth of weeds, maintaining proper moisture and reducing daily temperature fluctuations, mulch improves soil conditions for plant growth and development. The experiment was carried out between 2010 and 2012 at the Experimental Farm in Zawady as a split-block design with three replicates. The effect of the kind of straw (rye, corn, rape, buckwheat) and its dose (10 and 20 $t \cdot ha^{-1}$) applied as a mulch on the yield and quality of broccoli 'Milady F_1' cultivated for early harvest was investigated. The effect of straw was compared to a control plot without mulch. Weather conditions in the successive years of the study had a significant influence on the yield and quality of broccoli. The highest yields with the best parameters were obtained in 2010 and 2012, which were characterized by sufficient rainfall for broccoli. It was not found significant differences in the yield level and weight of head between particular kinds of straw, however, all kinds of straw investigated in the experiment, irrespective of dose, contributed to a significant increase in the yield and favourably influenced the biometric features of broccoli compared to that achieved from cultivation without straw. Soil mulching with corn straw was most favourable to yield and its parameters.

Keywords: yield, quality, broccoli, kind of straw, dose of straw.

INTRODUCTION

Soil mulching is one of the elements of ecological cultivation of vegetables. The effect of mulching on the soil properties and plant yield depends on the climatic and soil conditions, agrotechny and the kind of mulch. This treatment gives better results in less favourable soil conditions, in areas with a high abundance of weeds and at lower soil nutrients [Iwuafor, Kang 1994, Dobromilska et al. 1995, Zibilske, Makus 2009].

By limiting the growth of weeds, maintaining proper moisture and reducing the daily temperature fluctuations, mulch improves soil conditions for plant growth and development, resulting in a positive effect on vegetable yield [Swaider et al. 1992, Schonbeck, Evanylo 1998, Feldman et al. 2000, Ghosh et al. 2006, Dahiya et al. 2007, Nakhone, Tabatabai 2008, Saeed, Ahmad 2009]. According to Gill et al. [1996], the increase in yield due to mulching is higher for species grown for early harvest.

In the study by Jabłońska-Ceglarek et al. [2006] and Zaniewicz-Bajkowska et al. [2009], the application of rye straw as a mulch in cabbage and onion cultivation influenced the increase of total and marketable yield of these vegetables as compared to cultivation without mulch. An increased yield as the result of organic mulch application was found by Grassbaugh et al. [2004] as well as Rahman et al. [2006] (tomato), Law et al. [2006] (paprika), Jamil et al. [2005] (garlic), Adetunji [1990] (lettuce), Kar and Kumar [2007] (potato) and Johnson et al. [2004] (melon). In the study by Döring et al. [2005], mulch with straw applied to organically grown potatoes had no significant influence on the yield or tuber size fractions.

This study aimed to determine the effect of mulch with different kinds of straw on the yield and quality of broccoli.

MATERIALS AND METHODS

The experiment was carried out between 2010 and 2012 at the Experimental Farm in Zawady, which is located in central-eastern Poland (52°03'N, 22°33'E), 115 km east of Warsaw. According to the international system of FAO classification, the soil was classified as a Luvisol (LV) [WRB FAO 1998]. The soil organic matter content averaged 1.5% and its humus horizon reached a depth of 30–40 cm, the value of pH determined in H_2O was 5.4. The content of plant-available nutrients was lower than that specified by Sady [2000] for optimal content for broccoli (Table 1). In the autumn, soil liming at the rate of 2 t CaO·ha^{-1} (calcium carbonate fertilizer) was performed.

The experiment was established as a split-block design with three replicates. The effect of this kind of straw (rye, corn, rape, buckwheat) and its dose (10 and 20 t·ha^{-1}) applied as a mulch on the yield and the quality of broccoli 'Milady F_1' cultivated for early harvest was investigated. The effect of straw was compared to a control plot without mulch. The area of one plot for harvest was 12 m^2.

The forecrop for broccoli was triticale. In the autumn preceding broccoli cultivation, ploughing was performed. At the same time, farmyard manure at a rate of 30 t·ha^{-1} was incorporated. In the spring, two weeks before seedlings were planted, disc harrowing was applied to loosen the upper soil layer and prepare it for planting. After that, mineral fertilizers were applied in the amount of supplementary content to the optimal level for broccoli: 110 kg N, 98 kg P_2O_5, 220 kg K_2O per 1 ha. Mineral fertilizers were applied in the form of ammonium nitrate, triple superphosphate and 60% potassium salt. Directly before planting broccoli seedlings particular kind of straw in appropriate doses was application.

Broccoli seedlings were grown in a non-heated greenhouse. Seeds (10 g) were sown in the successive study years on the 19th, 18th and 20th of March in multi-trays (54 holes). The seedlings were produced using peat substrate. Before planting, seedlings were moved outdoors. Plants were planted on the 19th, 18th and 23th of April, at a spacing of 50×50 cm. Three weeks after planting, 50 kg N·ha^{-1} in the form of ammonium nitrate was applied (topdress).

Broccoli was harvested by hand on 30 June 2010, and the 28 June 2011 and 2012. During the harvest, the following were determined: the marketable yield (t·ha^{-1}); average weight of marketable head (kg); weight of individual heads per plant – edible part of plant (kg); length of broccoli arc (cm); and stalk diameter (cm).

The results of the experiment were statistically analysed by means of the analysis of variance following the mathematical model for the split-block design. Significance of differences was determined by the Tukey test at the significance level of $p = 0.05$.

Weather conditions in the study years varied (Table 2). Years 2010 and 2012 were characterized by a similar temperature during the growing period of the plant and favourable rainfall distribution for growing and development of broccoli. After seedlings were planted, there was abundant rainfall in the first and second 10-days of May. Also, two first 10-days of June were wet enough, what favoured broccoli heads formation and increased the yield. The least favourable conditions for broccoli

Table 1. Characteristic of soil conditions before experiment placing (available food components contents)

Years	pH	C–org.%	$N-NO_3$	$N-NH_4$	P	K	Ca	Mg
			mg·dm^{-3} air dry mass					
2010	5.7	1.7	8	7	32	70	220	24
2011	5.4	1.6	22	9	44	108	340	47
2012	5.0	1.3	13	6	27	70	220	37
Mean	5.4	1.5	14	7	34	83	260	36
Optimum content [Sady 2000]	6.2–7.0	–	105–120		50–60	160–190	1000–1500	45–55

Table 2. Mean air temperature and precipitation sums in the vegetation period of broccoli

Years	Decade	Temperature (°C)			Precipitation (mm)		
		April	May	June	April	May	June
2010	I	7.8	12.7	18.6	5.9	30.3	12.5
	II	9.7	14.8	16.7	2.4	41.2	47.3
	III	9.2	14.6	16.9	2.4	21.7	2.8
Mean air temperature and precipitation sum		8.9	14.0	17.4	10.7	93.2	62.6
2011	I	8.6	8.5	20.5	9.8	17.9	8.7
	II	7.9	14.9	16.7	18.5	5.2	18.8
	III	13.7	17.0	17.1	2.7	13.0	11.6
Mean air temperature and precipitation sum		10.1	13.4	18.1	31.0	36.1	39.1
2012	I	3.0	15.1	13.9	4.6	17.3	26.4
	II	8.9	12.2	17.6	21.1	33.0	37.7
	III	14.9	16.4	17.5	4.2	3.1	12.1
Mean air temperature and precipitation sum		8.9	14.6	16.3	29.9	53.4	76.2
Mean long-term air temperature and precipitation sum		7.2	13.2	16.2	29.4	54.3	69.3

growth were in 2011, which was characterized by quite high mean air temperatures compared with the other study years but very irregular rainfall distribution. A dry period during the last 10 days of April (2.7 mm rainfall) contributed to poor seedling growth. Not much rain fell during the growth and development period of plants (second and third 10-days of May) and in the period of head formation (first and second 10-days of June).

RESULTS AND DISCUSSION

Weather conditions in the successive study years had a significant influence on the marketable yield, average weight of marketable head and weight of individual heads per plant (Table 3–5). The highest, marketable yield (18.78 t·ha^{-1}) and average weight of head (0.55 kg) was obtained in 2010, the most favourable for broccoli cultivation. In the least favourable year, 2011, the marketable yield was lower on

Table 3. Marketable yield of broccoli (t·ha^{-1})

Years	Dose of straw (t·ha^{-1})	Kind of straw				Mean
		rye	corn	rape	buckwheat	
2010	no straw	15.18	19.89	15.67	15.87	16.65
	10	22.45	23.86	18.33	20.70	21.34
	20	16.30	20.30	17.55	19.28	18.36
	Mean	17.98	21.35	17.18	18.62	18.78
2011	no straw	9.29	9.55	8.96	10.02	9.46
	10	14.35	14.60	14.22	14.33	14.37
	20	18.22	16.78	14.65	15.81	16.37
	Mean	13.95	13.64	12.61	13.39	13.40
2012	no straw	9.47	9.31	9.81	9.56	9.54
	10	19.45	21.34	23.23	22.45	21.62
	20	17.91	22.98	15.74	17.60	18.56
	Mean	15.61	17.87	16.26	16.54	16.57
Mean	no straw	11.31	12.92	11.48	11.82	11.88
	10	18.75	19.93	18.59	19.16	19.11
	20	17.48	20.02	15.98	17.56	17.76
Mean for kind of straw		15.85	17.62	15.35	16.18	16.25

LSD0,05 for: years = 4.13; kind of straw = n.s.; straw dose = 1.87; years × kind of straw = n.s.; years × straw dose = 3.24; kind of straw × straw dose = 1.19.

average by 5.38 t·ha^{-1} and average weight of head by 0.16 kg (Table 3-4). A significantly higher weight of individual heads was found in 2010 and 2012 compared to 2011 (Table 5). A high influence of weather conditions on the broccoli yields was also reported by Kałużewicz et al. [2010] and Karistsapol et al. [2013]. According to Birch et al. [2000], broccoli is highly responsive to climatic conditions, especially air temperature. Temperature has a strong impact on the plant's transition from the vegetative development-phase to the generative phase and yield [Kałużewicz et al. 2002].

The marketable yield and weight of broccoli head in the particular study years depended on the straw dose applied to soil mulching (Table 3-4). In 2010, higher marketable yield of broccoli as well as weight of head was obtained from the plots mulched with straw at a dose of 10 t·ha^{-1} compared to control object without straw. Higher yield and heads with higher weight compared to control without straw were also obtained from the plots with mulching with straw at a dose of 20 t·ha^{-1}. However, the differences were not statistically significant. In 2011 and 2012, soil mulching with straw irrespective of dose caused a significant increase in yield and average weight of marketable head compared with control plots without straw. In 2011 more favourable for broccoli yield influenced soil mulching with straw at a dose of 20 t·ha^{-1}, whereas in 2012 higher yields was obtained in the plots with straw at a dose of 10 t·ha^{-1}. The yield achieved from the plots mulching with straw at a dose of 10 t·ha^{-1} in 2012 was more than 2.0 times higher compared to control without straw. Gail et al. [1994] in studies regarding corn cultivation found that, with increasing straw dose applied to soil mulching, there was increased average weight and number of cobs per plant. However, the highest yield of seeds was obtained at a straw dose of 5.1 t·ha^{-1} compared to the control and mulching with straw at a dose of 1.7 t·ha^{-1}, 3.4 t·ha^{-1} and 6.8 t·ha^{-1}. Similarly, the study by Uwah and Iwo [2011] showed that increasing mulch dose from 2 to 8 t·ha^{-1} increased the height and number of leaves per maize plants. However, the highest yield of grain and the highest mass of cob was achieved at a mulch dose of 6 t·ha^{-1}. The authors also found that the yield of grain obtained with mulch at a dose of 6 and 8 t·ha^{-1} was more than twice that without mulching. Döring et al. [2005] did not find a significant influence of mulch with straw on the potato yield and tuber size. According to these authors, this could be caused by too low dose of straw applied for soil mulching. This relationship was confirmed in the studies by Stoner et al. [1996] and Edwards et al. [2000].

The study results showed a significant influence of the interaction between the kind and dose

Table 4. Weight of marketable head (kg)

Years	Dose of straw (t·ha^{-1})	Kind of straw				Mean
		rye	corn	rape	buckwheat	
2010	no straw	0.47	0.54	0.45	0.43	0.47
	10	0.68	0.66	0.54	0.60	0.62
	20	0.53	0.62	0.52	0.51	0.55
	Mean	0.56	0.61	0.50	0.51	0.55
2011	no straw	0.32	0.30	0.28	0.31	0.30
	10	0.42	0.42	0.40	0.39	0.40
	20	0.49	0.49	0.43	0.46	0.47
	Mean	0.41	0.40	0.37	0.39	0.39
2012	no straw	0.35	0.34	0.35	0.33	0.34
	10	0.58	0.61	0.65	0.61	0.61
	20	0.57	0.62	0.52	0.59	0.57
	Mean	0.50	0.52	0.51	0.51	0.51
Mean	no straw	0.38	0.40	0.36	0.36	0.37
	10	0.56	0.56	0.53	0.53	0.55
	20	0.53	0.58	0.49	0.52	0.53
Mean for kind of straw		0.49	0.51	0.46	0.47	0.49

LSD0,05 for: years = 0.09; kind of straw = n.s.; straw dose = 0.05; years × kind of straw = n.s.; years × straw dose = 0.09; kind of straw × straw dose = n.s.

Table 5. Weight of individual heads per plant (kg)

Years	Dose of straw ($t \cdot ha^{-1}$)	Kind of straw				Mean
		rye	corn	rape	buckwheat	
2010	no straw	0.42	0.36	0.35	0,37	0.38
	10	0.52	0.47	0.44	0.42	0.46
	20	0.36	0.44	0.36	0.38	0.39
	Mean	0.43	0.42	0.38	0.39	0.41
2011	no straw	0.37	0.24	0.32	0.28	0.30
	10	0.37	0.33	0.33	0.40	0.36
	20	0.42	0.37	0.31	0.44	0.38
	Mean	0.39	0.32	0.32	0.38	0.35
2012	no straw	0.38	0.38	0.35	0.37	0.37
	10	0.48	0.46	0.47	0.48	0.47
	20	0.49	0.49	0.44	0.43	0.46
	Mean	0.45	0.44	0.42	0.43	0.44
Mean	no straw	0.39	0.33	0.34	0.34	0.35
	10	0.46	0.42	0.41	0.44	0.43
	20	0.42	0.43	0.37	0.42	0.41
Mean for kind of straw		0.42	0.39	0.37	0.40	0.40

LSD0,05 for: years = 0.04; kind of straw = n.s.; straw dose = 0.05; years × kind of straw = n.s.; years × straw dose = n.s.; kind of straw × straw dose = 0.04.

of straw applied to soil mulching on the yield level and weight of individual heads per plant (Table 3 and 5). It was found that all kinds of straw investigated in the experiment irrespective of dose contributed to a significant increase in marketable yield. Significantly, the lowest yield was obtained from the plots without straw. Soil mulching with rye, rape and buckwheat straw at a dose of 10 $t \cdot ha^{-1}$ contributed to a significant increase in the marketable yield compared to plots mulched with straw at a dose of 20 $t \cdot ha^{-1}$. Increase in the marketable yield amounted to 1.27 $t \cdot ha^{-1}$, 2.61 and 1.60 $t \cdot ha^{-1}$, respectively. All straws at a dose of 10 $t \cdot ha^{-1}$ caused a significant increase in weight of individual heads per plants. Similar to the study by Sinkevičienė et al. [2009], the yield of vegetables depended on the kind of mulch application to soil mulching. These authors found that the highest yield was from soil mulching with grass. Yields of onion, reed beet, cabbage and potato from plots with straw and peat substrate mulch were similar – however, a less useful as a mulch was sawdust. This was confirmed in the study by Jamil et al. [2005] regarding garlic cultivation, in which mulch with straw contributed to a higher yield of bulb and a higher bulb weight compared to plots with sawdust mulch and control without mulch. The yield of bulb was higher by 1.68 and 2.76 $t \cdot ha^{-1}$, and weight of bulb by 74.13 and 111.10 g, respectively. Olfati et al. [2008] found that all organic mulches in the experiment contributed to increased total yield and average weight of carrot roots compared to cultivation without mulch. A similar dependence was found by Grassbaugh et al. [2004] and Saeed and Ahmad [2009] in tomato cultivation as well as Derek et al. [2006] in paprika cultivation. In the studies by Jabłońska-Ceglarek et al. [2006], mulch with rye straw in head cabbage growing increased the total and marketable yield compared to cultivation without mulch. Similarly, in the study by Zaniewicz-Bajkowska et al. [2009], yields of cabbage and onion cultivated after mulch with straw were higher than that obtained in the control without mulch. In the study by Feldman et al. [2000], the highest yields of cabbage and melon were obtained after soil mulching with organic mulch compared to synthetic mulch and without mulch. In turn, Díaz-Pérez [2004] in the study regarding onion cultivation, found that yield with straw mulch was significantly lower compared to the control. Gajc-Wolska et al. [2005] found that yields of paprika cultivated on the mulch with straw compared to polypropylene fibre mulch were lower.

Weather conditions in the study years had a significant influence on the length of broccoli head arc and stalk diameter (Table 6 and 7). The most favourable parameters were in 2012 (warm and moist). In

Table 6. Length of broccoli arc (cm)

Years	Dose of straw (t·ha^{-1})	Kind of straw				Mean
		rye	corn	rape	buckwheat	
2010	no straw	22.85	24.80	26.20	22.85	24.18
	10	28.50	24.50	24.35	24.00	25.34
	20	25.00	24.80	23.70	21.00	23.63
	Mean	25.45	24.70	24.75	22.62	24.38
2011	no straw	23.25	24.09	23.05	24.29	23.67
	10	24.78	21.56	22.22	23.22	22.94
	20	22.56	22.67	22.56	23.78	22.89
	Mean	23.53	22.77	22.61	23.76	23.17
2012	no straw	24.43	23.63	24.10	24.63	24.20
	10	27.57	27.23	27.10	26.20	27.03
	20	27.43	26.57	26.67	28.80	27.37
	Mean	26.48	25.81	25.96	26.54	26.20
Mean	no straw	23.51	24.18	24.45	23.93	24.02
	10	26.95	24.43	24.56	24.47	25.10
	20	25.00	24.68	24.31	24.53	24.63
Mean for kind of straw		25.15	24.43	24.44	24.31	24.58

LSD0,05 for: years = 0.87; kind of straw = n.s.; straw dose = n.s.; years × kind of straw = 1.92; years × straw dose = 2.09; kind of straw × straw dose = 0.91.

Table 7. Stalk diameter (cm)

Years	Dose of straw (t·ha^{-1})	Kind of straw				Mean
		rye	corn	rape	buckwheat	
2010	no straw	4.19	3.72	4.72	4.59	4.30
	10	5.10	4.99	4.98	4.68	4.94
	20	4.67	4.64	4.17	4.79	4.56
	Mean	4.65	4.45	4.62	4.68	4.60
2011	no straw	4.32	4.31	4.46	4.23	4.33
	10	4.88	4.85	4.75	5.32	4.95
	20	4.77	4.99	4.95	4.54	4.81
	Mean	4.65	4.72	4.72	4.70	4.70
2012	no straw	3.73	3.67	3.60	3.80	3.70
	10	3.73	4.17	4.53	3.90	4.08
	20	4.70	4.63	4.40	4.33	4.52
	Mean	4.06	4.16	4.18	4.01	4.10
Mean	no straw	4.08	3.90	4.26	4.21	4.11
	10	4.57	4.67	4.76	4.63	4.66
	20	4.71	4.75	4.51	4.55	4.63
Mean for kind of straw		4.45	4.44	4.51	4.46	4.47

LSD0,05 for: years = 0.25; kind of straw = n.s.; straw dose = 0.43; years × kind of straw = n.s.; years × straw dose = n.s.; kind of straw × straw dose = 0.25.

this year, heads were characterized significantly by the highest length of arc and the lowest stalk diameter. The differences amounted to 1.82 and 0.50 cm compared to 2010, and 3.03 and 0.60 cm compared to 2011. Broccoli heads in 2010 were characterized by a significantly longer arc than in 2011.

The study results showed the effect of the kind and dose of straw in the successive study years on the length of broccoli arc (Table 6). In 2010, higher length of arc characterized heads from the plots mulched with rye, corn and rape straw compared to buckwheat straw. The dose of straw applied to soil mulching had an influence on the length of head arc in 2012: the heads from plots with straw at a dose of 10 and 20 $t \cdot ha^{-1}$ were characterized by a significantly higher length of arc compared to controls without straw. The differences amounted to 2.83 and 3.17 cm.

The results showed a significant interaction between the kind and dose of straw on the investigated parameters of broccoli heads (Table 6 and 7). Application of rye straw, irrespective of dose, contributed to increased length of broccoli head arc compared to without straw. Length of arc was higher by 3.44 cm with rye straw at a dose of 10 $t \cdot ha^{-1}$ and 1.49 cm with rye straw at a dose of 20 $t \cdot ha^{-1}$. A significant increase in the length of broccoli head arc in the combination with rye straw at a dose of 10 $t \cdot ha^{-1}$ was found compared to mulching with straw at a dose of 20 $t \cdot ha^{-1}$. In the remaining combinations, the differences were lower and were not statistically significant (Table 6).

Broccoli heads with the lowest stalk diameter were obtained from the controls without straw (Table 7). In the plots with rape straw significantly lower stalk diameter were characterized heads at a dose of 20 $t \cdot ha^{-1}$ compared to 10 $t \cdot ha^{-1}$. Olfati et al. [2008] showed that all organic mulches applied in the studies had a significant influence on the increased length of carrot roots compared to controls without mulch. However, the authors did not find significant differences in the height of plants or root diameter. This was confirmed by a study by Khan and Parvej [2010] regarding corn cultivation. The authors found that mulching enhanced the number of ears per plant, ear height, length and diameter, tassel length, number of seed rows and 1000-grains weight compared to cultivation without mulch.

CONCLUSIONS

1. Weather conditions in study years had a significant influence on the yield level and the quality of broccoli heads. The highest yield with the best parameters was obtained in 2010 and 2012, which were characterized by sufficient rainfall for broccoli.
2. It was found that all kinds of straw investigated in the experiment, irrespective of dose, contributed to a significant increase in the marketable yield, weight of marketable head and improved head quality compared to that achieved from cultivation without straw mulching.
3. Among the examined kinds of straw, the most favourable was corn straw at a dose of 20 $t \cdot ha^{-1}$.

Acknowledgments

Research supported by the Ministry of Science and Higher Education of Poland as the part of statutory activities of Department of Vegetable Crops, Siedlce University of Natural Sciences and Humanities.

REFERENCES

1. Adetunji I.A. 1990. Effect of mulches and irrigation on growth and yield of lettuce in semi-arid region. Biotronics 19: 93-98.
2. Birch C.J., Rickert K.G., Wearing A.H., Tan D.K.Y. 2000. Broccoli yield and quality can be determined by cultivar and temperature but not photoperiod in south-east Queensland. Aust. J. Exp. Agric. 39: 901-909.
3. Dahiya R., Ingwersen J., Streck T. 2007. The effect of mulching and tillage on the water and temperature regimes of a loess soil: experimental findings and modeling. Soil Till. Res. 96: 52-63.
4. Derek M.L., Rowell A.B., Snyder J.C. 2006. Weed control efficacy of organic mulches in two organically managed bell pepper production system. Hort. Tech. 16(2): 225-232.
5. Díaz-Pérez J.C. 2004. Effects of mulch and irrigation system on sweet onion: I. Bolting, plant growth and bulb yield and quality. J. Amer. Soc. Hort. Sci. 129(2): 218-224.
6. Dobromilska R., Orłowski M., Rekowska E., Słodkowski P. 1995. Ściółkowanie gleby w uprawie warzyw ciepłolubnych. Konf. Nauk. nt. „Nauka praktyce ogrodniczej". Lublin 13-15 września: 761-764.
7. Döring T.F., Brandt M., Heß J., Finckh M.R., Saucke H. 2005. Effect of straw mulch on soil nitrate dynamics, weeds, yield and soil erosion in organically grown potatoes. Field Crop Res. 94: 238-249.
8. Edwards L., Burney J.R., Richter G., MacRae A.H. 2000. Evaluation of compost and straw mulching on soil-loss characteristics in erosion plots of potatoes in Prince Edward Island, Canada. Agric. Ecosyst. Environ. 81: 217-222.
9. Feldman R.S., Holmes C.E., Blomgren T.A. 2000. Use of fabric and compost mulches for vegetable production in a low tillage, permanent bed system: Effects on crop yield and labor. Am. J. of Alternative Agric. 15(4): 146-153.
10. Gail A.W., Crutchfield D.A., Burnside O.C. 1994. Influence of wheat (*Triticum aestivum*) straw mulch and metolachlor on corn (*Zea mays*) growth and yield. Weed Sci. 42(1): 141-147.

11. Gajc-Wolska J., Zielony T., Radzanowska T. 2005. Ocena plonowania i jakość owoców nowych mieszańców papryki słodkiej (*Capsicum annuum* L.). Zesz. Nauk. AR Wrocław, LXXXVI, 215: 139-147.
12. Ghosh P.K., Dayal D., Bandyopadhyay K.K., Mohanty M. 2006. Evaluation of straw and polythene mulch for enhancing productivity of irrigated summer groundnut. Field Crop Reas. 99: 76-86.
13. Gill K.S., Gajri P.R., Chaudry M.R., Singh B. 1996. Tillage, mulch and irrigation effects on corn (*Zea mays* L.) in relation to evaporative demand. Soil & Till. Res. 39(3-4): 213-227.
14. Grassbaugh E.M., Regnier E.E., Bennett M.A. 2004. Comparison of organic and inorganic mulches for heirloom tomato production. Acta Hort. (ISHS) 638: 171-176.
15. Iwuafor E.N.O., Kang B.T. 1994. Soil conditions under conventional and zero tillage systems with and without mulch and fertilizers. Soil Tillage for Crop Production and Protection of the Environment, ISTRO Denmark: 1031-1041.
16. Jabłońska-Ceglarek R., Franczuk J., Rosa R., Zaniewicz-Bajkowska A., Kosterna E. 2006. Wpływ sposobów mulczowania gleby i rodzaju mulczu na plonowanie kapusty głowiastej 'Masada'. Acta Agrophys. 7(4): 885-894.
17. Jamil M., Munir M., Quasim M., Baloch J., Rehman K. 2005. Effect of different types of mulches and their duration on the growth and yield of garlic (*Allium Sativum* L.). Int. J. Agri. Biol. 7(4): 588-591.
18. Johnson J.M., Hough-Goldstein J.A., Vangessel M.J. 2004. Effects of straw mulch on pest insects, predators, and weeds in watermelons and potatoes. Environ. Entomol. 33: 1632-1643.
19. Kałużewicz A., Kordakow J., Czosnowski E., Knaflewski M. 2002. Effect of thermal conditions on the initiation of broccoli curd. EJPAU, Horticulture 5(2).
20. Kałużewicz A., Krzesiński W., Knaflewski M., Lisiecka J., Spiżewski T., Frąszczak B. 2010. The effect of temperature on the broccoli yield and length of the period from head initiation to harvest. Acta Sci. Pol., Hortorum Cultus 9(3): 167-174.
21. Kar G., Kumar A. 2007. Effects of irrigation and straw mulch on water use and tuber yield of potato in eastern India. J. Agricult. Water Manag. 94(109): 116.
22. Karistsapol N., Quanchitb S., Sompongb T. 2013. Effect of planting date and variety on growth and yield of broccoli during the dry season in southern Thailand. Int. J. of Plant, Animal and Environ. Sci. (IJPAES), 3(2): 121-124.
23. Khan M.A.H., Parvej M.R. 2010. Impact of conservation tillage under organic mulches on the reproductive efficacy and yield of quality protein maize. J. of Agric. Sci. 5(2): 52-63.
24. Law D.M., Rowell A.B., Snyder J.C. 2006. Weed control efficacy of organic mulches in two organically managed bell pepper production system. Hort. Tech., 16(2): 225-232.
25. Nakhone L.N., Tabatabai M.A. 2008. Nitrogen mineralization of leguminous crops in soils. J. Plant Nut. Soil S. 171: 231-241.
26. Olfati J.A., Peyvast Gh., Nosrati-Rad Z. 2008. Organic mulching on carrot yield and quality. Int. J. of Veg. Sci. 14(4): 362-368.
27. Rahman M.J., Uddin M.S., Bagum S.A., Mondol A.T.M.A.I., Zaman M.M. 2006. Effect of mulches on the growth and yield of tomato in the costal area of Bangladesh under rainfed conditions. Int. J. Sustain. Crop. Prod. 1: 6-10.
28. Sady W. 2000. Nawożenie warzyw polowych. Plantpress, Kraków.
29. Saeed R., Ahmad R. 2009. Vegetative growth and yield of tomato as affected by the application of organic mulch and gypsum under saline rhizosphere. Pak. J. Bot. 41(6): 3093-305.
30. Schonbeck, M.W., Evalylo G.E. 1998. Effects of mulches on soil properties and tomato production. I. Soil temperature, soil moisture, and marketable yield. J. of Sustainable Agric. 13: 55-81.
31. Sinkevičienė A., Jodaugienė D., Pupalienė R., Urbonienė M. 2009. The influence of organic mulches on soil properties and crop yield. Agron. Res. 7(1): 485-491.
32. Stoner K.A., Ferrandino F.J., Gent M.P.N., Elmer W.H., LaMondia J.A. 1996. Effect of straw mulch, spent mushroom compost, and fumigation on the density of Colorado potato beetle in potatoes. J. Econ. Ent. 89: 1267-1280.
33. Swaider, J.M., Ware G.W., McCollum J.P. 1992. Producing Vegetable Crops, 4th ed. Interstate Publishers, Inc., Danville, IL: 626.
34. Uwah D.F., Iwo G.A. 2011. Effectiveness of organic mulch on the productivity of maize (*Zea mays* L.) and weed growth. The Journal of Animal & Plant Sciences 21(3): 525-530.
35. World Reference Base for Soil Resources. 1998. World Soil Resource Reports FAO, ISRIC, ISSS, Rome.
36. Zaniewicz-Bajkowska A., Franczuk J., Kosterna E. 2009. Direct and secondary effects of soil mulching with straw on fresh mass and number of weeds, vegetable yield. Polish J. of Environ. Stud. 18(6): 1185-1190.
37. Zibilske L.M., Makus D.J. 2009. Black oat cover crop management effects on soil temperature and biological properties on a mollisol in Texas, USA, Geoderma 149: 379-385.

3

CO-DIGESTION OF SEWAGE SLUDGE AND MATURE LANDFILL LEACHATE IN PRE-BIOAUGMENTED SYSTEM

Agnieszka Montusiewicz[1]

[1] Faculty of Environmental Engineering, Lublin University of Technology, Nadbystrzycka 40B, 20-618 Lublin, Poland, e-mail: a.montusiewicz@wis.pol.lublin.pl

ABSTRACT

The study examined the effects of co-digestion of sewage sludge and mature landfill leachate at the volumetric ratio of 95:5% in primarily bioaugmented system. Bioaugmentation was carried out with the use of commercial product Arkea® in the volumetric dose of 5% and lasted three months prior to the co-digestion start-up. Co-digestion was undergone without bioaugmentation. The results indicated that in the first period (of three months) following bioaugmentation, co-digestion led to biogas/methane yields only 5-8% lower as compared to anaerobic digestion of sewage sludge, and the differences were not statistically significant. Moreover, a comparable value of volatile solids removal was obtained. However, the effects became worse over time, i.e. a lower organics removal efficiency of 16% as well as 9.5–13% decreases of biogas/methane yields were achieved by applying co-digestion for a further period (of the same duration). Co-digestion of sewage sludge and mature landfill leachate could be recognized as quite efficient in the system that was primarily bioaugmented with the use of Arkea®. However, the beneficial impact of bioaugmentation remained for the limited period of three months after its completion. To sustain the favourable effects a periodical, repeatable bioaugmentation of the co-digestion system is required.

Keywords: anaerobic co-digestion; primary bioaugmentation; sewage sludge; mature landfill leachate; biogas/methane yields.

INTRODUCTION

Landfill has been employed worldwide for municipal solid waste disposal due to low operational cost and cheap exploitation. Leachate generated there as a result of waste degradation in combination with percolating rainwater, is a high-strength contaminated liquid that may contain large concentrations of organic matter, inorganic macro components (i.e. ammonia nitrogen), heavy metals and xenobiotic compounds (i.e. aromatic hydrocarbons, phenols, chlorinated aliphatics and pesticides). Leachate composition depends on many factors, such as morphology and the type of waste disposed, compaction of waste layers, landfill hydrology, seasonal weather variation, site operations and management and, particularly, on landfill age and the corresponding waste decomposition stage [Kurniawan et al. 2006, Renou et al. 2008]. According to the latter factor, the leachate classification comprises three categories: young, intermediate and mature (old). With an increase of landfill age, leachate characteristics significantly differ. The mature one (classified as stabilized) is typically represented by relatively low chemical oxygen demand (COD < 3000–4000 mg·dm^{-3}), low biodegradability (BOD_5/COD < 0.1) and dominance of refractory, high molecular weight compounds such as humic and fulvic substances that prevail as components of leachate organic fraction. On the contrary, high concentration of ammonia nitrogen (NH_4-N> 400 mg·L^{-1}) and dissolved salts (notably chlorides, carbonates and sulphates) could be expected [Pi et al. 2009, Foo and Hameed 2009]. Other constituents are as follows: slightly basic pH 7.5–8.5, total organic carbon TOC/COD ratio > 0.5 and low heavy metals concentration (below 2 mg·dm^{-3}) [Li et al. 2010].

The composition of mature leachate indicates both biodegradability enhancement and ammonia nitrogen removal to be a critical issue of the treatment [Pi et al. 2009]. Due to the presence of bio-refractory and toxic contaminants [Eggen et al. 2010] suppressing the biodegradation, unassisted biological systems involving conventional aerobic activated sludge (CAS), aerobic or anaerobic sequencing batch reactors (SBR) and up-flow anaerobic sludge blanket reactors (UASB) did not assure a requisite efficiency, even if a multi-stage configuration was employed. In order to achieve the levels of purification required to reduce the environmental adverse effect, various frequently combined biological and physicochemical treatment of leachate has been applied [Wiszniowski et al. 2006, Abbas et al. 2009, Li et al. 2009, Cortez et al. 2011]. Among the physicochemical methods incorporated as an additional treatment stage, techniques such as coagulation-flocculation, ammonia-stripping, chemical precipitation, activated carbon adsorption, advanced oxidation (AOP) and membrane separation (MBR) have been widely investigated in the last decade [Kurniawan et al. 2006, Bohdziewicz et al. 2008, Foo and Hameed 2009, Li et al. 2010, Umar et al. 2010, Ahmed and Lan 2012]. Interestingly, two main approaches have been considered. One of them concerned the physicochemical pretreatment, converting initially biorecalcitrant compounds to more readily biodegradable components degraded during the subsequent biological stage. Post-treatment (last purification) could also be recommended when the removal of recalcitrant organic micro pollutants and toxicity is required.

Among many technological systems, co-digestion of sewage sludge and mature landfill leachate has also been studied [Lebiocka et al. 2010]. The findings indicated the method was practically insufficient due to the unacceptably low efficiency resulting probably from the presence of hardly biodegradable organic compounds as well as possible toxic influence of leachate on digester' biomass. However, in the last decade the new promising approach appeared to be incorporating a strategy that leads to improvement of several biological and chemical processes. This technique, bioaugmentation, consists in the application of a specific strain or a consortium of microorganisms to enhance a required biological activity in the system investigated. In anaerobic digestion it has been used more frequently to improve both process stability and biogas yields. Interestingly, Schauer-Gimenez et al. [2010] applied this method for reducing the recovery time of digesters exposed to toxic events. In order to obtain favourable results investigating co-digestion of sewage sludge and mature leachate, application of the organisms from the archaeal domain seems to be promising because of their adaptations to extreme habitats, including environments of high salt, high temperature, low pH and acute anoxia [McLain, 2007]. However, to make the process most cost-effective and minimize the operational costs, primary bioaugmentation should be considered.

In the present study, the effects of co-digestion of sewage sludge and mature landfill leachate have been examined in applying an anaerobic digestion system that was primarily bioaugmented with the use of the commercial product Arkea® containing organisms from the archaeal domain.

MATERIALS AND METHODS

Sewage sludge (primary and waste sludge after thickening) was obtained from the Puławy municipal wastewater treatment plant (Poland). Thickened sludge was sampled once a week in the WWTP, then transported and mixed at a volume ratio of 60:40 (primary waste : sludge) under laboratory conditions. The sludge samples were homogenized, screened through a 3 mm screen and stored at 4 °C in a laboratory fridge for a week at the longest. Sludge prepared in this manner fed the digester as mixed sewage sludge (SS). The main characteristics of SS during experiments are presented in Table 1.

Leachate was sampled once as a collected sample achieved from a storage tank of mature leachate in the Rokitno municipal solid waste landfill (the age of landfill exceeded 20 years). Under laboratory conditions it was homogenized and partitioned, then frozen and stored at 25 °C in a laboratory freezer. The leachate samples were thawed daily for 6 hours at 20 °C in indoor air. The mature leachate (L) composition is presented in Table 2.

The commercial product Arkea® containing organisms from the archaeal domain was used for prior bioaugmentation and prepared as a solution in continuous mode according to the procedure given by ArchaeaSolutions Inc. The average TS and VS of the Arkea® liquor were 0.47 and 0.042 $g \cdot kg^{-1}$, respectively, and pH was 7.16.

Table 1. Sewage sludge (SS) composition (the mean value and standard deviation are given)

Parameter	Unit	Experiment 1	Experiment 2
Chemical oxygen demand (COD)	kg m^{-3}	41.7 ± 4.6	44.9 ± 8.0
Total solids (TS)	g kg^{-1}	33.2 ± 4.31	36.7 ± 6.02
Volatile solids (VS)	g kg^{-1}	25.3 ± 3.41	26.3 ± 5.05
pH	–	6.89 ± 0.96	6.55 ± 0.33
Total nitrogen (TN)	g m^{-3}	2228 ± 468	3324 ± 327
Ammonia nitrogen (N-NH_4^+)	g m^{-3}	64 ± 43	49 ± 29
Total phosphorus (TP)	g m^{-3}	464 ± 131	582 ± 221
Phosphate phosphorus (P-PO_4^{3-})	g m^{-3}	122 ± 29	118 ± 26

Table 2. Leachate composition (the mean value and standard deviation are shown)

Parameter	Unit	Mean value
Chemical oxygen demand (COD)	kg m^{-3}	5.6 ± 0.2
BOD_5/COD	–	0.05 ± 0.01
Total solids (TS)	g kg^{-1}	25.4 ± 0.95
Volatile solids (VS)	g kg^{-1}	14.3 ± 0.62
Alkalinity	g m^{-3}	3950 ± 65
pH	–	7.95 ± 0.06
Volatile fatty acids (VFA)	g m^{-3}	3960 ± 225
Total nitrogen (TN)	g m^{-3}	7200 ± 123
Ammonia nitrogen (N-NH_4^+)	g m^{-3}	6915 ± 114
Total phosphorus (TP)	g m^{-3}	71 ± 15.1
Phosphate phosphorus (P-PO_4^{3-})	g m^{-3}	39 ± 8.4

The study was carried out in reactors operating at the temperature of 35 °C in semi-flow mode. The laboratory installation consisted of two completely mixed anaerobic digesters (with an active volume of 40 dm^3) working in parallel, equipped with a gaseous installation, an influent peristaltic pump and storage vessels. The gas system consisted of pipelines linked with a pressure equalization unit and a mass flow meter. The scheme of the installation is shown in Figure 1.

An inoculum for the laboratory reactors was obtained from the Puławy wastewater treatment plant as a digest collected from an anaerobic digester operating at mesophilic temperature. The adaptation of the digester biomass was achieved after 30 days. Then, the primary phase started: the first reactor was fed regularly once a day with sewage sludge (SS of 2 dm^3 volume) and operated without bioaugmentation, the second one was supplied with SS with Arkea® addition in the dose 0.1 dm^3. This phase lasted three months.

Actual investigations following by the primary phase were devoted to co-digestion of sewage sludge and mature leachate and undergone without bioaugmentation. The study was divided into two experiments taking into consideration the duration after the bioaugmentation: Experiment 1 was carried out immediately after, Experiment 2 appeared consequently in a further period to evaluate the system efficiency over time. Each of them consisted of two runs lasting 90 days (30 days for adaptation and 60 days for measurements) and were carried out simultaneously to the control in two parallel systems. The digesters were supplied regularly once a day with an applied volume of the SS or mixture of SS and L. The feed composition and operational parameters are shown in Table 3.

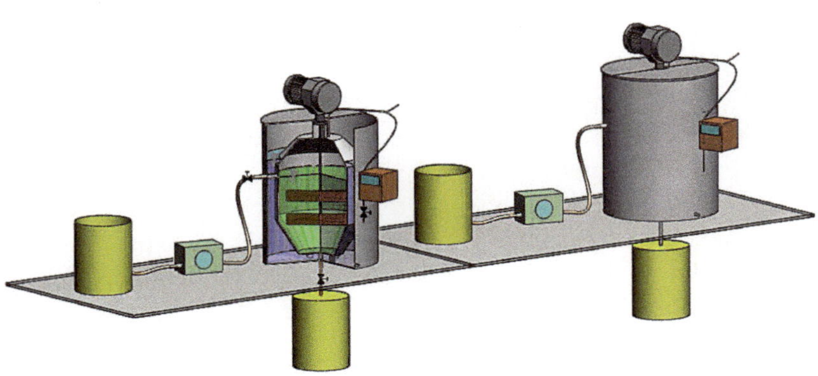

Figure 1. Laboratory installation of anaerobic digestion of waste in a wet system

Table 3. Feed composition and operational regime during experiments

Run	Feed composition	Feed volume	HRT*	OLR**
		dm^3	d	kg VS $m^{-3}d^{-1}$
Experiment 1				
R 1.1 (control)	SS (sewage sludge)	2.0	20	1.27
R 1.2	SS + L 95:5 v/v	2.0 + 0.1	19.1	1.30
Experiment 2				
R 2.1 (control)	SS	2.0	20	1.32
R 2.2	SS + L 95:5 v/v	2.0+0.1	19.1	1.35

* HRT – hydraulic retention time, ** OLR – organic loading rate

In the sewage sludge and its supernatant the following parameters were analyzed once a week: total chemical oxygen demand (COD), total solids (TS), volatile solids (VS), volatile fatty acids (VFA), alkalinity and pH level. Moreover, total nitrogen, ammonia nitrogen ($N\text{-}NH_4^+$), nitrite and nitrate nitrogen ($N\text{-}NO_x^-$), total phosphorus and ortho-phosphate phosphorus ($P\text{-}PO_4^{3-}$) were also determined. The supernatant samples were obtained by centrifuging the sludge at 4000 $r \cdot min^{-1}$ for 30 min.

The leachate composition was analyzed once after providing it to the laboratory. The values of the parameters were examined using the sewage sludge schedule besides biochemical oxygen demand (BOD_5).

In the digest and its supernatant, the specified parameters were determined twice a week, in accordance with the timetable accepted. Most analyses were carried out in accordance with Polish Standard Methods. The analyses of $N\text{-}NH_4^+$, $N\text{-}NO_x^-$ and $P\text{-}PO_4^{3-}$ were performed with FIASTAR 5000 using FOSS analytical methods. Ammonium was determined according to ISO 11732, nitrite and nitrate according to ISO 13395, and orthophosphate in accordance with ISO/FDIS 15681-1.

The anaerobic digestion efficiency was controlled by the daily evaluation of the biogas yield and its composition. Moreover, the volatile solids removal efficiency was evaluated according to Polish Standard Procedure. Biogas production was determined using digital mass flow meter Aalborg (USA). The composition of the biogas was measured using gas chromatograph Trace GC-Ultra coupled with a thermal conductivity detector (TCD) fitted with DVB packed columns. The Rt-Q-Bond column was used to determine CH_4 and CO_2 concentrations. The parameters used for the analysis were as follows – injector 50 °C and detector 100 °C. The carrier gas was helium with a flux rate of 1.5 $cm^3 \cdot min^{-1}$. Peak areas were determined by the computer integration program (CHROM-CARD).

RESULTS AND DISSCUSSION

The characteristics of feed (reactors' influent) and digest (reactors' effluent) for the specified experiments are listed in Tables 4 and 5 (the average value ± standard deviation are given). It should be noticed that feed conditions varied throughout the runs, which was attributed to the changes in sludge characteristics and the leachate addition.

The composition of feed supplemented by leachate (runs R 1.2 and R 2.2) did not essentially differ compared to SS (runs R 1.1 and R 2.1) for total solids, volatile solids and pH. TS and VS average values decreased to a little extent, respectively 1% and 2%, while pH was much the same. On the contrary, a clearly visible influence was found in relation to alkalinity and VFA (Table 5). These values were higher in the presence of leachate (as its composition indicated); the alkalinity increased by 17 and 23%, respectively for Experiment 1 and 2, and the increment of VFA concentration was of 27% (for R 1.2) and 16% (for R 2.2). The differences observed were statistically significant. Amazingly, the leachate's VFA concentration was untypically high regarding both the landfill age and low BOD_5/COD ratio.

The changes in feed volume (resulting from using leachate apart sewage sludge) led to small decreases in the hydraulic retention time (HRT) from 20 to 19.1 days, while organic loading rate (OLR) increased by about 2% with the highest value of 1.35 kg VS m^{-3} d for R 2.2 (Table 3). Anaerobic digestion proceeded in a stable way and pH remained in the range typical of methanogenesis 7.56–7.77 (with higher values corresponding with co-digestion runs). In Experiment 1, the digestion efficiency based on VS removal was comparable for both runs (Table 1), despite the shortened HRT. Contrarily, a decrease of η_{VS} appeared over time (Experiment 2), indicating the co-digestion response to pre-bioaugmentation being time-depen-

Table 4. Total solids and volatile solids, as well as their removal efficiency η for runs (TS and VS expressed in g kg^{-1}, efficiency reported as a percentage value)

Parameter	Experiment 1				Experiment 2			
	Runs							
	R 1.1		R 1.2		R 2.1		R 2.2	
	feed	digest	feed	digest	feed	digest	feed	digest
TS	33.2 ± 4.3	21.4 ± 0.6	32.8 ± 4.1	22.0 ± 1.1	36.7 ± 6.0	27.2 ± 2.7	36.2 ± 5.7	28.0 ± 4.1
η_{TS}	35.9 ± 9.8		34.1 ± 10.0		29.7 ± 6.7		29.6 ± 5.3	
VS	25.3 ± 3.4	14.5 ± 0.5	24.7 ± 3.2	14.3 ±	26.3 ± 5.1	17.5 ± 1.3	25.8 ± 4.8	17.4 ± 1.1
η_{VS}	42.7 ± 9.2		43.0 ± 9.7		38.0 ± 5.7		32.9 ± 13.7	

Table 5. Alkalinity, pH value and volatile fatty acids (VFA) for runs (concentrations reported in g m^{-3})

Parameter	Experiment 1				Experiment 2			
	Runs							
	R 1.1		R 1.2		R 2.1		R 2.2	
	feed	digest	feed	digest	feed	digest	feed	digest
Alkalinity*	853 ± 328	2916 ± 313	1001 ± 313	3333 ± 215	668 ± 68	2650 ± 87	824 ± 65	3034 ± 128
pH	6.89 ± 0.96	7.56 ± 0.13	6.91 ± 1.04	7.62 ± 0.16	6.55 ± 0.32	7.67 ± 0.16	6.56 ± 0.47	7.77 ± 0.20
VFA	591 ± 190	136 ± 47	751 ± 181	180 ± 124	894 ± 217	202 ± 74	1040 ± 206	215 ± 83

* Alkalinity given in g $CaCO_3$ m^{-3}

dent (the longer the duration after bioaugmentation, the minor the co-digestion efficiency).

The average biogas yields for the experiments are shown in Figure 2. It is worth noticing a similar, time-dependent tendency for the results obtained. In the first period following bioaugmentation (Experiment 1), the daily biogas production was a little lower for the run R 1.2 with leachate addition (18.1 ± 2.4 dm^3·d^{-1}; mean ± standard deviation is given) as compared to sewage sludge (18.7 ± 2.2 dm^3·d^{-1}). Similarly, biogas yields calculated per kg VS removed amounted to 0.87 ± 0.31 m^3·kg^{-1} and 0.93 ± 0.33 m^3·kg^{-1}, respectively. Analogous decreases were achieved per kg VS added (values of 0.35 ± 0.05 and 0.37 ± 0.05 m^3·kg^{-1}, respectively) as well as per kg TS added and removed (Figure 2a, c). In all the cases the difference was not statistically significant and did not exceed a few percent.

According to the author's previous investigations [Lebiocka et al., 2010], application of the same dose of same-sourced mature leachate as a co-substrate for sewage sludge anaerobic digestion (pre-bioaugmentation was not involved) led to much higher, statistically significant decreases in biogas yields of 40%. The results achieved in the present study seem to indicate a favorable impact of pre-bioaugmentation on the co-digestion process immediately following and lasting for three months. The Arkea® addition could improve the sludge digestion, probably due to enhanced activity of microorganisms involved in bioaugumenting systems and their resistance to toxic factors. This explanation is consistent with the research by Duran et al. [2006] regarding selected strains of the *Baccillus*, *Pseudomonas* and *Actinomycetes* species used for bioaugmentation.

Experiment 2, scheduled as a continuation of SS and L co-digestion in a further period (of the same duration), indicated the worsening of the results, despite the sustained operational conditions. A comparison of biogas yields, calculated both per kg VS and TS added and removed, showed much lower values in the presence of mature leachate as compared to SS (Figure 2b, d). Moreover, the yields achieved exceeded the levels reported for Experiment 1, although the differences still had no statistical significance. The daily biogas production unfavorably decreased from 23.2 ± 3.1 dm^3·d^{-1} for sewage sludge (R 2.1) to 22.0 ± 2.3 dm^3·d^{-1} for co-digestion process (R 2.2). The results were consistent with the study of Nielsen et al. [2007], which indicated bioaugmented anaerobic digestion of cattle manure to assure a high increase of methane yield in a two-stage thermophilic system, however for only a limited time after inoculation.

A similar average methane content was noted in biogas for runs applying leachate (53.4% ± 0.74 and 53.0% ± 1.01, respectively for R 1.2 and R 2.2), however it was a little lower compared to

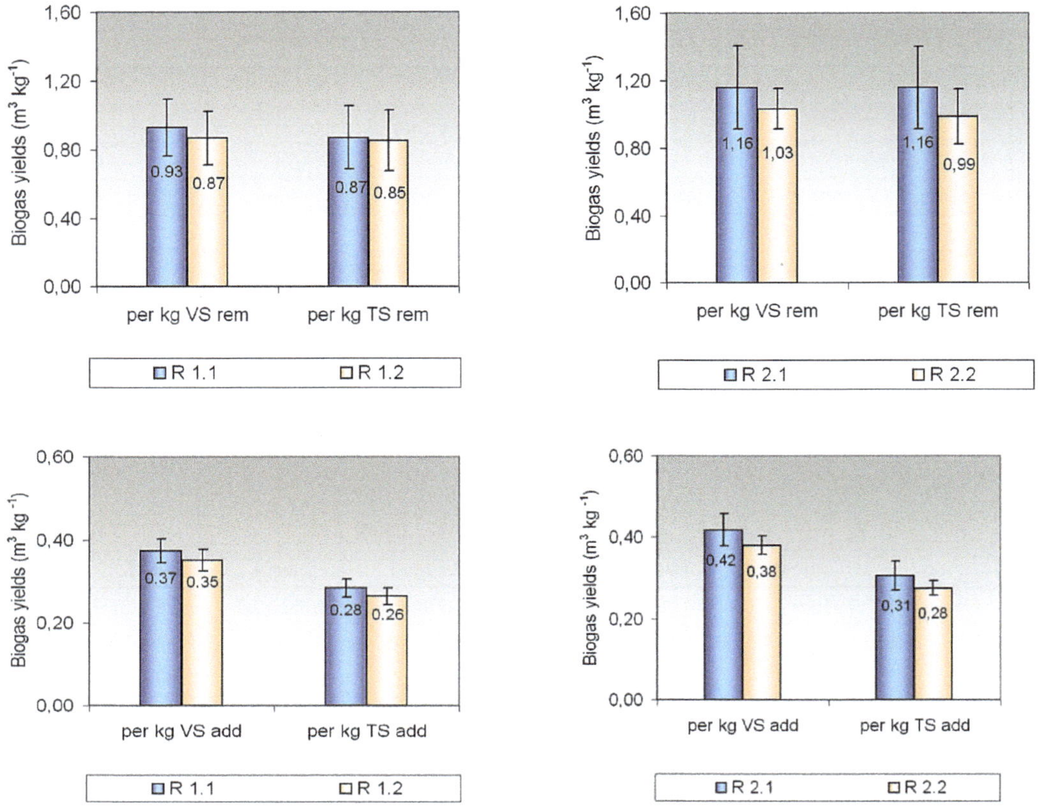

Figure 2. The biogas yields obtained in specified experiments and expressed per kg: a), b) of organic compounds removed; c), d) of organic compounds added to the system (error bars represent confidence levels, $\alpha = 0.05$)

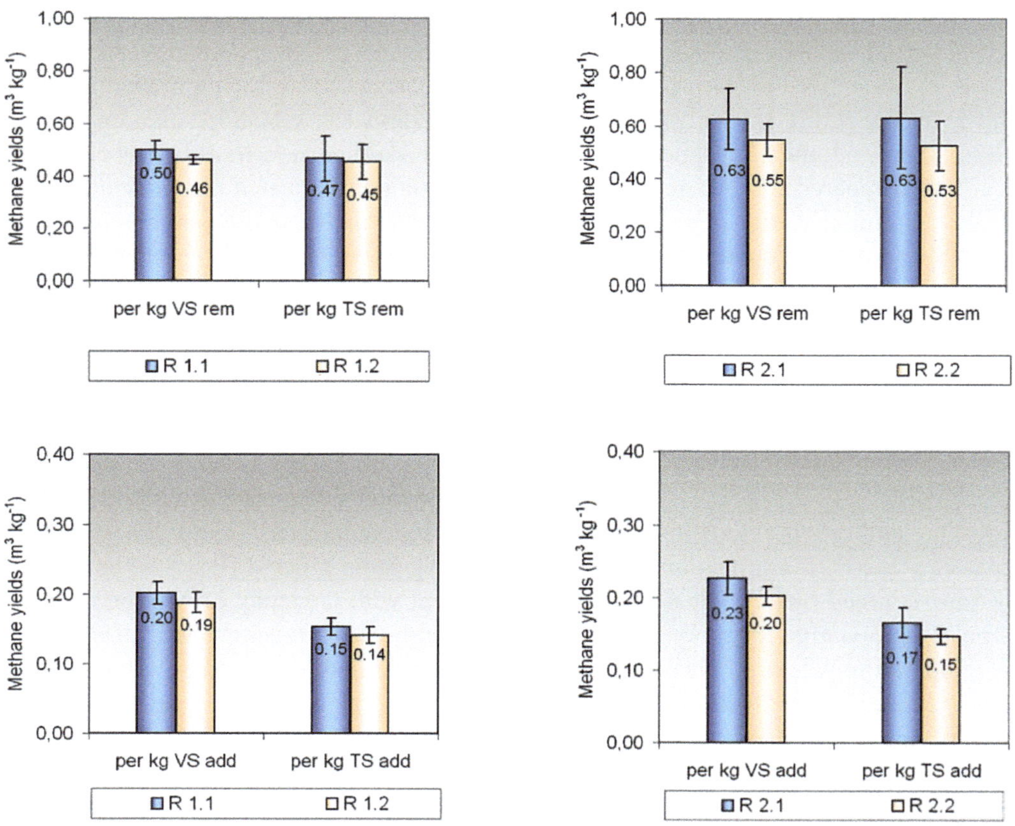

Figure 3. The methane yields obtained in specified experiments and expressed per kg: a), b) of organic compounds removed; c), d) of organic compounds added to the system (error bars represent confidence levels, $\alpha = 0.05$)

sewage sludge (54.0% ± 1.37 and 53.60 ± 1.01, respectively for R 1.1 and R 2.1). Consistently, the average methane yields decreased when co-digestion occurred (Figure 3).

Higher drops appeared in Experiment 2 that was carried out three months after the completion of bioaugmentation. The results seem to confirm the usefulness of pre-bioaugmentation with the commercial product Arkea® for enhancing co-digestion efficiency of sewage sludge and mature leachate, however the beneficial impact should be regarded as time-dependent. It remained for the limited period of three months after completion of bioaugmentation and then the results appeared to be worse. This indicates the requirement of periodical, repeatable bioaugmentation of the co-digestion system as the lifetime of the microorganisms added is limited in the reactor's conditions.

CONCLUSIONS

The results indicated that co-digestion of sewage sludge and mature landfill leachate mixed at the volumetric ratio of 95:5% could be quite efficient in the system that was primarily bioaugmented with the use of commercial product Arkea®. However, the beneficial impact of bioaugmentation remained only for the limited period of three months after its completion. To sustain the favourable effects a periodical, repeatable bioaugmentation of the co-digestion system is required.

Acknowledgements

The author thanks the National Centre of Science (Poland), No.7405/B/T02/2011/40 for financial support.

REFERENCES

1. Abbas A.A., Jingsong G., Ping L.Z., Ya P.Y., Al-Rekabi W.S. 2009. Review on landfill leachate treatments. Am. J. Appl. Sci., 6 (4), 672–684.
2. Ahmed F.N., Lan C.Q. 2012. Treatment of landfill leachate using membrane bioreactors: a review. Desalination, 287, 41–54.
3. Bohdziewicz J., Neczaj E., Kwarciak A. 2008. Landfill leachate treatment by means of anaerobic membrane bioreaktor. Desalination, 221, 559–565.
4. Cortez S., Teixeira P., Oliveira R., Mota M. 2011. Evaluation of Fenton and ozone-based advanced oxidation processes as mature landfill leachate pretreatments. J. Environ. Manage., 92 (3), 749–755.
5. Duran M., Tepe N., Yurtsever D., Punzi V.L., Bruno C., Mehta R.J. 2006. Bioaugmenting anaerobic digestion of biosolids with selected strains of *Bacillus*, *Pseudomonas*, and *Actinomycetes* species for increased methanogenesis and odor control. Appl. Microbiol. and Biotechnol., 73, 960–966.
6. Eggen T., Moeder M., Arukwe A. 2010. Municipal landfill leachates: a significant source for new and emerging pollutants. Sci. Total Environ., 408, 5147–5157.
7. Foo K.Y., Hameed B.H. 2009. An overview of landfill leachate treatment via activated carbon adsorption. J. Hazard. Mater., 171, 54–60.
8. Kurniawan T.A., Lo W., Chan G.Y.S. 2006. Physico-chemical treatments for removal of recalcitrant contaminants from landfill leachate. J. Hazard. Mater. B129, 80–100.
9. Lebiocka M., Montusiewicz A., Zdeb M. 2010. Anaerobic co-digestion of sewage sludge and old landfill leachate. Pol. J. Environ. Studies, Series of Monographs, vol. 2, 141–145.
10. Li H-s., Zhou S-q., Sun Y-b., Feng P., Li J-d. 2009. Advanced treatment of landfill leachate by a new combination process in a full-scale plant. J. Hazard. Mater., 172, 408–415.
11. Li W., Zhou Q., Hua T. 2010. Removal of organic matter from landfill leachate by advanced oxidation processes: a review. Int. J. Chem. Eng., vol. 2010, ID 270532, doi:10.1155/2010/270532.
12. McLain J.E.T. 2004. Archaea. In: Hillel D (ed.) Encyclopedia of Soils in the Environment, Elsevier Ltd., Oxford, UK, 88–94.
13. Nielsen H.B., Mladenovska Z., Ahring B.K. 2007. Bioaugmentation of a two-stage thermophilic (68°C/55°C) anaerobic digestion concept for improvement of the methane yield from cattle manure. Biotechnol. Bioeng., 97 (6), 1638–1643.
14. Pi K.W., Li Z., Wan D.J., Gao L.X., 2009, Pretreatment of municipal landfill leachate by a combined process. Process Saf. Environ., 87, 191–196.
15. Renou S., Givaudan J.G., Poulain S., Diarassouyan F., Moulin P. 2008. Landfill leachate treatment: review and opportunity. J. Hazard. Mater., 150, 468–493.
16. Schauer-Gimenez A., Zitomer D, Maki J., Struble C. 2010 Bioaugmentation for improved recovery of anaerobic digesters after toxicant exposure. Water Res., 44, 3555–3564.
17. Umar M., Aziz H.A., Yusof S. 2010. Trends in the use of Fenton, electro-Fenton and photo-Fenton for the treatment of landfill leachate. Waste Manage, 30, 2113–2121.
18. Wiszniowski J., Robert D., Surmacz-Gorska J., Miksch K., Weber J.V. 2006. Landfill leachate treatment methods: a review. Environ. Chem. Lett., 4, 51–61.

4

HOW DO DEGRADABLE/BIODEGRADABLE PLASTIC MATERIALS DECOMPOSE IN HOME COMPOSTING ENVIRONMENT?

Magdalena Vaverková[1], Dana Adamcová[1], Jan Zloch[1]

[1] Department of Applied and Landscape Ecology, Faculty of Agronomy, Mendel University in Brno, Zemědělská 1, 613 00 Brno, Czech Republic, e-mail: magda.vaverkova@uake.cz; dana.adamcova@mendelu.cz

ABSTRACT

This paper provides information about biodegradability of polymeric (biodegradable/degradable) materials advertised as 100%-degradable or certified as compostable, which may be a part of biodegradable waste, in home composting conditions. It describes an experiment that took place in home wooden compost bins and contained 9 samples that are commonly available in retail chains in the Czech Republic and Poland. The experiment lasted for the period of 12 weeks. Based on the results thereof it can be concluded that polyethylene samples with additive (samples 2, 4, 7) have not decomposed, their color has not changed and that no degradation or physical changes have occurred. Samples 1, 3 and 5 certified as compostable have not decomposed. Sample 6 exhibited the highest decomposition rate. Samples 8, 9 (tableware) exhibited high degree of decomposition. The main conclusion from this study is that degradable/biodegradable plastics or plastics certified as compostable are not suitable for home composting.

Keywords: biodegradation, degradability, degradable/biodegradable plastics, home composting, waste treatment.

INTRODUCTION

At present, waste management is becoming a global problem in developed countries due to the rapid collapse of landfills and the high impacts related to biowaste dumping. One of the major tasks of municipal waste management in European Union countries is the systematic reduction of waste that is removed and transported to landfills. This refers particularly to biodegradable waste [Suchowska-Kisielewicz et al. 2012]. The presence of such waste influences the amount of pollution emitted with leachate and biogas to the natural environment [Suchowska-Kisielewicz et al. 2012]. In view of these problems European Union (EU) published the Landfill Directive in 1999 [Directive 1999/31/63]. The Directive imposes on the member states the necessity of developing a three phase program to reduce the landfilling of organic waste, and explicitly defines the required degree of reduction in the volume of biodegradable waste going to landfills and the timetable for its implementation [Myszograj 2013]. However, The Directive does not specify how to achieve the necessary degree of reduction in the biodegradable fraction of waste [Myszograj 2013].

The overall annual food and garden waste included in mixed municipal solid waste in the European Union is within 76.5–102 Mt that represents 30–40% of the total annual municipal waste generation [Green Paper 2009, Martínez-Blanco et al. 2010]. The organic fraction of municipal solid waste (OFMSW), or biowaste, is mainly composed of food rejects of vegetable or animal origin and green waste, in an amount that depends on the region considered [Waste Statistics 2013, Puyuelo et al. 2013]. According to European recommendations and environmental considerations, these wastes need to be separately collected at home to be biologically treated through composting or anaerobic digestion to ensure the production of high quality compost, in accordance to new European regulations [Puyuelo et al. 2013].

Composting, which can be defined as aerobic biological degradation and stabilization of organic substrates under controlled, thermophilic and aerobic conditions [Martínez-Blanco et al. 2010, Haug 1993], has been presented as an environmental friendly and sustainable alternative to manage and recycle organic solid wastes, with the aim of producing a quality product known as compost, to be used as organic amendment in agriculture [Green Paper 2009, Martínez-Blanco et al. 2010]. Composting may considerably reduce the amounts of municipal waste that are presently directed to incineration facilities and landfills [Vasarevičius et al. 2011]. Composting is the keystone of sustainable development but, unfortunately, it is often not applied in a system of municipal waste [Vasarevičius et al. 2011]. For these reasons, exhaustive and systematic evaluations about its environmental performance are necessary. Potential environmental impacts, positive and negative, of municipal waste treatments should be considered including their potential pollution and their contributions to climate change, among other environmental impacts [Martínez-Blanco et al. 2010].

Home composting or backyard composting, which means self-composting of the biowaste as well as the use of the compost in a garden belonging to a private household [Green Paper 2009], presents some potential benefits when compared to the industrial process: it avoids the collection of the OFMSW; it considerably reduces the economic, material and energetic investments; and finally, it allows a direct control of the process and the organic materials input by avoiding or reducing the inclusion of impurities [McGovern 1997, Ligon and Garland 1998, Jasmin and Smith 2003, Martínez-Blanco et al. 2009]. However, home composting also presents some problems: compost obtained often is not homogeneous; odors and other pollutants such as methane, ammonia or nitrous oxide are emitted directly to the atmosphere during the decomposition process [Martínez-Blanco et al. 2010, Amlinger et al. 2008, Ansorena 2008], etc. The suitability of raw materials for biological treatment, as well as the efficiency of the processes of degradation depends on their chemical composition and structure. Besides organic compounds, the raw materials must also contain the appropriate proportions of other nutrients, and be characterized by adequate hydration and pH level [Myszograj 2013]. Moreover, in recent years, this kind of waste has been collected in single-use plastic bags. Manufacturers claim that these bags can be composted (inserted into compost pile). However, not all of these materials are truly biodegradable and therefore may pose problems within home composting.

The work reported within this paper documents observation and results obtained from monitoring home composting of biowaste from households. The aim of this paper is to provide information about biodegradability of polymeric (biodegradable/degradable) materials advertised as 100%-degradable or certified as compostable, which may be a part of biodegradable waste, in home composting conditions.

AIMS AND OBJECTIVES

In order to realize recovery and recycling via backyard composting, it must be clarified whether and how these different types of biodegradable/degradable materials or certified as compostable degrade under special conditions of domestic composting systems. In fact, the existing investigations concern almost exclusively aerobic degradation of biodegradable/degradable polymers under laboratory conditions or in technical composting facilities as certain standards are fixed for this way of recycling. Since the input material of backyard composting can differ considerably from that of a large plant, i.e. composition, amount, moisture etc., it must be examined how the composting process is influenced by inserting biodegradable/degradable polymers into the input material. Furthermore, the conditions of domestic composting can differ considerably from those in an industrial system. In home composting processes high temperature levels are partly not achieved, the material does not have optimum moisture contents and very wet biosolids with a bad structure can provoke anaerobic conditions. The objective of this work is to examine the degradability of biodegradable/degradable materials advertised as 100%-degradable or certified as compostable in two home or backyard composting systems thus providing data about the materials` degradation and to give recommendations concerning their treatment, or rather recycling in these systems under their special conditions.

MATERIAL AND METHODS

The research was aimed at verifying the degradation of biodegradable and compostable plastic materials at home composting conditions and in the environment of uncontrolled conditions – the soil conditions. The objective of the research was to find out how biodegradable or compostable plastics behave in conditions other than laboratory or real composting conditions (controlled conditions).

Description of the area

The site where the research was conducted is located in the Czech Republic, in a village called Borač that is situated in a valley on the border of the Vysočina Region and the South Moravian Region, with an altitude of 279 m. Location of the site is shown in Figure 1.

Figure 1. Location of the research

The experiment took place in home wooden compost bins (Figure 2). The bins were placed in an apple orchard under southern slope in the distance of about 10 m from a brick wall of a building. Direct sunlight was falling on the compost bins throughout the whole morning and after 2 pm the bins got into shade.

The following table (Table 1) shows data of annual precipitation [mm] and temperatures [°C] in the research area in 2012 according to the Czech Hydro-meteorological Institute (CHMI).

Table 1. Annual precipitation and temperatures at the site

Year 2012	Territorial annual precipitation [mm]	Territorial annual temperature [°C]
South Moravian Region	501	9.5
Vysočina Region	646	8

Characteristics of the samples and description of the compost bins

The experiment contained 9 samples that are commonly available in retail chains in the Czech Republic and Poland. These were plastic bags, single-use plastic bags and single-use compostable tableware (namely cups and trays). The manufacturers declared all samples biodegradable in nature or compostable. Photographic documentation of the samples at the beginning of the experiment is shown in Figure 3. The descriptions of the materials as well as the observed samples are listed in Table 2.

Following labeling (samples 1-9) and photographing of the samples, they were placed into compost pile of the home composting bins. The bins were made of wood, placed directly on the ground surface without any modification thereof. The first compost bin (compost bin 1) had a rectangular shape with dimensions of 160×130×100 cm,

Table 2. Description of the samples

Sample	Type	Description
1	N/A	Compostable 7P0202
2	N/A	BIO-D plastic material
3	Natural material	Compostable 7P0073
4	HDPE	Epi – 100% degradable
5	Starch and PCL	OK Compost AIB VINCOTTE
6	Starch	Compostable 7P0147
7	N/A	Degradable plastic material – d_2w
8	–	N/A
9	–	N/A

Figure 2. Home compost bins

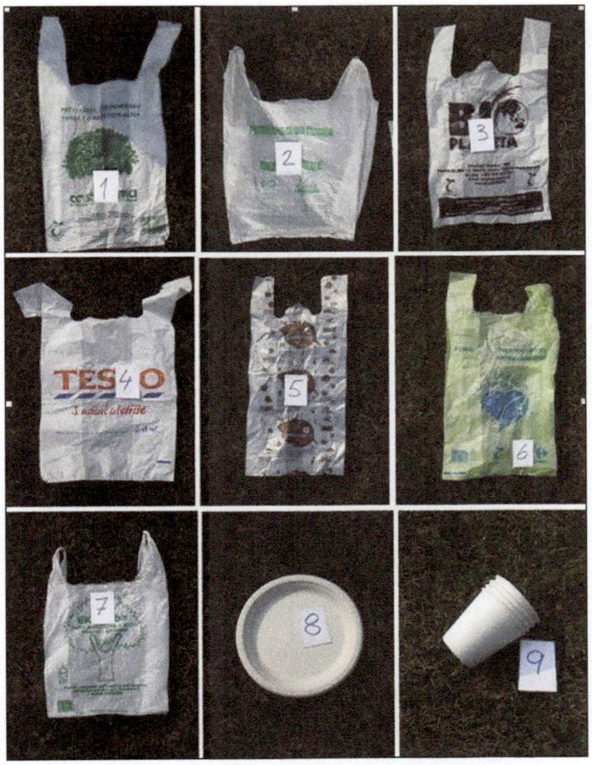

Figure 3. Photographs of the observed samples

the other bin (compost bin 2) had dimensions of 105×100×80 cm (see Figure 2).

Prior to the start of the experiment the compost pile had been already set up, the degradation process within the pile was thus in progress. The compost pile was dug over and refilled with new biowaste once during the monitored period of time. The compost pile consisted of rabbit manure, cut grass, biodegradable kitchen waste (tea bags, fruit rinds, coffee, etc.), as well as parts of plants, feathers or fur from farming animals and wood ashes. An elemental analysis of the compost pile wherein the investigated samples were put was carried out in laboratories of Mendel University in Brno. Parameters of the compost pile sample (incinerable substances, pH, lead, cadmium) are listed in Table 3.

Simultaneously, a test of degradation of the studied materials was performed outside the compost pile. All 9 samples were placed into soil in the orchard. The grassy area (the research site) was delineated with dimensions of 210×150 cm. Soil features based on the Estimated Pedologic-Ecological Units are stated in Table 4.

Table 3. Parameters of the compost pile sample

Parameters	Compost pile
Incinerable substances [%]	44.6
pH	8.74
Pb [mg/kg]	0.852
Cd [mg/kg]	0.186

The course of the experiment

The experiment was initiated on 1 September 2012 and continued for the period of 12 weeks. 7 samples containing plastic bags and single-use plastic bags made of degradable or compostable material were loosely placed into compost bin 1. The remaining 2 samples - disposable tableware (cup and tray) made of biodegradable plastic material were inserted into compost bin 2. The samples were placed into newly excavated spaces on the surface of the compost pile and subsequently covered with 20 cm layer of compost pile. Pictures documenting the initiation of the experiment are displayed in Figure 4.

The same samples that were put in home compost bins were evenly placed in grassy area of the size of 210×150 cm, which was located in the orchard. The samples were inserted into hollowed spaces with dimensions of 20×20×15 cm situated on the surface, then covered with the original layer of soil, see Figure 5. For the purpose of future visual comparison, the samples were photographically documented prior to their placement into the soil.

The compost pile was not dug over or otherwise interfered during the course of the research. The samples were not manipulated anyhow. After 12 weeks, in early December 2012, the samples were removed from the bins.

Table 4. Features of the soil at the research site based on Pedologic-Ecological Units

Pedologic-Ecological Units			
Climatic region	The main pedologic unit		
	Soil type	Soil substrate	Remark
Mildly dry and humid	Light to lighter medium heavy	Various substrates	Extreme inclination (up to 12 °C)
Inclination	Exposure	Skeletality	Depth of the soil
Considerable slope (12–17 °C)	South (SW-SE)	Medium to very skeletal	Shallow, medium deep and deep

Figure 4. Initiation of the experiment in compost bins

Figure 5. Initiation of the experiment in the orchard

RESULTS AND DISCUSSION

After 12 weeks the samples were removed and photographically documented. Visual evaluation of the decomposition of each sample material was conducted. The samples that were examined in compost bin 1 showed no visual modification, except for sample 6 (plastic bag, according to manufacturer made of starch, country of origin – Poland), see Figure 6. The sample exhibited partial degradation, separation into smaller fragments and colour changes. Samples 1–5 and 7 experienced no modifications compared to their original state (at the start of the experiment), see Figure 7. The following picture (Figure 7) shows photographs of samples 1–5 and 7 after the termination of the experiment.

The samples that were examined in compost bin 2, the disposable tableware made of degradable plastic material (samples 8 and 9) entirely degraded; no visible fragments were left, see Figure 8.

Figure 6. Sample 6 before and after the experiment – the compost pile

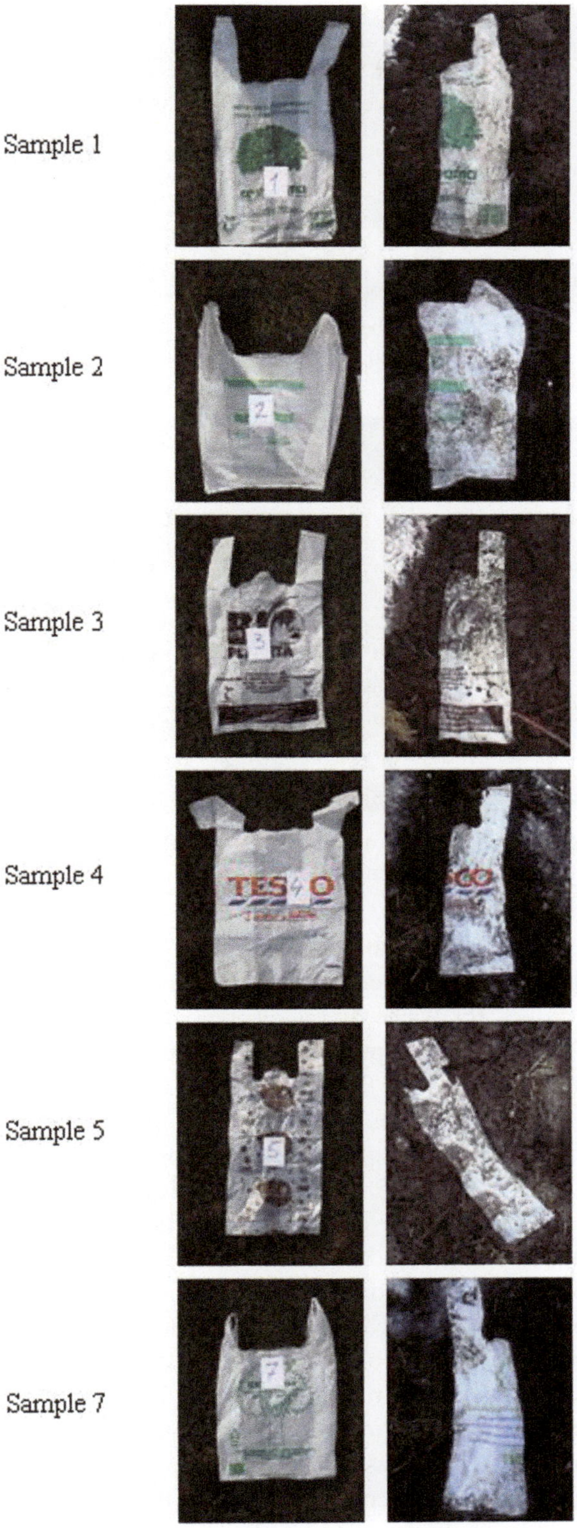

Figure 7. Samples 1-5, 7 before and after the experiment – the compost pile

The samples of plastic bags and single-use plastic bags (samples 2–7) stored in the soil of the grassy area of the orchard experienced no visual changes. Photographs of the samples before and after the experiment are shown in Figure 9.

Figure 8. Sample 8 and 9 before and after the experiment – the compost pile

Only sample 1 that was placed in the soil of the grassy area of the orchard showed signs of partial degradation into smaller pieces, see Figure 10.

The samples 8 and 9 (disposable tableware made of biologically degradable plastic material) placed in the grassy area of the orchard exhibited a high degree of decomposition. There were only several visible fragments left, see Figure 11.

Bio-based polymers have increasingly attracted attention from industry and consumers as one of the ways of reducing municipal solid waste and reducing the dependence on producing polymers from petroleum-based resources. These new polymers have the inherent advantage of being produced from renewable resources such as cellulose and starch, and if their end-of-life scenario is properly designed by using recycling or composting, their environmental footprint can be reduced. Most of them are blends, containing a certain percentage of mineral oil based biodegradable polymers. Marketable packing products are bags, trays and dishes for fruits, vegetables, bread, cakes, pastries and meat products as well as containers for dairy products, bin liners and compostable food ware/catering articles.

Unfortunately, there are now many plastic materials in the market claiming to be "biodegradable" or environmentally preferable in other ways. Often claims cannot be verified or are misleading. The companies selling those products are taking advantage of markets that are unaware of the difference between certifiable compostable and biodegradable products and those that are not. Oxo-degradable plastics for instance, do not meet any standards in place for biodegradability and should not be considered biodegradable or compostable.

While many bioplastics are certifiable as compostable in commercial compost facilities, not all can be home composted and not all are bio-

Sample 2

Sample 3

Sample 4

Sample 5

Sample 6

Sample 7

Figure 9. Samples 2-7 before and after the experiment – the orchard

degradable in the marine environment. Furthermore, a number of petrochemical-based polymers are certified biodegradable and compostable. Biodegradability is directly linked to the chemical structure, not to the origin of the raw materials.

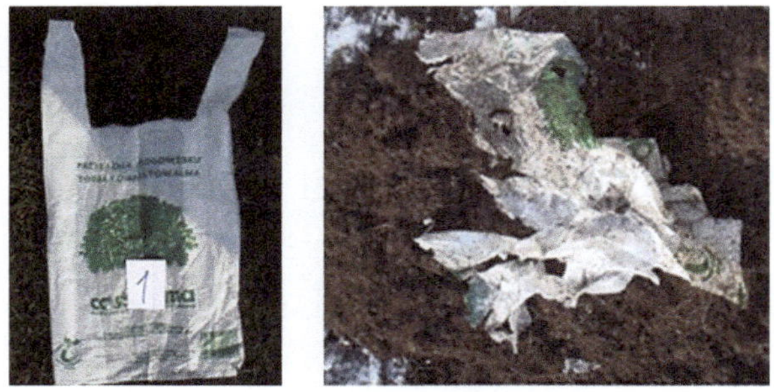

Figure 10. Sample 1 before and after the experiment – the orchard

Figure 11. Sample 8 and 9 before and after the experiment – the orchard

Products that can compost in a variety of backyard/home system greatly increase the opportunities for waste diversion, given that many regions of Czech Republic lack commercial composting facilities.

In real conditions of the Central Composting Plant in Brno, similar research was conducted in 2011 and 2012. It focused on verifying the degree of degradation of plastic bags labelled as compostable, degradable or 100% biodegradable in controlled composting conditions of the composting plant. The authors of that research discovered that the (observed) samples that carry a compostability certification, showed high degree of decomposition and some disintegrated completely. In contrast, samples made of PE with additives did not decompose, alike in the experiment in home composting conditions [Adamcová et al. 2013, Vaverková et al. 2013]. The results show that biodegradable plastic materials are not entirely suitable for domestic composting. The reason why the samples did not degrade may be the failure to achieve the required temperature inside the compost pile. Another aspect that may play a significant role in the decomposition and disintegration of the examined materials is the rain factor. In home composting conditions compost humidity depends on the amount of rainfall.

CONCLUSION

The monitoring undertaken has demonstrated that although PE with additive sacks are suitable for the collection of organic wastes, they are not compostable under home composting conditions as a determining feature of a compostable polymer is its ability to undergo biological degradation at a rate consistent with other known compostable materials, in addition to leaving no visually distinguishable residues within the final compost product. The plastics certified as compostable are not suitable for home composting and that in home compost bin they do not degrade.

Actually, the product may be backyard or home compostable if they are certified by Vinçotte OK Compost HOME. OK Compost HOME is the only certification that assesses compostability in typical backyard compost settings, which do not reach or maintain high temperatures of industrial compost facilities. Systematic research to investigate the effects of composting conditions on the degradability biodegradable/biodegradable plastics should be performed in the future.

REFERENCES

1. Suchowska-Kisielewicz M., Jędrczak A., Myszogra S. 2012. Kinetic constants of decomposition of the municipal solid waste prior to and after mechanical-biological processing field scale. Archives of Environmental Protection, 38(4), 71–86.
2. Council of the European Union 1999. Directive 1999/31/EC, on the Landfill of Waste: http://eurlex.europa.eu/LexUriServ/LexUriServ.do?uri=CELEX:31999L0031:EN:NOT (January 2010).
3. Myszograj S. 2013. Effects of the solubilisation of the cod of municipal waste in thermal disintegration. Archives of Environmental Protection, 39(2), 57–67.
4. European Commission 2009. Green Paper on the Management of Bio-waste in the European Union 2008. In: Official Journal of the European Communities: http://eurlex.europa.eu/LexUriServ/LexUriServ.do?uri=CELEX:52008DC0811:EN:NOT (January 2009).
5. Martínez-Blanco J., Colón J., Gabarrell X., Font X., Sánchez A., Artola A., Rieradevall J. 2010. The use of life cycle assessment for the comparison of biowaste composting at home and full scale, Waste Management, 30, 983–994.
6. Eurostat 2013. WasteStatistics: http://epp.eurostat.ec.europa.eu/statistics_explained/index.php/Waste_statistics> (January 2013).
7. Puyuelo B., Colón J., Martín P., Sánche A. 2013. Comparison of compostable bags and aerated bins with conventional storage systems to collect the organic fraction of municipal solid waste from homes. A Catalonia case study, Waste Management, 33, 1381–13.
8. Haug R. 1993. The Practical Handbook of the Compost Engineering, Laws Publishers, Boca Raton, Florida, USA.
9. Vasarevičius S., Baltrenas P., Baltrenaite E. 2011. Investigation and Evaluation of Green Waste, Composting Parameters, Polish Journal of Environmental Studies, 20(6), 1603–1609.
10. McGovern A. 1997. Home composting makes major impact, Biocycle, 38, 30.
11. Ligon P., Garland G. 1998. Analyzing the costs of composting strategies. BioCycle, 39, 30–37.
12. Jasmin S., Smith S. 2003. The Practicability of Home Composting for the Management of Biodegradable Domestic Solid Waste. Centre for Environmental Control and Waste Management, Department of Civil and Environmental Engineering, London, England.
13. Martínez-Blanco J., Colón J., Gabarrell X., Font X., Artola A., Rieradevall J., Sánchez A. 2009. Emissions and environmental assessment of the home composting of two domestic wastes. In: The Proceedings of the III International Conference of Life Cycle Assessment in Latinoamerica, Pucón, Chile.
14. Amlinger F., Peyr S., Cuhls C. 2008. Greenhouse gas emissions from composting and mechanical biological treatment, Waste Management and Research, 26, 47–60.
15. Ansorena J. 2008. Composting in Guipuzkoa. Retema, 34–4 (in Spanish).
16. www.maps.com, modified by Vaverková, Adamcová.
17. Adamcová D., Vaverková M., Toman F. 2013. Repeated research of biodegradability of plastics materials in real composting conditions, Acta Universitatis Agriculturae et Silviculturae Mendelianae Brunensis. In print.
18. Vaverková M., Adamcová D., Kotovicová J., Toman F. 2013. Evaluation of biodegradability of plastics bags in composting conditions. Ecological Chemistry and Engineering S. 2013. In print.

5

INFLUENCE OF SEWAGE SLUDGE ON THE CHOSEN SOIL PROPERTIES AND MICROBIOLOGICAL PARAMETERS OF URBAN GRASS MIXTURES RHIZOSPHERE

Urszula Wydro[1], Elżbieta Wołejko[1], Tadeusz Łoboda[1], Marzena Matejczyk[1], Andrzej Butarewicz[1]

[1] Division of Sanitary Biology and Biotechnology, Bialystok University of Technology, Wiejska 45E, 15-351 Białystok. Poland, e-mail: uwydro@gmail.com; elzbietawolejko@wp.pl; lobodat@vp.pl; m.matejczyk@pb.edu.pl; a.butarewicz@pb.edu.pl

ABSTRACT

The aim of this study was to analyse the influence of municipal sewage sludge on selected microbiological parameters in the rhizosphere of lawn grass mixtures. Four experiments on the lawns along the main roads of Bialystok were founded. The factors in the experiment were three doses of sewage sludge (0 – control; 7.5 and 15 $kg \cdot m^{-2}$) and two grass mixtures (Eko and Roadside). The studied parameters were monitored twice during 2011 vegetation season by determining the total number of soil microorganisms, the total number of *Gram*-negative bacteria, the number of bacteria of *Pseudomonas fluorescens* species, the number of amylolytic and proteolytic bacteria. Sewage sludge amended to soil resulted in a change of physical-chemical properties of soil. The sewage sludge application to soil influenced significantly the number of proteolytic and *Gram*-negative bacteria.

Keywords: sewage sludge, rhizosphere, urban soil, microorganisms.

INTRODUCTION

Urban soils and green spaces, including lawns, are a crucial component of urban ecosystems, contributing directly or indirectly to the general quality of life of city residents [Luo et al. 2012]. Urban soils are characterized by a distortion of the soil profile, increasing soil pH, salinity (especially street-adjacent areas), low water capacity and water permeability as well as weak biological activity and contamination with heavy metals [Wysocki, Stawicka 2005].

Sewage sludge is an increasingly important economic and environmental problem. According to Central Statistic Office [2013], in 2012 in Poland 533.3 thousand Mg of dry matter from municipal sewage sludge was produced. It is estimated that from 2018 over 700 thousand Mg dry matter of sewage sludge will be produced [Bień et al. 2011]. Therefore, it is necesary to look for methods and ways of their rational use.

One way of utilization of municipal sewage sludge is to use it for recultivation of low quality soils and degraded areas [Sanches-Monedero 2004]. It is justified by the fertilizer properties of the sewage sludge, such as high content of nitrogen, phosphorus, carbon, macro- and microelements, which are necessary to improve soil conditions e.g. respiration and enzymatic activity and plant growth [Singh, Agrawal 2008, Siuta 2003]. Therefore, a very promising future area of application of municipal sewage sludge is the land adjacent to roads. In addition, fertilizer components and organic matter from sludge return to the environment.

The rhizosphere zone is an area of intense biological activity, characterized by rich bacterial and fungal flora, which has a large impact on the uptake of nutrients by the roots [Jurkiewicz et al. 2004]. According to Wolińska [2010], microbial activity plays a very important role in the life of plants. Plants form soil microflora and the influence microbial metabolism, constantly take place in the environment.

The aim of this work is to investigate the impact of sewage sludge fertilization on the number of selected microbiological parameters of lawn grasses mixtures in the rhizosphere and linking these parameters with physico-chemical properties of the soil.

MATERIAL AND METHODS

Four experiments on the lawns along the main roads of Bialystok: Hetmanska Str., Piastowska Str., Popieluszki Str. and Raginisa Str. were conducted. The factors of the experiment were: three doses of sewage sludge (0.0 – control, 7.5 and 15.0 kg/m^2) and two mixtures of lawn grasses (Eko and Roadside).

Stabilized municipal sewage sludge from the Municipal Wastewater Treatment Plant in Sokolka was used. In autumn 2010 the test areas were fertilized with municipal sludge with two doses: 7.5 kg×m^{-2} (14.5 t/ha of dry matter) and 15.0 kg×m^{-2} and control plots (without sewage sludge) were founded. The doses of sewage sludge were established according to Kiryluk [2002], who found in several years study that the most effective doses for turfing of municipal waste disposal areas were those above 40 t/ha.

Before the establishment of the experiment both sewage sludge and soil from each combination were analyzed according to the Regulation of Ministry of Environment of July 13th, 2010 concerning municipal sewage sludge (Table 1 and 2). The obtained values of the studied parameters were compared with the values of the Regulation concerning non-agricultural land recultivation [Directive of… 2010].

In spring 2011, on the prepared plots two mixtures of lawn grasses were seeded, Eko from Nieznanice Plant Breeding Station which included 30% of *Lolium perenne* cv. Niga, 15% of *Poa pratensis* cv. Amason, 22.6% of *Festuca rubra* cv. Adio and 32.4% of *Festuca rubra* cv. Nimba, and grass mixture Roadside from Barenbrug which included 32% of *Lolium perenne* cv. Barmedia, 5% of *Poa pratensis* cv. Baron, 52% of *Festuca rubra rubra* cv. Barustic, 5% of *Festuca rubra commutata* cv. Bardiva (BE) and 6% of *Festuca rubra commutata* cv. Bardiva (NL).

Microbiological analyses

Rhizosphere soil samples for microbiological analyses were collected two times, in July and October 2011. A fresh sample of 5 g of grass mixture roots with the soil was suspended in 45 ml of sterile saline solution and shaken for 10 min and then subjected to dispersion. Subsequently, another ten-fold dilutions (10^{-1}–10^{-6}) and then the soil samples were inoculated on the surface of Petri dishes in triplicate on:

- 10% TSA (Tryptic Soy Agar) medium for determining the total number of bacteria [Galimska-Stypa et al. 1999];
- 10% TSA medium with crystal violet to determine the number of *Gram*-negative bacteria [Galimska-Stypa et al. 1999];
- King B medium with peptone for determining the number of bacteria of the species *Pseudomonas fluorescent* [Galimska-Stypa et al. 1999].
- Frazier medium for determining the number of proteolytic bacteria [Rodina 1968];
- Starch agar to determine the number of amylolytic bacteria [Rodina 1968].

The mean number of colonie was expressed as CFU·g^{-1} d.m. soil.

Physico-chemical analysis

pH in soil samples taken from a depth of 0–20 cm was determined by the potentiometric

Table 1. Selected physic-chemical properties of soils in research plots at Popieluszki, Hetmanska, Piastowska and Raginisa Streets, taken before the foundation of experience in autumn 2010 (the analyses were done by the Regional Chemical and Agricultural Station in Białystok)

Soil samples	pH	P_2O_5 [mg /100g d.m.]	Heavy metals content in mg/kg of DM of soil						
			Cd	Cu	Ni	Pb	Zn	Hg	Cr
Popiełuszki Street	7.6	22.0	1.7	17.9	5.9	26.5	82.4	0.14	13.1
Hetmanska Street	7.9	7.3	<0.5	9.5	4.5	12.4	40.9	0.06	10.9
Piastowska Street	7.7	18.4	<0.5	16.8	10.8	23.9	195	0.02	13.6
Raginisa Street	7.4	10.0	<0.5	8.8	4.6	12.1	36.6	0.05	8.3
Limited concentration for land reclamation for non-agricultural [Directive of …2010]			3	50	30	50	150	1	100

Table 2. Characteristics of municipal sewage sludge from sewage treatment plants in Sokolka used in experiments (the analyses were done by the Regional Chemical and Agricultural Station in Białystok)

The indicators determined in sludge							
pH	6.7	ammonium nitrogen (% DM)					0.1
Dry matter (%)	19.3	calcium (% DM)					5.5
Total nitrogen (% DM)	4.0	magnesium (% DM)					0.7
	Heavy metals content in mg/kg of DM of sewage sludge						
	Cd	Cu	Ni	Pb	Zn	Hg	Cr
Metal content in sludge	<0.5	194	22,0	23.5	1459	1.0	58
Limit value in the application of sludge to land reclamation for non-agricultural purposes [Directive of ...2010]	25	1200	400	1000	3500	20	1000
Viable helminth ova of *Ascaris sp., Trichris sp., Toxocara sp.*	not detected						
Bacteria of the genus *Salmonella* in 100 g of sludge	not detected						

method in distilled water. The total nitrogen was measured with the Kjeldahl method based on wet oxidation of organic matter using H_2SO_4 and conversion of organic nitrogen to the ammonium form. The available phosphorus (expressed as P_2O_5) was determinated by Egner-Riehm method in 0.04 solution of calcium lactate. Nitrogen and phosphorus were determined by UV-vis spectrophotometry. The organic carbon was measured by Tiurin method which is based on the oxidation of organic carbon by potassium dichromate in an acid medium and measurement of the excess dichromate using Mohr's salt [Ostrowska et al. 1991].

Statistical analysis

To test for the effects of the experimental factors (dose of sewage sludge, sampling time, grass mixture) in the variables analyzed, the tree-way analysis of variance ANOVA with interaction term was performed. In the case of significant F-tests, differences between group means were assessed by the Tukey post hoc test at significance level at $p < 0.05$. The correlation between characteristics was calculated using Spearman's rank correlation (for $n<50$) with the level of significance established at $p < 0.05$ by using Statistica 10.0.

RESULT AND DISCUSSION

The pH values for all samples tested ranged from 7.2 to 8.0, which means that at each point alkalinity of the soil occurred (Table 3).

The soil fertilization of sewage sludge caused an increase the content of organic carbon, total nitrogen and available phosphorus in urban soils (Table 3). The average content of organic carbon was smallest in soil without application of sewage sludge with Eko grass mixture (1.3% d.m.) and largest in the soil samples with double dose of sewage sludge and Eko grass mixture (2.2% d.m.). The average total content of total nitrogen ranged from 0.30% d.m. for the control plots with Eko grass mixture to 0.42% d.m. for the double dose of biosolids and Eko grass mixture. The average smallest content of available phosphorus was observed for soil samples for control plots with Eko mixture (13.3 mg/100 g d.m.) and largest for soil with sludge dose 15.0 kg/m^2 and Roadside grass mixture (36.5 mg/100 g d.m.).

The results of this study have shown that the fertilization with sewage sludge evoked changes in rhizosphere of microbiological activity of grass mixture (Figure 1).

The experimental factors did not have a significant effect on this parameter (Figure 1). The average total number of bacteria was the smallest in the soil without application of sewage sludge (13.0 CFU$\times 10^7$ g^{-1} DM of soil) with the mixture Eko in samples taken in July, while the largest average total number of bacteria was observed in the samples collected in October with the Eko mixture and the highest dose of biosolids (15 kg/m^2) (41.2 CFU$\times 10^7$ g^{-1} DM of soil). Generally, with increasing doses of sludge an increase in the total number of microorganisms was observed. However, soil samples taken in October where Roadside mixture was grown showed an increased number of microorganisms at a dose of 7.5 kg/m^2 (Figure 1).

Studies on the impact of different doses of sewage sludge on the number of *Gram*-negative bacteria revealed an increase in the number of these microorganisms in rhizosphere with an increase in the biosolids dose (Figure 1). The av-

Table 3. Selected physico-chemical parameters of soil after application different doses of sewage sludge (0, 7.5, 15 kg/m^2) and two lawn grass mixtures (Eko and Roadside) measured in October 2011 (average±standard deviation)

Sewage sludge dose [kg/m²]	pH_{H2O}		C_{org} [% d.m.]		N [% d.m.]		P_2O_5 [mg/100 g]	
	eko	roadside	eko	roadside	eko	roadside	eko	roadside
0.0 (n=8)	7.7±0.2	7.6±0.0	1.3±0.2	1.4±0.7	0.30±0.02	0.33±0.06	13.3±6.7	13.3±8.3
7.5 (n=8)	7.6±0.2	7.5±0.1	1.8±0.5	1.8±0.3	0.31±0.06	0.33±0.05	25.1±13.0	23.3±11.2
15.0 (n=8)	7,4±0.1	7.4±0.1	2.2±0.8	2.0±0.3	0.42±0.14	0.36±0.06	32.6±10.6	36.5±9.8
MIN (n=24)	7.3	7.2	0.6	0.5	0.23	0.28	4.9	4.3
MAX (n=24)	8.0	7.7	3.3	2.5	0.58	0.42	45.1	47.1

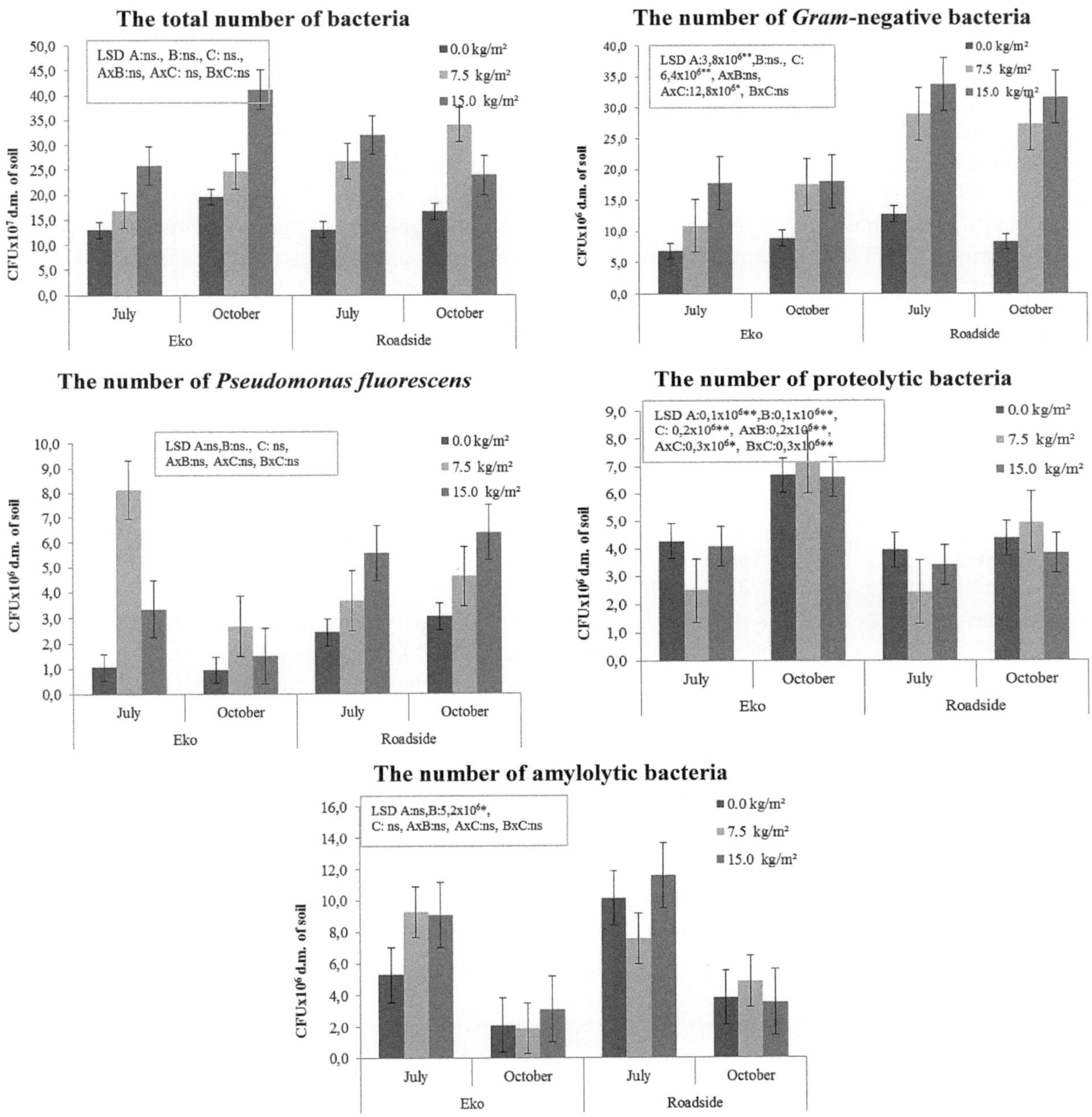

Figure 1. The number of selected bacteria in grass mixtures rhizosphere in samples collected in July and October 2011 after application of different doses of sewage sludge (0, 7.5, 15 kg/m^2) and two lawn grass mixtures (Eko and Roadside) (A: mixture of grass, B: sampling time, C: dose of sewage sludge, ns – not significant differences, ** significant differences for $p<0,01$; * for $p<0,05$)

erage largest number of this microbial parameter was found in samples where Roadside grass mixture was grown. The average number of *Gram*-negative bacteria was largest in the soil samples collected in July with the dose of 15.0 kg/m^2 and Roadside mixture (33.7 CFU×10^6 g^{-1} DM of soil), while the lowest results were obtained for the samples taken in July on the control plots with Eko mixture (6.9 CFU×10^6 g^{-1} d.m. of soil).

The genus *Pseudomonas* is an important component of the rhizosphere, which developed mechanisms that restrict pathogenic effect of fungi and bacteria. Species from genus of *Pseudomonas* are know as Plant Growth Promoting Rhizobacteria (PGPR). In addition to the protective function, *Pseudomonas* can also enrich soil with nutrients, plant hormones and vitamins synthesized and enhance their bioavailability for plants [Jankiewicz 2010].

We found that the number of *P. fluorescens* in soil varied (Figure 1). The experimental factors did not influence the studied parameter. The largest average number of *P. fluorescens* (6.4 CFU×10^6 g^{-1} d.m. of soil) was observed in the samples taken in October on plots with double dose of sewage sludge and Roadside grass mixture. The smallest number of these microorganisms were obtained in October for the control plots with Eko grass mixture (0.9 CFU×10^6 g^{-1} d.m. of soil).

Joniec et al. [2012], suggest that sewage sludge represents a rich source of proteins. The obtained results for the number of proteolytic bacteria in grass mixtures rhizosphere shows significant effect of the sewage sludge dose on this parameter (Figure 1). The average largest number of proteolytic bacteria was obtained in October on the plots with 7.5 kg/m^2 biosolids with Eko mixture, while the smallest were in July on the plots with 7.5 kg/m^2 with Roadside mixture (2.5 CFU×10^6 g^{-1} d.m. of soil).

The research study of Joniec and Furczak [2007], presented positive effects of sewage sludge on the number of proteolytic bacteria in the second year of the study. Smaller number of proteolytic bacteria in the soil amended with sewage sludge in the first research time was probably caused by the fact that bacteria brought with biosolids into the soil did not find suitable growth conditions in the soil and gradually decayed [Joniec et al. 2012].

The results obtained for the number of amylolytic bacteria show that the sewage sludge dose did not influence this microbial parameter (Figure 1). The largest number of amylolytic microbes was observed in July for the plots with double dose of sludge with Roadside grass mixture (11.6 CFU×10^6 g^{-1} d.m. of soil), while the lowest results were obtained in October for the plots with a single dose of sludge and Eko grass mixture (1.9 CFU×10^6 g^{-1} d.m. of soil).

Literature reports that climatic conditions, such as precipitation and temperature, can affect the microbial activity of soil. The amount of soil microorganisms may also have seasonal changes in vegetation period. Increased microbial growth occurs in spring when temperatures rise and in autumn after delivery of soil organic matter in a form of plant residues, which are food for microorganisms.

Rousk et al. [2009] and Su et al. [2004], suggest that soil pH is a very important parameter which directly determines the possibility of the development of microorganisms, plant growth, bioavailability of nutrients and heavy metals, the biological processes and creating other forms of acidity.

The correlation analysis showed a significant relationships between pH and the total number of bacteria and the number of *Gram*-negative bacteria in samples of soil (Table 4). It was found that the increase pH in soil caused decrease the number of these bacteria.

The preferred conditions for proliferation of bacteria in the soil could be close to neutral pH,

Table 4. The correlation coefficients between selected physic-chemical properties of soil and microbiological parameters (n=48)

The physic-chemical properties of soil	The total number of bacteria	The number of *Gram*-negative bacteria	The number of *P. fluorescens*	The number of proteolytic bacteria	The number of amylolytic bacteria
pH_{H2O}	-0.31*	-0.33*	-0.14	0.10	-0.08
C_{org}	0.09	0.18	-0.18	-0.12	0.01
Total N	-0.27	0.11	-0.08	-0.08	0.08
Available P	0.18	0.18	0.28	-0.03	0.28

* significant correlation coefficients for p<0.05.

providing large amounts of organic matter and minerals as well as small amounts of heavy metals, which do not affect the growth and development of bacteria. Biological properties have been suggested as indicators of the response of soil to a specific contaminants [Jorge-Mardomingo et al. 2013]. Ros et al. [2003] suggest that the microbial parameters are sensitive indicators of stress suffered by soil and its recovery after application on not composed sewage sludge, since microbial activity has a direct influence on the stability and fertility of ecosystem.

CONCLUSIONS

1. Application of sewage sludge to the soil caused changes in physic-chemical properties of urban soils. The content of organic carbon, available phosphorus and total nitrogen in the soils increased with the dose of sludge, but there was no significant correlation between the number of different groups of microorganisms in the rhizosphere zone.
2. The changes in the levels of the number of proteolytic and *Gram*-negative microorganisms in the rhizosphere depended on the dose of sludge and mixtures of grasses.
3. With increasing pH a decrease in of the total number of bacteria and the number of *Gram*-negative bacteria was observed.

Acknowledgement

This research was completed with financial support of the project number W/WBiIŚ/14/2014 and S/WBiIŚ/3/2011.

REFERENCES

1. Bień J., Neczaj E., Worwąg M., Grosser A., Nowak D., Milczarek M., Janik M. 2011. Kierunki zagospodarowania osadów w Polsce po roku 2013. Inżynieria i Ochrona Środowiska, 14, 4, 375–384.
2. Galimska-Stypa R., Małachowska-Jutsz A., Mrozowska J., Zabłocka-Godlewska, E. 1999. Laboratorium z Mikrobiologii Ogólnej i Środowiskowej. Mrozowska J. (Ed.). Wydawnictwo Politechniki Śląskiej, Gliwice.
3. Głażewska-Maniewska R., Maciejewska A., Melech A. 2004. Występowanie bakterii glebowych z rodzaju Arthrobacter spp. w uprawie żyta ozimego oraz ich enzymatyczne i antagonistyczne właściwości. Acta Scientiarum Polonorum Series Agricultura, 3, 1, 129–137.
4. Jankiewicz U. 2010. Bioaktywne metabolity ryzosferowych bakterii Pseudomonas, Woda-Środowisko-Obszary Wiejskie, 10, 2 (30), 83–92.
5. Jezierska-Tyś S., Frąc M. 2008. Effect of sewage sludge on selected microbiological and biochemical indices of soil fertility in view of domestic and world wide stusied: a review, Acta Agrophysica, 12, 2, 393–407.
6. Joniec J., Furczak J. 2007. Liczebność wybranych grup drobnoustrojów w glebie bielicowej pod uprawą wierzby użyźnionej osadem ściekowym po drugim roku jego działania, Annales Universitatis Mariae Curie-Skłodowska Sectio E Agricultura, 62, 1, 93–104.
7. Joniec J., Furczak J., Baran S. 2012. The importance of sludge microorganisms in nitrogen transformations in podzolic soil amended with sewage sludge, Archives of Environmental Protection, 38, 1, 35–47.
8. Jorge-Mardomingo I., Solver-Rovira P., Casermeiro M.A., de la Cruz M.T., Polo A. 2013. Seasonal changes in microbial activity in a semiarid soil after application of a high dose of different organic amendments, Geoderma, 206, 40–48.
9. Jurkiewicz A., Orłowska E., Anielska T., Godzik B., Turnau K. 2004. The influence of mycorrhiza and EDTA application on heavy metal uptake by different maize varieties. Acta Biologica Cracoviensa Series Botanica, 46, 7–18.
10. Kiryluk A. 2002. Mieszanki traw i osad ściekowy w procesie rekultywacji wysypiska odpadów komunalnych, AR Lublin PTG, Lublin, 85–86.
11. Luo X., Yu S., Zhu Y., Li X. 2012. Trace metal contamination in urban soils in China. Science of the Total Environment, 421-422, 17–30.
12. Ochrona Środowiska: Materiały i opracowania statystyczne. GUS. Warszawa 2013.
13. Ostrowska A., Gawliński S., Szczubiałka Z. 1991. Metody analizy i oceny właściwości gleb i roślin. Katalog. IOŚ, Warszawa.
14. Rodina A. 1968. Mikrobiologiczne metody badania wód. Państwowe Wydawnictwo Rolnicze i Leśne, Warszawa 1968.
15. Ros M., Hernandez M.T., Garcia C. 2003. Soil microbial activity after restoration of semiarid soil by organic amendments, Soil Biology and Biochemistry, 35, 463–469.
16. Rousk J., Brookes P.C., Baath E. 2010. Investigating the mechanisms for the opposing pH relationships of fungal and bacterial growth in soil, Soil Biology & Biochemistry, 42, 926–934.
17. Rozporządzenie Ministra Środowiska z dnia 13 lipca 2010 r. w sprawie komunalnych osadów ściekowych (Dz.U. 2010 nr 137 poz. 924).

18. Sanchez-Monedero M.A., Mondini C., de Nobili M., Leita L., Roig A. 2004. Land application of biosolids. Soil response to different stabilization degree of the treated organic matter. Waste Management, 24, 325–332.

19. Singh R.P., Agrawal M. 2008. Potential benefits and risks of land application of sewage sludge. Waste Management, 28, 347–358.

20. Siuta J. 2003. Uwarunkowania i sposoby przyrodniczego użytkowania osadów ściekowych. Inżynieria Ekologiczna, 9, 7–42.

21. Su D.C., Wong J.W.C., Jagadeesan H. 2004. Implication of rhizospheric heavy metals and nutrients for the growth of alfalfa in sludge amended soil. Chemosphere, 56, 957–965.

22. Tarrason D., Ojeda G., Ortiz O., Alcaniz J.M. 2010. Effect of different types of sludge on soil Microbial Poperties: a field experiment on degraded mediterraneand soils. Pedosphere, 20, 6, 681–681.

23. Wolińska A. 2010. Aktywność dehydrogenazowa mikroorganizmów glebowych i dostępność tlenu w procesie reoksydacji wybranych mineralnych gleb Polski. Acta Agrophysica, 180, Rozpr. Monogr. 3.

24. Wysocki C., Stawicka J. 2005. Trawy na terenach zurbanizowanych. Łąkarstwo w Polsce, 8, 227–236.

6

ANALYZING TRENDS IN REFERENCE EVAPOTRANSPIRATION IN NORTHWEST PART OF IRAN

Mohammadreza Azizzadeh[1], Khadije Javan[1]

[1] Department of Geography, Payam Noor University, PO Box 3697-19395, Tehran, Iran, e-mail: m_azizzadeh@pnu.ac.ir; kjavan20@yahoo.com

ABSTRACT

Evapotranspiration is one of the most important components of the hydrological cycle. It is essential in all of hydro climatological studies, irrigation and drainage calculations, water balance, and crop water requirements. The aim of this study is to investigate temporal trends of reference evapotranspiration (ETo) in the Northwest part of Iran. For this purpose, the meteorological data from 20 synoptic stations over a 22-year time period (1986–2007) were used. After the calculation of ETo using the Penman-Monteith FAO-56 method, the non-parametric Mann-Kendall test was used to investigate the temporal trends on monthly, seasonal and annual scales. The Sen's estimator method was used and to calculate the slope of the trend line. The results indicated that the ETo trend was increasing (positive) in various months, except for the Sarab station in December. On a monthly scale, the Maragheh station showed the highest positive slope in August, and the Bijar station showed the highest negative slope in May. 43.34% of the stations showed a significant trend, and 56.66% did not show a significant trend. In seasonal and annual time periods there was not a significant decreasing trend in any of the stations. The analysis of the results shows that in the spring 20%, in the summer 55%, in the fall 70%, in the winter 75%, and on an annual scale 60% of the stations under study showed an increasing trend. For modeling these changes, the spatial distribution of ETo trends (on monthly, seasonal, and annual scales) were mapped in ArcGIS. Final models determine that most stations in this study show an increasing trend on monthly, seasonal and annual scales.

Keywords: reference evapotranspiration, trend, Mann-Kendall test, Sen's estimator, Northwest of Iran.

INTRODUCTION

Evapotranspiration, following precipitation, is the most important component of the hydrological cycle. It is one of the main components of water balance, and proper irrigation scheduling programs in any location. Limited water resources and undesirable and uneconomic use of it are the main factors that limit agricultural development and food production in Iran. Therefore, it is essential to have sufficient information about the changes in evapotranspiration for accurate planning of agricultural irrigation and water resource management in any region.

In recent years, the issue of climate change and its impacts has created many challenges in international communities. Such that all nations attempt to determine the change in the components of climate and its process and direction. In addition to elements such as temperature and precipitation, which are often studied, recently other elements such as evapotranspiration have been of interest to researchers.

One of popular methods for analyzing hydro-meteorological time series is the study of the presence or absence of trends in them using statistical tests [Hajjam, 2008, p. 157]. Overall, two statistical methods –parametric and non-parametric– exist for the analysis of trends in time series. However, non-parametric statistical method is used more widely in hydro-meteorological studies. Parametric methods are often based on a regression relationship between a series of data. However, for series of data that do not fit any specific statistical distribution or have a high measure of skewness or kurtosis, the use of non-parametric methods is more suitable. Among the non-parametric tests, the Mann-Kendall test is the best method for the study of uniform data [Xu et al., 2003, p. 145]. This method is very suitable for time series without normal distribution and it is not sensitive to outlier values that are visible in most time series [Kumar et al., 2009, p. 172].

Several studies in relation to changes in evapotranspiration in different parts of the world have been performed. Some of them are as follows.

Thomas [1999] studied the trend in the potential evapotranspiration changes in China using a parametric method. His research showed a decreasing trend for the entire country and an increasing trend for northwest and southeast China. He reported the number of hours of sunshine as the main factor for the changes in potential evapotranspiration in China.

Bandyopadhyay et al. [2007] studied the trend in evapotranspiration in the reference crop in India using the Mann-Kendall test, with statistics from 133 stations during the statistical time period from 1971 to 2002. They determined a decreasing trend in the region. They stated that this decrease was due to a significant increase in relative humidity and also a decrease in wind speed.

Song et al. [2010] studied the time and regional distribution of ETo in the Northern valley in China during the statistical time period from 1961 to 2006. They determined the maximum yearly ETo occurred in central and western China, whereas the minimum occurred in eastern China.

Tabari et al. [2011] studied the trend in evapotranspiration in 20 synoptic stations in western Iran during the period from 1966 to 2005 using the Mann-Kendall method, and used the Sen's estimator method to determine the slope of the trend line. They concluded that the referential evapotranspiration had increasing and decreasing trends, where in 70 percent of the stations showed an increasing trend on an annual scale, and that the increasing trends occurred in the summer and winter.

Zhang et al. [2011] studied the changes in evapotranspiration potential in China using the Penman-Monteith method and the Mann-Kendall test. They used data from 590 meteorological stations, and concluded that seasonal and annual potential evapotranspiration in the east, south and northwestern China showed a decreasing change. The different meteorological variables are factors cause change in evapotranspiration potential, such that in eastern China net radiation changes, and in northwest China relative humidity are effective.

Dinpashoh et al. [2011] studied the trend in the evapotranspiration potential in Iran during the statistical period from 1965 to 2005, and concluded that the number of stations showing an increasing trend is more in Iran. Also, on a monthly time scale, they observed that increasing trends were more than decreasing ones.

Shirgholami et al. [2005] studied the trend in the potential evapotranspiration using data from 34 synoptic stations in Iran. They calculated the amount of potential evapotranspiration using the FAO-Penman-Monteith method, and used the Least-Squares method and the parametric t-test for testing the trend. They concluded that 12 percent of the stations showed a significant positive trend, 26% significant negative, and the rest showed no trend.

Sabzi-parvar et al. [2012] determined the trend of the reference evapotranspiration (ETo) in 13 synoptic stations in the country with warm climate in the period from 1957 to 2006 using the Mann-Kendall and the Age-Estimating method. This study showed that 53 percent of the stations in different seasons or on an annual scale had a significant trend, and 47 percent were insignificant.

Sabzi-Parvar and Shadmani [2012] determined the time variation of the parameter reference evapotranspiration in arid regions of Iran using the nonparametric Mann-Kendall and Spearman tests, and in order to reduce the impact of correlation on the results, they used the TFPW method. The results showed an increasing trend for some cities, and a decreasing trend for others. A comparison of the functionality and power of the two tests on a significant statistical level showed that in most cases a noticeable correspondence in revealing trends in reference evapotranspiration exists between the two tests.

Ismailpour and Dinpashoh [2013] determined the trend in evapotranspiration potential in 6 stations located in the south of the Aras river basin in the period from 1986 to 2008 using the Mann-Kendall test. The results showed that the trend is variable at different times. Among the stations studied, Khoy has experienced a significant increase on an annual scale.

Considering the studies performed relating to the subject of this research it was noted that so far an integral study on the trend in the changes in the evapotranspiration in the northwest of Iran has not been performed. Therefore, the main objectives of this study are:

- Estimating the reference evapotranspiration using the Penman-Monteith-FAO-56 in the Northwest.
- Studying the trend of its variations using the nonparametric Mann-Kendall method.
- Determining the slope of the significant trends using the Age-estimating method.

ARTICLES AND METHODS

Data

For studying the trend in the long term variations in the reference evapotranspiration (ETo) in northwestern Iran, monthly statistics from 20 select synoptic stations in the period from 1987 to 2007 were used. The parameters considered are:

- minimum temperature,
- maximum temperature,
- average temperature,
- minimum relative humidity,
- vapor pressure,
- wind speed,
- number of hours of sunlight.

To ensure the homogeneity of the data, Run test was used, and to fill the gaps in the data, the Averaging method was used. The specifications and the locations of the selected stations are given in Figure 1, and in Table 1.

Table 1. Specifications of the selected stations

Station name	Longitude °E	Latitude °N	Elevation
Ahar	47 04	38 26	1390.5
Ardebil	48 19	38 13	1335.5
Bijar	47 37	35 53	1883.5
Jolfa	45 40	38 45	736.2
Khalkhal	48 32	37 36	1797.4
Khoramdare	49 11	36 11	1575
Khoy	44 59	38 33	1198.7
Mahabad	45 42	36 45	1352
Makoo	44 23	39 22	1411.3
Maragheh	46 10	37 21	1344
Mianeh	47 42	37 27	1110
Oroomieh	45 03	37 39	1328
Parsabad	47 55	39 39	31.9
Piranshahr	45 08	36 41	1443.5
Saghez	46 16	35 15	1522.8
Sanandaj	47 00	35 20	1373.4
Sarab	46 17	37 56	1682
Tabriz	46 17	36 05	1361
Takab	47 06	36 23	1817.2
Zanjan	48 29	36 41	1663

Research method

- The Penman-Monteith-FAO-56 (PMF-56) Method.

This method, which has been introduced by Allen et al. [1998], was used to determine the reference evapotranspiration (ETo) in northwestern Iran. FAO has selected this method as the best for determining evapotranspiration. This method uses the following equation:

Figure 1. Geographic location of the region under study

$$ET_o = \frac{0.408\,\Delta(R_n - G) + \gamma\{\frac{900}{T+273}\}U_2(e_a - e_d)}{\Delta + \gamma(1 + 0.34U_2)} \quad (1)$$

In the above equation, the variables and parameters are defined as follows:

Variable	Description	Unit
ETo	Reference evapotranspiration	mm/day
R_n	Net radiation on the surface of the plants	MJ/m^2
T	Average temperature	˚C
U_2	Wind speed at 2m above the surface	m/s
e_a-e_d	Vapor pressure shortage at a height of 2m	KP
Δ	Slant of the vapor pressure	KP/˚C
γ	Humidity factor	KP/˚C
G	Heat transfer to the ground	MJ/m^2 per day

The values of ETo were calculated monthly using REF-ET software [Allen, 2000].

- The Mann-Kendall Test.

This test was originally introduced by Mann (1945), and was then expanded by Kendall [1975] [Serrano et al. 1999]. This test is used to determine trends and their significance. This test also gives information about the starting point of the trend and sudden changes in the climate [Karaca et al. 1995]. One of the strong points of this test is its suitability for time series that do not fit into any specific statistical distribution. Another advantage of this test is that it is marginally influenced by outlier data which is seen in many time series. The Null Hypothesis of this test is that the data series does not follow a trend and is accidental, and the acceptance of the Alternative Hypothesis means that a trend does exist in the data series. The steps involved in performing this test are as follows:

Calculating the difference between the individual observances, performing the *sgn* function, and determining the parameter *S*:

$$S = \sum_{k-1}^{n-1}\sum_{j=k+1}^{n} sgn(x_j - x_k) \quad (2)$$

In the above equation, *n* is the number of observances, x_j and x_k are the jth and the kth elements respectively.

The *sgn* function is determined as follows:

$$sgn(x) = \begin{cases} +1 & if \quad (x_j - x_k) > 0 \\ 0 & if \quad (x_j - x_k) = 0 \\ -1 & if \quad (x_j - x_k) < 0 \end{cases} \quad (3)$$

If $n>10$, variance (var) is determined as follows:

$$Var(s) = \frac{n(n-1)(2n+5) - \sum_{i=1}^{m} t(t-1)(2t+5)}{18} \quad (4)$$

In the above equation, *n* is the number of observances, *m* is the number of series with at least one repeated element. Parameter *t* indicates the abundance of elements with equal value.

Then, the statistic *Z* is determined as follows:

$$Z_{MK} = \begin{cases} \frac{S-1}{\sqrt{Var\ (S)}} & if \quad S = 0 \\ 0 & if \quad S = 0 \\ \frac{S+1}{\sqrt{Var\ (S)}} & if \quad S = 0 \end{cases} \quad (5)$$

In a two-sided test, the Null Hypothesis is accepted if the following inequality is satisfied:

$$|Z| \leq Z_{\alpha/2} \quad (6)$$

where, the parameter α determines the level of significance considered by the test, and Z_α is the standard normal distribution statistic in the α significance level. Due to the fact that this test is two-sided, $\alpha/2$ is used. In this test, a significance level of 95%, and 99% is considered. A positive value of the statistic Z indicates an increasing trend, whereas a negative Z indicates a decreasing trend [Hajjam et al. 2009, p. 160; Juraj et al. 2009, p. 475].

- The Sen's Estimator method.

Due to the non-parametric nature of the Mann-Kendall test, the Sen's estimating method was used to determine the slope of the trends. In this method, outlier data have a very small effect on the slope; this is its main advantage compared to the Linear Regression method [Novotny and Stefan, 2007, p. 336].

The slope between any pair of observances is calculated as follows:

$$Q = \frac{X_t - X_s}{t - s} \quad (7)$$

where X_t and X_s are the data at times *t* and *s* respectively, and *t* occurs one time unit after *s*. Applying this equation to all pairs of data, a series of slopes are calculated. The slope of the trend (Q_{med}) is then determined by finding the median of this series. A positive Q_{med} indicates an increasing trend, and a negative one indicates a decreasing trend [Hajjam et al. 2009, p. 160; Juraj et al. 2009, p. 475].

RESULTS AND DISCUSSION

Monthly time series

After calculating the amounts for the reference evapotranspiration, the Statistic of the

Mann-Kendall and the Sen's estimator methods for monthly time series were calculated, and the significance of the Statistic **Z** on a level of significance in of 95% and 99% was studied. The results of the Mann-Kendall test are shown in Table 2. As noted in this table, the reference evapotranspiration shows an increasing trend in various months, and it is only in the Sarab station in the month of December that the significant trend shows a decrease. In January, only the Takab station shows a significant increasing trend on a 99% level, and in the other stations being studied, a significant trend is not observed. In February, about half the stations show a significant trend.

In March, all stations, except the Sarab and the Saghez stations, show a significant increasing trend. In this month, on average, 90% of the region's stations show a significant trend. In April this number decreases and only eight stations in the region show a significant trend; in May this decline continues and it is only in the Tabriz and the Ardabil stations that a significant increasing trend is observed. From the month of June onward, the number of stations showing a trend increases, such that in August this number is 15, and in October it is 17. Again, in November and December the number of stations showing a significant trend, decreases. The highest amount of monthly increasing trend was seen in October in the Khoy station with a **Z** of 4.23, and the highest amount of monthly decreasing trend was seen in the Sarab station in December with a **Z** of -2.63.

After determining the amounts and the direction of the trends, the slope of the trends over the months of the year were calculated using the Sen's estimator method, as shown in Table 3. From the results, shown in Table 3, we can note that in the Bijar, Maragheh, Saghez, Tabriz, and Ahar stations have had a considerable slope compared to the other stations.

The highest amount of the increasing monthly reference evapotranspiration trend with a slope of 0.106 mm/day, occurred in the Maragheh station in August, and its highest decreasing trend occurred in the Bijar station in May with a slope of -0.045 mm/day (Figure 2).

In order to specify the scattering of the regions that have a significant trend, using the results of the Mann-Kendall test, the region was mapped out. The mapping of the region was carried out using the Kriging method. Figure 3 shows the regional distribution of the stations having a significant trend separated according to the monthly series.

Table 2. Mann-Kendall Statistic (Z) for reference evapotranspiration on a monthly scale

Station name	Jan	Feb	Mar	Apr	May	Jun	July	Aug	Sep	Oct	Nov	Des
Ahar	1.17	2.32*	2.00*	0.82	0.40	1.40	3.10**	3.41**	1.36	3.23**	0.77	1.38
Ardebil	1.82	2.84**	2.09*	2.09*	2.66**	0.99	0.96	2.06*	2.55**	2.48*	1.70	1.02
Bijar	1.09	1.27	3.23**	2.05*	-1.30	0.15	-0.18	-0.54	-0.18	1.30	-0.45	-0.21
Jolfa	0.26	1.17	2.49*	1.11	0.03	1.30	1.02	2.62**	0.93	2.11*	2.84**	0.58
Khalkhal	1.71	2.82**	2.96**	2.59**	0.39	0.76	1.42	2.99**	2.15*	2.71**	0.39	0.12
Khoramdare	1.51	2.99**	3.41**	2.17*	1.36	3.00**	1.33	2.03*	1.36	2.62**	0.62	1.00
Khoy	-0.79	1.40	2.55**	1.58	0.62	3.68**	3.75**	4.23**	3.62**	4.66**	2.35v	0.27
Mahabad	1.27	2.33*	3.30**	2.68**	0.54	2.51*	2.66**	3.44**	2.70**	2.76**	1.11	1.10
Makoo	-0.51	1.82	2.54*	1.16	1.07	1.80	1.34	2.91**	1.75	4.03**	2.90**	1.38
Maragheh	0.79	2.04*	2.85**	2.21*	1.68	3.33**	1.97*	3.64**	2.88**	4.29**	3.12**	2.37*
Mianeh	-0.13	1.60	2.08*	0.39	0.09	1.30	-0.15	2.08*	2.72**	1.72*	0.88	-0.80
Oroomieh	-0.79	1.00	2.31*	1.85	0.68	2.62**	2.79**	3.28**	2.67**	2.31*	0.00	-1.24
Parsabad	1.87	3.40v	2.67**	1.30	1.35	-0.06	0.51	-0.42	-0.11	3.53**	3.62**	1.99*
Piranshahr	0.79	2.28*	3.64**	2.83**	1.64	2.74**	2.38*	2.54*	1.89	2.89**	2.08*	2.08*
Saghez	-1.85	-0.35	1.30	0.28	-1.19	0.54	0.73	2.23*	1.75	2.45*	-0.83	-1.28
Sanandaj	1.18	1.53	3.11v	1.83	0.25	1.38	0.93	1.21	1.13	1.95	1.00	0.39
Sarab	-0.20	1.87	1.64	0.48	0.03	0.57	0.34	1.83	0.37	2.20*	-0.78	-2.63**
Tabriz	1.10	2.08*	2.65**	1.92	2.03*	2.06*	1.55	2.93**	3.98**	4.54**	1.61	1.92
Takab	3.24**	2.72**	2.91**	1.70	0.28	0.00	-1.33	-0.28	0.03	3.19**	1.42	2.66**
Zanjan	1.06	1.89	2.38*	2.49*	0.37	1.53	0.51	2.17	2.34*	2.85**	-0.09	-0.14

(*) trend at 95% and (**) trend at 99% level.

Table 3. The slope of the monthly reference evapotranspiration (mm per day)

Station name	Jan	Feb	Mar	Apr	May	Jun	July	Aug	Sep	Oct	Nov	Des
Ahar	0.015	0.032	0.028	0.009	0.004	0.024	0.020	0.053	0.019	0.035	0.012	0.010
Ardebil	0.012	0.026	0.021	0.019	0.041	0.016	0.016	0.045	0.038	0.042	0.015	0.007
Bijar	0.013	0.022	0.077	0.049	-0.045	0.010	-0.009	-0.029	-0.007	0.040	-0.008	-0.002
Jolfa	0.002	0.007	0.024	0.010	0.000	0.053	0.031	0.055	0.016	0.008	0.011	0.001
Khalkhal	0.009	0.020	0.021	0.036	0.007	0.011	0.024	0.055	0.026	0.023	0.003	0.000
Khoramdare	0.009	0.031	0.056	0.049	0.040	0.043	0.028	0.054	0.025	0.042	0.005	0.007
Khoy	-0.004	0.006	0.022	0.013	0.010	0.041	0.038	0.054	0.033	0.023	0.006	0.000
Mahabad	0.004	0.015	0.039	0.023	0.011	0.042	0.031	0.039	0.029	0.039	0.008	0.005
Makoo	-0.002	0.009	0.026	0.013	0.017	0.046	0.027	0.053	0.029	0.033	0.012	0.003
Maragheh	0.003	0.013	0.038	0.021	0.033	0.087	0.055	0.106	0.083	0.071	0.015	0.011
Mianeh	0.000	0.012	0.026	0.004	0.002	0.041	-0.001	0.038	0.034	0.025	0.003	-0.001
Oroomieh	-0.001	0.005	0.020	0.014	0.013	0.043	0.030	0.044	0.035	0.019	0.000	-0.004
Parsabad	0.009	0.014	0.022	0.023	0.016	-0.006	0.012	-0.006	-0.001	0.020	0.017	0.008
Piranshahr	0.007	0.017	0.053	0.043	0.034	0.054	0.051	0.054	0.040	0.040	0.022	0.013
Saghez	0.012	-0.004	0.023	0.003	-0.033	0.012	0.026	0.052	0.032	0.034	-0.006	-0.008
Sanandaj	0.003	0.003	0.035	0.022	0.010	0.037	0.030	0.035	0.014	0.019	0.004	0.003
Sarab	-0.001	0.009	0.014	0.007	0.000	0.020	0.001	0.040	0.005	0.024	-0.005	-0.006
Tabriz	0.008	0.028	0.049	0.035	0.039	0.071	0.040	0.076	0.085	0.069	0.018	0.014
Takab	0.007	0.015	0.031	0.017	0.010	0.004	-0.030	-0.010	0.001	0.030	0.003	0.006
Zanjan	0.004	0.018	0.037	0.029	0.013	0.046	0.009	0.049	0.028	0.038	-0.001	-0.001

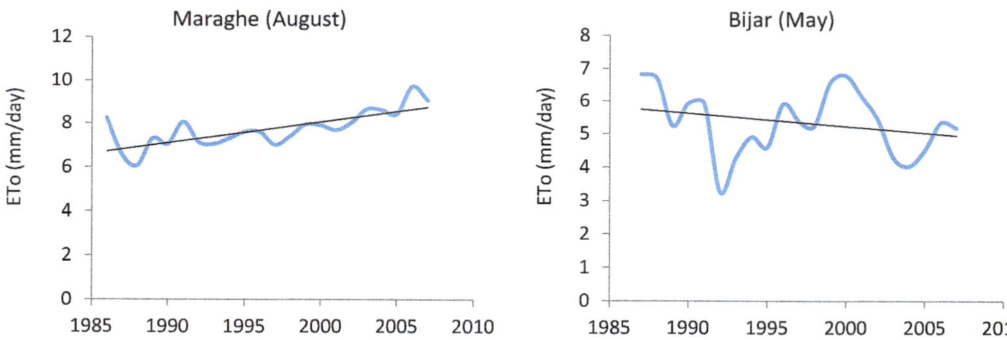

Figure 2. Monthly reference evapotranspiration in stations with the highest slope of trend

In January in all locations, other than central locations of the region and locations surrounding the Takab station, a significant trend is not seen. In February, the eastern border, the central location and a small area in the west of the region, show an increasing trend. In March these locations are expanded and a significant trend is seen in most locations of the region under study. Spatial distribution in April is similar to March; and in the month of May, it is shown only in the Tabriz and the Ardabil stations. In June, this location is expanded in the northwest and a small area in the southeast; and in July this situation continues in the northwest. In August and September, the locations having a significant trend is expanded, and it is seen in all locations other than the southern portion and the northeastern border. In October, locations having a trend is expanded to the northeast. In November and December the expanse of the locations having a trend is decreased, and it is seen only in a narrow border on the north and some areas in the west. However, in December the only location showing a significant decreasing trend is expanded around the Takab station.

Altogether, in analyzing the significance of monthly trends in the northwest of Iran, 42.92 percent of the cases showed an increasing trend, 0.42 percent showed a decreasing trend, and 56.66 percent of the cases showed no trend.

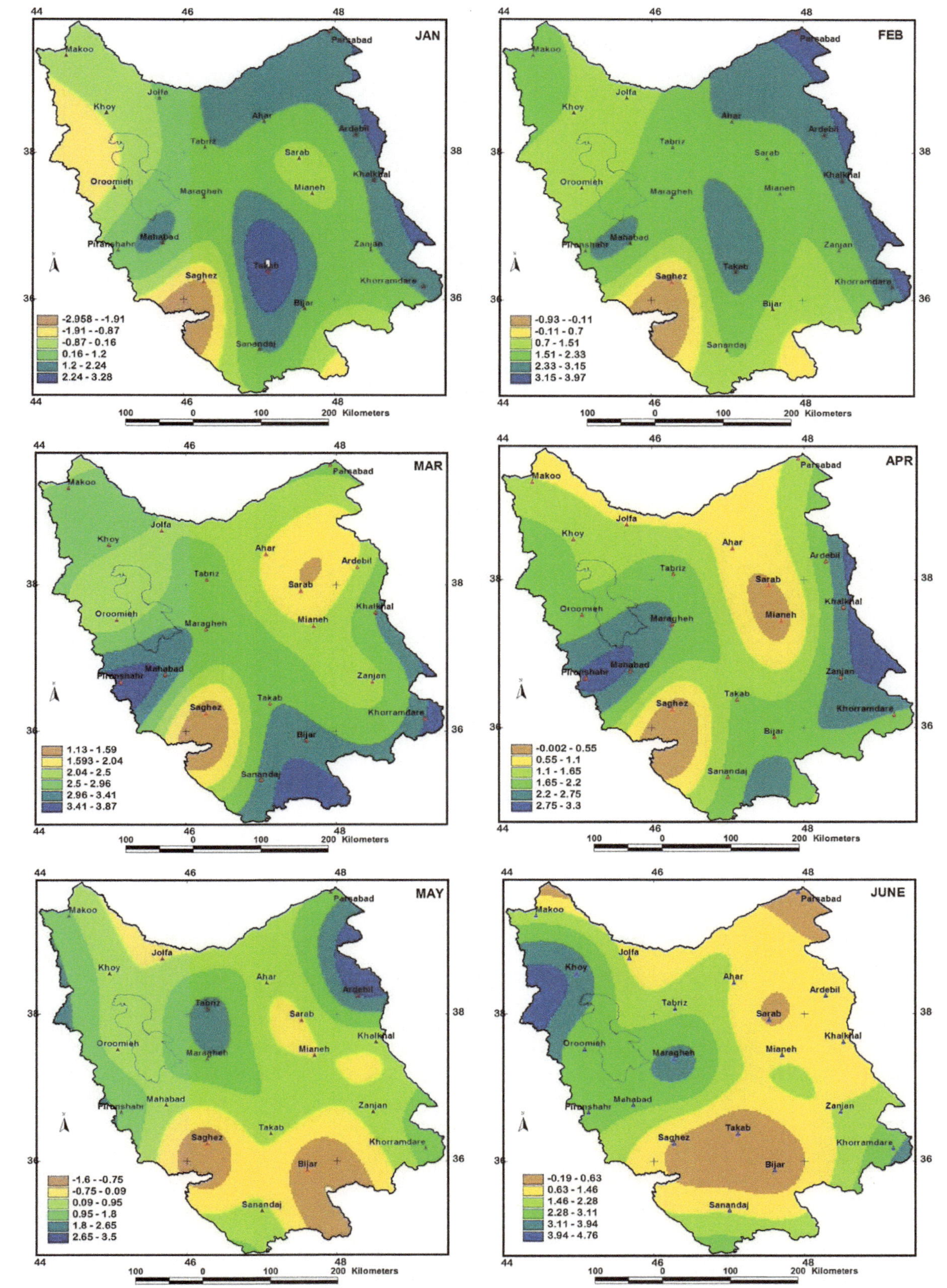

Figure 3. Regional distribution of the evapotranspiration in the northwest on a monthly scale

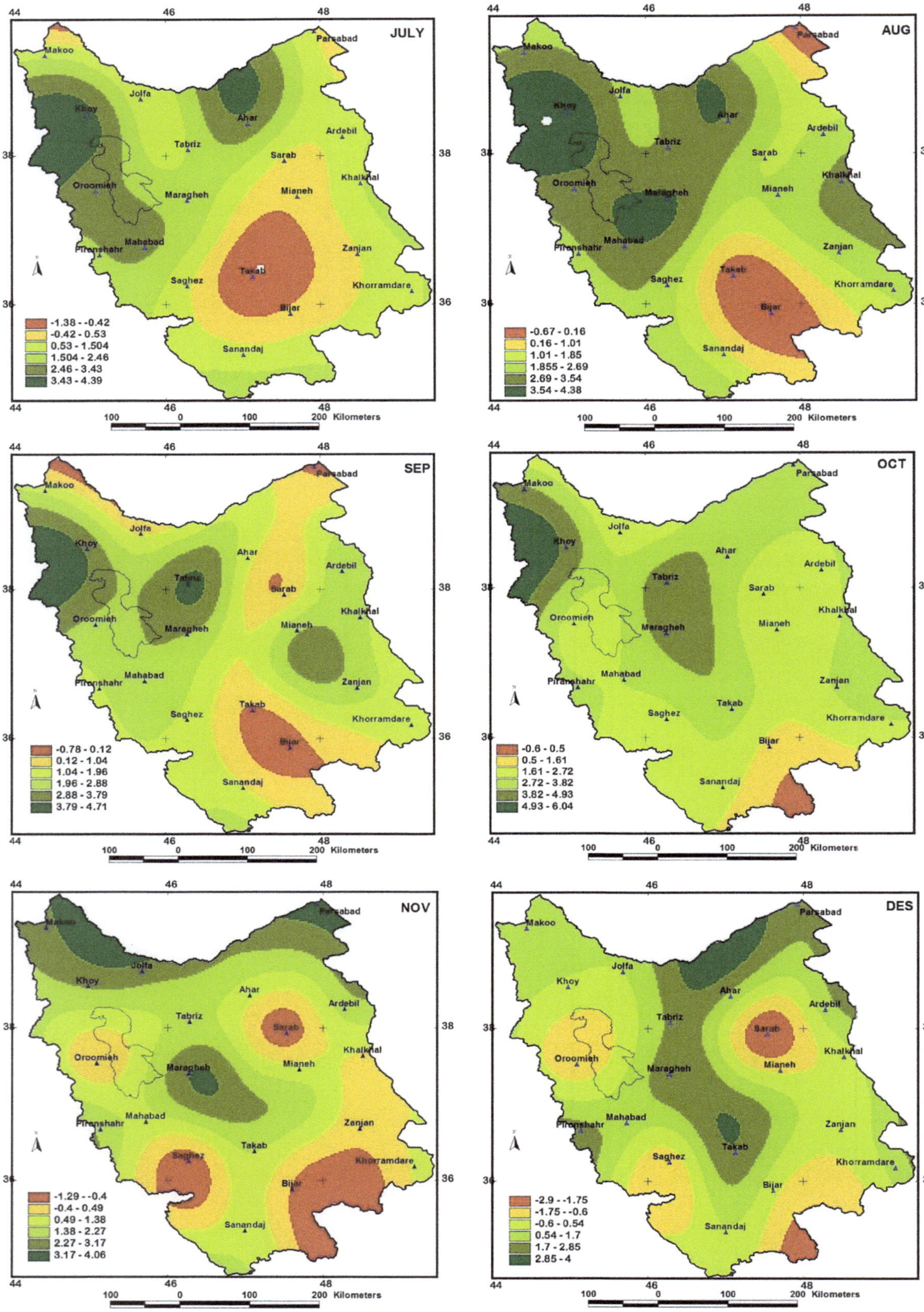

Figure 3. Regional distribution of the evapotranspiration in the northwest on a monthly scale

Seasonal and annual time series

The results of the Mann-Kendall test in seasonal and annual periods in all stations under study are shown in Table 4. As we can note from this table, in seasonal and annual time periods, none of the stations show a significant decreasing trend, and all the trends specified are increasing. Among the stations under study, the Sarab and the Saghez stations do not show a significant trend in seasonal and annual analyses. In the spring, only in four stations across the region (Maragheh, Piranshahr, Tabriz, and Ardebil), a significant increasing trend is seen; and in the winter, a significant trend is seen in most stations across the region. In this season, the number of significant trends is more than that in other seasons which shows that winter shows a higher sensitivity to the changes in the trends in ETo in relation to other seasons.

The highest trend was seen in the spring, fall, and also on an annual scale in the Maragheh station which were 2.93, 4.43, and 4.06 millimeters per year; and the fall shows the highest sensitivity. Furthermore, the highest trend in the summer occurred in the Khoy station with an amount of 4.17 millimeters per year, and in the winter it occurred in the Khalkhal station, with an amount of 3.36 millimeters per year.

In table 5, the slope of the annual and seasonal trends in the stations under study is presented. As we can notice from Table 5, among the stations, the Maragheh station in the spring and in the summer and on an annual scale, the Tabriz station in the Fall, and the Bijar station in the Winter have the highest increasing trend of the reference evapotranspiration. The analysis of the results shows that in the Spring 20%, in the Summer 55%, in the Fall 70%, in the Winter 75%, and on an annual scale 60% of the stations under study showed an increasing trend,

The spatial scattering of the stations having a significant trend, separated annually and seasonally, are shown in Figures 4 and 5. Studying Figure 4, we can observe that in the spring most locations, other that some central, northeastern, and western areas do not show a significant trend. However in the summer, in the entire region, except in the southern areas, and in small borders in the north and the east, a significant trend is observable. Spatial scattering of the stations in the fall and in the winter are similar, other than some small northwestern, southern, and central areas, the rest of the region shows an increasing trend. The spatial scattering of the annual reference evapotranspiration trend in the northwest of the country (Figure 5), shows a significant increasing trend in most stations.

Table 4. Mann-Kendall Statistic (Z) for reference evapotranspiration on a seasonal and annual scale

Station name	Spring	Summer	Autumn	Winter	Annual
Ahar	1.27	2.82**	2.28*	2.26*	2.82**
Ardebil	2.12*	2.31*	2.31*	2.71**	2.88**
Bijar	-0.18	-0.27	0.63	2.63**	0.60
Jolfa	1.33	1.92	3.11**	1.95	1.52
Khalkhal	1.10	3.02**	2.06*	3.36**	1.30
Khoramdare	1.92	1.61	2.26*	3.10**	2.65**
Khoy	1.95	4.17**	4.20**	1.91	3.95**
Mahabad	1.89	2.91**	2.28*	3.16**	3.10**
Makoo	1.95	2.17*	3.33**	2.21*	2.54*
Maragheh	2.93**	3.58**	4.43**	2.26*	4.06**
Mianeh	1.06	1.89	1.42	2.03*	1.84
Oroomieh	1.75	3.05**	1.13	1.66	2.71**
Parsabad	0.93	0.00	3.57**	2.90**	1.07
Piranshahr	2.51**	2.34**	2.93**	2.66**	2.82**
Saghez	-0.34	1.80	1.66	0.06	1.30
Sanandaj	1.35	1.30	2.23*	2.60**	2.43**
Sarab	0.28	1.02	0.87	1.55	0.68
Tabriz	2.43*	3.05**	3.68**	2.59**	2.76**
Takab	0.34	-0.34	3.11**	2.96**	0.56
Zanjan	1.47	2.09*	1.75	2.21*	2.28*

Table 5. Seasonal and annual slope of the reference evapotranspiration trend in the stations under study

Station name	Spring	Summer	Autumn	Winter	Annual
Ahar	0.016	0.031	0.016	0.027	0.038
Ardebil	0.023	0.037	0.023	0.018	0.029
Bijar	-0.007	-0.013	0.011	0.039	0.008
Jolfa	0.022	0.028	0.006	0.012	0.040
Khalkhal	0.015	0.030	0.009	0.016	0.033
Khoramdare	0.040	0.039	0.019	0.034	0.062
Khoy	0.020	0.047	0.009	0.010	0.045
Mahabad	0.028	0.037	0.017	0.017	0.045
Makoo	0.027	0.030	0.017	0.013	0.048
Maragheh	0.049	0.089	0.029	0.019	0.081
Mianeh	0.019	0.025	0.009	0.014	0.028
Oroomieh	0.020	0.038	0.005	0.008	0.016
Parsabad	0.015	0.001	0.015	0.015	0.014
Piranshahr	0.042	0.044	0.026	0.026	0.057
Saghez	-0.003	0.039	0.008	0.000	0.031
Sanandaj	0.014	0.021	0.011	0.017	0.029
Sarab	0.006	0.015	0.004	0.008	0.013
Tabriz	0.047	0.060	0.033	0.027	0.076
Takab	0.011	-0.005	0.014	0.018	0.009
Zanjan	0.031	0.029	0.012	0.016	0.040

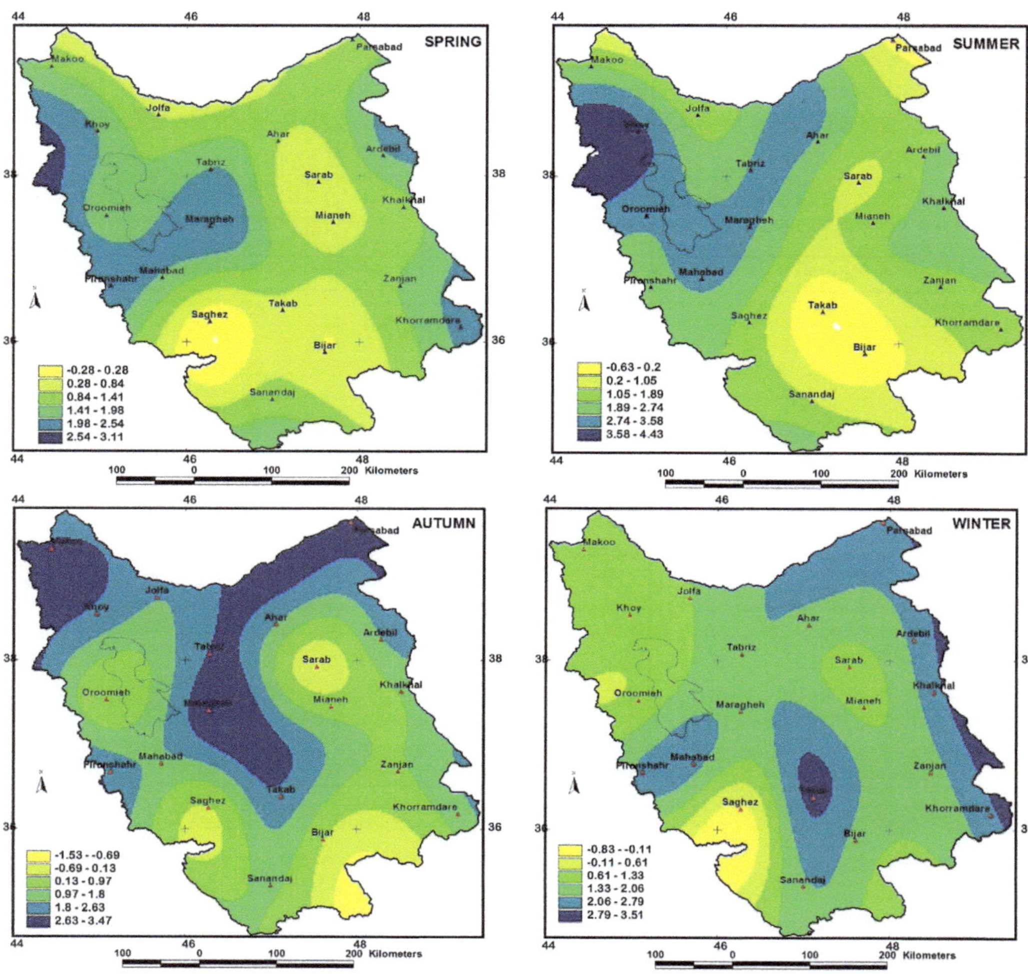

Figure 4. Spatial scattering of the reference evapotranspiration trend on a seasonal scale

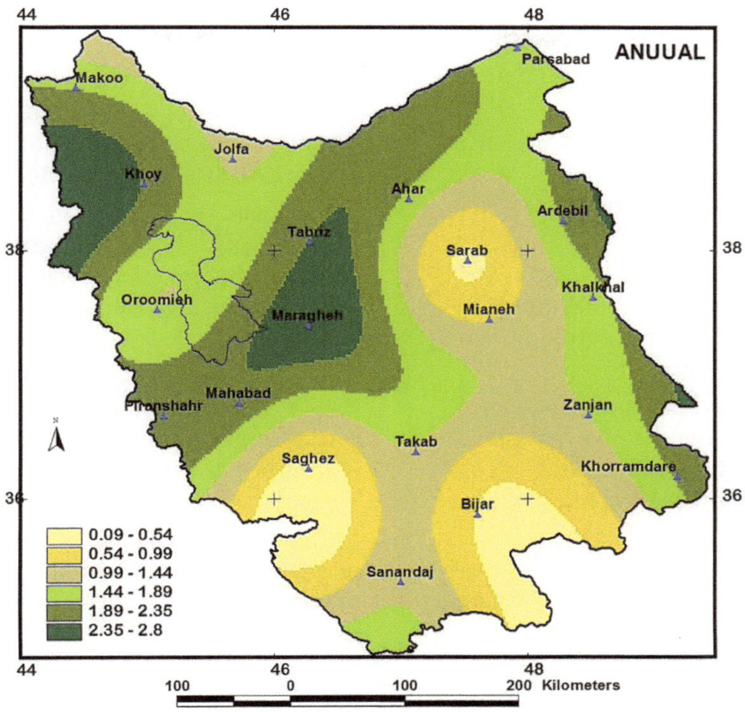

Figure 5. Locational scattering of the reference evapotranspiration trend on an annual scale

CONCLUSION

Evapotranspiration is one of the important elements of the hydrological cycle and plays an important role in agricultural studies, water resources management planning, irrigation and drainage design. The study of the reference evapotranspiration is essential for offering a proper cropping pattern and optimizing the use of existing water resources. In this study, using the Mann-Kendall test, the trend of reference evapotranspiration for 20 synoptic stations in northwest of Iran during the period from 1986 to 2007 was analyzed. After calculating the amount and direction of trends, the slope of the monthly trends were calculated using the Sen's estimator method. The results showed that the reference evapotranspiration had an increasing trend, and only the Sarab station in the month of December showed a significant decreasing trend. Overall, in the analysis of significance of the monthly trends in northwest Iran, 43.34 percent had a significant trend and 56.66 percent showed no trends.

On an annual and seasonal scale, none of the stations showed a significant decrease, all the specified trends were increasing. In the winter, in most stations in the region showed a significant trend, and the number of stations showing a significant trend in this season was more than other seasons. Analysis of the results shows that in the spring 20%, in the summer 55%, in the fall 70%, in the winter 75%, and on an annual scale 60% of the stations under study showed an increasing trend. Results obtained by other researchers [Dinpashoh, et al. 2011; Esmailpour and Dinpashoh, 2013] show an increasing trend of evaporation and transpiration.

Acknowledgement

This Work is supported by the Payame Noor university of I.R of Iran through the research program under Grant.

REFERENCES

1. Allen R.G., Pereira L.S., Raes D., and Smith M. 1998. Crop evapotranspiration: Guidelines for computing crop water requirements. FAO Irrigation and Drainage Paper, No. 56, FAO, Rome, Italy, 300 p.
2. Allen R.G. 2000. Manual REF-ET version Windows 2.0. Available online at www.kimberly.idaho.edu.
3. Bandyopadhyay A., Bhadra A., Raghuwanshi N.S, and Singh R. 2009. Temporal trends in estimates of reference evapotranspiration over India. Journal of Hydrologic Engineering, 14(5), 508–515.

4. Dinpashoh Y., Jhajharia D., Fakheri-Fard A., Singh V.P., Kahya E. 2011. Trends in reference crop evapotranspiration over Iran. Journal of Hydrology, 399, 422–433.
5. Esmailpour M. and Dinpashoh Y. 2013. Analysis of long-term potential evapotranspiration in the south of the Aras river basin. Geography and Environmental Planning, No. 47, 193–210.
6. Juraj M., Cunderlik Taha B., Ouarada M.J. 2009. Trends in the timing and magnitude of floods in Canada. Journal of Hydrology, No. 375, 471–480.
7. Hajjam S., Khoshkhu Y. and Shamsuddin-Vandi R. 2009. Analysis of changes in seasonal and annual precipitation of selected stations in central Iran using nonparametric methods. Geographical Studies, No. 64, 157–168.
8. Karaca M., Tayanc M., Toros H. 1995. Effects of urbanization on climate of Istanbul and Ankara. Atmospheric Environment, 29, 3411–3421.
9. Kumar S., Merwade V., Kam J. and Thurner K. 2009. Sreamflow trends in Indiana: Effects of long term persistence, precipitation and subsurface drains. Journal of Hydrology, 374, 171–183.
10. Novotny E.V., 2007. Stream flow in Minnesota: Indicator of climate change. Journal of Hydrology, 334, 319–333.
11. Sabzi-Parvar A. and Shadmani M. 2012. Analysis of reference evapotranspiration using Mann-Kendall test and Spearman in arid regions of Iran. Journal of Soil and Water, No. 4, 823–834.
12. Sabzi-Parvar A., Mir-Mousavi Sh. and Nazemosadat M. 2012. The study of long-term variations of reference evapotranspiration in some of the warm climates in the country. Research in Physical Geography, No. 75, 1–17.
13. Serrano A., Mateos V.L., and Garcia J.A. 1999. Trend Analysis of Monthly Precipitation over the Iberian Peninsula for the Period 1921-1995. Phys. Chem. EARTH (B), 24, 85–90.
14. Shirgholami H., Ghahreman B., Alizadeh F. and Bodagh-Jamali J. 2005. Evaluation of reference crop evapotranspiration in Iran. Journal of Agricultural Sciences and Natural Resources of the Caspian, 2 (3), 11–27.
15. Song Z.W., Zhang H.L., Snyder R.L., Anderson F.E., Chen F. 2010. Distribution and trends in reference evapotranspiration in the North China Plain. Journal of Irrigation and Drainage Engin., 136, 240–247.
16. Tabari H., Marofi S., Aeini A., Hosseinzadeh Talaeea P., and Mohammadi K. 2011. Trend analysis of reference evapotranspiration in the western half of Iran. Agricultural and Forest Meteorology, 151 (2), 128–136.
17. Thomas A., 1999. Spatial and temporal characteristics of potential evapotranspiration trends over China. International Journal of Climatology, 20, 16 p.
18. Turgay P. and Ercan K. 2005. Trend analysis in Turkish precipitation data. Hydrological processes. Published online in Wiley Inter Science (www.Interscience.wiley.com).
19. Xu Z.X., Takeuchi K. and Ishidiaira H. 2003. Monitoring trend step changes in Japanese precipitation. Journal of Hydrology, 279, 144–150.
20. Zhang Q., Xu CH.Y. and Chen X. 2011. Reference evapotranspiration changes in China: Natural processes or human influences? Theoretical and Applied Climatology, 103, 479–488.

7

THE ALLELOPATHIC INFLUENCE OF *TARAXACUM OFFICINALE* ON THE INITIAL GROWTH AND DEVELOPMENT OF *FESTUCA RUBRA* L.

Jolanta Jankowska[1], Grażyna Anna Ciepiela[1], Kazimierz Jankowski[1], Roman Kolczarek[1], Jacek Sosnowski[1], Beata Wiśniewska-Kadżajan[1]

[1] Department of Grassland and Green Areas Creation, Siedlce University of Natural Sciences and Humanities, B. Prusa 14. 08-110 Siedlce, Poland, e-mail: laki@uph.edu.pl

ABSTRACT

Common dandelion (*Taraxacum officinale* F.H. Wigg) is perennial plant often stepping out in very large quantity in the sward of natural meadows. This species expands, enlarging its population very quickly. In the literature there is the lack of data related to influences of the common dandelion on plants growing in it's neighbourhood. It is also unknown why this species creates large clusters. It may compete with different plants through the allelopatic influence. Therefore, the aim of this work was to test of the influence of water extracts from leaves and from the roots of common dandelion (*Taraxacum officinale*) as well as soil extracts from the radicular layer of this species on the germination of the seeds and the initial growth of Red fescue (*Festuca rubra* L.). The investigative material came from leaves and roots of *Taraxacum officinale* and soil coming from the radicular layer of this species. It was applied most often in biotest experiments on germination. Red fescue was the tested plant. The germination energy of red fescue was the most braked through the plant extracts prepared from roots and leaves of *Taraxacum officinale*.

Keywords: common dandelion, red fescue, allelopathy.

INTRODUCTION

In plant communities, such as natural grassland, predominance of these species which are particularly well adapted to the suitable environmental conditions occurs e.g., drought, excess of moisture or to the shortage of nutritive components. The botanical composition of natural ecosystems can be largely considered as the result of varied competitive relations [Callaway et al., 2005; Goslee et al., 2001; Ridenour et al., 2001; Weidenhamer 2006]. One may also encounter the occurrence of allelopathic influences [Basis et al., 2003; Spyreas et al., 2001).

According to many authors [Oleszek 1996], allelopathy is the process consisting of slowed down or stimulated development of one plant by another one. It happens as the result of the secretion of various chemicals, which have allelopathic proprieties, to the environment. These chemicals in high concentrations can act as inhibitors of growth and development, however in low concentrations they can be the stimulants of these processes [Inderjit and Duke 2003; Rene et al., 2004].

One of the reasons for meadow sward degradation is the allelopatical substances secreted to the soil by some species of plants. These substances are called allelocompounds [Perry et al., 1005; Birkett et al., 2001; Sinkonen 2003]. They can be released to the environment through evaporating and leaching from the underground parts and by the exudation from the root system or in the process of the decompose of the organic matter [Einhelling 1995a, 1995 b].

Plants of various species and their different organs are the source of the allelopatical substances [Lipińska and Harkot 2007]. They are in aboveground parts (generative and vegetative) and underground plants both alive and dead, which are on the surface of soil, and also in the soil [Bertin et al., 2003]. Many allelopatical compounds in aboveground parts of plants (letter, stems, flowers) dissolve in water, therefore, they can be washed by rain, fog or drops of dew [Beyschlag et al., 1996]. The important source of relationships allelopatical compounds is also decomposing plant residues [Pratley et al., 2000]. Specific products of the secondary metabolism of plants are secreted by root systems of many species of meadow plants, which have allelopatical proprieties that is why they can influence on the development of plants growing in their environment [Suman et.al., 2002; Callaway et. al 2000; Julita et.al.,2002].

According to many authors [Harkot and Lipińska 1997] high allelopatical activity among others are demonstrated by such species as: *Elymus repens, Festuca rubra, Festuca arundinacea, Bromus inermis*, *Bromus hordeaceus*, and also *Plantiago lanceolata, Achilea millefolium* and *Hieracium pilosella.* These plants affect negatively the germination and growth of different plants.

Common dandelion (*Taraxacum officinale* F.H. Wigg) is perennial plant often stepping out in very large quantity in the sward of natural meadows. This species expands, enlarging its population very quickly. In the literature there is the lack of data relating to influences of common dandelion on plants growing in its neighborhood. It is also unknown, why this species creates large clusters. It may compete with different plants through the allelopatic influence.

So, the aim of this work was to test the influence of water extracts from leaves and from the roots of common dandelion (*Taraxacum officinale*) as well as soil extracts from the radicular layer of this species on the seeds germination and the initial growth of red fescue (*Festuca rubra* L. sensu lato).

MATERIALS AND METHODS

The investigative material came from leaves and roots of *Taraxacum officinale* and soil coming from the radicular layer of this species. The samples of the material were taken in October 2006 from the natural meadow community, in which the large quantity *Taraxacum officinale* stepped out. The plant and soil material after drying in room temperature and crumbling was used for preparing water extracts. For this purpose samples of leaves and root with the mass about 25, 50 and 75 grams and the samples of the soil with the mass about 500, 1000 and 1500 grams were prepared. The weighed out material was placed in Erlenmayer butts and flood with distilled water in the following quantity:

- 500 cm^3 – for leaves or roots,
- 1000 cm^3 – for the soil.

Prepared samples were left in the room's temperature for 24 hours. The received extract was infiltrated by filter-paper, and then use to the performance of biotests. During the whole experience, extracts were kept in a refrigerator. Most often experimented biotest with germination were applied. *Festuca rubra* was the tested plant. For this purposethe authors used Petry patern with the diameter 15 cm, lay out with two layers of filter-paper and placed 25 seeds, watering them every three days with 5 ml of one prepared extracts. In the next days, it was watered with distilled water. The experience was put in three replicants with the control object (distilled water).

Following combinations were applied in this experiment:

- control (distilled water)
- water extracts from leaves – (25, 50 and 75 g of leaves)
- water extracts from roots – (25, 50 and 75 g of roots)
- water extracts from the soil – (500, 1000 and 1500 g of the soil)

In the carried experiment the seeds germination of *Festuca rubra* was in room temperature in daylight.

Energy and the ability of seeds germination of this plant were estimated. The measurement of the energy of germination was done after 7 days from the date of the experiment foundation however the germination ability after 21 days from this date. Moreover, three times in the three-days' interval lengths of leaf sheaths and the leaves length of *Festuca rubra* were estimated and once was estimated the radicular lengths of this plants. First measurement of the length of leaf sheath and the leaves length of *Festuca rubra* was conducted the next day after the estimation of germination ability. During the last measurement the roots length of this plant was also measured.

Variation analysis was used in the statistical study of the results for one factors experiences. Significance of the differentiation the means characterizing the investigative factor was estimated by the Tukey'a test for the level of significance $\alpha \leq 0{,}05$.

RESULTS AND DISCUSSION

The germination energy of *Festuca rubra*

The significant parameter in the biological estimation of grasses species is germination energy of the seeds.

The results obtained in the work show that the germination energy of *Festuca rubra* seeds was differentiated in dependence on the kind of solution applied to watering of these seeds, as also on the mass of sample of plant and soil material used to preparing of extract (Table 1, 2, 3 and Figure 1).

The highest energy of germination was affirmed in the conditions of applying soil extracts (from 52 to 60%). Extracts from leaves the mostly braked the germination of seeds, and only single germination of seeds was observed after seven days from their sowing.

It should also be emphasized that generally all tested water extracts significantly reduced the energy of germination in the relation to the control.

The germination ability of *Festuca rubra* L.

The germination ability of *Festuca rubra* seeds was very high on the control object, where the value amounted to 96%. In the case of soil extracts (Figure 2) this ability in comparison with control object (96%) reached slightly lower values for the successive concentrations, getting suitably values 88%, 80% and 72%.

The germination ability of *Festuca rubra* seeds after applying water extracts from roots was considerably lower both in relation to the control object and to the soil extracts. The germination ability carried out respectively 36%, 16% and 12% in dependence on the applied increasing of concentrations of these solutions.

However, the studied feature had the lowest values after applying water extracts from leaves of *Taraxacum officinale* (from 1 to 40%), what can testify to their big allelopathy influence. It should be noticed, that water extracts both from the soil and from the plants (leaves, roots) significantly reduced germination ability of *Festuca rubra* seeds in relation to the control object (Table 1, 2, 3). Also stronger influence of all the applied water solutions was observed on the energy of

Table 1. Influence of the water solution of the soil extract on initial growth and development of *Festuca rubra*

Studied feature	Control	Mass of the sample (g)		
		500	1000	1500
Germination energy	64a	60ab	56bc	52c
Germination ability	96a	88b	80c	72d
Length of the leaf [mm]	42a	52b	45a	40a
Length of the sheath [mm]	7a	9a	9a	8a
Length of radicle [mm]	25a	21ab	20ab	16b

Comment: Means in lines estimated with the same letters do not differ significantly.

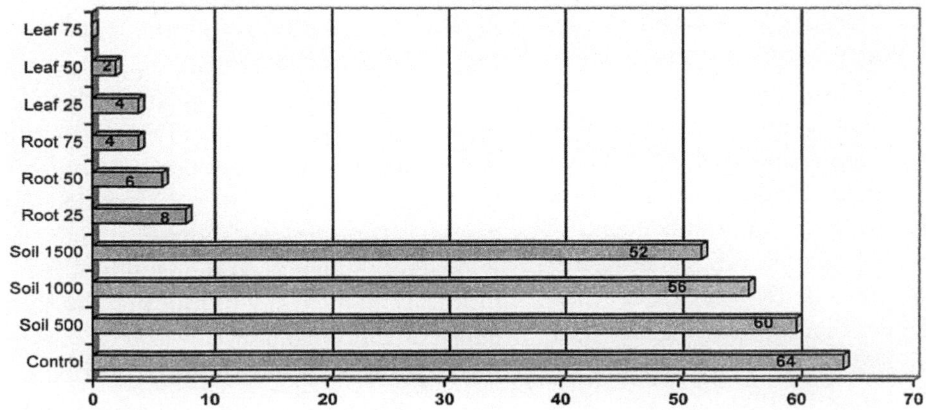

Figure 1. The germination energy of seeds of *Festuca rubra* L. in dependence on various concentrations of studied extracts

Table 2. Influence of the water solution of the extract from *Taraxacum officinale* roots on initial growth and development of *Festuca rubra*

Studied feature	Control	Mass of the sample (g)		
		25	50	75
Germination energy	64a	8b	6b	4b
Germination ability	96a	36b	16c	12c
Length of the leaf [mm]	42ab	48b	42a	36a
Length of the sheath [mm]	7a	13a	12a	7a
Length of radicle [mm]	25a	6b	3bc	2c

Comment: Means in lines estimated with the same letters do not differ significantly.

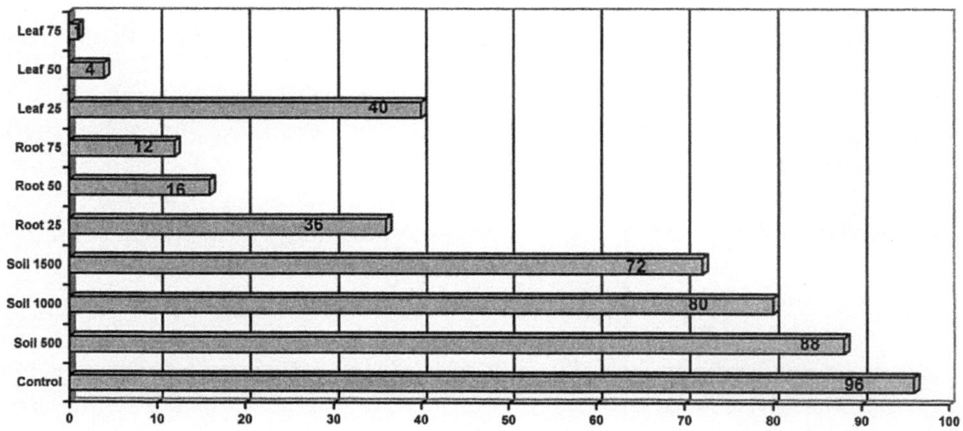

Figure 2. The germination ability of seeds *Festuca rubra* L. in dependence on various concentrations of studied extracts

germination than on the germination ability of *Festuca rubra*.

Biometric parameters of *Festuca rubra* L. seedlings

In the conducted experiment the measurements of the length of leaves, leaf sheaths and root's length of *Festuca rubra* were done.

In these investigations (Table 1) *Festuca rubra* developed the longest leaves under the influence of soil extracts, but the largest value – 52 mm – was obtained at the lowest concentration (500 grams of the soil/litre). With the increase of the concentrations of soil extracts, the seedlings developed slightly shorter leaves achieved the value 45 mm at the concentration 1000 grams of the soil/litre and 40 mm at the highest concentration. The strongest allelopathic influence on the studied feature was recorded for extracts from leaves (Table 3). Under their influence the length of leaves of *Festuca rubra* reached the lowest values, and by highest concentration the leaves did not develop. Similar tendency in the change of the length of leaves was affirmed under the influence of water extracts from roots of *Taraxacum officinale*.

Similarly to the case of the lengths of leaves, the studied solutions influenced the length of leaf sheaths of the tested grass. The results obtained in the work (Table 3) show, that the extracts from leaves influenced the most stopped the growth of leaf sheaths. Only under the influence of the applying lowest concentration of the extract from leaves, were leaf sheaths created, however they were short with the length of 3 mm. After using higher concentrations of extracts from leaves leaf sheaths did not create at all. extracts from the roots smallest allelopathic influence on the length of the leaf sheath (Table 2). In this case the longest sheaths were created, which reached the length suitably 13 mm, 12 mm and 7mm. The extracts from the soil showed slightly larger allelopathic influence on the length of leaf sheaths (Table 1), in the result of this, they were shorter at approximately 2,0 mm.

In the allelopathic influence of extracts from *Taraxacum officinale* their influence on the length of seedlings root's of *Festuca rubra* is also important. After applying the soil extract with the con-

Table 3. The influence of the water solution of the extract from leaves of *Taraxacum officinale* on initial growth and development of *Festuca rubra*

Studied feature	Control	Mass of the sample (g)		
		25	50	75
Germination energy	64a	4b	2b	0b
Germination ability	96a	40b	4c	1c
Length of the leaf [mm]	42a	27b	11c	0d
Length of the sheath [mm]	7a	3ab	0b	0b
Length of radicle [mm]	25a	1b	0b	0b

Comment: Means in lines estimated with the same letters do not differ significantly.

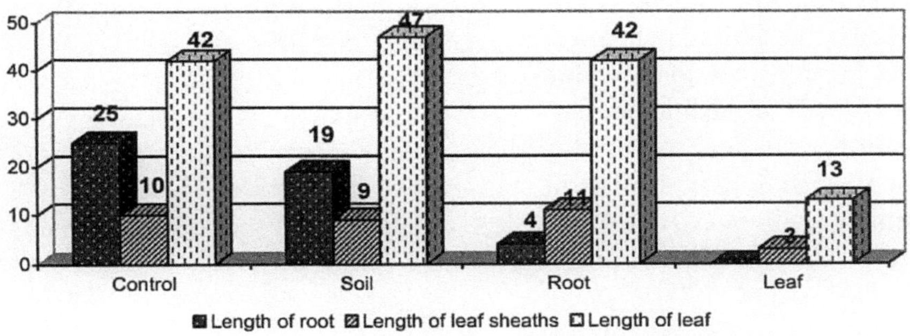

Figure 3. Length of studied parts of plants in regard to the kind of the applied extracts (mean for studied solutions)

centration of the solution 1500 grams of the soil/litre *Festuca rubra* had the longest roots of 21 mm. Together with the growth of the concentrations especially the extracts from roots (Table 2) clear inhibitory influence on the studied feature in such considerable degree was observed(they were very short). However, the extracts from leaves, showed the strongest allelopathic influence (Table 3). Under their influence, only in the case of the lowest concentration were created the roots with length 1mm, however under the influence of remaining concentrations they did not create at all.

The conducted investigations (Figure 3) show strong influence of plant extracts from leaves of Taraxacum officinale on initial growth and development of *Festuca rubra.* Slightly weaker allelopathic influence showed the extracts from the roots of this plant.

The inhibitory effect of allelocompounds [Harkot and Lipińska 1997] can already appear while imbibition of seeds. The disorders to which comes then, dealy the plants germination, break the growth of creation of the roots and cause their deformation and decay. Many cases of allelopathic influences of plants were already proved during the seeds germination. One of them is allelopatic influence of *Hieracium pilosella* L. on initial growth and development of *Lolium perenne* L. and *Festuca rubra* L. [Jodełka et al., 2003].

Extracts from leaves broke the germination of seeds most effectively, and only single germination of seeds was observed after seven days from their sowing. Investigations led by many authors (Harkot and Lipińska 1996) showed similar dependence in the case of *Poa pratensis*, where substances relived from decomposed leaves showed larger allelopatic activity, than the ones from decomposed roots of this species.

Stronger influence of the all applied water solutions was observed on the energy of germination than on the germination ability in *Festuca rubra*. Also definitely larger influence of water solutions (plant, soil and root's, secretions) on the germination energy than on the germination ability of *Bromus inermis* and *Poa pratensis* seeds was proved by many authors [Lipińska 2002; Harkot and Lipińska 1995].

Their investigations [Harkot and Lipińska 1996] proved that germination of *Dactylis glomerata* L. and *Phleum pratense* L. on the soil from under sward with *Poa pratensis* L. domination was indeed worse in comparison with germination on the control objects however germination of *Festuca rubra* were similar in all objects. The investigations conducted by the other authors

[Lipińska 2002] showed stronger influence of water extracts from the soil under the sward of *Lolium perenne* L. on the extension of the roots than on the length of seedlings of *Phleum pratense* L. and *Poa pratensis* L.

Conducted investigations show the strong influence of plant extracts from leaves of *Taraxacum officinale* on initial growth and development of *Festuca rubra.* Slightly weaker allelopathic influence was showed the extracts from the roots of this plant.

Some danger, connected with the interpretation of the importance of individual compounds in allelopathic affecting is the physico-chemical complexity of the soil, its ability to bind organic compounds, and the existence of the microorganisms which can transform individual plant products to more toxic compounds or to destroy them in such a degree that they will not show any activity in the relation to tissues of the higher plants [Bertin et al., 2003; Callaway et al., 2000]. So, the results of the influence of the soil on the germination, height of seedlings or length of radicula of the tested plants cannot be attributed exclusively to the proprieties of the allelopathic substance presented by Suman at al. [2002]. In field conditions, it is also allelopathic compounds that have influence on their interactions connected with various botanical compositions of various composition agrophytocenosis [Callaway et al., 2005; Weidenhamer 2006]. These effectis are so closed that the control of their influence is often symbolic. The results of experimental investigations cannot be transferred directly onto conditions existing in the nature. However, this does not change the fact that the phenomenon of allelopathy is one of several basic ecological factors, which create the conditions of environment [Basis et al., 2003; Inderjit and Duke 2003].

CONCLUSIONS

1. The germination energy of *Festuca rubra* independently on the applied soil or plant extract, underwent break together with the growth of the concentration of applied extracts prepared from *Taraxacum officinale.*
2. The higher concentrations of *Taraxacum officinale* secretions broke lengthening of the roots and the growth of the leaf sheath as well as the leaves of *Festuca rubra.*
3. On the all studied features the strongest allelopathic influence independently on the rare of the concentration of studied extracts, had the extract from the leaves of *Taraxacum officinale* in the contrast to the soil extract, which did not show such affecting.

REFERENCES

1. An M., Pratley J.E. and Haig T. 2000. Phytotoxicity of Vulupia residues: IV. Dynamics allelochemicals during decomposition of vulpia residues and their corresponding phytotoxicity. J. of Chem. Ecol., 26, 2603–2617.
2. Basis H.P., Vepachedu S., Callaway G.R.M. and Vivanco J.M. 2003. Allelopathy and exotic plant invasio: from molecules and genes to species interactions. Science, 30, 1377-1380.
3. Bertin C., Paul R.N., Duke S.O. and Weston L.A. 2003. Laboratory assessment of the allelopathic effects of fine leaf fescues. J. Chem. Ecol., 29, 1919-1937.
4. Beyschlag W., Ryel R.J., Ullmann I. and Eckstein J. 1996. Experimental studies on the competitive balance between two Central European roadside grasses with different growth forms. 2 Controlled experiments on the influence of soil depth, salinity and allelopathy. Bot. Acta, 109, 6, 449–455.
5. Birkett MA. Chamberlain K., Hooper A.M. and Pickett JA. 2001. Does allelopathy offer real promise for practical weed management and for explaining rhizosphere interactions involving higer plants?. Plant and Soil, 232, 31–39.
6. Callaway R.M., Ridenour W.M., Laboski T., Weir T. L. and Vivanco J.M. 2005. Natural selection for resistance to the allelopathic effects of invasive plants. J. Ecol., 93, 576–583.
7. Callaway R.M. and Aschehoug E.T. 2000. Invasive plants versus their mew and old neighbors: A mechanism for exotic invasion. Science, 290, 521–523.
8. Einhelling F.A. 1995a. Allelopathy: Current status and future goals. In: Allelopathy, Organisms, Proceses and Aplications. Red. Inderjit, K.M.M. Dakashini, F.A. Einhelling. Am. Chem Soc., Washington, 1–24.
9. Einhelling F.A. 1995b. Mechanizm of action of allelochemicals in allepathy. In: Allelopathy, Organisms, Proceses and Aplications. Red. Inderjit, K.M.M. Dakashini, F.A. Einhelling. Am. Chem Soc., Washington, 96–116.
10. Goslee S.C., Peters D.P.C. and Beck K.G. 2001. Modeling invasive weeds in grasslands: the role of allelopathy in Acroptilon repens invasion. Ecolog. Modelling, 139, 31–45.

11. Harkot W. and Lipińska H. 2005. Allelopathic effects of water leachates of *Poa pratensis* leaves. Allelopathy Jour., 16, 251-260.

12. Harkot W. and Lipińska H. 1996. Wpływ wydzielin korzeni siewek niektórych gatunków traw i koniczyn na kiełkowanie ich nasion In: Teoretyczne i praktyczne aspekty allelopatii. Mat. Konf. IUNG Puławy, 147-153.

13. Harkot W. and Lipińska H. 1997. Allelopatyczny wpływ stokłosy bezostnej na kiełkowanie, początkowy wzrost i rozwój niektórych gatunków traw i motylkowatych. Zesz. Post. Nauk Roln., 452,185-197.

14. Renne I.J., Rios B.G., Fehmi J.S. and Tracy B.F. 2004. Low allelopathic potential of invasive forage grass on native grassland plants: a cause for encouragement?. Basic and Applied Ecology 5, 261-269.

15. Inderjit, and Duke S.O. 2003. Ecophysiological aspects of allelopathy. Planta, 217, 529–539.

16. Jodełka J., Jankowski K. and Ciepiela G.A. 2003. Wpływ allelopatyczny jastrzębca kosmaczka na początkowy rozwój życicy trwałej i kostrzewy czerwonej. Biul. IHAR, 225, 353-358.

17. Julita H.M. and Et Grace J.B. 2002. Effects of disturbance on germination and seedling establishment in a costal prairie grassland: A test of the competitive release hypothesis. J. of Ecol., 90, 291–302.

18. Lipińska H., and Harkot W. 2007. Allelopatia w zbiorowiskach trawiastych. Post. Nauk Roln., 1, 49-61.

19. Lipińska H. 2002. Allelopatyczne oddziaływanie *Lolium perenne* na wybrane gatunki traw. Łąkarstwo w Polsce, 5, 137-144.

20. Oleszek W. 1996. Allelopatia – rys historyczny i definicje, Materiały Konferencji: Teoretyczne i praktyczne aspekty allelopatii, IUNG, Puławy, K(10).

21. Perry L.G., Thelen G.C., Ridenour W.M., Weir T.L., Callaway R.M, Paschke M.W. and Vivanco J.M. 2005. Dual role for an allelochemical: catechin from Centaurea maculosa root exudates regulates conspecific seedling estabilishment. J. of Ecol., 93, 6: 1126-1135.

22. Ridenour W.M. and Callaway R.M. 2001.The relative importance of allelopathy in interference: the effects of an invasive weed on a native bunchgrass. Oecologia, 126, 444-450.

23. Sinkkonen A. 2003. A model describing chemical interference caused by decomposing residues at different densities of growing plants. Plant and Soil, 250, 315–322.

24. Spyreas G., Gibson D.J. and Et Middleton B.A. 2001. Effects of endophyte infection in tall fescue (Festuca arundinaac: Poaceae) on community diversity. In. J. of Plant Sci., 162, 1237–1245.

25. Suman A., Shahi H.N., Pushpaa S. and Gaur A. 2002. Allelopathic influence of Vigna mungo (black gram) seeds on germination and radical growth of some crop plants. Plant Growth Regulation, 38, 69–74.

26. Weidenhamer J.D. 2006. Distinguishing allelopathy from resource competition: The role of density. [In:] Allelopathy: A physiological process with ecological implications, Reigosa M., Pedrol N. and González L. (eds), Springer, Dordrecht (Neth.) pp. 85-103.

IMPACT OF GROWTH INHIBITOR BERCEMA ON SHOOTS REGROWTH OF PERENNIAL RYEGRASS AND COCKSFOOT

Beata Wiśniewska-Kadźajan[1], Kazimierz Jankowski[1], Elżbieta Malinowska[1], Jacek Sosnowski[1], Roman Kolczarek[1]

[1] Department of Grassland and Green Areas Creation, Siedlce University of Natural Sciences and Humanities, B. Prusa 14, 08-110 Siedlce, Poland, e-mail: laki@uph.edu.pl

ABSTRACT

In a pot experiment conducted in a greenhouse, the effect of CCC Bercema growth inhibitor to reduce vegetative shoot growth of perennial ryegrass and cocksfoot was studded. The experiment was set up in triplicate. Growth regulator was applied once, in a form of an aqueous solution. The following experimental objects were specified: control object, Bercema with a concentration of 10%, 20% and 30%. Ten measurements were performed every 3 days. The study was conducted under conditions of 12-hour artificial light daily. The results were statistically analyzed using two-factorial variance analysis . It was found that the greatest shoot growth reduction of ryegrass due to the highest inhibitor concentration (B- 30%), but for cocksfoot under the influence of the minimal concentration (B- 10%), compared to the control object. It has been noted that cocksfoot is more sensitive to the applied growth regulator, resulted in significantly greater inhibition of shoot regrowth than for ryegrass. In the experiment with the perennial ryegrass significant differences in the growth of shoots were found under the influence of inhibitor application at a concentration of 30%, compared to the control object. In the case of cocksfoot, all the concentration increases of Bercema inhibitor significantly differentiated the shoots regrowth.

Keywords: growth regulator Bercema, perennial ryegrass, cocksfoot.

INTRODUCTION

Plants' physiological processes can be modified by organic compounds, called growth regulators [10, 11, 13, 6]. The application of growth regulators can affect: inhibition of cell elongation in stems, stalks' thickening, increased protein content, chlorophyll, minerals up-taking, increased resistance to weather stress [8, 5, 9]. The use of growth inhibitors in agricultural crops, horticultural crops and fruit can contribute significantly to the reduction of cultivation cost, which is an important economic element [1, 2]. One of the most common and earliest used in the cultivation of cereals is Bercema. Exogenous inhibitor containing chloride 2-chloroectylothreemehlamine with a short period of activity, is used primarily as antiloder. In the cultivation of grasses fertilized with nitrogen lodging it is a big problem. Thus, the selection of appropriate substance for regulating the shoots' regrowth of grasses is the subject of many studies.

The aim of the study was to determine the effect of the Bercema inhibitor used in various concentrations on the regrowth of vegetative shoots of perennial ryegrass and cocksfoot.

MATERIALS AND METHODS

The experiment was conducted in 2008, in the laboratory conditions. From the permanent grassland the soil samples with two grass species: perennial ryegrass and orchard grass. The grasses were collected in development phase BBCH 32 in quantities of 24 units. The soil has characterized by low abundance in phosphorus (64 mg $P_2O_5 \cdot kg^{-1}$ of soil), and potassium (88 mg $K_2O \cdot kg^{-1}$ of soil), and the average of magnesium (41MgO·kg^{-1} of soil), with KCl pH = 5.44. Plastic vases were filled with cut turf, about 2 kg of soil·$vase^{-1}$. In the experiment, CCC Bercema growth inhibitor containing an active substance 2-chloroectylothreemehlamine chloride was used. Growth regulator was used once, in a form of aqueous solution, on 7 May 2008. The experiment was established as follows: control object, Bercema with a concentration of 10% (B-10%), 20% (B-20%) and 30% (B-30%). The experiment was carried out in triplicate. Ten measurements were performed every 3 days, in the period from 7 May to 6 June 2008. The study was conducted under conditions of 12-hour artificial light daily with an average intensity of 4000 lux. On the first day of the study (May 7) tested grasses were cuted to a height of 10 cm – perennial ryegrass and 12 cm – orchard grass, and then growth regulator was applied. The results were statistically analyzed using analysis of variance for two-factors experiment. The significance of differences was verified using Tukey's test at a significance level of $p \leq 0.05$.

RESULTS AND DISCUSSION

In the experiment with the use of growth inhibitor Bercema the differences of perennial ryegrass regrowth under the influence of the largest 30% concentrations were significant, as compared to the control object (Table 1).

A similar relationship in their research indicate [7] for the growth of ryegrass under the influence of Flordimex inhibitor. Statistical analysis showed a significant effect of all the Bercema inhibitor concentrations on the regrowth of cocksfoot, compared to the control, except for the first measurement term (after 3 days). Changes in regrowth of vegetative shoots of grasses tested fluctuated on different measurement dates.

In the experiment with perennial ryegrass a considerable reduction of shoots re-growth was found under the influence of the preparation Bercema concentrations used (Figure 1). With the increase of inhibitor concentration inhibition of regrowth of the tested grass was observed most.

The greatest reduction of shoot regrowth of perennial ryegrass was found between 5 and 15 days of the experiments after inhibitor application at a concentration of 30% and it amounted to, between 6.3 and 6.8 cm respectively. Inhibitor concentration of 20% and 10% at the beginning of the measurements resulted in a much smaller decrease in regrowth of perennial ryegrass in 9 day of measurements amounted to between 4.6 cm and 3.1 cm respectively. At a subsequent measurement, starting from 18th day of the experiment there was a much smaller differences in the reduction of vegetative shoot growth of perennial ryegrass in relation to the control object, than during the initial period. Inhibitor Bercema 30% during this period of reduced regrowth an average about 4 cm, B – 20% about 0.8 cm, B – 10% about 1.2 cm. In the last period of the measurement, after

Table 1. The shoots regrowth of perennial ryegrass and cocksfoot (cm) depending on the concentration of growth inhibitor in various periods of measurement

Inhibitor concentration	Dates of measurements – growth of shoots (cm)											Mean for growth
	1	3	5	9	12	15	18	21	24	27	30	
	Lolium perenne											
0%	–	7.7	12.3	17.8	21.5	23.7	26.0	29.3	31.0	31.8	32.7	23.4
B 10%	–	6.0	10.3	14.7	19.7	22.3	23.8	29.0	29.8	31.0	31.3	21.8
B 20%	–	5.5	9.7	13.2	17.0	20.5	25.0	29.0	30.0	31.0	33.3	21.4
B 30%	–	4.0	6.0	11.0	14.7	17.4	22.3	24.7	26.7	27.3	28.5	18.3
Mean	–	5.8	9.6	14.2	18.2	21.0	24.3	28.0	29.4	30.3	31.5	21.2
$LSD_{0.05}$ for: A – concentration A = 4.01 B – dates of measurements B = 25.6												
	Dactylis glomerata											
0%	–	11.3	18.7	26.0	28.8	29.5	31.7	33.3	33.5	33.7	34.0	28.1
B 10%	–	8.7	11.8	13.7	14.3	14.0	13.0	14.7	14.7	13.7	14.0	13.3
B 20%	–	7.0	11.7	14.8	16.0	17.0	18.0	19.3	18.3	19.0	20.3	16.1
B 30%	–	6.0	9.0	11.3	12.7	14.3	15.2	15.3	15.8	15.3	16.3	13.1
Mean	–	8.3	12.8	16.5	18.0	18.7	19.5	20.7	20.6	20.4	21.2	17.7
$LSD_{0.05}$ for: A – concentration A = 6. 6 B – dates of measurements B = 13.2												

30 days, the acting of this inhibitor at 20% dose resulted in a higher regrowth than the tested grass on the control object. The highest concentration of Bercema (30%) resulted in inhibiting perennial ryegrass of the regrowth. The most intense impact takes place within two weeks after its application. Jankowski et.al [6] demonstrated the inhibitory effect of the regulator trinexsapac-ethyl on the growth of several grass species including perennial ryegrass. Grzyś et.al, Stachecki and Praczyk [4, 12] confirmed the inhibitory effect of retardants on the blades regrowth of spring wheat in their study.

Bercema inhibitor at a concentration of 30% inhibited the regrowth of cocksfoot shoots to 12 days of the experiment most (Figure 2).

At a subsequent measurement terms the strongest inhibitory effect was observed when the tested concentration of the growth regulator was 10%, producing an average of 18.6 cm of growth.

In the last days of the experiment (in 27 and 30 days) the greatest reduction of shoots regrowth of cocksfoot was under the influence of the smallest concentration of the inhibitor, amounting to 20 cm. However, after application of 20% Bercema, the regrowth inhibition of the tested grass, compared with the control object increased, at the beginning of the experiment and from 12th day, remained at the same level, average 13.8 cm. For this reason, grass species the lowest concentration of inhibitor (B – 10%) most effectively inhibited the growth of the vegetative shoots, with respect to the control object. The strongest limiting effect the growth of cocksfoot occurred in two weeks after the application of the tested inhibitor (B – 10%).

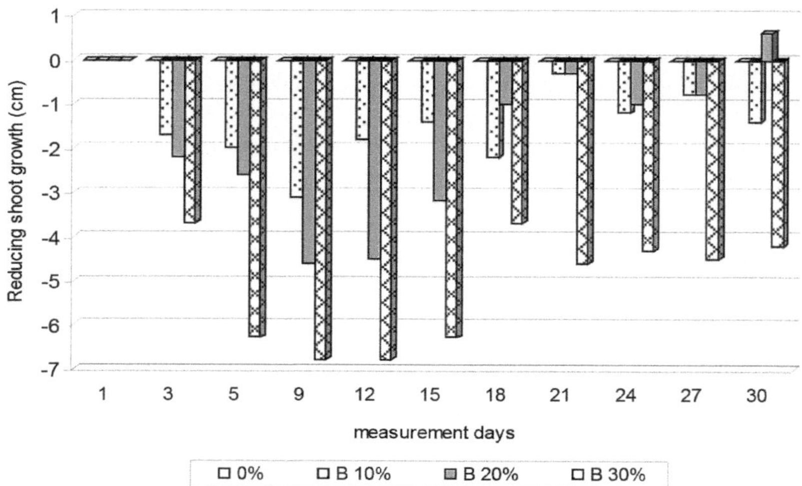

Figure 1. Inhibition in the growth of shoots of perennial ryegrass, depending on the concentration of the growth inhibitor, in relation to the control objects in the individual measurement dates

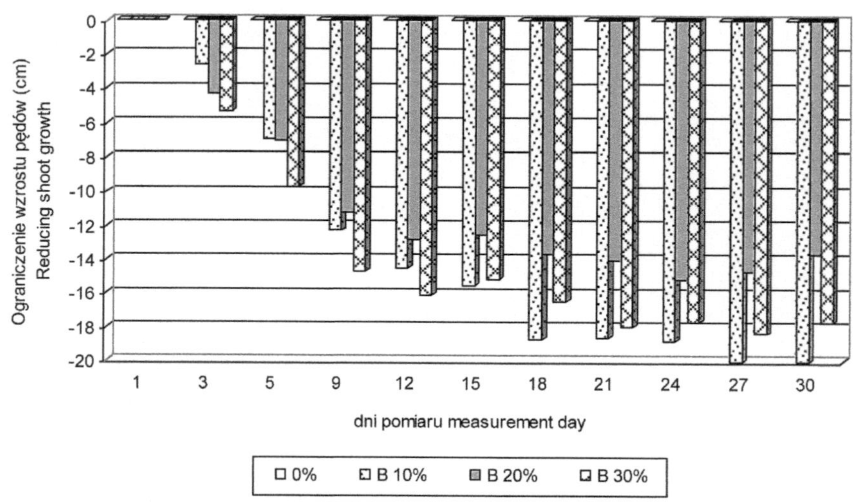

Figure 2. Inhibition in the growth of shoots in cocksfoot, depending on the concentration of the growth inhibitor in relation to the control of the individual measurement dates

In plants the cultivation, in synthetic versions of natural plant hormones is very often used [10]. According to many authors [3, 4] retardant treated the plants maintain a constant photosynthetic efficiency, photochemical activity and hydration of the cells and are more resistant to lodging. The use of inhibitors limits the frequency of mowing lawns, causing intense sodding. According to [6] the use of growth regulators influenced lawns in maintaining in a good condition especially in the summer, when the extent of their use is highest.

CONCULSIONS

1. In the experiment the use of different concentrations of Bercema inhibitor the largest regrowth inhibition of perennial ryegrass stems were stated under the influence of the largest concentrations (B – 30%), but for cocksfoot the minimal concentration (B – 10%) in comparison to the control object.
2. Cocksfoot was noted to be more sensitive to the application of growth regulator, what resulted in significantly greater inhibition of shoot regrowth than for ryegrass.
3. In the experiment with the perennial ryegrass significant differences in the growth of shoots were found under application of an inhibitor at a concentration of 30%, compared with the control object. In case of cocksfoot, all Bercema inhibitor concentrations significantly differentiated shoots regrowth, as compared to the control object.

REFERENCES

1. Adams R.E., Kerber K., Pfister K., Weiler E.W. 1992. Studies on the action of a new growth retardant CGA 163'935 (Cimectacarb). Progress in plant growth regulation. [In:] Proceedings of the 14th International Conference on Plant Growth Substances (C. Karssen., L. Van Loon., D. Vreugdenhill, eds). Dordrecht, Netherlands, Kluwer Academic Publishers: 818-827.
2. Bragg P., Rubino P., Henderson F.K.G., Fielding, W.J., Cannel R.Q. 1984. A comparison of the root and shoot growth of winter barley and winter wheat, and the effect of an early application of chloromequat. J. Agri. Sci., Cambridge 103: 257-264.
3. Feng Y., Stoeckel D.M., von Santen E., Walker R.H. 2002. Effects of subsurface aeration and trinexapec-ethyl application on soil microbial communities in a creeping bentgrass putting green. Biol. Fertil. Soils 36, 456-460.
4. Grzyś E., Demczuk A., Sacała E., Kulczycki G. 2012. Effect of retardants on blade growth and content of microelements (Fe, Mn, Zn, Cu) in the grain and stalks of spring wheat cultivars. Zesz. Nauk. UP Wrocław, Rolnictwo II, 588, 65-73.
5. Jankiewicz L.S. 1997. Retardanty i niektóre inne syntetyczne inhibitory wzrostu oraz wybrane substancje modyfikujące wzrost. [In:] Regulatory wzrostu i rozwoju roślin. Jankiewicz L.S. (ed.) PWN, Warszawa: 108-123.
6. Jankowski K., Czeluściński W., Jankowska J., Kolczarek R., Sosnowski J. 2012. The influence of the growth regulator trinexapac-ethyl on the regrowth rate of lawn grasses. Acta Sci. Pol. Hortorum Cultus 11(4), 67-76.
7. Jankowski K, Kolczarek R, Jankowska J, Sosnowski J, Wiśniewska-Kadżajan B, Truba M, Kaczorek A. 2013. Effect of different concentrations of the Flordimex inhibitor on the shoot growth of perennial ryegrass and cocksfoot. Environmental Protection and Natural Resources. Vol. 24, No. 4(57): 1-4.
8. Koziara W., Czajka M., Sobiech S. 1993. Wpływ deszczowania, nawożenia azotowego i stosowania Bercemy CCC na plonowanie pszenżyta jarego. Biologia i uprawa pszenżyta, Międzyzdroje, p. 31.
9. Malinowska E., Kalembasa S. 2012. The yield and content of Ti, Fe, Mn, Cu in celery leaves (*Apium graveolens l. var. dulce mill. pers.*) as a result of Tytanit application. Acta Sci. Pol., Hortorum Cultus, 11(1), 69-80.
10. Mikos-Bielak M. 2005. Egzogenne regulatory wzrostu w uprawie ziemniaka. Annales UMCS, Sec. E, 6, 281-292.
11. Silva T.R.B., Schmidt R., Silva C.A.T., Nolla A., Faver F., Poletine J.P. 2011. Effect of Trinex –apacethyl and nitrogen fertilization on wheat growth and yield. J. Food Agric. Environ. 9(1), 596-598.
12. Stachecki S., Praczyk T. 2004. Biologiczna aktywność chlorku chlormekwatu (CCC) stosowanego z adiuwantami w pszenicy ozimej. Post. Ochr. Roślin 44(1), 414-422.
13. Starczewski K., Affek-Starczewska A., Jankowski K. 2011. Wpływ wybranych regulatorów wzrostu na kolorystykę muraw mieszankowych na bazie życicy trwałej. Łąkarstwo w Polsce 14, 137-145.

9

THE ASSESSMENT OF ATMOSPHERIC DROUGHT DURING VEGETATION SEASON (ACCORDING TO STANDARDIZED PRECIPITATION INDEX SPI) IN CENTRAL-EASTERN POLAND

Elżbieta Radzka[1]

[1] Department of Agrometeorology and Land Reclamation, University of Natural Sciences and Humanities, B. Prusa 14, 08-110 Siedlce, Poland, e-mail: elzbieta.radzka@uph.edu.pl

ABSTRACT

The paper presents an assessment of atmospheric drought during vegetation season defined on the basis of standardized precipitation index (SPI). The data used in this paper come from nine IMWM stations from central-eastern region of Poland, and they were registered in 1971–2005. The frequency of occurrence of vegetation season's months was determined in particular drought classes. Spatial distribution of SPI index values was shown in all of the vegetation season's months on the area examined. The direction and significance of values changes tendency of the analyzed index during the vegetation season were also defined. It was noticed that extreme droughts appeared four times less frequently than the normal months. Very dry months were noted most frequently in September while moderately dry – in August. The analysis of the frequency of spatial distribution of particular drought classes showed that extreme dry and very dry months occurred most frequently in western part of the area examined, while the moderately dry months also in south-eastern part. On the basis of the linear trend analysis it can be said that the SPI index values were slightly decreasing year by year.

Keywords: standardized precipitation index (SPI), time trend, central-eastern Poland.

INTRODUCTION

Drought is an atmospheric as well as soil and hydrological phenomenon. It is related to the period with no precipitation or to the repeated less than medium precipitation. To be able to fight negative consequences of drought effectively one should possess proven and reliable coefficients to define drought. One of them is SPI – standardized precipitation index [Bąk and Łabędzki 2004, Lana 2001, Lloyd-Hughes and Saunders 2002, Turke's and Tatlı 2009, Wu et al. 2005]. Coefficient's value shows precipitation's deviation from the average value in a multi-year. It is used to detect periods of drought and to assess its intensity on a variety of time scales. SPI index is used in many countries for operational monitoring of meteorological drought, e.g. in USA in National Drought Mitigation Center (NDMC) [Łabędzki 2006, McKee et al. 1993]. The analysis of frequency of occurrence, intensity and duration of drought during vegetation season is very important from the viewpoint of agriculture because it is an important factor that limits yielding of crops.

The aim of this paper was to determine, on the basis of SPI index, the frequency and intensity of drought occurrence during vegetation season in central-eastern Poland. Moreover, the attention was paid to the scope of variation of the examined coefficient during the time series analyzed. Statistical significance of directional coefficients of linear trends was also examined.

MATERIALS AND METHODS

The data used in this paper concerning daily sums of precipitation come from nine Institute of Meteorology and Water Management stations from central-eastern region of Poland from the years 1971–2005 (Table 1).

The assessment of drought was made on the basis of standardized precipitation index (SPI).

$$SPI = \frac{f(P) - \mu}{\sigma} \quad (1)$$

where: SPI – standardized precipitation index,
$f(P) = \sqrt[3]{P}$ – transformed precipitation sum,
μ – average value of normalized precipitation series,
σ – average standard deviation of normalized precipitation series.

Table 1. Geographic coordinates of synoptic and climatic IMGW stations in central-eastern Poland

Station	Geographic coordinates		H_s
	φ°	λ°	m a.s.l.
Ostrołęka	53° 05'	21° 34'	95
Białowieża	52° 42'	23° 51'	164
Włodawa	51° 33'	23° 32'	163
Szepietowo	52° 51'	22° 33'	150
Legionowo	52° 24'	20° 58'	93
Biała Podlaska	52° 02'	23° 05'	133
Sobieszyn	51° 37'	22° 09'	135
Pułtusk	52° 44'	21° 06'	95
Siedlce	52° 11'	22° 16'	146

Explanations: φ°– geographic latitude, λ° – geographic longitude, H_s – elevation above sea level.

In practice the following fact is commonly used: for random variable X with gamma distribution variable $Z = \sqrt[3]{X}$ has approximately a normal distribution. Using that fact an approximate method of calculating SPI index can be suggested through standardization of random variable $Z = \sqrt[3]{X}$ [Gąsiorek et al. 2012].

Table 2 presents ranges of values (drought classes) for each of the vegetation season's months.

The frequency of occurrence of drought classes was examined in particular months of vegetation season in nine stations in central-eastern Poland and spatial distribution of the frequency of their occurrence was presented as well. The frequency of drought occurrence in a specific month was calculated as the ratio of the number of cases of drought in a given month over the multi-year (1971–2005) to the number of 35 years being analyzed.

Table 2. Drought classes according to standardized precipitation index SPI

SPI index	Type of period
≤–2	extremely dry
From –1 to –1.50	very dry
From –1.49 to –0.49	moderately dry
From –0.48 to 0.49	normal

The basic characteristics of the distribution have been determined: the arithmetic mean, the minimum and the maximum. In order to express the dynamics of changes of the coefficient during the period examined, the standard deviations were calculated. The direction and significance of changes tendency were determined on the basis of the linear trend equations. The significance of directional coefficient of the trend was estimated with t-Student test where $\alpha = 0.05$.

RESULTS AND DISCUSSION

During the multi-year analyzed in central-eastern Poland the lowest value of SPI index was noted in June in Białowieża (-5.94) and in August in Ostrołęka (-5.77), Białowieża (-5.71) and BiałaPodlaska (-5.00) (Table 3).

While the highest values occured in June in Legionowo (4.21) and in August also in Legionowo (4.63) and Siedlce (4.00).

Extremely dry months occurred most frequently in western part of the region examined (frequency about 10%), while the least frequently in central and eastern part (frequency about 8%) (Figure 1).

Very dry months were noted with the highest frequency in western and southern part of the region examined (frequency above 14%) and moderately dry in western and south-eastern part (frequency above 30%). Whereas optimally humid (normal) vegetation seasons occurred most frequently in southern part (frequency in Sobieszyn – 36%).

In central-eastern Poland in all of the stations during vegetation season extremely dry and very dry months occurred most rarely, while normal and moderately dry months were most common. In the multi-year analyzed normal months occurred with the frequency from 30% (September) to 35% (May and July), moderately dry months from 26% (May) to 34% (September), very dry months from 9% (May) to 14% (September), extremely dry months from 6% (August) to 9% (April, May and July) (Table 4).

Łabędzki and Bąk [2004] defining SPI index variety in Poland in years 1954–1998 concluded that dry periods with different intensity

Table 3. Average, minimum and maximum values of standardized precipitation index SPI during the vegetation season in central-eastern Poland between the years 1971–2005

Station		Month					
		IV	V	VI	VII	VIII	IX
Ostrołęka	min	-3.38	-2.08	-3.46	-2.96	-5.77	-3.03
	max	2.42	2.07	2.28	2.54	2.31	3.37
	average	-0.23	-0.23	-0.28	-0.23	-0.32	-0.29
Białowieża	min	-2.93	-4.14	-5.94	-3.88	-5.71	-6.33
	max	2.27	1.82	2.27	2.23	2.88	1.79
	average	-0.23	-0.24	-0.30	-0.25	-0.28	-0.27
Włodawa	min	-2.08	-3.37	-4.66	-3.30	-4.00	-2.88
	max	2.99	2.28	3.17	2.13	1.66	1.86
	average	-0.07	-0.17	-0.24	-0.22	-0.22	-0.18
Szepietowo	min	-2.42	-4.42	-3.11	-3.43	-2.70	-2.57
	max	1.61	2.92	3.00	2.81	2.59	2.14
	average	-0.21	-0.20	-0.23	-0.26	-0.14	-0.26
Legionowo	min	-2.90	-2.83	-3.89	-2.93	-3.10	-2.51
	max	2.46	1.82	4.21	2.70	4.63	2.66
	average	-0.25	-0.09	-0.13	-0.18	-0.29	-0.29
BiałaPodl.	min	-2.60	-3.53	-3.51	-2.31	-5.00	-3.09
	max	2.62	2.58	2.09	2.73	2.02	2.27
	average	-0.21	-0.12	-0.24	-0.21	-0.32	-0.13
Sobieszyn	min	-2.37	-2.45	-2.06	-3.53	-3.37	-2.25
	max	2.20	2.62	2.53	2.60	2.46	1.98
	average	0.00	-0.05	-0.20	-0.19	-0.21	-0.18
Pułtusk	min	-3.10	-2.35	-3.03	-2.69	-3.91	-3.02
	max	2.58	2.80	2.65	2.36	3.00	2.24
	average	-0.30	-0.13	-0.25	-0.23	-0.23	-0.25
Siedlce	min	-2.57	-2.38	-2.96	-2.30	-4.72	-1.80
	max	2.80	2.50	3.58	2.67	4.00	2.01
	average	-0.03	0.00	0.09	-0.04	0.00	-0.04

Table 4. Average territorial frequency of occurrence (%) of particular humidity classes (type of month) during the vegetation season (IV-IX) in central-eastern Poland between the years 1971–2005

Droughtclass/month	IV	V	VI	VII	VIII	IX	Average IV-IX
Extremely dry	9	9	8	9	6	7	8
Very dry	12	9	13	12	13	14	12
Moderately dry	27	26	30	29	34	31	29
Normal	32	35	32	35	33	30	33

of precipitation deficiency lasted for about 1/3 part of the whole vegetation season. Among the driest months were mentioned May and July, while the least dry – August. Extreme droughts occurred most frequently in August, very dry periods in May and moderately dry in May and July. Łabędzki's studies [2008] during the years 1961–2006 from the station Bygdoszcz – IMUZ show that the frequency of the months in which drought was identified was 40–45% for various time scales. On the other hand Kanecka-Geszke and Smarzyńska [2007] claimed that according to SPI during the whole 35 years period (1970–2004) in the four out of five stations examined (Bydgoszcz, Kraków, Suwałki, Szczecin) there were 16 (46%) vegetation periods with drought, and in one (Wrocław) – 17 vegetation periods with drought.

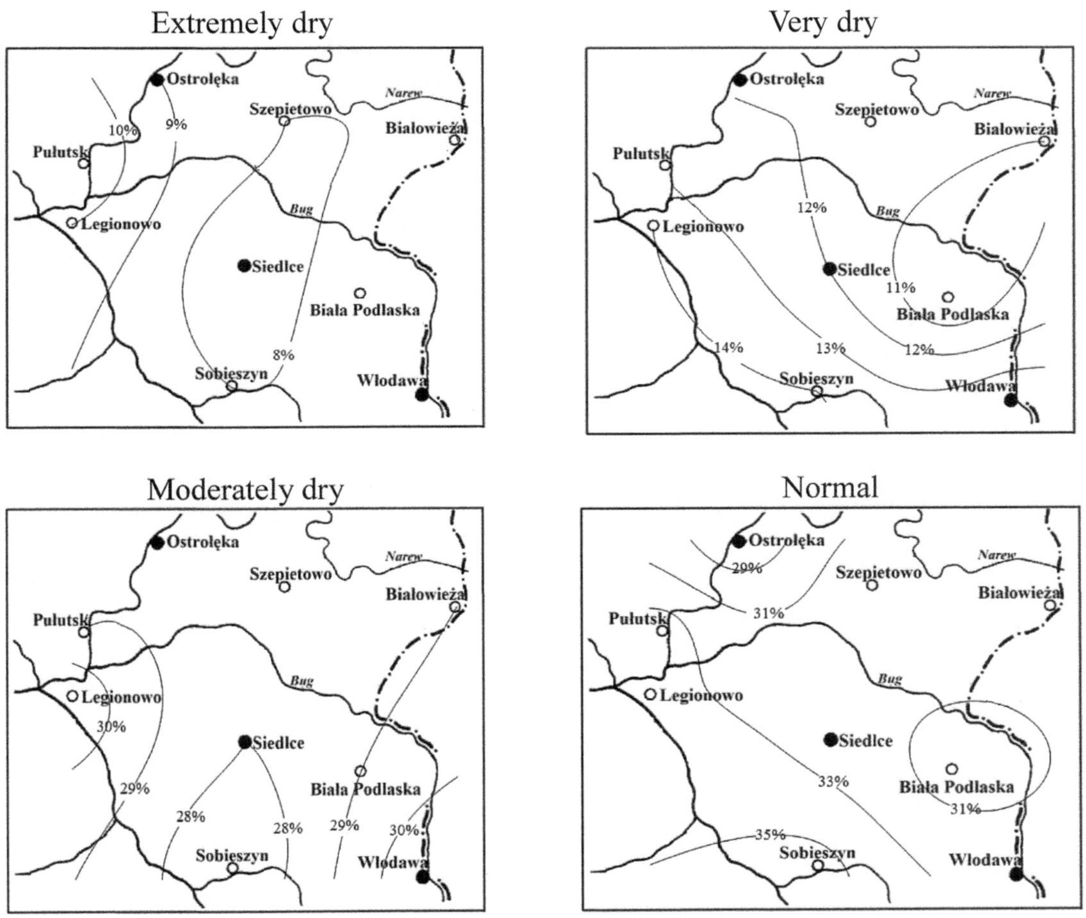

Figure 1. Frequency of occurrence (%) of particular humidity classes during the vegetation season (IV-IX) in stations of central-eastern Poland between the years 1971–2005

Variability of average SPI index values during vegetation period (IV–IX) in all of the stations was on a similar level. Standard deviation value balanced from 0.34 in Włodawa to 0.48 in Legionowo (Table 5).

Table 5. Values of directional coefficients of the trend and of coefficients of variation of Sielianinow's coefficient during vegetation season (IV-IX) in central-eastern Poland between the years 1971–2005

Station	Directional coefficient of the trend (a)	Standard deviation
Ostrołęka	-0.006	0.41
Białowieża	-0.013	0.46
Włodawa	-0.010	0.34
Szepietowo	-0.014*	0.41
Legionowo	-0.014	0.48
Biała Podlaska	-0.008	0.40
Sobieszyn	-0.014	0.45
Pułtusk	-0.006	0.44
Siedlce	-0.014*	0.39

Explanations: * significant at $\alpha = 0.05$.

On the basis of the analysis of directional coefficients of the trend values it can be stated that SPI index values were slightly decreasing year by year. Whereas statistically significant changes were noted only in Szepietowo and Siedlce. SPI index in those stations was decreasing about 0.14 within 10 years. Different tendencies of precipitation changes are observed in particular months in different regions of Poland [Banaszkiewicz et al. 2007, Musiał and Rojek 2007].

CONCLUSIONS

1. Variability of average SPI index values during vegetation season (1971–2005) in central-eastern Poland remained on a similar level.
2. During the vegetation season of the multi-year analyzed extreme droughts occurred four times less frequently than normal months, with at the same time similar frequency of occurrence in separate months. Whereas very dry months

were noted most frequently in September and moderately dry – in August.

3. The analysis of frequency of spatial distribution of particular drought classes showed that extremely dry and very dry months occurred most frequently in western part of the area examined while moderately dry also in south-eastern part.
4. On the basis of the linear trend analysis it can be stated that SPI index values were slightly decreasing year by year. Statistically significant changes were noted only in Szepietowo and Siedlce (about 0.14 within 10 years).
5. Standardized precipitation index SPI allows to evaluate the atmospheric drought. It can be used to monitor the atmospheric drought as well but in relation to the agricultural needs, further analysis of the soil and hydrological effects is a must. Its value depends in a large extent on precipitation variability.

REFERENCES

1. Banaszkiewicz B., Grabowska K., Suchecki S. 2007. Characteristics of Precipitation Conditions in Suwałki Lake district in the years 1971–2000. Scientific Review Engineering and Environmental Sciences 3 (37), 23–32.
2. Bąk B., Łabędzki L. 2002. Assessing drought severity with the relative precipitation index (RPI) and the standardized precipitation index (SPI). J. Water Land Develop. 6, 89–105.
3. Bąk B., Łabędzki L. 2004. Standardized climatic water balance as drought index. Acta Agrophisica 3(1), 275–283.
4. Kanecka-Geszke E., Smarzyńska K. 2007. Assessing meteorological drought in some agro-climatic regions of Poland by using different indices. Acta Sci. Pol., Formatio Circumiectus 6 (2), 41–50.
5. Gąsiorek E., Grządziel M., Musiał E., Rojek M., 2012. Determination of relative precipitation index based on standardized precipitation index for monthly precipitation sums. Infrastructure and Ecology of Rural Areas, 3/III, 197–208.
6. Lana X., Serra C., Burgueno A. 2001. Patterns of monthly rainfall shortage and excess in terms of the standardized precipitation index. International Journal of Climatology 21, 1669–1691.
7. Lloyd-Hughes B., Saunders M. 2002. A drought climatology for Europe.International Journal of Climatology, 22, 1571–1592.
8. Łabędzki L. 2006. Agricultural drought – an outline of the problems and methods of monitoring and classification. Water-Environment-Rular Areas, 17, pp. 107.
9. Łabędzki L. 2008. Estimation of different duration drought frequency using the StandardizedPrecipitation Index SPI. Zeszyty Problemowe Postępu Nauk Rolniczych, 526, 105–112.
10. Łabędzki L., Bąk B. 2004. Differentiation of the atmospheric drought index SPI in the vegetation period in Poland. Water-Environment-Rular Areas 4, 2a(11), 111–122.
11. McKee T.B., Doesken N.J., Kleist J. 1993. The relationship of drought frequency and duration to time scales. Proc. of the 8th Conference of Applied Climatology, 17-22 January 1993, Anaheim, California, 179–184.
12. Musiał E., Rojek M.S. 2007. Potential evapotranspiration and precipitation in Wrocław-Swojec during the period 1965-2000. Acta Agrophysica 9(3), 685–698.
13. Turke's, M., Tatlı H. 2009. Use of the standardized precipitation index (SPI) and a modified SPI for shaping the drought probabilities over Turkey. International Journal of Climatology 29, 2270–2282.
14. Wu H., Hayes M.J., Wilhite D.A., Svoboda M.D. 2005. The effect of the length of record on the standardized precipitation index calculation. International Journal of Climatology 25, 505–520.

10

CONTENT AND UPTAKE OF SELECTED TRACE ELEMENTS BY WEEDS IN POTATO TO CULTIVATION UNDER DIFFERENT CONDITIONS OF SOIL TILLAGE AND WEED CONTROL METHODS

Krystyna Zarzecka[1], **Marek Gugała**[1], **Alicja Baranowska**[2]

[1] Chair of Plant Cultivation, University of Natural Sciences and Humanities in Siedlce, B. Prusa 14, 08-110 Siedlce, Poland, e-mail: kzarzecka@uph.edu.pl

[2] Pope John Paul II State School of Higher Education in Biala Podlaska, Sidorska 95/97, 21-500 Biala Podlaska, Poland

ABSTRACT

The study utilized data from a field experiment carried out at the Experimental Station in Zawady owned by the University of Natural Sciences and Humanities in Siedlce in the years 2005–2007. The experimental factors included two soil tillage systems and seven weed control methods in potato. Iron, copper and zinc in weeds were determined with the AAS method. The trace element content in weed dry matter before row closure of potato depended significantly on soil tillage methods (excluding Cu), weed control methods and weather conditions, and, prior to tuber harvest, on soil tillage and weather conditions during the growing season. The uptake of Fe, Cu, and Zn by weeds from the area of 1 hectare depended on the experimental factors, weed biomass and weed chemical composition.

Keywords: weeds, trace elements, content, uptake, potato.

INTRODUCTION

Potato yield losses resulting from weed infestation are estimated at 20–80% [Hashim 2003; Jaiswal, Lal 1996; Knezevic et al. 1995]. According to many authors [Boligłowa, Gleń 2003; Boydston, Vaughin 2002; Dorado, Lopez-Fando 2006] an application of soil tillage simplifications increases weed infestation, reduces potato tuber yield and reduces its quality. Negative effects of simplified soil tillage may be alleviated by an application of accurately chosen herbicides. An application of herbicides and their mixtures in potato cultivation reduces weed infestation by 40 to 99%, compared with mechanical operations [Gugała et al. 2013; Guttieri, Eberlein 1997; Mišovic et al. 1997; Shah et al. 2003; Zarzecka et al. 1999]. When weed infestation is moderate (1.5 t dry matter of weeds) use of nutrients reaches the level of 150 kg NPK. Also, the uptake of macroelements is significant. Under average conditions this amount of nutrients is sufficient to produce over 10 tonnes of potato tubers [Domańska 1980]. Increased weed infestation is followed by an increased uptake of nutrients by weeds, and a reduced uptake by the crop plant [Lehoczky et al. 2003; Żurawski, Sienkiewicz 1981]. The existing literature on the subject of microelement content in segetal weeds is quite abundant whereas the literature pertaining to the microelement content is much scarcer. Also, research on an impact of plant protection agents, being the second most important industrial production means, on accumulation of microelements in weeds is lacking [Czuba, Wróbel 1983; Lozowicka, Konecki 2011].

The research aimed at determining, under various conditions of soil tillage, the content and uptake of selected trace elements by weeds which occur in the cultivation of potato including different herbicides.

MATERIALS AND METHODS

The research included a field experiment conducted over the years 2005–2007 at the Experimental Farm in Zawady. The soil of the experimental site was the rye very good complex with a pH of 5.6 to 6.5 (1 mol KCl dm^3), organic matter content of 11.3 to 11.5 g kg^{-1}, and Fe, Cu and Zn contents determined in the tilled layer, amounting to, respectively: 755–1300, 3.8–6.2 and 8.4–11.0 mg kg^{-1}. The experiment was designed as randomised sub-blocks, and it included two factors:

a) factor I included two methods of soil tillage:
   - conventional (reversing ploughing, winter ploughing, harrowing, cultivating, harrowing),
   - simplified (reversing ploughing, cultivating);
b) factor II included weed control methods:
   - control treatment – mechanical weeding prior to and after potato stand establishment,
   - Plateen 41,5 WG (metribuzin + flufenacet) 2.0 kg ha^{-1},
   - Plateen 41,5 WG (metribuzin + flufenacet) 2.0 kg ha^{-1} + Fusilade Forte 150 EC (fluazyfop-P-butyl) 2.5 dm^3 ha^{-1} (mixture),
   - Plateen 41,5 WG (metribuzin + flufenacet) 1.6 kg ha^{-1} + Fusilade Forte 150 EC (fluazyfop-P-butyl) 2.0 dm ha^{-1} + adjuvant Atpolan 80 EC 1.5 dm^3 ha^{-1} (mixture),
   - Barox 460 SL (bentazone + MCPA) 3.0 dm^3 ha^{-1},
   - Barox 460 SL (bentazone + MCPA) 3.0 dm^3 ha^{-1} + Fusilade Forte 150 EC (fluazyfop-P-butyl) 2.5 dm^3 ha^{-1} (mixture),
   - Barox 460 SL (bentazone + MCPA) 2.4 dm^3 ha^{-1} + Fusilade Forte 150 EC (fluazyfop-P-butyl) 2.0 dm^3 ha^{-1} + adjuvant Atpolan 80 EC 1.5 dm^3 ha^{-1} (mixture).

The experimental site received a mineral and organic fertilizer application of 90 kg N, 32.9 kg P, 112.1 kg K, and 25.0 t ha^{-1} farmyard manure, respectively. Botanic and weight analyses of weeds were made twice – before and after row closure. Weeds from the area of 1 m^2 were collected on each plot and, after drying, their air dry weight was determined. Iron, copper and zinc contents were determined in weed air dry matter by the method of atomic absorption spectrophotometry (AAS). Digestion of weeds (1.0 g samples) was performed using a mixture of HNO_3 (7 ml) + H_2O_2 (1 ml). It was mineralized in a laboratory microwave oven Ethos plus. Distilled water was added to the sample and it was mixed. The sample was diluted to 50 ml with distilled water. Element concentration were expressed on a dry weight in mg kg^{-1}. An uptake of elements from an area of 1 ha was calculated on the basis of weed dry weight and weed chemical composition. The results obtained were statistically analysed with the analysis of variance and the significance of differences was determined using Tukey test at the significance level of $p=0.05$. Weather conditions over the growing seasons of studies varied. The year 2007 was most favourable as far as rainfall and temperature were concerned. In contrast, the season of the year 2006 was the least favourable due to a heavy drought that took place then.

RESULTS AND DISCUSSION

The average three-year air dry weight of weeds was the highest in the mechanically-cultivated control treatment (Table 1). The herbicides applied reduced the weed weight at the start of vegetation and prior to tuber harvest by 40.1 to 80.3%, and 33.2 to 61.9%, respectively. The works of other authors [Mišovic et al. Shah et al., 2003; Zarzecka et al. 1999; Mirabelli et al. 2005] support this finding.

Mineral composition and nutrient uptake of weeds are a function of at least several factors. They include: stage of weed growth, soil fertility of the site, the length of time of competition between plants, application of fertilizers and herbicides, and thermal and moisture conditions throughout the growing period [Johansen et al. 2005; Ali et al. 2006; Baćmaga et al. 2007; Królak 2003; Kirchmann et al. 2005].

Table 1. Weed control of air dry matter of weeds (2005–2007)

Weed control methods	Weed control [%]	
	before row closing	before tubers harvest
1.	0 (23.7 g m^2)	0 (98.9 g m^2)
2.	55.7	36.9
3.	75.9	48.7
4.	80.3	61.9
5.	40.1	33.2
6.	55.8	42.1
7.	59.5	50.0
Mean	61.2	45.5

* Explanation as in materials and methods.

The analysis of the results obtained in the present work showed that iron, copper and zinc contents in weeds sampled at the beginning of vegetation significantly depended on soil tillage methods, cultivation methods and growing conditions over the years of study (Table 2). More Fe and Zn were accumulated by weeds sampled from the simplified tillage treatment than from the conventionally-tilled plots. An application of herbicides reduced the iron content and increased the copper and zinc contents compared with the mechanically-tilled control treatment. The differences in the contents of the elements between herbicide treatments were not statistically proven, which means that the effect of individual herbicides was not pronounced.

The contents of the elements determined in older weeds (prior to tuber harvest) were lower than in younger weeds, and depended significantly on the tillage methods and weather conditions over the years of study. Also Parylak [1994] found that the contents of elements in weeds decreased as they matured. Fe, Cu and Zn concentrations in weeds were similar to the values obtained by other authors [Ali et al. 2006; Johansen et al. 2005; Kalny et al. 2007; Kirchmann et al. 2005].

Trace element uptake by weeds (at both dates of determination) was significantly affected by the experimental factors of the trial (Table 3). Higher uptake of elements was recorded in simplified tillage treatments. Herbicides reduced the volume of weed biomass prior to the crop plant harvest by 45.5%, on average. As a result, the uptake of the elements by weeds decreased almost twice (Fe – from 116.76 in the control treatment to 70.33 g ha^{-1} in herbicide treatments; Cu – from 12.558 to 7.570; Zn – from 34.820 to 21.538 g ha^{-1}). Kirchmann et al. [2005] and Trąba [2001] also found that uptake of elements by weeds from a one-hectare area depended on the concentration of the elements in weeds, and on their biomass.

In the studies discussed, a large effect of weather conditions on Fe, Cu and Zn contents and accumulation was found. Under favourable moisture conditions of the year 2004 (precipitation was higher than in the multi-year period, and the temperature was close to the average) the uptake of the elements was several times higher than in the year 2006 when a heavy drought took place, which was confirmed in the studies by Dziekanowski et al. [1992].

Some authors [Mirabelli et al. 2005; Ali et al. 2006; Trąba 2001] believe that chemical composition of weeds is influenced by the species which dominate in a weed community. In the research carried out the dominating were *Chenopodium album* and *Echinochloa crus-galli*, and the Fe, Cu and Zn contents were higher in white goosefoot than barnyard grass (Figure 1). Higher Cu and Zn

Table 2. Content of Fe, Cu and Zn of dry matter of weeds (mg kg^{-1})

Experimental factors		Harvest times of weeds					
		I*			II**		
		Fe	Cu	Zn	Fe	Cu	Zn
Tillage systems	1.	125.5	13.01	36.44	114.8	12.99	35.69
	2.	128.8	12.94	36.74	118.1	12.95	36.16
	$LSD_{0.05}$	0.7	n.s.	0.15	0.5	0.02	0.29
Weed control methods	1.	127.6	12.85	36.29	116.9	12.94	35.68
	2.	127.0	12.93	36.59	116.9	12.98	35.85
	3.	126.8	13.05	36.56	116.1	12.99	36.14
	4.	127.0	13.04	36.61	116.6	12.96	35.94
	5.	127.8	13.04	36.68	116.2	13.01	36.06
	6.	127.3	12.98	36.75	116.0	12.96	35.91
	7.	127.3	13.02	36.66	116.6	12.98	35.91
	$LSD_{0.05}$	0.6	0.47	0.19	n.s.	n.s.	n.s.
Years	2005	126.9	12.57	36.34	115.7	12.60	35.39
	2006	125.8	12.78	36.95	114.1	13.66	35.92
	2007	128.9	13.60	36.48	119.7	12.66	36.47
	$LSD_{0.05}$	1.1	0.34	0.24	0.8	0.03	0.44
Mean		127.2	12.98	36.59	116.5	12.97	35.93

* before row closing of potato, ** before tubers harvest
n.s. – non-significant differences

Table 3. Uptake of Fe, Cu and Zn by weeds (g ha^{-1})

Experimental Factors		Harvest times of weeds					
		I*			II**		
		Fe	Cu	Zn	Fe	Cu	Zn
Tillage systems	1.	11.8	1.3	3.4	68.6	7.6	21.5
	2.	18.3	1.8	5.1	85.3	8.9	25.3
	$LSD_{0.05}$	2.2	0.2	0.6	7.6	0.8	n.s.
Weed control methods	1.	30.2	4.0	8.6	116.8	12.6	34.8
	2.	16.1	1.6	4.4	81.2	8.7	24.6
	3.	7.1	0.7	2.0	61.8	6.8	19.2
	4.	5.3	0.5	1.5	47.3	5.1	14.43
	5.	19.9	2.0	5.6	87.3	9.4	26.7
	6.	14.0	1.4	3.9	77.4	8.3	23.9
	7.	12.6	1.3	3.5	66.9	7.2	20.5
	$LSD_{0.05}$	3.4	0.4	0.9	15.8	1.7	1.3
Years	2005	7.9	0.8	2.3	47.8	5.3	14.7
	2006	4.1	0.4	1.1	19.5	2.3	6.1
	2007	33.1	3.4	9.3	163.6	17.2	49.5
	$LSD_{0.05}$	3.4	0.4	0.9	11.7	1.2	3.7
Mean		15.0	1.5	4.2	77.0	8.3	23.4

* before row closing of potato, ** before tubers harvest
n.s. – non-significant differences

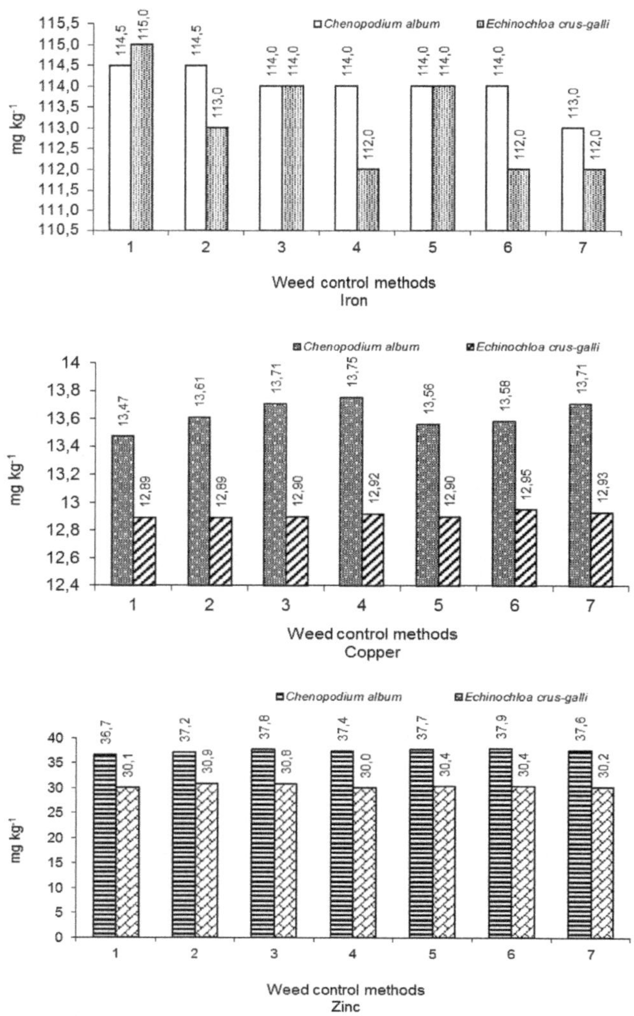

Figure 1. Content of iron, copper and zinc in *Chenopodium album* and *Echinochloa crus-galli* before tubers of potato harvest (mg kg^{-1})

concentrations in dicotyledonous weeds than in monocotyledonous weeds were recorded by other workers [Czuba, Wróbel 1983].

CONCLUSIONS

1. Iron, copper and zinc contents in weed dry weight at the beginning of vegetation and prior to potato harvest depended, respectively, on tillage methods (excluding Cu), weed control, and growth conditions, and on tillage methods and whether conditions over the research years.
2. Herbicides significantly reduced Fe content and increased Cu and Zn contents in young weeds, which indicates that plant protection agents influenced the chemical composition of segetal vegetation.
3. The amount of trace element uptake from a one-hectare area depended on experimental factors, weed biomass and weed chemical composition.

REFERENCES

1. Ali M.F., Heng L.Y., Ratman W., Nais J., Ripin R. 2006. The content and accumulation of arsenic and heavy metals in medicinal plants near mamut river contaminated by copper-mining in Sabah, Malaysia. Fresenius Envir. Bull., 15(10), 1316–1321.
2. Baćmaga M., Kucharski J., Wyszkowska J. 2007. Impact of crop protection chemicals on plants and animals. J. Elementol., 12(2), 135–148.
3. Boligłowa E., Gleń K. 2003. Yielding and quality of potato tubers depending on the kind of organic fertilization and tillage method. Electronic Journal of Polish Agricultural Universities, S. Agronomy Vol. 6, Iss. 1, 1–8. www.ejpau.media.pl/series/volumne 6/issue1/agronomy/art-03.html
4. Boydston R.A., Vaughin S.F. 2002. Alternative weed management systems control weeds in potato (*Solanum tuberosum*). Weed Technology, 16(1), 23–28.
5. Czuba R. and Wróbel S. 1983. Ocena roli chwastów jako konkurentów w pobieraniu składników pokarmowych przez rośliny uprawne. Roczniki Gleboznawcze, 34(3), 175–184.
6. Domańska H. 1980. Chwasty i ich zwalczanie. PWRiL Warszawa.
7. Dorado J., Lopez-Fando C. 2006. The effect of tillage system use of a paraplow on weed flora in a semiarid soil from central Spain. Weed Research, 46(5), 424–431.
8. Dziekanowski A., Ciećko Z., Nowak G. 1992. Zawartość podstawowych makro- i mikroskładników w bulwach ziemniaka w zależności od poziomu nawożenia potasem Acta Acad. Agricult. Tech. Olst., Agricultura, 54, 117–126.
9. Gugała M., Zarzecka K., Sikorska A. 2013. Ocena skuteczności działania herbicydów i ich wpływ na plon handlowy ziemniaka. Biul. IHAR, 270, 75–84.
10. Guttieri M.J., Eberlein C.V. 1997. Preemergence weed control in potatoes with rimsulfuron mixtures. Weed Technology, 11(4), 755–761.
11. Hashim S. 2003. Chemical weed control efficiency in potato *(Solanum tuberosum* L.) under agro-climatic conditions of Peshawar, Pakistan. Pakistan J. Weed Science Research, 9(1–2), 105–110.
12. Jaiswal V.P., Lal S.S. 1996. Efficacy of cultural and chemical weed control methods in potato *(Solanum tuberosum).* Indian J. Agronomy, 41(3), 454–456.
13. Johansen A., Bakken A. K., Synnes O.M. 2005. Green fodder crop and dicotyledonous weeds as sources for micronutrients in ruminant diet, in: NJF Sem. No. 370 "Essential trace elements for plants, animals and humans", Reykjavik, Iceland, 15–17 August: pp. 61–63.
14. Kalny P., Fijałek Z., Daszczuk A., Ostapczuk P. 2007. Determination of selected microelements in polish herbs and their infusions. Science of the Total Environment, 381(1–3), 99–104.
15. Kirchmann H., Thorvaldsson G., Björnsson H., Mattson L. 2005. Trace elements in crop from Swedish and Icelandic long-term experiment, in: NJF Sem. No. 370 "Essential trace elements for plants, animals and humans", Reykjavik, Iceland, 15–17 August: pp. 30–33.
16. Knezevic M., Durkic M. and Samota D. 1995. Chemical and mechanical weed control in potatoes. Fragmenta Phytomedica et Herbologica, 23(2), 61–67.
17. Królak E. 2003. Accumulation of Zn, Cu, Pb and Cd by Dandelion (*Taraxacum officinale* Web.) in Environments with Various Degrees of Metalic Contamination. Pol. J. Environ. Stud. 12(6), 713–721.
18. Lehoczky É., Dobozi M., Gyüre K. 2003. Competition between weeds and potato with special regard to competition for nutrients. Magyar Gyomkutatás és Technológia, 4(1), 19–30.
19. Łozowicka B., Konecki R. 2011. Selected aspects of chemical protection of agricultural crop in northern-eastern Poland. Acta Sci. Pol. Agricultura, 10(4), 107–119.
20. Mirabelli C.,Colla A., Fiorill A., Cardarelli M., Rouphael Y., Paolini R. 2005. The effect of mechanical weed control technique and irrigation method on yield, tuber quality and wed suppression in organic potato. Acta Horticulturae, 684, 127–134.

21. Mišovic M.M., Brocic Z. A., Momirovic N.M., Šinzar B.C. 1997. Herbicide combination efficacy and potato yield in agro-ecological conditions of Dragacevo. Acta Horticulturae, 462, 363–368.

22. Parylak D. 1994. Pobieranie składników pokarmowych przez chwasty i pszenżyto ozime w różnych okresach jego rozwoju. Zeszyty Naukowe AR w Szczecinie, Rolnictwo, LVIII, 162, 185–188.

23. Shah N.H., Hassan G., Khan I.A., Azim A. 2003. Management of grassy and broadleaf weeds in potato in the hills of Hazara, Pakistan. Pakistan J. Weed Science Research, 9 (1–2), 111–116.

24. Trąba Cz. 2001. Konkurencyjność chwastów wobec owsa w warunkach nawożenia mineralnego i organicznego. Progress in Plant Protection/Postępy w Ochronie Roślin, 41(2), 941–944.

25. Trąba Cz., Wiater J. 2000. Pobranie niektórych mikroelementów przez chwasty w uprawie lędźwianu siewnego, w warunkach nawożenia mineralnego i odpadami organicznymi. Zesz. Probl. Post. Nauk Rol., 471, 803–809.

26. Zarzecka K., Ceglarek F., Gąsiorowska B., Grużewska A. 1999. Impact of weed control on potato infestation and yielding. Electronic Journal of Polish Agricultural Universities, S. Agronomy Vol. 2, Iss. 2, 1–8. www.ejpau.media.pl/series/volumne2/issue2/agronomy/art-07.html

27. Żurawski, H., Sienkiewicz, J. 1981. Wpływ uproszczeń w uprawie roli i zróżnicowanego nawożenia na plony roślin i pobranie składników pokarmowych. Pamiętnik Puławski, 74, 73–84.

11

ANALYSIS OF HYDROTHERMAL CONDITIONS AND THEIR IMPACT ON EARLY POTATO YIELDS

Elżbiata Radzka[1], Katarzyna Rymuza[1], Tomasz Lenartowicz[2]

[1] University of Natural Sciences and Humanities in Siedlce, Prusa 14, 08-110 Siedlce, Poland, e-mail: elzbieta.radzka@uph.edu.pl

[2] The Research Centre for Cultivar Testing in Słupia Wielka, 63-022 Słupia Wielka, Poland

ABSTRACT

The work is based on data on monthly air temperatures, monthly sums of atmospheric precipitation and early potato yields in 2000–2013 obtained from seven COBORU (Research Centre for Cultivar Testing) stations situated in east-central Poland. Hydrothermal conditions during the growing season (April–July) of early potato were described by means of the Sielianinow's coefficient. The relationships between potato yields and the Sielianinow's hydrothermal coefficient of the growing season were examined using multiple regression equations generated by means of the step-wise regression procedure followed by polynomial regression procedure. The average Sielianinow's coefficient values ranged from 1.43 in April to 1.66 in May during the early potato growing season in east-central Poland. The applied statistical methods revealed that the hydrothermal factors impacted on early potato yields, particularly in the second stage of potato growth and development.

Keywords: Sielianinow's coefficient, growing season, east-central Poland, yield, early potato.

INTRODUCTION

Potato is one of the most widely cultivated crop plants [Chotkowski and Rembeza 2005]. It is grown predominantly on lighter soils, which makes potato yields dependent on evenly distributed precipitation. Importance of water has been stressed in many works published so far [Chmura 2001, Dzieżyc et al. 1987, Nowak 2001, Szutkowska and Lutomirska 2002]. The main meteorological factors shaping the living conditions of organisms are precipitation and thermal conditions. These two meteorological (and in multiyear – climatic) parameters are so important because they condition water circulation as well as mineral and organic substances circulation in the environment [Bochenek 2012].

The Sielianinow's hydrothermal coefficient is a useful tool for determining the water status of the environment [Kapuściński and Nowak 2003, Bartoszek and Banasiewicz 2007, Hutorowicz et al. 2008]. The relationship of crop plant development and yield with hydrothermal conditions is based on current or predicted temperature and atmospheric precipitation [Ziernicka-Wojtaszek 2009]. The list of crop plant cultivars encompasses many potato cultivars belonging to different maturity categories. The categories have got different development rhythm, time of growth to maturity and, first and foremost, different usability characteristics. As a result, they can be used in many ways in practice. Many cultivars cease to be registered on the Polish National List in a natural way and are replaced by new ones. Both new and already admitted cultivars respond differently to changeable cultivation conditions [Dmowski et al. 2004]. The already published works on potato precipitation needs are of help while assessing irrigation requirements of this crop at different times and in different areas of the country as potato is very responsive to improved moisture conditions [Mazurczyk et al. 2009, Nowak 2006, Rolbiecki et al 2009, Żarski 2011]. The effect of precipitation on early potato yields seems to be obvious but statistical methods of calculation used by many authors do not always confirm this [Biniak et al. 2007]. Useful as they are for analysing and determining quantitative relationships, the methods have got limitations, which stimulates researches to look for new ways of analysing the results [Kalbarczyk Kalbarczyk 2009].

The objective of this work was to describe hydrothermal conditions in east-central Poland and determine their effect on early potato yields.

MATERIALS AND METHODS

The work is based on average monthly air temperatures and monthly precipitation sums obtained from seven COBORU stations in east-central Poland in 2000–2013. The second data group consisted of average yields of early potato cultivars grown at seven stations included in the hydrothermal analysis. Values of Sielianinow's hydrothermal coefficient (*k*), which is a measure of rainfall efficiency in a given month, were calculated for the growing season (April-July). Months were assigned to classes according to the following criterion:

- extremely dry (ed) $k < 0.40$
- very dry (vd) $0.41 < k < 0.70$
- dry (d) $0.71 < k < 1.0$
- quite dry (qd) $1.01 < k < 1.30$
- optimum (o) $1.31 < k < 1.60$
- quite wet (qw) $1.61 < k < 2.0$
- wet (w) $2.01 < k < 2.5$
- very wet (ww) $2.51 < k < 3.0$
- extremely wet (ew) $k > 3.01$ [Skowera and Puła 2004].

Spatial distribution of average monthly values of Sielianinow's coefficient in the study area are presented using isolines. The relationship between potato yields and Sielianinow's hydrothermal coefficient was examined by means of multiple regression equations. Multiple regression coefficients were obtained using the stepwise regression procedure. Moreover, to check the significance of regression coefficients, t-Student test values were calculated and determination coefficients were computed (R^2). The determination coefficient is a measure of goodness of fit which explains what portion of the total variation of variable *Y* is explained by linear regression on independent variables [Trętowski, Wójcik 1991]. In order to more fully describe the phenomenon, the potato growing season was divided into two periods with individual months including the most important developmental stages of plants (Stage1: April–May and Stage2: June–July). Equations of multiple regression were used to construct models describing the relationship of potato yield with Sielianinow's coefficient in individual stages. Significance of regression coefficients was checked using t-Student test at $p \leq 0.05$. All calculations were performed in Statistica 10.0.

RESULTS AND DISCUSSION

The transitional nature of Polish climate manifests itself in substantial variability of values of individual climate components, including air temperature and precipitation. Thus, it is important to examine hydrothermal conditions taking into accounts agricultural economy needs. Analysis of hydrothermal conditions demonstrated that, during the growing season, the average value of Sielianinow's coefficient in the study area ranged from 1.43 in April to 1.66 in May. Skowera [2014] analysed the hydrothermal conditions in Poland from 1971 to 2010 and noted that the average value of hydrothermal coefficient was the lowest in August ($k = 1.3$). Kalbarczyk [2003] claims that this value ranges from 1.29 to 1.67 during the growing season (April-September), the lowest average values occurring in September and highest in April. This work demonstrated that the southern part of the east-central Poland was the wettest in all the months of the growing season (Figure 1).

The driest months were the most common of all the classes to which months were assigned (on average, 2.9 times in 14 years). Extreme droughts occurred in May and June in the study area, being most frequently observed in Krzyżew and Uhnin. Extremely wet months were recorded most often in April but were less frequent than extreme droughts (on average, twice during 14 years). In turn, in Krzyżew and Nowy Lubliniec, April was extremely wet 3 times in the study period. July was the month which was most frequently very wet (on average, 0.9 times during 14 years). Months assigned to this class most often occurred in Cicibór Duży and Nowy Lubliniec and were the rarest in Marianów. Skowera and Puła 2004] analysed extreme pluviometric conditions in Poland in the spring from 1971 to 2000 and concluded that average Sielianinow's coefficients differed regionally throughout the country. According to

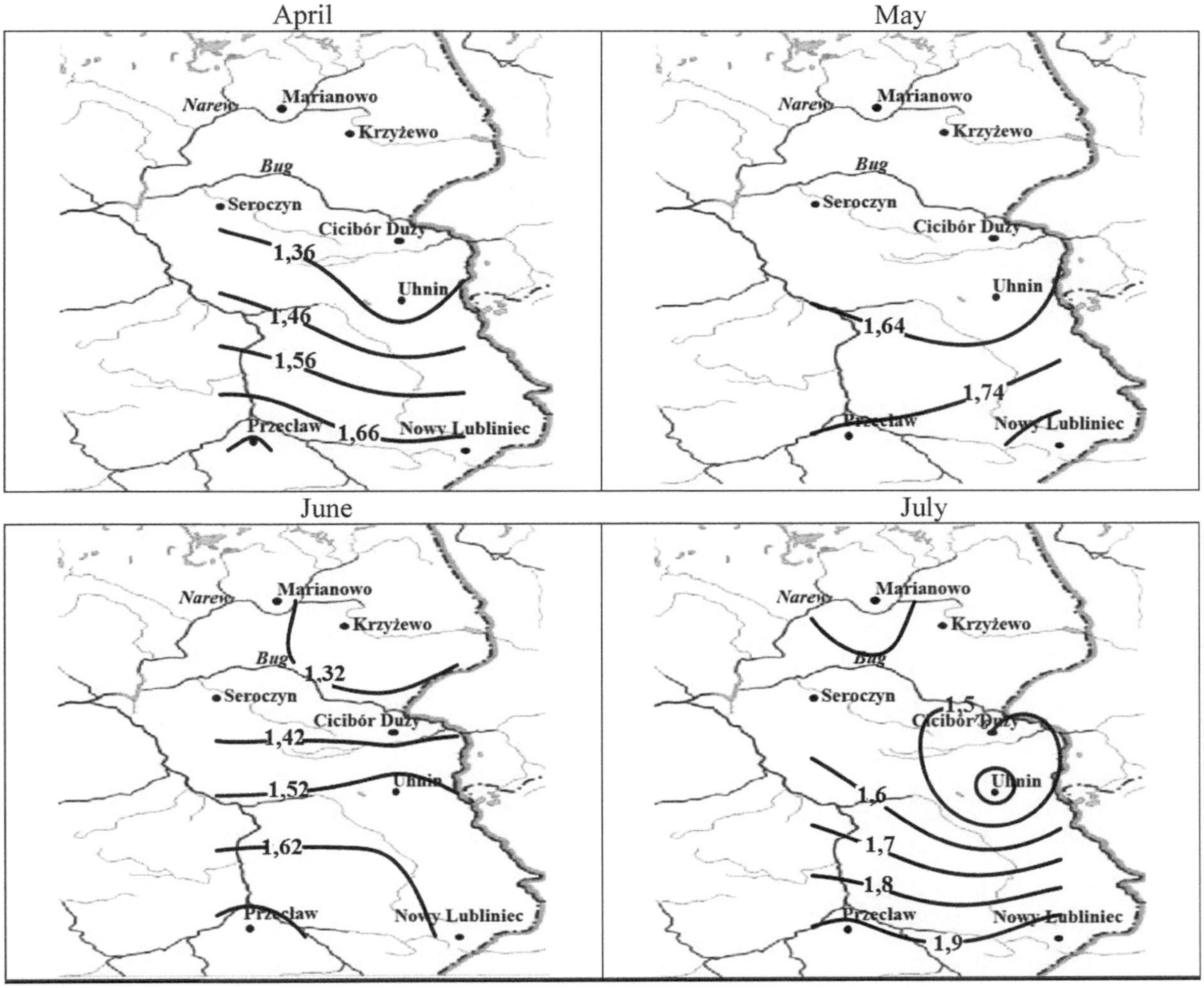

Figure 1. Spatial distribution of average values of Sielianinow's hydrothermal coefficient during the growing season (April–June) in east-central Poland in 2000–2013

the aforementioned authors, the south and north of Poland are the wettest regions in April whereas in May quite dry conditions prevail in most of the country. Quite wet and wet conditions were observed in southern Poland only. Throughout June optimum conditions predominate, quite wet and wet weather occurring in the north, south and north-east. Many authors point to an unfavourable effect of drought on plant growth and development due to decreased yields and poorer quality Radzka et al. 2010, Szwejkowski et al. 2008].

Values of the Sielianinow's hydrothermal coefficient in 38% determined potato yields during the growing season (Table 1). The model revealed that potato yields were significantly affected by hydrothermal coefficient values in June only. The coefficients for April and May appeared in the model but their influence was not significant.

A more detailed description and Sielianinow's hydrothermal coefficient in individual stages on potato yields was obtained using multiple regression (Table 1 and 2).

Sielianinow's coefficient values in the second growth stage (June-July) significantly impact on early potato yield in east-central Poland. The relationship is parabolic in character. The maximum value of 450 dt can be obtained when the Sielianinow's coefficient is 2.3. Adverse weather conditions affect the length of phonological stages and, sometimes, may extend the growing season, which, according to Jefferies and Lawson [1991], Sawicka [1991] as well as Sawicka and Krochmal-Marczak [2005], may directly influence potato yield and quality.

When the weather after potato planting is warm, plants establish more quickly. Plant water demand during this period is at its minimum. From emergence to flowering, plants require quite warm weather and moderate, evenly distributed precipitation. The greatest water demand

Table 1. Regression equations for the relationship between early potato yields and Sielianinow's hydrothermal coefficient during the growing season

Regression equations on Sielianinow's coefficient (R^2 = 0.384)			
variables in the model	regression coefficients	t test values	p level
Free term	348.43	7.321	0.00
Sielianinow's coefficient in June	-41.82*	2.784*	0.00
Sielianinow's coefficient in April	-21.53	-1.401	0.179
Sielianinow's coefficient in May	17.08	1.005	0.318

* – significant at p ≤ 0.,05.

Table 2. Components of orthogonal regression equation, t-Student values and graphs presenting early potato yields depending on Sielianinow's coefficient at two stages of growth and development

Equation component	Stage 1			Stage 2		
	free term	linear component	quadratic component	free term	linear component	quadratic component
value	421.772	-8.928	1.708	320.07*	107,35*	23.078*
t-Student	11.534	-0.224	0.174	8.709	2.515	-2.258
p	0.000	0.823	0.863	0.000	0.013	0.026
Graph	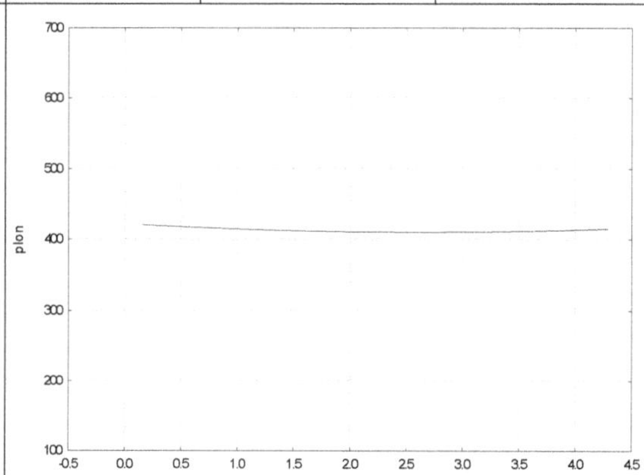			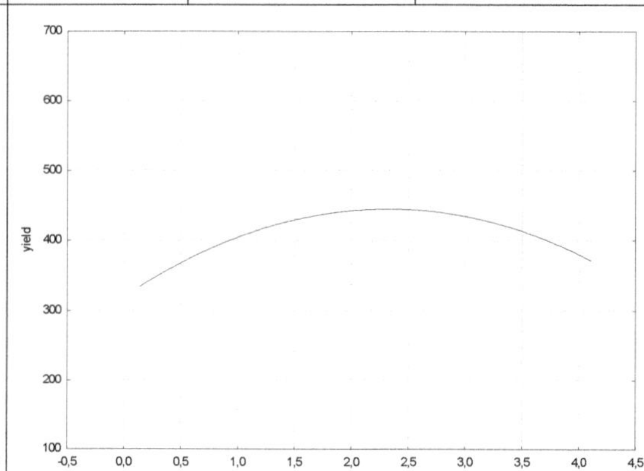		

is at tuber set and intensive growth. At the final stage of tuber growth, warm, sunny weather with moderate and evenly distributed precipitation contributes to dry matter and starch accumulation in tubers. On the contrary, excessive precipitation during this period results in tuber infection with diseases and poorer storage quality in winter [Lutomirska 2005, 2006, 2008].

CONCLUSIONS

Average values of Sielianinow's hydrothermal coefficient in east-central Poland during the growing season of early potato ranged from 1.43 in April to 1.66 in May. The lowest k values in all the months of the growing season were obtained in the southern part of the study area.

The regression equation demonstrated that the yield was significantly affected by the Sielianinow's hydrothermal coefficient value in June.

The statistical methods applied in this work demonstrated that hydrothermal conditions exerted more impact on early potato yields particularly during the second part of the growing season. Maximum tuber yields of 460 and 450 dt can be achieved when Sielianinow's coefficient is 2.3 respectively.

REFERENCES

1. Bartoszek K., Banasiewicz I., 2007. Agrometeorological characteristics of the vegetation period in 2005 against the background of the period of 1951-2005 in the Lublin region. Acta Agroph. 9(2), 275–283.
2. Bochenek W. 2012. Evaluation of precipitation at the IG&SO pas research station in Szymbark during 40-year period (1971–2010)and its impact on the variability of ater runofffrom the Bystrzanka stream basin]. Woda Środ. Obsz. Wiej. 2(38), 29–44.

3. Biniak M., Kostrzewa S., Żyromski A., 2007. Thermal and rainfall determinants of water demand in the Wrocław area as exemplified by medium-late potatoes. Zesz. Probl. Post. Nauk Rol. 519, 31–45.
4. Chmura K., 2001. Natural and agrotechnological determinants of potato cultivation in south-western Poland. Zesz. Nauk. AR in Wrocław, Dissertations 410.
5. Chotkowski J., Rembeza J., 2005. Consumer preferences and changes in potato consumption. Rocz. Nauk. Stow. Ekon. Rol. i Agrobiznesu 7(3), 42–49.
6. Dmowski Z., Nowak L., Chmura K., 2004. Reaction of varying in the length of a vegetation period to differentiated water-fertilization conditions. Biul. Inst. Hod.i Aklim. Rośl. No 232. 141–148.
7. Dzieżyc J., Nowak L., Panek K., 1987. Decade indices of precipitation requirements by crops in Poland. Zesz. Probl. Post. Nauk Roln., 314, 11–33.
8. Głuska A., 2000. Effect of cultivation technology on the quality of potato yield. Biul. IHAR 213, 173–178.
9. Hutorowicz H., Grabowski J., Olba-Zięty E., 2008. Frequency of occurrence of dry spells and droughts in two mesoregions of Masurian Lakeland. Acta Agroph. 12(3), 663–673.
10. Jakubowski T., 2007. Potato tuber cropping and susceptibility to mechanical damage related to hydrothermal coefficient. Inżynieria Rol. 8(96), 87–93.
11. Jefferies R.A., Lawson H.M., 1991: A key for the stages of development of potato (*Solanum tuberosum*). Ann. Appl. Biol., 119, 387–389.
12. Kalbarczyk R., 2003. Thermal and precipitation conditions in relation to the yielding of potato in Poland. Annales UMCS. Sectio E. Vol. 58. 35–44.
13. Kalbarczyk R., Kalbarczyk E., 2009. The requirements and deficiency in atmospheric precipitation in cultivated mid-late and late potato in Poland. Infrastructure and Ecology of Rural Areas, Nr 3. 129–140.
14. Kapuściński J., Nowak R., 2003. The frequency of the occurrence of droughts and post-droughts periods in mid-west Poland on the example of Poznań, Wałcz and Wieluń. In: Management and conservation of the forest environment. Climate vs forest (Ed. Miler A.) Agricultural University in Poznań, 76–88.
15. Lutomirska B., 2005. The effect of temperature in mid-growing season on potato yield. Ziemniak Polski No 4. Bonin. 14–15.
16. Lutomirska B., 2006. Accelerating very early potato harvest. Ziemniak Polski No 1, 9–11.
17. Lutomirska B., 2008. The effects of cultivars and meteorological factors on variability of the potato phenophases. Ziemniak Polski No 247. Bonin. 119–134.
18. Mazurczyk W., Wierzbicka A., Wroniak J., 2009. Influence of optimisation of irrigation and nitrogen fertilisation on some growth parameters and yield of early potato variety. Infrastructure and Ecology of Rural Areas, No 3, 91–100.
19. Nowak L., 1989. Water requirements of rainfed tuber and root crops. Zesz. Probl. Post. Nauk Roln., 343, 27–42.
20. Nowak L., 2001. The effect of irrigation of medium-early potato in the Wrocław area. Fragm. Agron. 1, 68–75.
21. Nowak L., 2006. Water requirements of tuber and root crops. In: Plant Irrigation (Eds Kaczmarczyk S., Nowak L., PWRiL, Poznań, 368–372.
22. Radzka E., Jankowska J., Koc. G, Rak J., 2010. The effect of droughts on potato yields in east-central Poland. Fragm. Agron. 27(4), 111-118.
23. Rolbiecki S., Rzekanowski Cz., Rolbiecki R., 2009. Estimation of needs and results of irrigation of medium-early potato in the vicinity of Bydgoszcz in the years 2005-2007. Acta Agroph., 13(2), 463–472.
24. Sawicka B., 1991. A study of variation in selected characteristics and degeneration of different potato cultivars in the Biała Podlaska area. Publishing Centre of Agricultural University in Lublin, Rozpr. Nauk., 141, 1–75.
25. Sawicka B, Krochmal-Marczak B., 2005. Influence of agrometeorological factors on length of developmental phases of very early and early potato cultivars. Acta Agroph. 6(1), 225–236.
26. Skowera B., 2014. Changes of hydrothermal conditions in the Polish area (1971–2010). Fragm. Agron. 31(2), 74–87.
27. Skowera B., Puła J., 2004. Pluviometric extreme conditions in spring season in Poland in the years 1971-2000. Acta Agroph., 3(1), 171–177.
28. Szutkowska M., Lutomirska B., 2002. Influence of environmental conditions and agriculture factors on tubers infection with *Streptomyces scabies*. Biul. IHAR 221, 153–167.
29. Szwejkowski Z., Dragańska E, Suchecki S., 2008. Forecast of influence of expected global warming in year 2050 on crop yielding in north-eastern Poland. Acta Agroph. 12(3), 791–800.
30. Trętowski J., Wójcik A. R., 1991. Methodology of Agricultural Experiments. WSRP Siedlce, 79–94.
31. Ziernicka-Wojtaszek A., 2009. Verification of agro-climatic regionalisation types in Poland in the light of contemporary climate change. Acta Agroph., 13(3), 803–812.
32. Żarski J., 2011. Trends in changes of climatic indices for irrigation needs of plants in the region of Bydgoszcz. Infrastructure and Ecology of Rural Areas, 5, 29–37.

12

THE CONTENT OF MACRO- AND MICROELEMENTS IN THE SHOOTS OF *GLYCERIA MAXIMA* OF THE SŁUPIA RIVER

Agnieszka Parzych[1], Małgorzata Cymer[2]

[1] Environmental Chemistry Department, Institute of Biology and Environmental Protection, Pomeranian University, Arciszewskiego 22b St., 76-200 Slupsk, Poland, e-mail: parzycha1@op.pl

[2] Department of Land Reclamation and Environmental Management, University of Warmia – Mazury, Plac Łódzki 2, 10-719 Olsztyn, Poland

ABSTRACT

The study shows the content of macro- and microelements in the leaves and rhizomes of *Glyceria maxima* of the Słupia River in Słupsk. The content of macro- and microelements in *G. maxima* was analyzed for each component separately and in an integrated way, and comparing the demand for nutritions. The largest quantity of nitrogen, phosphorus, potassium, magnesium and calcium were found in the leaves, meanwhile zinc, iron, manganese, nickel and copper in the *G. maxima* rhizomes. The amount of Zn and Cu, in all of the tested positions, within the range of the limit for the plants, and the concentrations of Ni and Mn exceed physiological needs. The U Mann Whitney test showed a number of statistically significant differences in the concentration of the analyzed elements in leaves – rhizomes, leaves – bottom sediment and rhizome - bottom sediment relation. The relations between designated heavy metals formed a following series: Mn>Fe>Zn>Ni>Cu in leaves and Fe>Mn>Zn>Ni>Cu in rhizomes. The *G. maxima* shoots accumulated 985.8 do 1441.4 $mmol_c \cdot kg^{-1}$ in all the analyzed components, and the lower value of the sum of the accumulated macro- and microelements were found in rhizomes. The sum of ion comospition of the macronutrients in the leaves and rhizomes was similar. The content of nitrogen was 51.7–53.7% of this amount, 5.8–8.6% phosphorus, potassium 22.8 – 26.6%, 4.9-5.8% of magnesium, calcium, 8.2–11.9%, and trace elements were 0.58% in total, in the case of leaves and 8.70% in the rhizomes.

Keywords: macrophytes, macroelements, trace elements, accumulation, pollution.

INTRODUCTION

The pollution of urban ecosystems with the anthropogenic substances, including heavy metals, significantly affects the increase in the concentration of the macro-and microelements in the bottom sediments and the coastal vegetation of many rivers [Buszewski et al. 2000, Yang et al. 2001, Dąbkowska-Naskręt and Różański 2009]. The macrophytes, which next to the phytoplankton, macrozoobenthos and fish, are one of the basic elements of the assessment of the ecological status of rivers, are used to bioindicate aquatic ecosystems [Skorbiłowicz 2003, Obolewski and Glińska-Lewczuk 2006, Rabajczyk and Jóźwiak 2009]. The state of water reservoirs pollution, with both macro- and microelements, is very well reflected by the aquatic plants [Aksoy et al. 2005, Letachowicz et al. 2006, Sasmaz et al. 2008, Baldantoni et al. 2009, Klink et al. 2013]. Currently, a significant impact on the chemical composition of the bottom sediments has the intensity of catchment area use and sewage inflow [Misztal et al. 1996, Salati and Moore, 2009]. This leads to an increase in the content of many elements in river sediments, which are important part of the water environment. Most of the heavy metals entering the river are connected and transported with the slime, which leads to the rise of sediments [Kruopiene 2007] and may be a secondary source of pollution, which despite the cut-off of primary sources continues to threaten aquatic ecosystems [Grynkiewicz et al. 2006]. Most of the chemical elements accumulated in the river sediments, is bioaccumulated in plant and animal tissue, what causes the increase of the risk of their poisoning within the trophic chain, on the top of which stands a man. The absorption and bioaccumulation of the components needed by the plants results from the physiological needs and is a part of the natural cycle [Kabata-Pendias and Szteke 2005]. The proper functioning of the plants requires basic macroelements (N, P, K, Ca, Mg), as well as small amounts of microelements (Fe, Zn, Cu, Ni, Mn), that are natural components of the ecosystems [Wolf and Gworek 2009]. However, their excessive concentration in the environment is harmful, because of their sufficiently high concentration interferes with the functioning of the ecosystems, threat to living organisms [Gruca-Królikowska and Wacławek 2006]. Controlling the chemical composition of the coastal vegetation and bottom sediments in urban areas is necessary because it allows the identification of existing and potential risks arising from the toxic effects of heavy metals on the aquatic environment and human health. A good indicator of the quality of bottom sediments is *Glyceria maxima* (Hartm.) Holmb.), [Rabajczyk and Jóźwiak 2009, Teuchies et al. 2013].

The aim of the study was to compare the ability of accumulation of the above and belowground shoots of the *Glyceria maxima* nutrients and selected heavy metals. The contents of macro-and microelements in *G. maxima* were analyzed considering each component separately and in an integrated way – by comparing the demand for nutrients. The study takes into account the effect of the bottom sediments to the tested macrophyt.

MATERIAL AND METHODS

The study area

The study was conducted on a section of the Słupia River within the city of Słupsk. The Słupia River is located in the central part of Pomerania (northern Poland). The lowland watercourse is 138,6 km long and the catchment occupies the area of 620 km^2. From the north, the catchment is bordered by the Baltic Sea catchment, from the west by the Wieprza River catchment, from the south by the Brda River catchment, from the east by the of Leba and Łupawa catchment. The sources of the Słupia River are located in the Kashubian Lake District, near Sierakowska Huta, at an altitude of 178 m above sea level. The width of the riverbed ranges from 7 m in the upper part of the river to 40 m at the estuary, where the average flow is 15,5 m^3·s^{-1}. The area of the city of Słupsk includes 8 km section of the Słupia River with varying degrees of transformation of the riverbed (low, medium or high) [Obolewski 2010], with shores covered with numerous macrophytes (Table 1).

The methodology

The study was conducted in July 2013, within the 10 stations located throughout the city. Samples of the aboveground (leaves) and belowground (rhizomes) shoots of *Glyceria maxima*, collected in the river side zone of the Słupia River. Within each of the experimental stations sev-

Table 1. The characteristics of the research stations

Research stations	The degree of the riverbed transformation	Type ground	Dominant macrophytes			
			Glyceriamaxima	*Typhalatifolia*	*Phragmitesaustralis*	*Phalarisarundinacea*
1	low	mud	+	+	+	
2	low	mud	+		+	+
3	low	mud	+		+	
4	medium	sand	+	+	+	+
5	high	gravel	+	+		+
6	high	gravel	+	+	+	+
7	high	gravel	+	+	+	+
8	medium	sand	+			+
9	medium	gravel	+		+	
10	low	mud	+			

eral shoots were taken. Then the mixed samples were formed separately from the aboveground and belowground shoots. After transportation to the laboratory the plant material was cleaned from mineral soil portion, rinsed in distilled water and dried to constant weight at 65 °C and homogenized in a grinder. The total contents of nitrogen was determined by Kiejldahl method, and the phosphorus by the molybdate method, after mineralization in the mixture of concentrated H_2SO_4 and 30% H_2O_2. In order to determine the metallic elements, the plant samples were mineralized wet in a closed system, in the mixture of concentrated HNO_3 and 30% H_2O_2 (Sobczyński et al. 1996). To the resulting solution, the contents of K, Mg, Ca, Zn, Cu, Ni, Mn, and Fe by atomic absorption spectrometry (Aanalyst 300, PerkinElmer). The analysis were carried in air-acetylene flame of the Ostrowska et al. [1991]. The study used the original standard solutions (Merck KGaA, 1g/1000 ml).

The desription of the results

In order to characterize and compare the concentration of selected macro- and microelements in the shoots above ground and belowground *Glyceria maxima*, medium, minimum and maximum values were calculated, standard aberration and coefficient of variation (CV). The distribution of the content of the analyzed elements was tested by the Shapiro-Wilk test. The significance of the differences in the macro- and microelements in the shoots of *Glyceria maxima* was verified by non-parametric Mann Whitney U test. For the calculations Statistica (7.1) was used. *G. maxima's* demand for nutrients was described by the ANE (Accumulation Nutrient Elements) method [Ostrowska 1987]. The sum of the components (Y) $mmol_c \cdot kg^{-1}$ was calculated from the formula:

$$Y = \sum_{i=1}^{i} (Z : z),$$

where: Z – content of the element in $mg \cdot kg^{-1}$,
z – atomic weight/ion valency.

After the calculation of Y, the percentage (X) of each element in the sum of:

$$X = \frac{(Z : z) \cdot 100}{Y}$$

was calculated.

The study contained the effect of the sediments on the tested macrophyt (Table 2).

RESULTS AND DISCUSSION

The sediments of the Słupia River were characterized by a low content of organic matter and erythrocyte and a slightly alkaline reaction (Table 2). The concentration of most elements in the studied sediments was low and remained within the limits of the geochemical background for most of the analyses [Kabata-Pendias and Pendias 1999]. Only in the case of zinc, copper and nickel the concentration occasionally exceeded the geochemical background for stations in the central part of the city. According to the classification of Lawa [1998] and Wolska and Mędrzycka [2009], the Słupia River sediments were classified as class I, due to the presence of

Table 2. The physical and chemical features of the bottom sediments of the Słupia River

Parameter	Average ± SD	Minimum – maximum	CV, %
Organic matter %	2.1 ± 2.5	0.6 – 7.9	117.0
pH (H_2O)	7.8 ± 0.4	7.0 – 8.2	4.6
pH (KCl)	7.7 ± 0.4	6.7 – 8.1	5.2
N, $mg \cdot kg^{-1}$	969.5 ± 558.1	560.0 – 2030.0	57.6
P, $mg \cdot kg^{-1}$	609.9 ± 241.2	396.2 – 1197.7	39.6
K, $mg \cdot kg^{-1}$	506.7 ± 191.7	296.3 – 852.7	37.8
Mg, $mg \cdot kg^{-1}$	842.9 ± 309.7	316.3 – 1328.7	36.7
Ca, $mg \cdot kg^{-1}$	2268.8 ± 1471.5	848.3 – 5836.7	64.9
Zn, $mg \cdot kg^{-1}$	34.1 ± 28.0	12.4 – 104.7	82.3
Cu, $mg \cdot kg^{-1}$	8.36 ± 4.4	2.6 – 15.1	52.8
Ni, $mg \cdot kg^{-1}$	11.9 ± 2.5	8.6 – 16.4	20.8
Mn, $mg \cdot kg^{-1}$	152.1 ± 111.5	58.1 – 376.7	73.3
Fe, $mg \cdot kg^{-1}$	9293.5 ± 4203.6	4008.0 – 15503.3	45.2

SD – standard deviation, CV – coefficient of variation.

heavy metals assayed, as the non-contaminated sediments. The content of the tested components in the sediments of the Słupia River does not affect living organisms (Table 2).

The content of macronutrients in the shoots of *Glyceria maxima* reflects the supply of nutrients. The largest amount of nitrogen, phosphorus, potassium, magnesium and calcium were in the leafs and the largest amount of zinc, iron, manganese, nickel and copper the *G. maxima* rhizomes (Figures 1 and 2).

These results are confirmed by the research conducted by Vardanyan and Ingol [2006] and Klink et al. [2013] on the ecology of aquatic plants. According to Sharma et al. [2006], the higher concentration of macroelements in above-ground shoots is determined by the processes of photosynthesis. The research conducted by Baldontini et al. [2009] also confirm the presence of high concentrations of macroelemnts in above-ground shoots rather than in the below-ground water vegetation. The results also indicate that the rhizomes of the *G.maxima* contain significant amounts of macroelements and are a valuable reservoir of ingredients necessary for their growth and development.

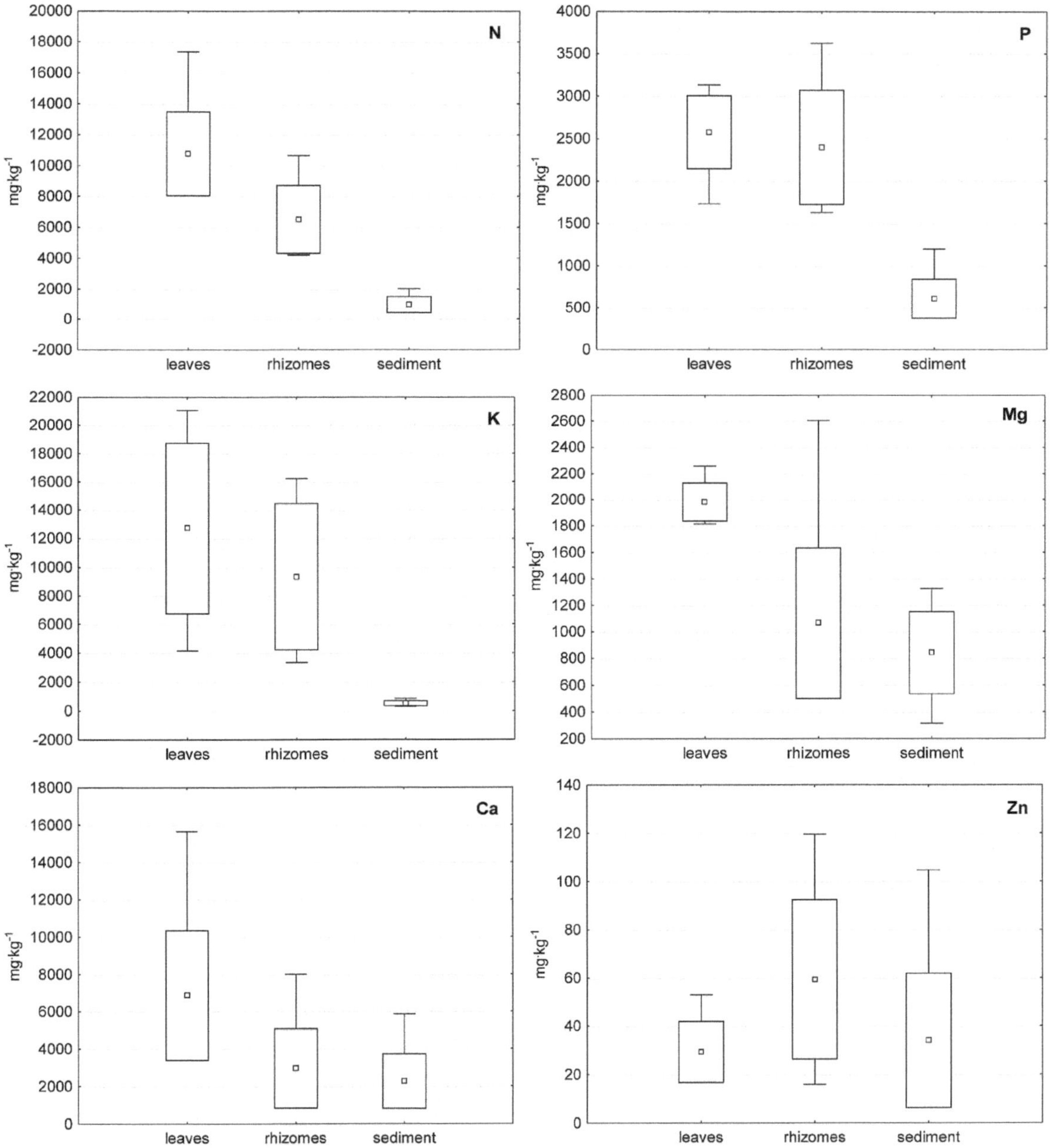

Figure 1. The content of nitrogen, phosphorus, potassium, magnesium, calcium and zinc in the *G. maxima* sediments and shoots. The dot (average value), rectangle (standard deviation), mustache (minimum-maximum).

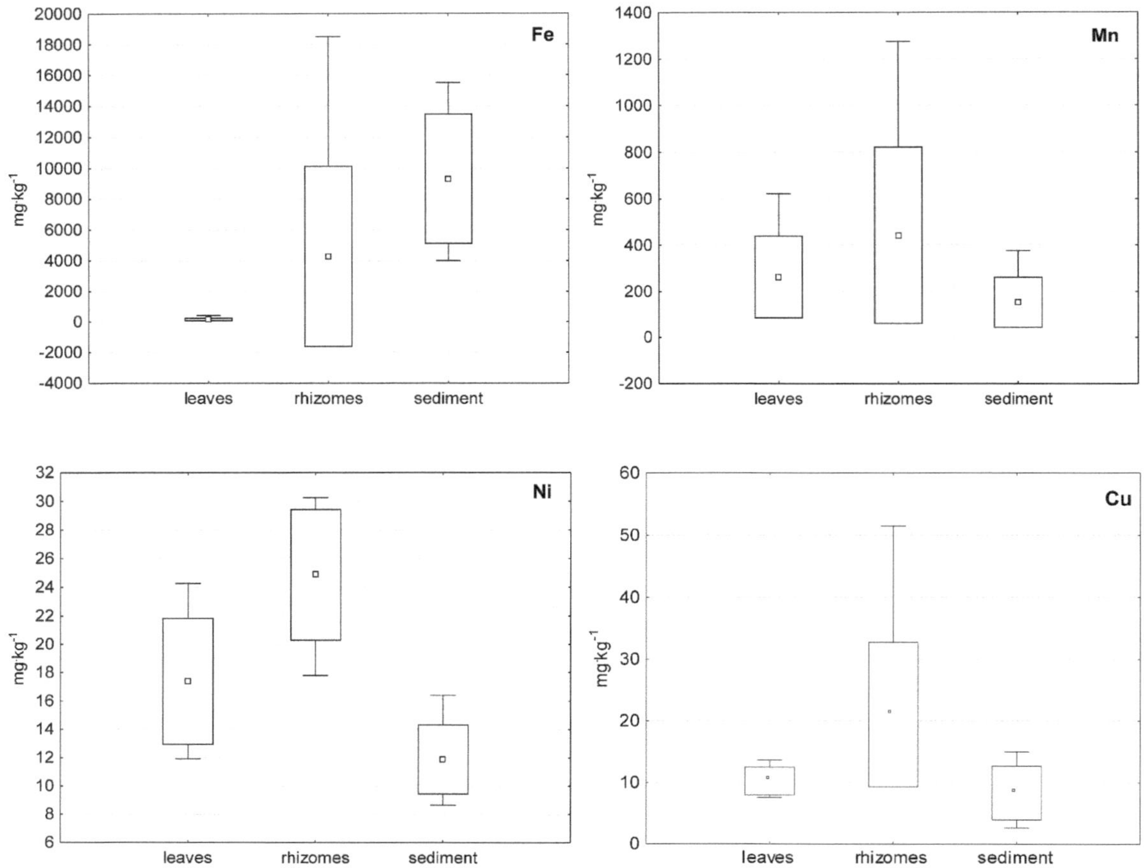

Figure 2. The content of iron, manganese, nickel and copper in the *G. maxima* sediments and shoots (average value), rectangle (standard deviation), mustache (minimum-maximum)

The content of the heavy metals in the water plants often exceeds its content in the surrounding aqueous environment by a number of times [Sadler 1998, Aksoy et al. 2005]. The wide range of variation in the concentration of the heavy metals is caused by biology and ecology of the individual species. In the case of Zn, Fe, Mn, Ni and Cu the content was larger in the belowground stems (rhizomes) than in the aboveground (leaves). Similar relations in the distribution of microelements in the shoots of other macrophytes were shown in the studies by Aksoy et al. [2005], Vardanyan and Ingol [2006] and Klink, et al. [2013]. The levels of zinc and copper in the *G.maxima* shoots stays in the limit of values for the plants, respectively Zn: 10-70 mg·kg^{-1} and Cu: 5-20 mg·kg^{-1} (Figure 2) [Kabata-Pendias and Pendias 1999, Ostrowska, Porębska 2002]. In the case of Mn, the increased content was found both in the aboveground and the belowground shoot of the *G. maxima*. However, at any position manganese level did not exceed the value recognized as toxic (>500 mg·kg^{-1}) [Kabata-Pendias and Pendias 1999]. The increased content of Mn in the shoots, in relation to the physiological demand, indicates a significant share of this macrophyt in the purification of the water and sediments of the Słupia River from manganese compounds available for the plants. Fe content in the leaves of *G. maxima* was within the range of limit values for the plants (50–250 mg·kg^{-1}), and in the rhizomes it reached a very high level (average 4259.9 mg·kg^{-1}), Figure 2. Such a high concentration of iron in the rhizomes may indicate accumulating properties of this species in relation to Fe. At the same time, small concentrations of this element in the leaves indicate a protective barrier that restricts the movement of the iron compounds from below to aboveground shoots. The content of Ni in both leaves and rhizomes of the *G. maxima*, exceeded the physiological requirements (0.1–5.0 mg·kg^{-1}) of the most plants, [Kabata-Pendias and Pendias 1999], which indicates the little contamination of the aquatic environment with that element. Nickel is easily taken up by the plants and transported to the aboveground parts and it's excess is accumulated in the roots [Ostrowska et al. 1991]. In the areas such as agglomerations, it's values are usually higher than the physiological demand, because nickel is easily bioaccumulated especially by the water plants [Sarosiek and Natkaniec-Wożakowska 1993]. The relations between the designated heavy metals are arranged in the following series: Mn>Fe>Zn>Ni>Cu and in the leaves Fe >Mn>Zn>Ni>Cu in the rhizomes.

The results of the non-parametric Mann Whitney U test showed a number of statistically significant differences in the content of macro- and microelements within 10 research stations in the city of Słupsk, in relations: leaves – rhizomes, leaves – bottom sediment and rhizome – bottom sediment (Table 3).

In order to assess the content of macro- and microelements in the *Glyceria maxima* shoots the ANE (Accumulation Nutrient Elements) method was used [Ostrowska 1987]. The sum of the components reflects the whole so-called nutritional factor. The shoots of *G. maxima* accumulated from 985.8 to 1441.4 $mmol_c \cdot kg^{-1}$ of all the analyzed components, and the lower value of the sum of the accumulated macro- and microelements found in the rhizomes (Table 4).

Macronutrients were collected in 61.4% of the leaves and 38.6% of the rhizomes. Micronutrients were characterized by a different distribution, 8.9% was accumulated in the leaves, and 91.1% in the rhizomes. The composition of the macronutrients in the leaves and rhizomes was similar. The share of nitrogen was 51.7–53.7% of this amount, 5.8–8.6% phosphorus, potassium 22.8–26.6%, 4.9–5.8% of magnesium, calcium, 8.2–11.9%, and trace elements were 0.58% in total in the case of leaves and 8.70% in the rhizomes.

The similar relations between the measured amount of ingredients in the various plant species were shown by Porębska and Ostrowska [2002] in their study. A large amount of manganese and iron in all of the measured components in the leaves and iron the rhizomes show excessive absorption from the bottom sediments, which favored slightly alkaline reaction of the bottom sediments (Table 2). The *G. maxima* leaves accumulated more nitrogen, calcium, magnesium, zinc, copper, nickel and manganese, since the rhizomes accumulated more of phosphorus, potassium and iron.

Table 3. The results the Mann-Whitney test – the importance of the diversity of the content of macro- and microelements in the *G. maxima* leaves and rhizomes and the bottom sediments of the river

Elements [$mg \cdot kg^{-1}$]	Leaves – rhizomes	Leaves – bottom sediments	Rhizomes – bottom sediments
N	+++	+++	+++
P	–	+++	+++
K	–	+++	+++
Mg	++	+++	–
Ca	–	+++	–
Zn	–	–	–
Cu	+++	–	++
Ni	++	++	+++
Mn	–	–	+
Fe	+++	+++	++

– no differences, + statistically significant differences p<0,05, ++ statistically significant differences p<0,01, +++ statistically significant differences p<0,001.

Table 4. The average accumulation of components in the above and belowground shoots of the *G. maxima**

Elements	*Glyceria maxima*	
	Leaves	Rhizomes
Σ macro [$mmol_c \cdot kg^{-1}$]:	1433	900
% N in Σ	53.7	51.7
% P in Σ	5.8	8.6
% K in Σ	22.8	26.6
% Ca in Σ	11.9	8.2
% Mg in Σ	5.8	4.9
Σ mikro [$mmol_c \cdot kg^{-1}$]:	8.4	85.8
% Zn in Σ	5.3	1.1
% Cu in Σ	1.9	0.4
% Ni in Σ	3.5	0.5
% Mn in Σ	56.5	9.4
% Fe in Σ	32.8	88.6
Σ macro + Σ mikro	1441.4	985.8

* Expressed as the sum of the components and their share in the total.

CONCLUSIONS

The results indicate that the greatest amount of nitrogen, phosphorus, potassium, magnesium and calcium were in the leaves of *Glyceria maxima*. The rhizomes also accumulated significant amounts of macroelements so that they can act as a backup body.

The largest amounts of heavy metals were found in belowground stems (rhizomes). The content of zinc and copper at all tested locations ranged within the limit for the plant. The increased content of manganese was shown, both in the leaves and the rhizomes, which indicates a considerable share of the *G. maxima* in the purification of water and sediment river from the compounds of manganese. Fe content in the leaves maintained at an acceptable level, and in the rhizomes it reached a very high level, which may indicate the accumulating feature of this species relative to Fe. The content of Ni in both the above and belowground *G.* shoots exceeded the physiological need, indicating the little contamination of the aquatic environment of the Słupia River by this element.

The results of the non-parametric Mann Whitney U test showed a number of statistically significant differences in the content of macro- and microelements in the relation: leaves - rhizomes, leaves - bottom sediment and rhizome - bottom sediment.

Shoots of *G.maxima* accumulated 985.8 to 1441.4 $mmol_c \cdot kg^{-1}$ in all analyzed components, where lower value of the accumulated sum of macro- and microelements were found in the rhizomes. The sum of the ionic composition of macronutrients in the leaves and rhizomes were similar. The content of nitrogen was 51.7–53.7% of this amount, 5.8–8.6% phosphorus, potassium 22.8–26.6%, 4.9–5.8% of magnesium, calcium, 8.2–11.9%, and trace elements were 0.58% in the leaves and 8.70% in the rhizomes.

REFERENCES

1. Aksoy A., Demirezen D., Duman F. 2005. Bioaccumulation, detection and analyses of heavy metal pollution in Sultan marsh and its environment. Water Air Soil Poll. 164, 241.
2. Baldantonid D., Ligrone R., Alfania A. 2009.Macro- and trace-element concentration in leavesand roots of *Phragmitesaustralis* in volcanic lake in Southern Italy. J. Geochem. Explor. 101, 166.
3. Buszewski B., Jastrzębska A., Kowalkowski T., Górna-Binkul A., 2000. Monitoring of selected heavy metals uptake by plants and soils in the area of Toruń, Poland. Pol. J. Environ. Stud. 9, 6, 511–515.
4. Dąbkowska–Naskręt H., Różański S. 2009. Formy połączeń Pb i Zn w glebach miasta Bydgoszczy, Ochrona Środowiska i Zasobów Naturalnych 41, 489–496.
5. Gruca-Królikowska S., Wacławek W. 2006. Metale w środowisku. Cz. II. Wpływ metali ciężkich na rośliny, Chemia · Dydaktyka · Ekologia · Metrologia 11, 1–2, 41–56.
6. Grynkiewicz M., Dembska G., Wisniewski S., Aftanas B. 2006. Ocena jakości osadów dennych pobieranych przy strefie brzegowej Zatoki Gdańskiej, Chem. Inż. Ekol. 13, 284–289.
7. Kabata-Pendias A., Pendias H. 1999. Biogeochemia pierwiastków śladowych, PWN, Warszawa.
8. Kabata-Pendias A., Szteke B. 2005. Pierwiastki śladowe w układzie gleba-roślina, Inż. Ekol. 26, 28–29.
9. Klink A., Wisłocka M., Musiał M., Krawczyk J. 2013. Macro- and trace-elements accumulation in *Typhaangustifolia* L. and *Typhalatifolia* L. organs and their use in bioindication, Pol. J. Environ. Stud. 22, 1, 183–190.
10. KruopieneJ. , 2007. Distribution of heavy metals in sediments of the Nemunas River (Lithuania). Pol. J. Environ. Stud. 16, 5, 715–722.
11. Lawa, 1998. Landesarbeitsgemeinschaft Wasser: Beurteilung der Wasserbeschaffen-heit von Fließgewässern in der Bundesrepublik Deutschland – chemische Gewässergüteklassifikation, Zielvorgaben zum Schutz oberirdischer binnengewässer – Band 2, Berlin.
12. Letachowicz B., Krawczyk J., Klink A., 2006. Accumulation of heavy metals in organs of *Typhalalifolia* L., Pol. J. Environ. Stud. 15 (2a), 407.
13. Misztal M., Smal H., Ligęza S. 1996.Zawartość wybranych makro- i mikropierwiastków w osadach dennych rzeki Bystrzycy. Zesz. Prob. Post. Nauk Rol. 437, 291–298.
14. Obolewski K. 2010. Ocena jakości wód powierzchniowych na obszarach zurbanizowanych z wykorzystaniem makrobezkręgowców bentosowych na przykładzie rzeki Słupi, Ochr. Środ. 32, 2, 35–42.
15. Obolewski K., Glińska-Lewczuk K. 2006. Contents of heavy metals in bottom sediments of oxobow lakes and the Słupia River. Pol. J. Environ. Stud. 15, 2a, 440–444.
16. Ostrowska A., 1987. Application of ANE value- and shares of individual elements in this value for determining the difference between various plant species. Genetic aspects of plant mineral nutrition, 27 43, MartinusNijhoff Pub. (Plant and Soil).

17. Ostrowska A., Gawliński S., Szczubiałka Z., 1991. Metody analizy i oceny właściwości gleb i roślin. Katalog,IOŚ, Warszawa.

18. Ostrowska A., Porębska G., 2002. Skład chemiczny roślin, jego interpretacja i wykorzystanie w ochronie środowiska. IOŚ, Warszawa.

19. Rabajczyk A., Jóźwiak M.A. 2009. Możliwości wykorzystania makrofitów jako bioindykatorów metali ciężkich zdeponowanych w osadach dennych, Monitoring Środowiska Przyrodniczego 9, 19–26.

20. Sadler P., 1998. Wetlands for mine water treatment Workshop. W: Constructed Wetlands Biofiltration. University of Wales, Cardiff.

21. Salati S., Moore F. 2009. Assessment of heavy metal concentration in the Khoshk River water and sediment, Shiraz, Southwest Iran. Environmental Monitoring and Assessment 164, 677–689.

22. Sarosiek J., Wożakowska-Natkaniec H. 1993. Chromi nikiel w roślinach z rodziny *Lemnaceae* i w ich środowisku. [W.] Chrom nikiel i glin – problemy ekologiczne i metodyczne. [red.] Kabata-Pendias A., Z. PAN Kom. Człowiek i środowisko.

23. Sharma P., Asaeda T., Manatunge J., Fijino T., 2006. Nutrient cycling in a natural stand of *Typhaangustifolia* J. Freshwater Ecol. 21, 431.

24. Sasmaz A., Obek E., Hasar H. 2008. The accumulation of heavy metals in *Typhalatifolia* L. grown in a stream carrying secondary effluent. Ecol. Eng. 33, (3–4), 78.

25. Skorbiłowicz E., 2003. Heavy metal content in bottom deposits and water plants in the Ełk River on urbanized area, Ecological Chemistry and Engineering 10, 9, 937–945.

26. Sobczyński T., Elbanowska H., Zerbe J., Siepak, J., 1996.Digestion of samples of bottom sediments prior to the determination of total contents of heavy metals. Gospodarka Wodna, 1996, 6, 570, pp. 173.

27. Teuchies J., Jacobs S., Oosterlee L., Bervoets L., Meire P., Role of plants in metal cycling in a wetland: Impications for phytoremediation. Science of Total Environment 445–446, 146–154.

28. Vardanyan L.G., Ingole B.S., 2006. Studies on heavy metal accumulation in aquatic macrophytes from Sevan (Armenia) and Carambolim (India) lake systems. Environ. Int. 32, 208–218.

29. Wilk M., Gworek B. 2009.Metale ciężkie w osadach ściekowych, Ochrona Środowiska i Zasobów Naturalnych 39, 40–59.

30. Wolska L., Mędrzycka K. 2009. Ocena ekotoksyczności osadów dennych z portów morskich w Gdańsku i Gdyni. Ochrona Środowiska 31, 1, 49–52.

31. Yang Y., Peterson E., Cambell C. 2001. Accumulation of heavy metals in urban soils and impacts on microorganism. Huan Jing KeXue 22, 3, 44–48.

DIATOMS (*BACILLARIOPHYCEAE*) IN RIVERS AND STREAMS AND ON CULTIVATED SOILS OF THE PODKARPACIE REGION IN THE YEARS 2007–2011

Teresa Noga[1], Natalia Kochman[1], Łukasz Peszek[2], Jadwiga Stanek-Tarkowska[3], Anita Pajączek[3]

[1] Department of Biological Foundations of Agriculture and Environmental Education, Faculty of Biology and Agriculture, University of Rzeszów, Ćwiklińskiej 2, 35–601 Rzeszów, Poland, e-mail: teresa.noga@interia.pl; kochman_natalia@wp.pl

[2] International Doctoral Studies in Natural Sciences, Faculty of Biology and Agriculture, University of Rzeszów, Ćwiklińskiej 2, 35–601 Rzeszów, Poland, e-mail: lukaspeszek@gmail.com

[3] Department of Soil Studies, Environmental Chemistry and Hydrology, Faculty of Biology and Agriculture, University of Rzeszów, Ćwiklińskiej 2, 35–601 Rzeszów, Poland, e-mail: jagodastanek@wp.pl; chuanita66@poczta.fm

ABSTRACT

Diatoms *Bacillariophyceae* are a specialized, systematic group of algae, living in different types of ecosystems, and they are very good indicators of water quality. Recently, detailed studies have been carried out in the territory of the Podkarpacie Province on the diversity of diatoms in running waters. The area of study covers mostly the catchment of the biggest rivers of the province (mainly the Wisłok River) and cultivated soils. Running waters in the Podkarpacie Province are characterized by a huge species richness of diatoms, 738 different taxa have been found so far. The highest number of taxa was noted in the Wisłok River (400), while the Żołynianka and Różanka streams and the Mleczka were also characterized by huge diversity (from 200 to over 350 taxa). In the majority of studied sites diatoms of alkaline (pH>7) and nutrient rich (eutrophic) waters dominated: *Navicula lanceolata*, *N. gregaria*, *N. capitatoradiata*, *Nitzschia palea*, *N. dissipata* ssp. *dissipata*, *Achnanthidium minutissimum* var. *minutissimum*, *A. pyrenaicum*, *Amphora pediculus*, *Diatoma vulgaris*, *Rhoicosphenia abbreviata*, *Planothidium lanceolatum*, *Cocconeis pediculus* and *C. placentula* var. *lineata*. About 150 taxa were indentified on cultivated soils, among which the largest populations were formed by *Hantzschia amphioxys*, *Mayamaea atomus*, *Nitzschia palea*, *N. pusilla*, *Pinnularia obscura* and *Stauroneis thermicola*.

Keywords: diatoms *Bacillariophyceae*, diversity, taxonomy, Podkarpacie Region.

INTRODUCTION

Diatoms are a specialized, systematic group of algae occurring in almost all water ecosystems and other wet habitats, such as: soils, rocks, caves, sea ice, springs (cold, hot and salty). It includes above 10 000 species, from which above 3000 taxa (sea and freshwater) have been recorded on the territory of contemporary Poland. They are a systematic group used in assessment of water quality. Their huge richness and diversity causes the degree of recognition of diatoms to be unequal [Siemińska 1977, Rakowska 2001, Wołowski 2003, Siemińska et al. 2006]. Rivers and streams in southern Poland are the best recognized due to the studies of Siemińska, Kawecka and Kwandrans [Siemińska 1990, Siemińska, Pająk 1992, Kwandrans 1989, Kawecka 2012].

Until recently the Podkarpacie Province had not been studied in algological terms. Only fragmentary studies have been conducted in the upper course of the San River in connection with the appearance of the *Didymosphenia geminata* diatom, which developed massively below reservoirs in Solina and Myczkowce [Kawecka, Sanecki 2003].

Since 2007, studies have been conducted on the diversity of diatoms in running waters of the Podkarpacie Province [Mucha et al. 2009, Noga, Siry 2010, Tambor, Noga 2011, Noga 2012, Pajączek et al. 2012], and since 2009, studies have also been carried out on cultivated soils [Stanek-Tarkowska, Noga 2011, 2012].

The aim of studies on running waters and soils of the Podkarpacie Province is, first of all, to recognize the species diversity of diatoms, while pointing out both endangered and rare species. The present work is a summary of current studies and their results, gathered over 5 years (2007–2011).

STUDY AREA

Three separate physiographic regions can be singled out on the territory of the Podkarpacie Province: the Sandomierska Valley (in the north), the Carpathian Foothills (the middle part), the Low Beskid and the Bieszczady mountains in the south [Kondracki 2001]. Just a small south-eastern part of the Province (about 1.3%) belongs to the catchment area of the Diestr River, and it is drained by Strwiąż, Mszaniec and Lechnawa rivers. The Wisła River makes the primary drainage system, with two big inflows: the San and the Wisłok River, which belongs to the Baltic Sea Basin. The San River, about 443.4 km long, is the richest water of the Carpathian Region, with its sources located on the territory of the Ukraine. The longest inflow of the San River is the Wisłok River, about 228.5 km, which, similarly to the Wisłoka River (163.6), begins in the Low Beskids. All the biggest rivers of the voivodeship have a mountainous character in their upper courses, and their catchment areas are extremely valuable in natural terms.

Generally conducted studies on the territory of the Podkarpacie Province include rivers and streams in the valley of the Wisłok River (the Carpathian Foothills). However, single study sites are located in valleys of the Wisłoka and the San rivers, and also on the afore-mentioned rivers themselves (Figure 1).

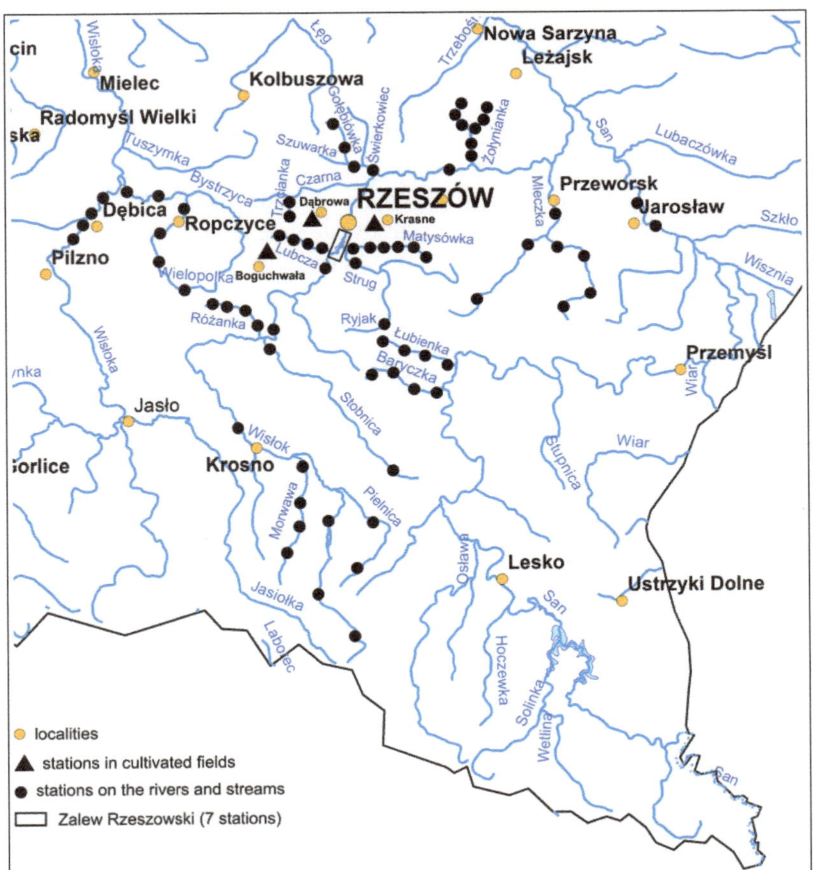

Figure 1. Distribution of studies stations in Podkarpacie Region

MATERIALS AND METHODS

Materials for study were taken, over a period of 5 years (2007–2011), from the territory of the Podkarpacie Province. The majority of study sites were located in the catchment area of the Wisłok River. The number of study sites on rivers and streams were: Wisłok – 8 sites, and in the Rzeszów Reservoir – 7 sites, Morwawa – 4 sites, Pielnica – 2 sites, Stobnica – 2 sites, Mleczka – 7 sites, Lubcza – 4 sites, Strug – 1 site and streams: Różanka – 4 sites, Ryjak – 1 site, Szuwarka – 1 site, Gołębiówka – 2 sites, Świerkowiec – 1 site, Żołynianka – 8 sites, Trzcianka – 2 sites, and Matysówka – 6 sites. Studies were also conducted on the Wisłoka River (4 study sites) and its inflow: Wielopolka (6 sites), and also on the San River (2 sites) and its two inflow streams: Łubienka (4 sites) and Baryczka (4 sites). Also, since 2009, soil diatoms have been studied, mainly on cultivated soils in Krasne (2 study sites), Boguchwała (3 study sites) and Dąbrowa (2 study sites) (Figure 1).

Material for studies in rivers and streams was taken from rocks, silt and water plants, in seasons from spring to autumn within a year in every studied current, and it was then prepared and processed according to the methods applied in this type of research [Kawecka 2012]. At the same time pH values and conductivity were measured. Collected material was preserved in a 4% formalin solution. Soil samples were taken once during the autumn season on study sites in Dąbrowa and Boguchwała and on the study site in Krasne once a month (from spring to autumn). In order to obtain pure valves of diatoms, part of the obtained material was subjected to maceration in a mixture of sulphuric acid and potassium dichromate at a proportion of 3:1, and rinsed in a centrifuge (at 2500 rev×min^{-1}). Cleaned diatoms were embedded in Pleurax synthetic resin (refractive index 1.75).

Soil samples were taken from stratum 0–3 cm to Petri dishes (3 samples from each study site), and next the material was subjected to maceration in acids (similarly as in the case of samples taken from water).

Diatoms were identified using a light microscope, "Nikon ECLIPSE 80i" at a magnification of 1000x, according to the keys: Krammer and Lange-Bertalot [1986–1991], Lange-Bertalot [2001], Krammer [2000] and Hofmann et al. [2011].

The number of noted species was obtained through calculating valves in randomly chosen fields of view of the microscope, until a total number of 400 cells was obtained. The most numerous species were those, whose participation in a given community was above 5%.

Based on the Red List of Algae, created for Poland [Siemińska et al. 2006], species, which were rare and endangered to different degrees, were singled out and assigned to the following categories: E – endangered, V – vulnerable, R – rare, I – indeterminate.

RESULTS

Measurements of pH and conductivity conducted in the open field indicated that all studied currents were characterized by alkaline or close to neutral water reactions (pH: 7–9), whereas conductivity was between above 100 to above 700 μS/cm.

The existence of 738 taxa of diatoms was identified during studies in 2007–2011 on the territory of the Podkarpacie Province. The largest diversity was encountered in the Wisłok River, in which 400 taxa were found on 8 sites. Huge species richness was noted also on many inflows of the Wisłok River, in which the Żołynianka stream was characterized by the largest number (365 taxa on 8 studying sites). Over 200 taxa were also identified in the Mleczka, Stobnica, Pielnica, Morwawa and Ryjak streams and on the San River (253), on two study sites near Jarosław city (Table 1).

Diatoms characteristic for alkaline (pH>7) and fertile, i.e. eutrophic waters, dominated on the majority of studied rivers and streams in the Podkarpacie Province. The most frequent were: *Navicula lanceolata*, *N. gregaria*, *N. capitatoradiata*, *Nitzschia palea*, *N. dissipata* ssp. *dissipata*, *Achnanthidium minutissimum* var. *minutissimum*, *A. pyrenaicum*, *Amphora pediculus*, *Diatoma vulgaris*, *Rhoicosphenia abbreviata*, *Planothidium lanceolatum*, *Cocconeis pediculus* and *C. placentula* var. *lineate* (Plate 1: 1–30).

Achnanthidium minutissimum var. *minutissimum* and *A. pyrenaicum* formed the most numerous populations in the upper sections of rivers and streams (30–50%), whereas in the middle and lower sections the most numerous were *Navicula lanceolata* and *N. gregaria* (above 30%).

Among 738 diatom taxa, 80 were endangered and rare, appearing in the Polish Red List

Table 1. The list of diatom taxa found at studies sites in the Podkarpacie Region

Taxa / Site	1	2	3	4	5	6	7	8	9	10	11	12	13	14	15	16	17	18	19	20	21	22	0
Achnanthes coarctata (Bréb.) Grun.		+								+	+	+		+	+		+		+			+	R
Achnanthes expressa Carter		+											+	+									
Achnanthes exigua var. *exigua* Grun.		+																			+		V
Achnanthes rupestoides Hohn		+																					
Achnanthes cf. *expressa* Carter										+													
Achnanthes sp.					+																		
Achnanthidium affine (Grun.) Czarn.																					+		
A. eutrophilum (Lange-Bert.) Lange-Bert.																					+		
Achnanthidium minutissimum var. *jackii* (Rabenh.) Lange-Bert.		+																				+	
Achnanthidium minutissimum (Kütz.) Czarn. var. *minutissimum*	+	+	+	+	+	+	+	+	+	+	+	+	+	+	+	+	+	+	+	+	+	+	
Achnanthidium pyrenaicum (Hust.) Kobayasi	+	+	+		+				+	+	+	+	+	+	+	+	+		+	+	+	+	
Achnanthidium saprophilum (Kobayasi & Mayama) Round & Bukht.					+				+			+	+	+		+	+				+		
Achnanthidium subatomus (Hust.) Lange-Bert.		+							+														
Adlafia brockmannii (Hust.) Bruder & Hinz		+										+	+		+						+	+	
Adlafia bryophila (Petersen) Moser, Lange-Bert. & Metzeltin																						+	
Adlafia minuscula (Grun.) Lange-Bert. var. *minuscula*	+	+		+																			
Adlafia minuscula var. *muralis* (Grun.) Lange-Bert.																	+						
Amphipleura pellucida (Kütz.) Kütz.		+								+	+	+	+	+	+	+	+						R
Amphipleura sp.						+																	
Amphora alpestris Levkov																					+		
Amphora cimbrica Østrup		+	+		+						+	+	+	+		+	+			+			R
Amphora copulata Ehrenb.	+	+	+	+	+	+	+	+	+	+	+	+	+	+	+	+	+	+	+	+	+	+	
Amphora hemicycla Stoermenr & Yang																					+		R
Amphora inariensis Krammer		+	+	+	+						+	+	+	+	+	+	+			+	+	+	
Amphora indistincta Levkov																						+	
Amphora micra Levkov												+									+		
Amphora minutissima? W. Smith												+											
Amphora ovalis (Kütz.) Kütz.		+	+	+	+				+	+	+	+	+	+	+	+	+		+	+	+	+	
Amphora pediculus (Kütz.) Grun.	+	+	+	+	+	+		+	+	+	+	+	+	+	+	+	+	+	+	+	+	+	
Amphora sp. 1		+											+										
Amphora sp. 2		+																					
Amphora sp. 3		+																					
Amphora sp. 4	+																						
Anomoeoneis sphaerophora (Ehrenb.) Pfitzer		+			+				+	+			+		+						+		
Asterionella formosa Hassal	+	+	+		+					+	+		+	+	+	+	+		+	+	+	+	
Aulacoseira ambigua (Grun.) Simonsen	+	+			+									+		+	+			+	+		
Aulacoseira distans (Ehrenb.) Simonsen	+	+			+						+		+			+	+				+		
Aulacoseira granulata (Ehrenb.) Simonsen	+	+	+		+					+	+		+	+	+	+	+	+		+	+		
Aulacoseira muzzanensis (Meister) Krammer																					+		
Aulacoseira cf. *muzzanesis* (Meister) Krammer		+																					
Aulacoseira sp. 1		+														+	+					+	
Aulacoseira sp. 2		+																					
Brachysira brebissonii Ross																						+	
Brachysira vitrea (Grun.) Ross		+				+						+											E
Caloneis aerophila Bock													+				+					+	R
Caloneis amphisbaena (Bory) Cleve		+	+		+		+	+	+	+	+	+	+	+	+	+	+	+	+	+	+	+	
Caloneis amphisbaena var. *subsalina* (Donkin) Huls				+																			
Caloneis bacillum (Grun.) Cleve		+	+	+	+	+				+	+	+	+	+	+	+	+		+	+	+		R
Caloneis fontinalis (Grun.) Lange-Bert. & Reichardt												+									+	+	
Caloneis hyalina Hust.				+																			R
Caloneis lancettula (Schulz) Lange-Bert. & Witkowski		+			+						+	+								+	+	+	
Caloneis leptosoma (Grun.) Krammer																					+		
Caloneis macedonica (?) Hust.															+								
Caloneis molaris (Grun.) Krammer		+							+	+	+	+	+	+	+		+	+			+	+	R
Caloneis pulchra Messikommer			+																				
Caloneis schumanniana (Grun.) Cleve		+			+						+		+		+		+						R
Caloneis silicula (Grun.) Cleve	+	+	+	+	+	+	+	+	+	+	+	+	+	+	+	+	+		+		+	+	

Table 1 contin.

Taxa / Site	1	2	3	4	5	6	7	8	9	10	11	12	13	14	15	16	17	18	19	20	21	22	0
Caloneis tenuis (Greg.) Krammer																					+	+	
Caloneis vasileyavae Lange-Bert., Genkal & Vekhov																						+	
Caloneis cf. *lancettula* (Schulz) Lange-Bert. & Witkowski	+										+												
Caloneis cf. *tenuis* (Greg.) Krammer													+										
Caloneis sp.												+										+	
Campylodiscus hibernicus Ehrenb.																	+						
Cavinula cocconeiformis (Gregory) D.G.Mann & Stickle												+											
Cavinula lapidosa (Krasske) Lange-Bert.												+											
Cavinula sp.													+										
Chamaepinnularia krookii (Grun.) Lange-Bert. & Krammer		+																			+		R
Chamaepinnularia soechrensis (Krasske) Lange-Bert. & Krammer												+											
Chamaepinnularia submuscicola (Krasske) Lange-Bert.																					+	+	
Chamaepinnularia cf. *krookii*			+																				
Cocconeis disculus (Schum.) Cleve		+																					R
Cocconeis neodiminuta Krammer		+											+										
Cocconeis pediculus Ehrenb.	+	+	+	+	+				+	+	+	+	+	+	+	+	+	+	+	+	+	+	
Cocconeis placentula var. *euglypta* (Ehrenb.) Grun.	+	+	+	+	+	+		+	+	+	+	+	+	+	+	+	+	+	+	+	+	+	
Cocconeis placentula var. *lineata* (Ehrenb.) Van Heurck	+	+	+	+	+	+	+	+	+	+	+	+	+	+	+	+	+	+	+	+	+	+	
Cocconeis placentula Ehrenb. var. *placentula*		+	+	+	+						+		+	+	+		+		+		+		
Cocconeis pseudolineata (Geitler) Lange-Bert.		+			+					+	+	+	+	+	+		+				+		
Cocconeis sp.		+									+		+		+		+						
Craticula accomoda (Hust.) D.G. Mann		+	+	+	+						+		+	+	+	+	+		+	+	+		
Craticula ambigua (Ehrenb.) D.G. Mann		+	+	+	+				+	+	+	+	+	+	+	+	+		+	+	+	+	
Craticula cuspidata (Kütz.) D.G. Mann		+	+	+	+	+		+		+	+		+	+	+	+	+	+	+		+		
Craticula halophilla (Grun.) D.G. Mann		+				+			+						+								
Craticula molestiformis (Hust.) D.G. Mann		+	+	+	+				+	+	+	+	+	+	+	+	+			+	+	+	
Craticula sp. cf. *molestiformis* (Hust.) D.G. Mann	+																						
Craticula sp.		+													+								
Cyclostephanos dubius (Fricke) Round	+	+			+							+	+		+				+	+	+		
Cyclostephanos invistatus (Hohn, Hellerman) Stroermer, Theriot & Håkansson	+				+																+		
Cyclotella atomus Hust.		+	+	+	+	+			+	+	+		+	+	+	+	+	+	+		+		
Cyclotella bodanica var. *affinis* Grun.		+																					
Cyclotella distinguenda Hust.																							
Cyclotella distinguenda Hust. var. *distinguenda*		+										+											
Cyclotella iris Brun		+																					
Cyclotella meneghiniana Kütz.	+	+	+	+	+	+	+	+	+	+	+	+	+	+	+	+	+	+	+	+	+	+	
Cyclotella ocellata Pantocsek		+			+				+	+	+			+	+	+	+						
Cyclotella planktonica Brunnthaler																					+		
Cyclotella cf. *cyclopuncta* Håkansson & Carter	+											+									+		
Cyclotella sp. cf. *distinguenda* Hust.	+																						
Cyclotella sp. cf. *glabriuscula* (Grun.) Håkansson																					+		
Cyclotella sp. cf. *ocellata* Pantocsek		+																					
Cyclotella sp. 1		+																					
Cyclotella sp. 2		+																					
Cyclotella sp. 3											+				+								
Cylindrotheca gracilis (Bréb.) Grun.												+			+	+	+			+		+	
Cymatopleura elliptica (Bréb.) W. Smith		+	+		+					+	+	+	+	+	+	+	+		+		+		
Cymatopleura solea var. *apiculata* (W. Smith) Ralfs	+	+	+	+	+		+	+	+	+	+	+	+	+	+	+	+		+		+	+	
Cymatopleura solea (Bréb.) W. Smith var. *solea*		+	+	+	+				+	+	+	+	+	+	+	+	+	+	+	+	+	+	
Cymbella affinis Kütz.	+	+	+		+				+	+	+		+	+	+	+	+	+		+			
Cymbella aspera (Ehrenb.) Cleve		+			+	+				+	+		+	+	+	+					+	+	V
Cymbella cistula (Ehrenb.) Kirchner		+	+	+	+	+				+	+	+	+	+	+	+	+	+				+	I
Cymbella cistuliformis Krammer																					+		
Cymbella compacta Østrup					+																	+	
Cymbella cymbiformis Ag.																					+		
Cymbella cymbiformis var. *nonpunctata* Fontell									+														
Cymbella excisa Kütz.												+									+	+	

Table 1 contin.

Taxa / Site	1	2	3	4	5	6	7	8	9	10	11	12	13	14	15	16	17	18	19	20	21	22	0
Cymbella helvetica Kütz.		+	+		+					+	+		+	+	+			+				+	R
Cymbella hustedtii Krasske																					+		
Cymbella laevis Naegeli																					+		
Cymbella lanceolata (Ehrenb.) Van Heurck			+		+				+										+		+		R
Cymbella neocistula Krammer																					+		
Cymbella neocistula var. *lunata* Krammer																					+		
Cymbella parva (W. Smith) Kirchner												+									+	+	
Cymbella proxima Reimer																					+		V
Cymbella subcistula Krammer												+									+	+	
Cymbella tumida (Bréb.) Van Heurck		+	+	+	+			+		+	+	+	+	+	+	+		+			+		
Cymbella tumidula var. *lancetula* Krammer		+																					
Cymbella sp. 1		+																					
Cymbella sp. 2		+																					
Cymbopleura amphicephala Naegali		+	+		+						+	+	+	+	+		+		+			+	
Cymbopleura anglica (Lagerstedtt) Krammer																					+		
Cymbopleura cuspidata Kütz.		+																			+		
Cymbopleura inaequaliformis Krammer																					+		
Cymbopleura inaequalis (Ehrenb.) Krammer													+								+		V
Cymbopleura lata (Grun.) Krammer																					+		
Cymbopleura naviculiformis (Auerswald) Krammer		+			+	+	+	+	+			+	+	+			+	+			+		
Cymbopleura subaequalis Grun.		+													+						+		
Denticula subtilis Grun.																						+	
Denticula tenuis Ag.		+	+		+					+	+	+	+	+	+			+				+	
Diadesmis contenta (Grun.) D.G. Mann		+	+							+	+	+	+	+	+	+	+			+	+	+	
Diadesmis paracontenta Lange-Bert. & Werum var. *paracontenta*												+											
Diadesmis perpusilla (Grun.) D.G. Mann	+	+										+	+	+			+				+	+	
Diadesmis sp. cf. *brekkaensis* (Petersen) D.G. Mann												+									+		
Diadesmis sp. 1														+								+	
Diadesmis sp. 2			+																				
Diadesmis sp. 3	+											+											
Diatoma ehrenbergii Kütz					+													+				+	
Diatoma mesodon (Ehrenb.) Kütz.		+	+	+	+					+													
Diatoma moniliformis Kütz.	+	+			+					+	+	+	+	+	+	+	+	+		+	+		
Diatoma tenuis Kütz.	+	+	+		+						+		+		+						+		
Diatoma vulgaris Bory	+	+	+	+	+				+	+	+	+	+	+	+	+	+		+	+	+	+	
Diatoma sp. cf. *moniliformis* Kütz.		+	+		+						+				+								
Didymosphenia geminata (Lyngbe) M. Schmidt		+			+																		
Diploneis elliptica (Kütz.) Cleve	+	+			+							+					+			+	+		
Diploneis fontanella Lange-Bert.												+									+		
Diploneis fontium Reichardt																						+	
Diploneis krammeri Lange-Bert. & Reichardt												+								+	+	+	
Diploneis minuta Petersen		+								+	+				+							+	
Diploneis modica Hust.		+																					
Diploneis oblongella (Naegeli) Cleve-Euler		+	+	+	+				+	+	+	+	+	+	+	+	+		+	+		+	R
Diploneis ovalis (Hilse) Cleve		+								+		+			+					+	+	+	R
Diploneis oculata (Bréb.) Cleve		+	+		+						+		+	+	+	+	+				+		
Diploneis parma Cleve		+																			+		E
Diploneis petersenii Hust.		+																				+	
Diploneis pseudoovalis Hust.		+											+										R
Diploneis puella (Schum.) Cleve		+											+								+	+	
Diploneis separanda Lange-Bert.												+									+	+	
Diploneis subovialis Cleve				+																			
Diploneis sp.																+							
Discostella pseudostelligera (Hust.) Houk & Klee		+			+					+	+	+	+	+	+	+	+		+	+	+	+	
Discostella stelligera (Cleve & Grun.) Houk & Klee		+			+							+		+									
Discostella woltereckii (Hust.) Houk & Klee																					+		
Discostella cf. *woltereckii* (Hust.) Houk & Klee	+																						
Encyonema caespitosum (Kütz.) D.G. Mann		+	+		+				+	+	+		+		+	+	+		+		+	+	
Encyonema minutum (Hilse) D.G. Mann	+	+	+	+	+	+	+	+	+	+	+	+	+	+	+	+	+		+	+		+	

Table 1 contin.

Taxa / Site	1	2	3	4	5	6	7	8	9	10	11	12	13	14	15	16	17	18	19	20	21	22	0
Encyonema perpusillum (Cleve) D.G. Mann		+											+	+							+		
Encyonema prostratum (Berk.) Kütz.		+	+		+				+	+	+	+	+	+	+	+	+		+		+	+	
Encyonema reichardtii (Krammer) D.G. Mann		+		+										+							+		
Encyonema silesiacum (Bleisch.) D.G. Mann		+			+	+					+	+	+	+	+		+		+		+	+	
Encyonema ventricosum (Ag.) Grun.												+									+	+	
Encyonema cf. *perpusillum* (Cleve) D.G. Mann					+																		
Encyonopsis cesatii (Rabenh.) Krammer																						+	
Encyonopsis microcephala (Grun.) Krammer		+			+						+	+	+	+	+	+	+	+				+	
Encyonopsis subminuta Krammer & Reichardt		+																					
Eolimna minima Lange-Bert.	+	+	+		+					+	+	+	+	+	+	+	+			+	+	+	
Eolimna subminuscula (Manguin) Moser, Lange-Bert. & Metzeltin	+	+	+	+	+					+	+	+	+	+	+	+	+			+	+	+	
Epithemia adnata (Kütz.) Bréb.	+	+			+					+	+	+	+	+	+						+		
Epithemia sorex Kütz.													+								+		
Epithemia turgida (Ehrenb.) Kütz.																					+		
Eucocconeis austriaca (Hust.) Lange-Bert.																						+	
Eucocconeis flexella (Kütz.) Meister		+																					
Eucocconeis laevis (Østrup) Lange-Bert.		+	+								+		+	+	+								
Eunotia bilunaris (Ehrenb.) Scharschmidt	+	+	+		+	+	+			+	+	+		+	+	+	+					+	
Eunotia botuliformis Wild, Nörpel & Lange-Bert.												+											
Eunotia curtagrunowii Nörpel-Schempp & Lange-Bert.					+						+												
Eunotia diodon Ehrenb.												+											
Eunotia exigua (Bréb.) Rabenh.											+	+		+								+	
Eunotia flexuosa (Bréb.) Kütz.												+											
Eunotia formica Ehrenb.																					+		V
Eunotia implicata Nörpel-Schempp, Lange-Bert. & Alles						+	+																
Eunotia incisa Greg.																					+		
Eunotia islandica Østrup												+											
Eunotia meisteri Hust.										+											+	+	I
Eunotia minor (Kütz.) Grun.		+	+							+		+									+		
Eunotia muscicola v. *tridentula* (Grun.) Nörpel-Schempp & Lange-Bert.		+										+											
Eunotia mucophilla (Lange-Bert. & Nörpel) Lange-Bert.						+	+		+			+		+	+							+	
Eunotia praerupta Ehrenb.												+											
Eunotia septentrionalis Østrup		+								+													E
Eunotia soleirolii (Kütz.) Rabenh.																					+		
Eunotia sp. cf. *crista-galli* Cleve												+											
Eunotia sp. cf. *groenlandica* (Grun.) Nörpel & Lange-Bert.												+											
Eunotia sp. 1		+										+							+				
Eunotia sp. 2		+																					
Eunotia sp. 3																	+						
Eunotis sp. 4										+													
Fallacia egregia (Hust.) D.G. Mann																					+		
Fallacia insociabilis (Krasske) D.G. Mann		+										+		+	+	+	+			+		+	
Fallacia lenzii (Hust.) Lange-Bert		+	+						+	+	+		+	+	+	+	+					+	E
Fallacia monoculata (Hust.) D.G. Mann	+	+	+	+						+	+	+	+	+	+	+	+		+	+	+	+	
Fallacia omissa (Hust.) D.G. Mann		+													+								
Fallacia pygmaea (Kütz.) Stricke & D.G. Mann	+	+	+	+	+			+	+	+	+	+	+	+	+	+	+		+	+	+		
Fallacia subhamulata (Grun.) D.G. Mann		+	+	+	+			+	+	+	+	+	+	+	+	+	+		+	+	+	+	R
Fallacia sublucidula (Hust.) D.G. Mann					+						+		+	+		+							
Fistulifera pelliculosa (Bréb.) Lange-Bert.																						+	
Fistulifera saprophila (Lange-Bert.& Bonik) Lange-Bert.	+				+																+		
Fragilaria acus (Kütz.) Lange-Bert.	+	+	+		+	+	+	+	+	+	+	+	+	+	+	+	+		+		+	+	
Frafilaria austriaca (Grun.) Lange-Bert.					+																	+	
Frafilaria biceps (Kütz.) Lange-Bert.		+			+							+	+								+		
Frafilaria bicapitata A. Mayer		+					+														+		
Frafilaria bidens Heiberg		+																					R
Frafilaria brevistriata Grun.					+																+		R
Frafilaria capucina Desmazieres var. *capucina*	+	+	+		+						+	+	+	+	+	+	+			+	+	+	
Frafilaria construens f. *binodis* (Ehrenb.) Hust.		+			+																+		

Table 1 contin.

Taxa / Site	1	2	3	4	5	6	7	8	9	10	11	12	13	14	15	16	17	18	19	20	21	22	0
Frafilaria construens f. *construens* (Ehrenb.) Grun.	+	+	+		+	+	+	+													+		
Frafilaria construens f. *subsalina* (Hust.) Hust.																					+		R
Frafilaria construens f. *venter* (Ehrenb.) Hust.					+																+		
Frafilaria crotonensis Kitton	+	+	+		+																+		
Frafilaria dilatata (Bréb.) Lange-Bert.					+																+		V
Frafilaria distans (Grun.) Bukht.																					+		
Frafilaria elliptica Schumann		+																			+		
Frafilaria exigua Grun.																					+		I
Frafilaria famelica (Kütz.) Lange-Bert. var. famelica		+																			+		
Frafilaria fasciculata (Ag.) Lange-Bert.	+	+	+		+	+	+		+														
Frafilaria gracilis Østrup	+	+	+	+	+	+					+	+	+	+		+	+			+	+	+	
Frafilaria lapponica Grun.																					+		E
Frafilaria leptostauron (Ehrenb.) Hust. var. *leptostauron*											+								+		+		
Frafilaria mesolepta Rabenh.		+	+		+							+	+								+		
Frafilaria nanana Lange-Bert.	+																				+		V
Frafilaria oldenburgiana Hust.																					+		R
Frafilaria pararumpens Lange-Baert., Hofmann, Mann & Werum												+									+		
Frafilaria parasitica (W. Smith) Grun. var. *parasitica*	+	+	+		+						+	+	+	+	+	+	+				+	+	
Frafilaria parasitica var. *subconstricta* Grun.		+	+	+	+	+		+	+	+	+	+	+	+	+	+	+		+		+		
Frafilaria perminuta (Grun.) Lange-Bert.		+			+																		
Frafilaria pinnata Ehrenb.	+	+			+	+					+		+	+	+				+		+	+	
Frafilaria pinnata var. *intercedens* (Grun.)Hust.		+																			+		
Frafilaria pulchella (Ralfs) Lange-Bert.		+																			+		
Frafilaria pseudoconstruens Marciniak																					+		
Frafilaria radians (Kütz.) Lange-Bert.		+											+			+	+				+		
Frafilaria robusta (Fusey) Manguin																					+		R
Frafilaria tenera (W. Smith) Lange-Bert.																					+		V
Frafilaria ulna var. *oxyrhynchus* (Kütz.) Lange-Bert.		+																					I
Frafilaria ulna var. *ulna* (Nitzch.) Lange-Bert.	+	+	+	+	+	+	+	+	+	+	+	+	+	+	+	+	+		+	+	+	+	
Frafilaria vaucheriae (Kütz.) Petersen	+	+			+					+	+	+	+	+	+	+	+		+	+	+	+	
Frafilaria virescens Ralfs		+										+		+									E
Frafilaria cf. *capensis* Grun.		+																					
Fragilaria sp. 1		+																					
Fragilaria sp. 2		+	+													+							
Fragilaria sp. 3		+																					
Fragilaria sp. 4										+													
Frustulia crassinerva (Bréb.) Ross										+												+	
Frustulia rhomboides (Ehrenb.) De Toni						+	+		+										+				
Frustulia saxonica Rabenh.		+																					
Frustulia vulgaris (Thwait.) De Toni	+	+	+	+	+	+	+	+	+	+	+	+	+	+	+	+	+		+	+	+	+	
Geissleria decussis (Østrup) Lange-Bert. & Metzeltin		+			+	+	+														+		R
Geissleria ignota var. *palustris* (Krasske) Lange-Bert. & Metzeltin														+									
Gomphonema acuminatum Ehrenb.	+	+	+	+	+	+	+	+		+	+		+	+	+	+	+			+	+	+	
Gomphonema affine Kütz.		+													+						+		R
Gomphonema augur Ehrenb.		+	+						+	+	+	+	+	+	+	+			+		+		
Gomphonema angustatum (Kütz.) Rabenh.																						+	
Gomphonema angustum Ag.									+											+			
Gomphonema bohemicum Reichelt & Fricke		+																					E
Gomphonema clavatum Ehrenb.																						+	
Gomphonema clevei Fricke																					+		
Gomphonema cymbeliclinum Reichard & Lange-Bert.												+			+						+		
Gomphonema elegantissimum Reichardt & Lange-Bert.																						+	
Gomphonema exilissimum (Grun.) Lange-Bert. & Reichardt												+									+	+	
Gomphonema gracile Ehrenb.		+	+		+						+		+	+	+		+				+	+	
Gomphonema hebridense Gregory																						+	
Gomphonema insigne Greg.		+									+				+				+				
Gomphonema micropus Kütz.	+	+	+	+	+	+	+		+	+	+	+	+	+	+	+	+		+	+	+	+	
Gomphonema minutum (Ag.) Ag.	+	+	+	+	+				+				+	+	+	+			+	+	+	+	

Table 1 contin.

Taxa / Site	1	2	3	4	5	6	7	8	9	10	11	12	13	14	15	16	17	18	19	20	21	22	0
Gomphonema minutum f. *pachypus* Lange-Bert. & Reichardt		+																					
Gomphonema olivaceum var. *minutissimum* Hust.		+												+									
Gomphonema olivaceum var. *olivaceoides* (Hust.) Lange-Bert.		+		+										+	+						+		
Gomphonema olivaceum (Horn.) Bréb. var. *olivaceum*	+	+	+	+	+				+	+	+	+	+	+	+	+	+		+	+	+	+	
Gomphonema olivaceolacuum (Lange-Bert. & Reich.) Lange-Bert. & Reichardt					+							+											
Gomphonema parvulum (Kütz.) Rabenh. var. *parvulum*	+	+	+	+	+	+	+	+	+	+	+	+	+	+	+	+	+		+	+	+	+	
Gomphonema parvulum (Kütz.) Rabenh. var. *parvulum* f. *saprophilum*												+											
Gomphonema parvulius (Lange-Bert & Reichardt) Lange-Bert & Reichardt																					+	+	
Gomphonema procerum ? Reichardt & Lange-Bert.		+																					
Gomphonema productum (Grun.) Lange-Bert. & Reichardt		+							+			+	+	+	+	+				+	+	+	
Gomphonema pseudoaugur Lange-Bert.			+																				
Gomphonema pumilum (Grun.) Reichardt & Lange-Bert.	+					+						+								+	+	+	
Gomphonema reichardtii Lange-Bert.		+																					
Gomphonema sarcophagus Greg.		+	+		+					+	+	+	+	+	+	+	+		+	+	+	+	V
Gomphonema subclavatum (Grun) Grun.	+	+	+	+	+				+	+	+	+	+	+	+	+	+			+	+	+	
Gomphonema tergestinum (Grun.) Fricke	+	+	+	+	+					+	+	+	+	+	+	+	+		+	+	+	+	I
Gomphonema truncatum Ehrenb.		+	+		+	+	+	+	+	+	+		+	+	+	+					+	+	
Gomphonema utae Lange-Bert. & Reichardt		+										+		+	+							+	
Gomphonema cf. *angustum* (Kütz.) Rabenh.	+	+	+							+	+		+	+	+	+	+				+	+	
Gomphonema cf. *bavaricum* Reichardt & Lange-Bert.											+												
Gomphonema cf. *bohemicum* Reicheld & Fricke											+												
Gomphonema cf. *hebridense* Gregory							+																
Gomphonema cf. *minutum* (Ag.) Ag.																	+						
Gomphonema cf. *pumilum* (Grun.) Reichardt & Lange-Bert.		+	+		+					+	+		+	+	+	+	+		+			+	
Gomphonema sp. cf. *clevei* Fricke		+																					
Gomphonema sp. cf. *dichotomum* Kütz.		+																					
Gomphonema sp. 1		+										+	+	+			+					+	
Gomphonema sp. 2		+																				+	
Gomphonema sp. 3		+																				+	
Gomphonema sp. 4													+										
Gomphonema sp. 5		+																					
Gomphonema sp. 6		+																					
Gomphonema sp. 7		+									+		+		+		+						
Gomphonema sp. 8		+																					
Gomphonema sp. 9													+										
Gomphonema sp. 10													+										
Gomphonema sp. 11																					+		
Gyrosigma acuminatum (Kütz.) Rabenh.	+	+	+	+	+	+		+	+	+	+	+	+	+	+	+	+		+	+	+	+	
Gyrosigma attenuatum (Kütz.) Rabenh.		+	+	+	+		+		+	+	+	+	+	+	+	+	+		+		+	+	
Gyrosigma elongatum (W. Smith) Griffith & Henfrey									+														
Gyrosigma obtusatum (Sulivant & Wormley) Boyer	+	+	+	+	+			+	+	+	+	+	+	+	+	+	+		+	+		+	
Gyrosigma sciotonense (Sulivant) Cleve		+	+	+	+				+	+	+	+	+	+	+	+	+		+			+	
Gyrosigma spencerii (W. Smith) Cleve										+													
Gyrosigma sp.												+											
Halamphora montana (Krasske) Levkov	+	+	+	+	+					+	+	+	+	+	+	+	+		+	+	+	+	
Halamphora normannii (Rabenh.) Levkov		+		+						+		+	+	+	+	+	+			+	+	+	
Halamphora veneta (Kütz.) Levkov	+	+											+		+			+		+	+	+	
Halamphora cf. *veneta* (Kütz.) Levkov				+																			
Hannaea arcus (Ehrenb.) Patrick					+																		
Hantzschia abundans Lange-Bert.	+		+							+	+	+			+						+	+	
Hantzschia amphioxys (Ehrenb.) Grun.	+	+	+	+	+					+	+	+	+	+	+	+	+		+	+	+	+	
Hippodonta capitata (Ehrenb.) Lange-Bert., Metzeltin & Witkowski	+	+	+	+	+	+	+	+	+	+	+	+	+	+	+	+	+		+	+	+	+	
Hippodonta costulata (Grun.) Lange-Bert., Metzeltin & Witkowski												+			+								

Table 1 contin.

Taxa / Site	1	2	3	4	5	6	7	8	9	10	11	12	13	14	15	16	17	18	19	20	21	22	0
Hippodonta cf. *lueneburgensis* (Grun.) Lange-Bert., Metzeltin & Witkowski															+								
Karayevia clevei (Grun.) Round & Bukht.		+																					
Karayevia laterostrata (Hust.) Round & Bukht.						+		+						+									
Kobayasiella subtilissima (Cleve) Lange-Bert.											+				+								
Kolbesia ploenensis (Hust.) Round & Bukht.		+																			+		
Lemnicola hungarica (Grun.) Round & Bassed	+	+	+		+	+	+	+			+	+	+	+	+	+	+				+		
Luticola acidoclinata Lange-Bert.		+								+	+	+		+		+	+				+	+	R
Luticola cohnii (Hilse) D.G. Mann	+										+												
Luticola dismutica (Hust.) D.G. Mann																+						+	
Luticola goeppertiana (Bleisch.) D.G. Mann		+	+		+					+	+		+	+	+	+	+		+		+	+	
Luticola kotschyi (Grun.) D.G. Mann		+			+					+	+	+	+	+	+	+	+			+	+	+	
Luticola mutica (Kütz.) D.G. Mann	+	+	+	+	+	+				+	+	+	+	+	+	+	+		+	+	+	+	
Luticola nivalis (Ehrenb.) D.G. Mann	+	+	+			+				+	+	+	+	+	+	+	+		+	+		+	
Luticola paramutica var. *binodis* Bock (Van de Vijver)		+			+								+	+	+		+						
Luticola paramutica Bock (Van de Vijver)	+	+										+			+							+	
Luticola sueoconum var. *dismutica* (Hust.) D.G. Mann		+																					
Luticola ventriconfusa Lange-Bert.												+											
Luticola ventricosa (Kütz.) D.G. Mann	+	+		+	+		+			+	+	+	+	+	+	+	+		+	+	+	+	
Luticola cf. *cohnii* (Hilse) D.G. Mann															+								
Luticola ?		+																					
Mayamaea agrestis (Hust.) Lange-Bert.	+	+	+		+								+		+		+		+				
Mayamaea asellus (Weinhold) Lange-Bert.		+																		+			
Mayamaea atomus var. *alcimonica* (Reichardt) Reichardt					+							+									+	+	
Mayamaea atomus (Kütz.) Lange-Bert. var. *atomus*	+	+	+	+						+	+	+	+	+	+	+	+			+	+	+	
Mayamaea atomus var. *permitis* (Hust.) Lange-Bert.		+	+		+					+	+	+	+	+	+	+	+			+	+	+	
Mayamaea excelsa (Krasske) Lange-Bert.	+																						
Mayamaea fossalis (Krasske) Lange-Bert. var. *fossalis*		+												+		+	+				+		
Mayamaea fossalis (Hust.) Lange-Bert. var. *obsidialis*																						+	
Mayamaea fossalloides (Hust.) Lange-Bert.					+																		
Mayamaea cf. *atomus* var. *permitis* (Hust.) Lange-Bert.	+																						
Mayamaea cf. *agrestis* (Hust.) Lange-Bert.										+		+			+								
Mayamaea cf. *recondita* (Hust.) Lange-Bert.	+																						
Mayamaea sp. 1		+												+	+							+	
Mayamaea sp. 2	+																						
Melosira lineata (Dilwyn) Ag.											+												
Melosira varians Ag.	+	+	+	+	+	+	+	+	+	+	+		+	+	+	+	+		+	+	+	+	
Meridion circulare Ag. var. *circulare*	+	+	+	+	+	+	+	+	+	+	+	+	+	+	+	+	+		+	+	+	+	
Meridion circulare var. *constrictum* (Ralfs) Van Heurck		+	+		+	+	+	+	+	+	+	+	+	+	+	+	+		+	+	+	+	
Muelleria gibbula Spaulding & Stoermer		+																				+	
Muelleria sp. cf. *gibbula* Spaulding & Stoermer	+																						
Navicula amphiceropsis Lange-Bert. & Rumrich																						+	
Navicula antonii Lange-Bert.	+	+	+	+	+				+	+	+	+	+	+	+	+	+		+	+	+	+	
Navicula arvensis var. *major* Lange-Bert.		+												+								+	
Navicula associata Lange-Bert.	+	+	+		+								+	+	+	+	+						
Navicula aquaedurae Lange-Bert.											+	+											
Navicula bacilloides Hust.	+									+		+			+	+	+				+	+	
Navicula capiata Ehrenb.																			+				
Navicula capitatoradiata Germain	+	+	+	+	+	+			+	+	+	+	+	+	+	+	+		+		+		
Navicula cari Ehrenb.		+			+				+	+	+	+		+	+	+	+			+	+	+	
Navicula cariocincta Lange-Bert.															+								R
Navicula caterva Hohn & Hellerman		+																					
Navicula cincta (Ehrenb.) Ralfs		+	+		+					+		+	+	+	+	+	+		+	+	+	+	
Navicula cryptocephala Kütz.	+	+	+	+	+	+	+	+	+	+	+	+	+	+	+	+	+		+	+	+	+	
Navicula cryptotenella Lange-Bert.	+	+	+	+	+	+	+	+	+	+	+	+	+	+	+	+	+		+	+	+	+	
Navicula cryptotenelloides Lange-Bert.	+	+	+		+					+	+	+	+	+	+	+	+		+	+	+	+	
Navicula digitoradiata (Gregory) Ralfs									+														R
Navicula digitoconvergens Lange-Bert.																					+		
Navicula erifuga Lange-Bert.		+	+		+					+	+		+	+	+	+	+			+		+	

Table 1 contin.

Taxa / Site	1	2	3	4	5	6	7	8	9	10	11	12	13	14	15	16	17	18	19	20	21	22	0
Navicila exigua Gregory																						+	
Navicula germainii Wallace		+			+																		
Navicula gregaria Donkin	+	+	+	+	+	+	+	+	+	+	+	+	+	+	+	+	+		+	+	+	+	
Navicula harderii Hust.														+		+	+					+	
Navicula integra (W. Smith) Ralfs	+	+			+	+	+	+											+		+		
Navicula lanceolata (Ag.) Kütz.	+	+	+	+	+	+	+	+	+	+	+	+	+	+	+	+	+		+	+	+	+	
Navicula libonensis Schoeman					+								+		+	+	+				+	+	
Navicula longicephala Hust.		+													+						+		
Navicula medioconvexa ? Hust.															+								R
Navicula microcari Lange-Bert.															+								
Navicula moskalii Metzeltin, Witkowski & Lange-Bert.	+																				+		
Navicula novaesiberica Lange-Bert.		+		+	+				+		+	+	+	+	+								
Navicula oblonga (Kütz.) Kütz.		+												+	+						+		R
Navicula oppugnata Hust.		+			+																+		R
Navicula phyllepta Kütz.				+																			
Navicula radiosa Kütz.		+	+	+	+	+	+	+		+	+	+	+	+	+	+	+		+		+	+	
Navicula radiosafallax Lange-Bert.		+																					
Navicula recens (Lange-Bert.) Lange-Bert.		+							+		+	+	+	+	+					+	+	+	
Navicula reichardtiana Lange-Bert.	+	+	+	+	+					+	+	+	+	+	+	+	+	+	+	+	+	+	
Navicula reinhardtii (Grun.) Grun.		+	+		+																+		
Navicula rhynchotella Lange-Bert.		+			+																+		R
Navicula rhynchocephala Kütz.		+				+	+	+	+					+		+					+		
Navicula rostellata (Kütz.) Cleve		+	+		+	+				+	+			+	+	+					+	+	
Navicula salinarium Grun. var. *salinarium*		+	+		+				+														
Navicula salinarium var. *rostrata* Hust.					+																		
Navicula simulata Manguin																						+	
Navicula sinuata Schumann																						+	
Navicula slesvicensis Grun.	+	+	+		+				+	+	+		+	+	+	+	+		+	+	+	+	
Navicula splendicula Land		+																					V
Navicula tenelloides Hust.		+				+				+	+	+	+	+	+	+	+		+	+	+	+	
Navicula tripunctata (O.F. Müller) Bory	+	+	+	+	+				+	+	+	+	+	+	+	+	+		+	+	+	+	
Navicula trivialis Lange-Bert.	+	+	+	+	+	+			+	+	+	+	+	+	+	+	+		+	+	+	+	
Navicula trophicatrix Lange-Bert.																					+		
Navicula upsaliensis (Grun.) Peragallo	+	+	+	+	+	+	+	+	+	+	+	+	+	+	+	+	+		+	+	+	+	R
Navicula vandamii Schoeman & Archbald var. *vandamii*		+			+								+	+	+		+		+		+		
Navicula veneta Kütz.	+	+	+		+					+	+	+	+	+	+	+	+		+	+	+	+	
Navicula ventralis Krasske		+																					
Navicula vilaplanii (Lange-Bert. & Sabather) Lange-Bert. & Sabather		+								+		+	+	+	+	+	+		+	+	+	+	
Navicula viridula (Kütz.) Ehrenb.	+	+	+	+	+				+	+	+	+	+	+	+	+	+	+	+		+	+	
Navicula viridulacalcis Lange-Bert.		+																					
Navicula vulpina Kütz.		+																					
Navicula wiesneri Lange-Bert.		+												+	+	+	+					+	
Navicula sp. cf. *microdigitoradiata* Lange-Bert.																		+					
Navicula sp. cf. *moskali* (?) Metzeltin, Witkowski & Lange-Bert.		+																					
Navicula cf. *parsura* Hust.															+								
Navicula sp. cf. *symmetrica* Patrick		+																					
Navicula sp. cf. *streckerae* Lange-Bert. & Witkowski					+																		
Navicula sp. cf. *suchlandtii* Hust.												+											
Navicula cf. var. *vilaplanii* Lange-Bert. & Sabather				+																			
Navicula sp. 1		+																				+	
Navicula sp. 2		+											+	+	+							+	
Navicula sp. 3		+																					
Navicula sp. 4			+																				
Naviculadicta stauroneioides? Lange-Bert.		+																					
Navicymbula pusilla (Grun.) Krammer					+																		
Neidium affine (Ehrenb.) Pfitzer		+			+					+	+	+	+	+	+	+					+		
Neidium affine var. *longiceps* (Greg.) Cleve															+								
Neidium ampliatum (Ehrenb.) Krammer		+	+	+	+	+	+		+	+	+	+	+	+	+		+	+	+		+	+	V

Table 1 contin.

Taxa / Site	1	2	3	4	5	6	7	8	9	10	11	12	13	14	15	16	17	18	19	20	21	22	0
Neidium apiculatum Reimer				+																			
Neidium binodis (Ehrenb.) Pfitzer													+										
Neidium binodeforme Krammer		+	+				+			+	+	+	+	+	+	+	+		+		+	+	
Neidium bisulcatum var. *subampliatum* Krammer					+								+		+							+	
Neidium dubium (Ehrenb.) Cl.	+	+			+			+		+	+	+	+	+	+	+	+		+		+		I
Neidium hercynicum ? A. Mayer															+								
Neidium productum (W. Smith) Cleve						+	+								+						+	+	
Nitzschia abbreviata Hust.																					+		
Nitzschia acicularis (Kütz.) W. Smith	+	+	+	+	+				+	+	+	+	+	+	+	+	+	+	+	+	+	+	
Nitzschia acicularoides Hust.					+																		
Nitzschia acidoclinata Lange-Bert.	+	+									+	+		+	+				+	+	+	+	
Nitzschia acula (Kütz.) Hantzsch																					+	+	
Nitzschia alpina Grun.												+											
Nitzschia amphibia Grun.	+	+	+	+	+	+				+	+		+	+	+	+	+		+	+	+	+	
Nitzschia angustata Lange-Bert.		+			+								+	+	+	+		+	+		+		
Nitzschia angustatula Lange-Bert.		+	+		+					+	+	+	+		+		+						
Nitzschia archibaldii Lange-Bert.	+	+	+		+						+	+	+	+	+	+	+			+	+		
Nitzschia bacillum Hust.		+	+		+								+	+		+							
Nitzschia brevissima Grun.				+					+			+											
Nitzschia brunoi Lange-Bert.												+											
Nitzschia bulheimana (Rabenh.) Smith		+	+							+		+		+		+	+						R
Nitzschia calida Grun.		+		+	+				+	+	+	+	+	+	+	+	+		+	+		+	
Nitzschia capitellata Hust.	+	+	+	+	+	+	+		+	+	+	+	+	+	+	+	+		+	+	+	+	
Nitzschia clausii Hantzsch																+	+						
Nitzschia communis Rabenh.	+	+										+	+	+	+	+	+				+	+	
Nitzschia constricta (Kütz.) Ralfs		+	+	+	+				+	+	+	+	+	+	+	+	+	+	+	+	+	+	
Nitzschia debilis (Arnott) Grun.	+	+	+	+	+	+	+	+	+	+	+	+	+	+	+	+	+	+	+	+	+	+	
Nitzschia delognei Grun.					+																		
Nitzschia denticula Grun.		+											+					+					
Nitzschia desertorum Hust.		+								+					+								
Nitzschia dissipata (Kütz.) Grun. ssp. *dissipata*	+	+	+	+	+	+	+		+	+	+	+	+	+	+	+	+	+	+	+	+	+	
Nitzschia dissipata var. *media* (Hantzsch) Grun.		+	+		+				+	+	+	+	+	+	+		+		+		+	+	
Nitzschia dravelliensis Coste & Ricard		+				+																	
Nitzschia dubia W. Smith		+	+	+	+			+	+	+	+	+	+	+	+	+	+		+	+		+	
Nitzschia filiformis (Smith) Van Heurck				+																			
Nitzschia fonticola Grun.	+	+	+		+	+	+		+	+	+		+	+	+	+	+	+		+	+		
Nitzschia frequens Hust.												+											
Nitzschia frustulum (Kütz.) Grun. var. *frustulum*	+	+	+	+	+	+		+		+	+	+	+	+	+	+	+	+	+	+		+	
Nitzschia frustulum var. *inconspicua* (Grun.) Grun.		+	+		+					+	+		+	+	+	+	+		+	+	+	+	
Nitzschia graciliformis Lange-Bert. & Simonsen		+	+									+			+								
Nitzschia gracilis Hantzsch	+	+	+		+	+				+	+	+			+		+				+	+	
Nitzschia hantzschiana Rabenh.																						+	
Nitzschia heufleriana Grun.		+	+		+					+	+	+	+	+	+	+	+		+	+	+	+	
Nitzschia homburgiensis Lange-Bert.	+	+										+		+		+	+				+		
Nitzschia hungarica Grun.	+	+	+	+	+				+	+	+	+	+	+	+	+	+		+	+	+	+	
Nitzschia intermedia Hantzsch		+	+		+				+			+	+	+	+	+		+					
Nitzschia lacuum Lange-Bert.																						+	
Nitzschia levidensis (W. Smith) Grun.		+	+	+	+	+	+	+	+	+	+		+	+	+	+	+		+	+			
Nitzschia linearis (Ag.) Smith	+	+	+	+	+				+	+	+	+	+	+	+	+	+	+	+	+	+	+	
Nitzschia microcephala Grun.		+	+								+												
Nitzschia ovalis Arnott																+	+						
Nitzschia palea (Kütz.) W. Smith	+	+	+	+	+	+	+	+	+	+	+	+	+	+	+	+	+		+	+	+	+	
Nitzschia palea var. *debilis* (Kütz.) Grun.																						+	
Nitzschia paleacea (Grun.) Grun.		+	+		+						+							+			+		
Nitzschia perminuta Grun.	+	+	+	+					+	+	+	+	+	+	+	+	+				+	+	
Nitzschia pura Hust.					+							+											
Nitzschia pusilla Grun.	+	+	+		+					+	+	+	+	+	+	+	+		+	+	+	+	
Nitzschia pumila Hust.		+																					
Nitzschia recta Hantzsch	+	+	+	+	+				+	+	+	+	+	+	+	+	+	+	+	+	+	+	

Table 1 contin.

Taxa / Site	1	2	3	4	5	6	7	8	9	10	11	12	13	14	15	16	17	18	19	20	21	22	0
Nitzschia sigma (Kütz.) W. Smith		+	+		+	+	+			+	+	+	+	+	+	+	+	+	+	+	+	+	
Nitzschia sigmoidea (Nitzsch) W. Smith		+	+	+	+	+	+	+	+	+	+	+	+	+	+	+	+		+	+	+	+	
Nitzschia salinarum Grun.		+			+							+	+	+		+	+				+	+	
Nitzschia solgensis Cleve-Euler		+	+									+	+	+	+	+	+				+	+	
Nitzschia sociabilis Hust.		+	+		+				+	+	+	+	+	+	+	+	+		+			+	
Nitzschia solita Hust.			+												+								
Nitzschia subacicularis Hust.		+	+		+				+	+	+	+									+		
Nitzschia sublinearis Hust.																					+		
Nitzschia subtilis Grun.		+			+					+	+	+	+	+	+	+	+		+	+		+	
Nitzschia supralitorea Lange-Bert.		+			+						+	+							+		+	+	
Nitzschia tabellaria Grun.	+		+		+						+			+	+			+			+		
Nitzschia tenuis W. Smith	+	+	+	+	+	+	+	+		+	+	+	+	+	+	+	+	+	+	+	+	+	
Nitzschia terrestris (Perersen) Hust.														+									
Nitzschia tubicola Grun.															+				+			+	
Nitzschia trybionella Hantzsch		+	+		+					+			+		+						+		
Nitzschia umbonata (Ehrenb.) Lange-Bert.		+		+					+			+			+	+						+	
Nitzschia vermicularis (Kütz.) Hantzsch		+	+	+	+				+	+	+	+	+	+	+	+	+	+	+	+	+	+	
Nitzschia vermicularoides Lange-Bert.																						+	
Nitzschia sp. cf. *agnita* Hust.											+												
Nitzschia sp. cf. *bergii* Cleve-Euler																							
Nitzschia sp. cf. *commutata* Grun.		+																					
Nitzschia sp. cf. *fasciculata* (Grun.) Grun.										+													
Nitzschia sp. cf. *ovalis* Arnott												+											
Nitzschia sp. cf. *paleacea* (Grun.) Grun.				+	+																		
Nitzschia cf. *supralitorea* Lange-Bert.													+										
Nitzschia sp. 1																							
Nitzschia sp. 2						+																	
Nitzschia sp. 3	+																						
Nitzschia sp. 4	+																						
Parlibellus crucicula (Smith) Witkowski, Lange-Bert. & Metzeltin		+			+																+		R
Parlibellus protracta (Grun.) Witkowski, Lange-Bert. & Metzeltin		+		+					+			+	+	+							+		
Parlibellus protractoides (Hust.) Witkowski, Lange-Bert. & Metzeltin		+	+	+		+	+	+	+		+	+	+	+	+	+	+	+	+		+		R
Pinnularia acuminata W. Smith																					+		
Pinnularia appendiculata (Ag.) Cleve		+								+		+	+	+	+	+				+	+		
Pinnularia borealis Ehrenb. var. *borealis*	+	+										+		+			+		+	+	+	+	
Pinnularia borealis var. *sublinearis* Krammer		+									+					+	+						
Pinnularia borealis var. *scalaris* (Ehrenb.) Rabenh.										+													
Pinnularia borealis var. *rectangularis* Carlson		+													+								
Pinnularia brebissonii (Kütz.) Rabenh.		+	+	+	+		+		+	+		+	+	+	+	+	+		+	+	+	+	
Pinnularia clevei var. *minor* (Hust.) Krammer																					+		
Pinnularia divergentissima var. *triundulata* Krammer																						+	
Pinnularia dubitabis var. *minor* Krammer														+									
Pinnularia esoxiformis Fusey																					+		
Pinnularia flexuosa Cleve																					+		
Pinnularia frauenbergiana Reichardt														+						+			
Pinnularia gentilis (Donkin) Cleve																					+		
Pinnularia gibba Ehrenb.						+	+					+									+		
Pinnularia globiceps Greg.			+	+					+			+		+						+		+	
Pinnularia grunowii Krammer		+	+			+					+	+	+	+	+		+			+	+		
Pinnularia iselana Krammer												+	+	+	+		+				+	+	
Pinnularia islandica ? Østrup		+													+								
Pinnularia legunem Ehrenb.													+										
Pinnularia lundii Hust.		+	+	+	+				+	+		+	+	+	+	+	+		+	+	+		
Pinnularia major (Kütz.) Cleve		+																					
Pinnularia marchica Schönfelder												+										+	
Pinnularia microstauron (Ehrenb.) Cleve	+	+	+	+	+							+	+					+			+		V
Pinnularia neomajor Krammer											+												
Pinnularia nodosa (Ehrenb.) W. Smith												+											

Table 1 contin.

Taxa / Site	1	2	3	4	5	6	7	8	9	10	11	12	13	14	15	16	17	18	19	20	21	22	0
Pinnularia nodosa var. *robusta* (Foged) Krammer																					+		
Pinnularia obscura Krasske	+	+		+	+					+	+	+	+	+	+	+	+			+	+	+	
Pinnularia obscuriformis Krammer					+								+	+			+				+		
Pinnularia oriunda Krammer		+	+		+							+	+	+	+		+			+	+	+	
Pinnularia parvulissima Krammer																					+		
Pinnularia perirrorata Krammer												+									+	+	
Pinnularia renata Krammer																							
Pinnularia rhenohassiaca Krammer & Lange-Bert.																					+		
Pinnularia rhombarea var. *udulata* Krammer										+													
Pinnularia rhombarea var. *variarea* Krammer																					+		
Pinnularia rupestris Hantzsch		+	+	+					+	+	+		+	+	+		+				+		E
Pinnularia schonfelderi Krammer												+			+	+	+				+	+	E
Pinnularia schroederii (Hust.) Krammer												+											
Pinnularia silvatica Petersen												+										+	
Pinnularia similis Hust.										+			+		+								
Pinnularia sinistra Krammer	+	+										+	+	+			+			+	+		
Pinnularia septentrionalis Krammer												+											
Pinnularia subcapitata Greg.												+		+	+						+		
Pinnularia subcapitata var. *subrostrata* Krammer																				+			
Pinnularia subcapitata var. *elongata* Krammer		+				+	+					+											
Pinnularia subcommutata Krammer														+			+				+	+	
Pinnularia subcommutata var. *nonfasciata* Krammer		+									+	+			+					+	+	+	
Pinnularia subgibba Krammer																					+	+	E
Pinnularia subgibba var. *undulata* Krammer																					+		
Pinnularia subrupertris Krammer					+							+	+	+	+	+	+			+	+	+	E
Pinnularia subundulata Østrup																					+		
Pinnularia undula (Schumann) Krammer													+								+		
Pinnularia viridiformis Krammer		+	+	+	+	+	+			+	+	+	+	+	+	+	+			+	+	+	E
Pinnularia viridiformis var. *minor* Krammer																					+		
Pinnularia viridis (Nitzsch) Ehrenb.		+			+	+			+	+		+	+	+	+		+		+	+	+	+	
Pinnularia cf. *appendiculata* (Ag.) Cleve														+									
Pinnularia cf. *cherryfieldiana* Krammer																					+		
Pinnularia cf. *frauenburgiana* Reichardt															+								
Pinnularia cf. *kuetzingii* Krammer														+								+	
Pinnularia cf. *oriundiformis* Krammer																					+		
Pinnularia cf. *schoenfelderii* Krammer		+																					
Pinnularia cf. *silvatica* Petersen																							
Pinnularia cf. *subrombica* Krammer																						+	
Pinnularia cf. *viridiformis* Krammer												+											
Pinnularia sp. 1		+										+											
Pinnularia sp. 2		+																					
Pinnularia sp. 3													+										
Pinnularia sp. 4																					+		
Placoneis abiscoensis (Hust.) Lange-Bert. & Metzeltin																					+		
Placoneis anglica (Ralfs) Cox																					+		
Placoneis clementis (Grun.) Cox		+		+	+	+	+	+	+		+	+	+	+	+	+	+		+		+		
Placoneis constans (Hust.) Cox												+									+		
Placoneis dicephala (Ehrenb.) Mereschkowsky									+										+		+		
Placoneis elginensis (Greg.) Cox		+	+	+	+	+						+			+				+		+	+	
Placoneis exigua var. *signata* (Hust.) Hawarth & Kelly																					+		
Placoneis gastrum (Ehrenb.) Mereschkowsky	+				+																+		
Placoneis paraelginensis Lange-Bert.		+			+						+	+		+	+					+	+	+	
Placoneis placentula (Ehrenb.) Cox					+						+				+	+					+		
Placoneis pseudoanglica Lange-Bert.						+								+							+		
Placoneis subplacentula (Hust.) Cox																							
Placoneis undulata (Østrup) Lange-Bert.																					+		
Placoneis cf. *abiskoensis* (Hust.) Lange-Bert. & Metzeltin																			+				
Placoneis sp.														+		+						+	
Platessa conspicua (Mayer) Lange-Bert.		+			+					+			+	+	+						+		

Table 1 contin.

Taxa / Site	1	2	3	4	5	6	7	8	9	10	11	12	13	14	15	16	17	18	19	20	21	22	0
Planothidium delicatulum (Kütz.) Round & Bukht.		+											+										
Planothidium ellipticum (Cleve) Round & Bukht.		+							+	+	+		+	+	+	+	+		+			+	
Planothidium frequentissimum (Lange-Bert.) Round & Bukht.	+	+			+	+	+	+	+		+	+	+	+	+	+	+			+	+	+	
Planothidium lanceolatum (Bréb.) Round & Bukht.	+	+	+	+	+	+	+	+	+	+	+	+	+	+	+	+	+		+	+	+	+	
Planothidium rostratum (Østrup) Lange-Bert.		+			+	+	+	+						+							+		
Psammothidium biorettii (Germain) Bukht. & Round		+				+	+	+	+					+							+		
Psammothidium grischunum (Wuthrich) Bukht. & Round		+									+	+		+	+								
Psammothidium helveticum (Hust.) Bukht. & Round		+																					
Psammothidium laueunburgianum (Hust.) Bukht. & Round		+									+	+	+	+		+	+				+		V
Psammothidium subatomoides (Hust.) Bukht. & Round		+				+	+				+	+			+						+		V
Psammothidium cf. *grischunum* (Wuthrich) Bukht. & Round				+																			
Puncticulata radiosa (Lemmermann) Håkansson		+	+		+					+						+							
Reimeria sinuata (Greg.) Kociolek & Stoermer	+	+	+		+					+	+	+	+	+	+		+				+	+	
Reimeria uniseriata Sala, Guerrero & Ferrario		+	+		+					+	+	+	+	+	+	+	+		+			+	
Rhoicosphenia abbreviata (Ag.) Lange-Bert.	+	+	+	+	+	+	+	+	+	+	+	+	+	+	+	+	+	+	+	+	+	+	
Rhopalodia brebissonii Krammer																+							
Rhopalodia gibba (Ehrenb.) Müller		+									+					+	+	+			+	+	
Rhopalodia gibberula (Ehrenb.) Müller		+	+								+		+	+	+	+	+					+	
Rhopalodia rupestris (W. Smith) Krammer															+								R
Rhopalodia cf. *rupestris* (W. Smith) Krammer										+													
Rossithidium petersenii (Hust.) Round & Bukht.						+																	
Scoliopleura peisonis Grun.												+											
Sellaphora americana (Ehrenb.) D.G. Mann																					+		R
Sellaphora bacillum (Ehrenb.) D.G. Mann		+	+		+					+	+	+	+		+	+				+	+		V
Sellaphora hustedtii (Krasske) Lange-Bert. & Werum															+							+	V
Sellaphora joubaudii (Germain) Aboal	+	+	+								+	+	+	+	+	+	+			+	+	+	
Sellaphora levissima (Kütz.) D.G. Mann.						+															+		
Sellaphora mutata (Krasske) Lange-Bert.					+										+								
Sellaphora nana (Hust.) Lange-Bert.	+																			+			
Sellaphora pseudopupula (Krag.) Lange-Bert.												+							+		+	+	E
Sellaphora pupula (Kütz.) Mereschk.		+	+	+	+	+	+	+	+	+	+	+	+	+	+	+	+	+	+	+	+	+	
Sellaphora seminulum (Grun.) D.G. Mann		+	+		+						+	+	+	+	+	+	+			+	+	+	
Sellaphora stroemii (Hust.) D.G. Mann		+											+									+	
Sellaphora cf. *stroemii* (Hust.) D.G. Mann												+											
Simonsenia delognei (Grun.) Lange-Bert.		+									+	+	+	+	+	+	+			+	+	+	
Stauroneis acidoclinata Lange-Bert. & Werum												+										+	
Stauroneis acuta W. Smith																					+		
Stauroneis amphicephala Kütz.																						+	
Stauroneis anceps Ehrenb.	+	+	+		+				+	+	+	+	+	+	+	+	+		+	+	+	+	
Stauroneis borrichii (Petersen) Lund	+																				+	+	
Stauroneis gracilior (Ehrenb.) Reichard															+						+		
Stauroneis gracilis Ehrenb.												+									+		V
Stauroneis kriegerii Patrick		+	+			+				+	+	+	+	+	+						+		
Stauroneis laterostrata Hust.		+																					
Stauroneis lauenburgiana Hust.																					+		
Stauroneis legumen (Ehrenb.) Kütz.						+														+	+	+	
Stauroneis leguminopsis Lange-Bert. & Krammer												+										+	
Stauroneis optusa Lagerstedt											+										+		
Stauroneis parathermicola Lange-Bert.	+											+									+	+	
Stauroneis phoenicenteron (Nitzsch) Ehrenb.		+	+	+	+	+	+	+	+	+	+	+	+	+	+		+		+	+	+	+	V
Stauroneis producta Grun.																						+	
Stauroneis reichardtii Lange-Bert.												+										+	
Stauroneis separanda Lange-Bert. & Werum		+	+		+					+	+		+	+	+	+	+		+	+	+	+	
Stauroneis smithii Grun.		+	+	+	+	+	+	+	+	+	+	+	+	+	+	+	+	+	+	+	+	+	
Stauroneis subgracilis Lange-Bert. & Krammer												+									+		
Stauroneis tackei (Hust.) Krammer, Lange-Bert., Kuesber & Metzeltin		+									+	+	+	+	+					+	+	+	

Table 1 contin.

Taxa / Site	1	2	3	4	5	6	7	8	9	10	11	12	13	14	15	16	17	18	19	20	21	22	0
Stauroneis thermicola (Petersen) Lund	+	+			+					+		+		+	+	+	+			+	+	+	R
Stauroneis undata Hust.		+														+							
Stauroneis cf. *agrestis* Petersen	+																						
Stauroneis cf. *borrichii* (Petersen) Lund	+	+													+	+	+						
Stauroneis sp. 1		+																					
Stauroneis sp. 2		+																					
Stauroneis sp. 3											+				+								
Stenopterobia sp.													+	+	+	+	+						
Stephanodiscus hantzschii Grun.	+	+	+		+				+	+	+	+	+	+	+	+	+	+	+	+	+	+	
S. minutulus (Kütz.) Cleve & Moller		+			+						+	+	+	+	+	+	+				+		
Surirella angusta Kütz.	+	+	+	+	+	+	+	+	+	+	+	+	+	+	+	+	+	+	+	+	+	+	
Surirella bifrons Ehrenb.																					+		V
Surirella biseriata Bréb.		+	+		+						+	+	+	+	+	+	+		+		+		
Surirella brebissonii Krammer & Lange-Bert. var. *brebissonii*		+	+	+	+	+	+	+	+	+	+	+	+	+	+	+	+	+	+	+	+	+	R
Surirella brebissonii var. *kuetzingii* Krammer & Lange-Bert.	+	+	+		+			+		+	+	+	+	+	+		+		+	+	+	+	
Surirella brebissonii var. *punctata* Krammer									+													+	
Surirella crumena Bréb.		+			+						+		+	+	+	+							R
Surirella elegans Ehrenb.		+							+						+						+		V
Surirella helvetica Brun												+									+		
Surirella linearis W. Smith			+		+					+			+	+	+	+	+		+	+	+	+	
Surirella minuta (Bréb.) Kütz.	+	+	+	+	+	+			+	+	+	+	+	+	+	+	+	+	+	+	+	+	
Surirella ovalis Bréb.		+		+	+				+	+	+	+	+	+	+	+	+			+	+	+	
Surirella robusta Ehrenb.			+																				V
Surirella spirallis Kütz.																						+	
Surirella splendida (Ehrenb.) Kütz.		+	+	+	+					+	+	+	+		+						+		
Surirella tenera Greg.		+	+		+			+	+				+	+	+	+	+						
Surirella tericola Lange-Bert. & Alles		+	+		+				+	+	+	+	+	+	+	+	+		+	+	+	+	
Surirella visurgis Hust.	+																				+		
Surirella cf. *terricola* Lange-Bert.	+											+											
Surirella sp. 1		+			+									+	+								
Surirella sp. 2	+																					+	
Tabellaria fenestrata (Lyngbe) Kütz.		+				+	+																V
Tabellaria flocculosa (Roth) Kütz.		+				+	+			+		+									+	+	
Thalassiosira duostra Pienaar					+																+		
Thalassiosira pseudonana Halse & Heimal	+	+			+							+	+	+	+	+					+		
Thalassiosira weissflogii (Roth) Kütz.		+	+		+						+			+	+		+						
Thalassiosira cf. *weissflogii* (Roth) Kütz.																							
Thalassiosira cf. *aestiwalis* Grun.		+													+								
Thalassiosira cf. *duostra* Pienaar		+																					
Thalassiosira sp.																					+		
The sum of diatom taxa	147	400	197	121	253	100	71	61	126	192	227	279	255	261	286	213	226	53	147	155	365	271	–

Explanations: Column 0: E = endangered; V = vulnerable; R = rare; I = indeterminate (according to Siemińska et al. 2006). Column 1 – soil. Columns 2–22 are a names of rivers and streams: 2 – Wisłok, 3 – Zalew Rzeszowski, 4 – Łubienka, 5 – San (near Jarosław), 6 – Gołębiówka, 7 – Szuwarka, 8 – Świerkowiec, 9 – Lubcza, 10 – Różanka, 11 – Morwawa, 12 – Baryczka, 13 – Pielnica, 14 – Stobnica, 15 – Mleczka, 16 – Strug, 17 – Ryjak, 18 – Wisłoka (near Dębica), 19 – Wielopolka, 20 – Trzcianka, 21 – Żołynianka, 22 – Matysówka.

of Algae (11.5% micro-flora). In this group of 80 taxa there were: 40 of the most rare taxa (R), of which 23 were vulnerable (V), 11 endangered (E) and 6 indeterminate (I) – Table 1, Plate 1: 31–40.

Studies conducted on cultivated soils in Boguchwała, Dąbrowa and Krasne in 2009–2011 indicated the presence of 147 diatom taxa, of which the most numerous were – *Hantzschia amphioxys*, *Mayamaea atomus*, *Nitzschia palea*, *N. pusilla*, *Pinnularia obscura* and *Stauroneis thermicola*. Along with the noted diatoms there were 6 taxa from the Red List, including: *Stauroneis thermicola,* which occurs numerously and is a dominant species, frequently reaching above 20% of the total number.

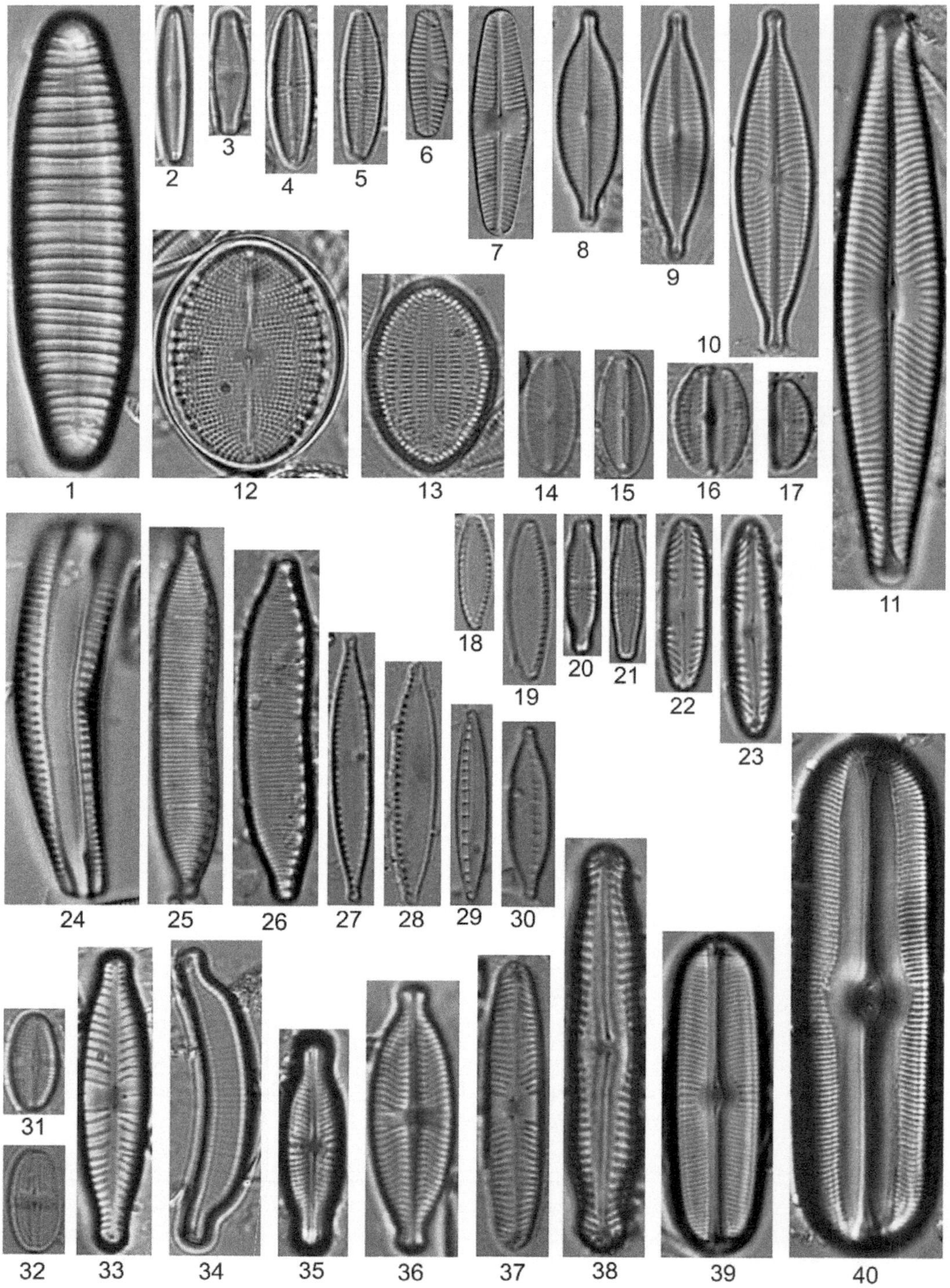

Plate 1. Selected aquatic and soil diatoms taxa: 1 – *Diatoma vulgaris* Bory, 2-3 – *Achnanthidium minutissimum* (Kütz.) Czarn. var. *minutissimum*, 4-5 – *Achnanthidium pyrenaicum* (Hust.) Kobayasi, 6-7 – *Planothidium lanceolatum* (Bréb.) Round & Bukht., 8-9 – *Navicula gregaria* Donkin, 10 – *N. capitatoradiata* Germain, 11 – *N. lanceolata* (Ag.) Kütz., 12 – *Cocconeis pediculus* Ehrenb., 13 – *C. placentula* var. *lineata* (Ehrenb.) Van Heurck, 14-15 – *Mayamaea atomus* (Kütz.) Lange-Bert. var. *atomus*, 16-17 – *Amphora pediculus* (Kütz.) Grun., 18-19 – *Nitzschia pusilla* Grun., 20-21 – *Stauroneis thermicola* (Petersen) Lund, 22-23 – *Pinnularia obscura* Krasske, 24 – *Rhoicosphenia abbreviata* (Ag.) Lange-Bert., 25-26 – *Hantzschia amphioxys* (Ehrenb.) Grun., 27-28 – *Nitzschia palea* (Kütz.) W. Smith, 29-30 – *N. dissipata* (Kütz.) Grun. ssp. *dissipata*, 31 – *Psammothidium lauerburgianum* (Hust.) Bukht. & Round, 32 – *P. subatomoides* (Hust.) Bukht. & Round, 33 – *Gomphonema sarcophagus* Greg., 34 – *Eunotia septentrionalis* Østrup, 35 – *Parliberllus protractoides* (Hust.) Witkowski, Lange-Bert. & Metzeltin, 36 – *Geissleria decussis* (Østrup) Lange-Bert. & Metzeltin, 37 – *Pinnularia schoenfelderi* Krammer, 38 – *P. nodosa* (Ehrenb.) W. Smith, 39 – *Sellaphora bacillum* (Ehrenb.) D.G. Mann, 40 – *S. americana* (Ehrenb.) D.G. Mann (magnification1000×)

DISCUSSION

Running waters and cultivated soils on the territory of the Podkarpacie Province are characterized by a huge richness of diatom species (Table 1). The amounts of taxa found (738 taxa in waters and 147 taxa in soils) during 5 years of studies are prooved.

Most of the studies were conducted on inflows of the Wisłok River [Mucha et al. 2009, Tambor, Noga 2011, Pajączek et al. 2012] and on the Wisłok itself [Noga 2012], in which there were 400 taxa diatom identified. In 1970–1995 Rakowska [2001] recorded a similar number of diatoms in the rivers of central Poland (the highest number was identified in the Rawka River – 444 taxa). During studies conducted in 1993–2007 in the valley of the Kobylanka stream, on the Krakowsko–Częstochowska Upland, the existence of 307 diatom taxa was noted [Wojtal 2009]. In both cases these were studies over periods of many years and they involved a huge amount of studied samples. Hence the species richness of diatoms in rivers and streams of the Podkarpacie Province presented is very impressive.

The studied rivers and streams of the territory of the Podkarpacie Province were characterized by a similar domination structure of diatoms, especially in the middle and lower courses, where there are anthropogenical transformations of the environment. On the majority of study sites the following diatoms dominated: *Navicula lanceolata*, *N. gregaria*, *N. capitatoradiata*, *Nitzschia palea*, *N. dissipata* ssp. *dissipata*, *Achnanthidium minutissimum* var. *minutissimum*, *A. pyrenaicum*, *Amphora pediculus*, *Diatoma vulgaris*, *Rhoicosphenia abbreviata*, *Planothidium lanceolatum*, *Cocconeis pediculus*, and *C. placentula* var. *lineata*. These diatoms prefer alkaline waters (pH>7) and eutrophic, that is fertile, waters, of the second and third classes of quality [Van Dam et al. 1994, Hofmann et al. 2011].

The most often recorded dominants on the majority of study sites were: *Navicula lanceolata* and *N. gregaria*, the latter being a halophytic species. *N. lanceolata* growth the best in waters, which are more mesotrophic and eutrophic [Krammer, Lange-Bertalot 1986–1991, Lange-Bertalot, Steindorf 1996, Hofmann et al. 2011].

The upper sections of the majority of studied rivers and streams were characterized by the dominance of: *Achnanthidium minutissimum* var. *minutissimum* and *A. pyrenaicum*. *Achnanthidium minutissimum* var. *minutissimum* has a wide ecological amplitude and develops from oligo- to eutrophic conditions, whereas *A. pyrenaicum* is determined as an alkaliphilic species [Van Dam et al. 1994, Lange-Bertalot, Steindorf 1996]. According to Hofmann [1994] it is known as an indifferent species, with respect to pH values. Both taxa have a wide range of tolerance in relation to the trophy [Hofmann 1994], whereas, according to Van Dam et al. [1994], *A. pyrenaicum* is determined as a mesotrophic species. These taxa were also registered numerously in the streams of the Tatra Mountains [Kawecka 2012].

According to different authors of studies of benthic diatoms in slightly or highly polluted rivers there are similar dominant species, from which the most frequent are: *Planothidium lanceolatum*, *Cocconeis placentula*, *Navicula lanceolata*, *N. gregaria*, *N. tripunctata*, *Nitzschia palea*, etc. [Wasylik 1985, Kawecka 1986, Ligowski 1988, Rakowska 2001].

In the diatom flora of the Podkarpacie Province, 11.5% taxa are endangered and rare, located on the Polish Red List of Algae [Siemińska et al. 2006]. The places of occurrence of these diatoms are, among others, the upper sections of rivers and streams, where the endangered species of *Pinnularia*: *P. rupestris*, *P. schoenfelderii*, *P. subrupestris*, *P. subgibba* and *P. undula* occur. These are mainly species, which prefer waters from oligo- to mesosaprobous, with low and moderate values of conductivity [Krammer 2000, Hofmann et al. 2011]. Some species occur in the majority of rivers and streams in the Podkarpacie Province, others only in the form of individual specimens [Noga 2012, Pajączek et al. 2012].

The area of the Podkarpacie Province is still weakly known with respect to algological studies. The whole valley of the San River has hardly been studied, with the exception of the section of the river around Jarosław city and two inflows in the Dynowski Foothills – the Łubienka and Baryczka streams [Noga, Siry 2010]. At the moment studies have begun on diatoms in the valley of the upper San, which promise to be unusually interesting [Noga, unpublished data]. Similar studies on soil diatoms have been conducted for two years, so far on a small amount of study sites. The introductory results presented indicate a huge species richness of diatoms, which in the future, as a result of other studies conducted, will surely increase in number.

REFERENCES

1. Hofmann G. 1994. Aufwuchs-Diatomeen in Seen und ihre Eignung als Indikatoren der Trophie. Bibliotheca Diatomologica 30, pp. 241. J. Cramer, Berlin – Stuttgart.
2. Hofmann G., Werum M. & Lange-Bertalot H. 2011. Diatomeen im Süβwasser – Benthos vom Mitteleuropa. Bestimmungsflora Kieselalgen für die ökologische Praxis. Über 700 der häugfisten Arten und ihre Ökologie. [In:] H. Lange-Bertalot (eds), pp. 908. A.R.G. Gantner Verlag K.G.
3. Kawecka B. 2012. Diatom diversity in streams of the Tatra National Park (Poland) as indicator of environmental conditions, pp. 213. Szafer Institute of Botany, Polish Academy of Sciences, Kraków.
4. Kawecka B. 1986. The effect of light deficiency on communities of sessile algae in the Olczycki stream (Tatra Mts, Poland). Acta Hydrobiol. 28: 379–386.
5. Kawecka B., Sanecki J. 2003. *Didymosphenia geminata* in running waters of S Poland – Symptom of change in water quality. Hydrobiol. 495: 193–201.
6. Kondracki J. 2001. Geografia regionalna Polski, pp. 463. PWN, Warszawa.
7. Krammer K. 2000. The genus *Pinnularia.* [In:] Diatoms of Europe 1. H. Lange-Bertalot (eds), pp. 1–703. A.R.G. Ganter Verlag K.G., Vaduz.
8. Krammer K., Lange-Bertalot H. 1986–1991. Bacillariophyceae. [In:] Süsswasserflora von Mitteleuropa 2(1–4). H. Ettl, J. Gerloff & H. Heyning (eds), G. Fischer Verlag, Stuttgart – New York – Jena.
9. Kwandrans J. 1989. Ekologiczna charakterystyka zbiorowisk glonów osiadłych w Białej i Czarnej Wisełce, źródłowych potokach rzeki Wisły (Beskid Śląski, Polska Południowa). Acta Hydrobiol. 31(1–2): 43–74.
10. Lange-Bertalot H. 2001. Diatoms of the European inland waters and comparable habitats. *Navicula* sensu stricto, 10 Genera Separated from *Navicula* sensu lato, *Frustulia.* [In:] Diatoms of Europe 2. H. Lange-Bertalot (ed.), pp. 1–526. A.R.G. Gartner Verlag. K.G., Vaduz.
11. Lange-Bertalot H., Steindorf A. 1996. Rote liste der limnischen Kieselalgen (Bacillariophyceae) Deutschlands. Schrittenreihe für Vegetationskunde, 28: 633–677.
12. Ligowski R. 1988. Sessile algae of the Widawka river. Acta Univ. Lodz., Folia Limnol. 3: 9–49.
13. Mucha Ł., Musiałek M., Noga T., Pajączek A., Paśko M., Pelczar J. 2009. Różnorodność glonów w wodach Wielopolki, Różanki, Mleczki i Morwawy. *Homo Naturalis* przyrodnicze, społeczne i ekonomiczne aspekty rozwoju zrównoważonego, pp. 203–210. WORD–PRESS, Katowice.
14. Noga T. 2012. Diversity of diatom communities in the Wisłok River (SE Poland). [In:] K. Wołowski, I. Kaczmarska, J. M. Ehrman & A.Z. Wojtal (eds), Phycological Reports: Current advances in algal taxonomy and its applications: phylogenetic, ecological and applied perspective, pp. 109–128. Institute of Botany Polish Academy of Sciences, Krakow.
15. Noga T., Siry. 2010. Różnorodność flory okrzemek w potoku Łubienka (Pogórze Dynowskie, Polska SE). Zeszyty naukowe PTIE i PTG, 12: 75–86.
16. Pajączek A., Musiałek M., Pelczar J., Noga T. 2012. Diversity of diatoms in the Mleczka River, Morwawa River and Różanka Stream (tributaries of the Wisłok River, SE Poland), with particular reference to threatened species. [In:] K. Wołowski, I. Kaczmarska, J.M. Ehrman & A.Z. Wojtal (eds), Phycological Reports: Current advances in algal taxonomy and its applications: phylogenetic, ecological and applied perspective, pp. 129–152. Institute of Botany Polish Academy of Sciences, Krakow.
17. Rakowska B. 2001. Studium różnorodności okrzemek ekosystemów wodnych Polski niżowej, pp. 77. Wydawnictwo Uniwersytetu Łódzkiego, Łódź.
18. Wołowski K. 2003. Różnorodność gatunkowa – glony. [In:] R. Andrzejewski & A. Weigle (eds), Różnorodność biologiczna Polski, pp. 37–48. Narodowa Fundacja Ochrony Środowiska, Warszawa.
19. Siemińska J. 1977. Stopień poznania flory glonów Polski i jej zmiany pod wpływem działalności człowieka. Wszechświat, 9: 222–224.
20. Siemińska J. 1990. Polska Bibliografia Fykologiczna. Bibliografie Botaniczne, 3, pp. 464. Polska Akademia Nauk, Instytut Botaniki im. Szafera, Kraków – Wrocław.
21. Siemińska J., Pająk J. 1992. Polska Bibliografia Fykologiczna za lata 1981–1990. Bibliografie Botaniczne, 6, pp. 181. Polska Akademia Nauk, Instytut Botaniki im. Szafera, Kraków.
22. Siemińska J., Bąk M., Dziedzic J., Gąbka M., Gregorowicz P., Mrozińska T., Pełechaty M., Owsiany P. M., Pliński M., Witkowski A. 2006. Red list of the algae in Poland. [In:] Z. Mirek, K. Zarzycki, W. Wojewoda & Z. Szeląg (eds), Red list of plants and fungi in Poland, pp. 37–52. Szafer Institute of Botany, Polish Academy of Sciences, Kraków.
23. Stanek-Tarkowska J., Noga T. 2011. The influence of two tillage systems on diatom communities in wheat plantations. [In:] J. Matuła, D. Richter & J. Urbaniak (red.), Book of abstract XXX International Conference of the Polish Phycological Society, pp. 187. Wrocław – Pawłowice, 19–21.05.2011.
24. Stanek-Tarkowska J., Noga T. 2012. Zbiorowiska okrzemek rozwijające się na glebach pyłowych pod uprawą kukurydzy w rejonie Podkarpacia. Fragm. Florist. Geobot. Polon., 19(2): 525–536.

25. Tambor A., Noga T. 2011. Różnorodność flory okrzemek w rzece Lubcza i potoku Lubenia (Podgórze Rzeszowskie, Polska SE). Rocznik Przemyski, 47(3): 105–118.

26. Van Dam H., Martens A., Sinkeldam J. 1994. A coded checklist and ecological indicator values of freshwater diatoms from the Netherlands. Netherlands J. Aquatic Ecol., 28: 117–133.

27. Wasylik K. 1985. Diatom communities in pure and polluted waters in the Biała Przemsza river basin (Southern Poland). Acta Hydrobiol., 25/26: 287–315.

28. Wojtal A. 2009. Diatom flora of the Kobylanka stream near Kraków (Wyżyna Krakowsko-Częstochowska Upland, S Poland). Polish Bot. Journal, 54(2): 129–330.

ASSESSMENT OF IMPACT OF COHERENT LIGHT ON RESISTANCE OF PLANTS GROWING IN UNFAVOURABLE ENVIRONMENTAL CONDITIONS

Małgorzata Śliwka[1]

[1] Department of Environmental Engineering and Mineral Processing, Faculty of Mining and Geoengineering, AGH University of Science and Technology in Krakow, 30 Al. Mickiewicza Ave., 30-059 Krakow, Poland, e-mail: sliwka@agh.edu.pl

ABSTRACT

The results of experiments on the effect of the coherent light emitted by lasers on plant material show that properly selected laser stimulation parameters, such as: wavelength, power, time and type of exposure, allow to obtain a greater growth of plant biomass, changes in the content of elements in the biomass and increasing plant resistance to unfavorable environmental conditions. The aim of this study was to determine the effect of laser stimulation on selected plant species (*Iris pseudoacorus* L., *Lemna minor* L.) to increase their resistance to low temperatures and the ability to adapt to an environment polluted by mining activities (*Phelum pratense* L.). Plants from experimental groups (*Iris pseudoacorus* L., *Phelum pratense* L., *Lemna minor* L.) were stimulated with coherent light with specific characteristics. To irradiate plants from experimental groups different algorithms of stimulation parameters, differentiating the method and time of exposure were used. Plants group without the stimulation, were the reference group. The article discusses the results of preliminary experiments carried out on a laboratory scale and pot experiments.

Keywords: laser stimulation, biostimulation, plants resistance, stress factors, ecological valence.

INTRODUCTION

Visible light is responsible for the stimulation of important physiological and biochemical processes and biophysical phenomena such as: photosynthesis, phototropism, photomorhogenesis, which are based on a series of photo-biological reactions induced by quantas energy of electromagnetic radiation in the visible range 380–780 nm. The results of scientific research [Inyushin et al. 1981, Suppan 1997, Kopcewicz, Lewak 2007] on the laser stimulation of biological material confirm a significant effect of radiation on bioenergetic processes of cells. Depending on the irradiation parameters: the amount of energy supplied, the method of exposure and the type of biological material, you can get different effects of exposure: a phototermel effect, a photochemical effect, photoionization or effect of biostimulation. The effect of stimulation is the effect of laser irradiation, accompanied by a local increase of temperature (0.5-1 °C), and the changes on the cellular level does not respond to the stress. Laser radiation is absorbed by the biologically active compounds or cellular organelles, and can lead to changes in the metabolism of the cells, rather than to its destruction [Inyushin 1977, Bryszewska 1997, Dobrowolski 1999, Drozd, Szajsner 2001, Drozd et al. 2003, Popp 2006, Szajsner et al. 2008]. Scientific research [Inyushin 1977] also showed that the effect of stimulation is closely related to the properties of coherent light, such as coherence and polarization, which allow better penetration of the radiation into the tissue and cause photochemical reactions through resonant absorption and provide energy charge corresponding to a specific wavelength. Incoherent light does not show comparable to laser effects on bioenergetic processes [Karlander 1968, Friedman, Lubart 1993, Cenian et al. 2005, Liedtke, Popp 2006].

The practical use of the effect of laser stimulation of biological material were used, among other things in medicine [Inyushin et al. 1981, Dobrowolski 1993, Sieroń, Cieślar 1994, Dobrowolski, Różanowski 1995, Fiedor 1995]. It was found the impact of coherent light on bone marrow cells and blood and the activity of some enzymes. These studies have been used in the elimination of cancer cells, stimulation of the bone marrow and tissue regeneration.

There are also known studies [Gosh, Bupp 1992, Wilczek et al. 2006, Aladjadjiyan 2007, Wilczek, Ćwintal 2009, Ćwintal 2010] relating to stimulation crops before their sowing, in order to improve their germination, growth and yield, as well as alternative preparations of seed dressing against fungal diseases. An interesting direction of the use of laser stimulation is to use this technology in the practice of environmental engineering, primarily to increase the efficiency and effectiveness of biological reclamation of bio- and phytoremediation, enhance biological wastewater treatment, accelerate the growth of energy crops and increase the resistance of plants growing on soils contaminated with heavy metals and saline soils [Śliwka 2007, Śliwka, Jakubiak 2009, Śliwka, Jakubiak 2010, Dobrowolski et al. 2012, Jakubiak 2012]. Research on the proecological use of laser stimulation has already proposed in the 70's of last century by Dobrowolski.

The objective of these experiments was to determine the effect of the laser simulation of selected plant species for their ability to adapt to unfavorable environmental conditions, such as climate and pollution.

METHODS

Selected plant species were chosen as an experimental material: duckweed (*Lemna minor* L.) and yellow iris (*Iris pseudoacorus* L.) used in the wastewater treatment plant and species of grass *Phleum pratense* L. recommended for use in the reclamation treatments: fallow, contaminated land, landfills of waste and ash [Góral 2001].

In the following experiments was used a coherent light source:

- a laser diode emitting light with a wavelength corresponding to red color (660 nm) with a power of 20 mW,
- a laser diode emitting light with a wavelength corresponding to blue color (473 nm) with a power of 20 mW,
- an argon laser emitting light with a wavelength corresponding to celadon color (514 nm) with a power of 21 mW,
- impulse medical laser emitting light with a wavelength corresponding to red color (670 nm) with a power of 20 mW.

The article discusses two preliminary experiments.

Experiment I (2005–2007) was conducted in the field conditions. In this experiment were used duckweed (*Lemna minor* L.) and the yellow iris (*Iris pseudoacorus* L.) as experimental material.

In the case of duckweed were irradiated whole plants. In each experimental group were 90 plants. After this, plants were placed in three ponds with the same volume (160 dm^3) filled with wastewater from a municipal sewage treatment Kraków Płaszów, diluted (1:1) with a marked chemical composition [Śliwka 2007]. Irradiated rhizomes of yellow iris (about the same weight: 400 g) were also placed in each of the ponds.

The rhizomes of yellow iris and duckweed (whole plants) were irradiated using an argon laser (514 nm) and laser diode (660 nm). For irradiation was used parameters of stimulation, which have been selected under laboratory conditions (wavelength, exposure time and method of exposure), in terms of the expected stimulation effect: increased biomass growth, the resistance of plants [Śliwka 2007].

For duckweed were used the following parameters: wavelength 514 nm exposure time 3 times 3 seconds and a wavelength of 660 nm at the time of exposure 3 times 3 seconds. For yellow iris were selected the following parameters: wavelength 514 nm exposure time 3 times 30 seconds and a wavelength of 660 nm at the time of exposure 3 times 30 seconds. In order to compare the results prepared unexposed control groups of plants. Plants were irradiated with perpendicular light beam from a distance of 20 cm.

Experiment II was conducted in laboratory conditions.

Timothy grass (*Phleum pratense* L.) was sown in Petri dishes on the soil collected from the heaps area of KWK Jankowice (0–20 cm depth), contaminated with Cd (38 $mg \cdot kg^{-1}$

d.m.) [Rozporządzenie... 2002] according to the scheme: 30 seeds × 3 repetitions for each of the irradiation algorithm. In the case of timothy grass as an optimal (to increase the capacity of germination, germination rate) on the basis of preliminary experiments were chosen following parameters: wavelength of 473 nm for the second time 3×1 and the wavelength of 670 nm for 3×9 seconds of exposure time. Additionally, three control groups of unexposed plant (30 seeds) as control groups were prepared. Observations continued for 14 days. Plants were irradiated with perpendicular light beam from a distance of 20 cm.

Following parameters were measured: biomass growth and plant condition at the end of each of the three growing seasons (2005–2007) for duckweed (*Lemna minor*) and for yellow iris (*Iris pseudoacorus*) and ability to germinate, germination rate by Pieper method, growth of roots and shoots of *Phleum pratense*.

The programs: Statistica 7.1 and MS Excel were used for statistical evaluation of the results. The Student t-test and ANOVA ranks Kruskal-Wallis were used to assess the statistical significance of the results and also used in the analysis of the test of normality the Shapiro-Wilk.

Discussed below relate to the results of preliminary observations and will be prerequisite for the continuation of these studies.

RESULTS AND DISCUSSION

Results of experiments and preliminary observations confirm the significant impact of the coherent light on biological material. It was found, among other things, a positive and statistically significant effect of laser stimulation on the growth of plant biomass and increasing ecological valence unfavorable environmental factors.

Pot experiments on duckweed (*Lemna minor*) showed a significant effect of laser stimulation on the growth of biomass. As the most optimal algorithm of irradiation (wavelength: 660 nm and exposure time: 3 times 3 seconds), at the end of the first growing season 300% greater increase of biomass in compared to the control group of unirradiated plants was obtained a [Śliwka 2010, Śliwka 2011, Śliwka 2012]. Similar effects were noted in case of *Salix viminalis* [Jakubiak, Gdowska 2013].

It is worth noting that the effect of stimulation was maintained during the following observations in the next growing seasons, which indicates its durability. These conclusions apply only to plant reproducing vegetatively, and the effect of stimulation can carry on the progeny plants. In the case of plants which reproduce generatively, a similar relationship was not found.

Duckweed (*Lamna minor*) exposed to laser diode 660 nm were characterized by a higher resistance to temperature drop, also exhibited a higher viability (in winter) in subsequent years of observation (preliminary observations) (Figure 1).

Significant differences in the condition of individual plants of the experimental groups were found (Figure 2).

Experiments concerning the influence of coherent light on the growth of yellow iris (*Iris pseudoacorus*) showed significant differences in the increase of the biomass of leaves for groups irradiated with argon laser (514 nm) with an exposure time 3 times 30 seconds. These groups of plants were characterized as more abundant and flowered earlier (by observation). For the group irradiated with diode laser (660 mn) re-growth of leaves after collecting the biomass at the end of the vegetation season was found (end of October).

The results of laboratory experiments on the influence of coherent light on seed germination

Figure 1. Comparison of plants condition from each experimental groups of duckweed (*Lemna minor*) after ponds frozen a) control group, b) plants irradiated with laser diode (660 nm), irradiation time: 3 times 3 seconds, c) plants irradiated with argon laser (514 nm), irradiation time: 3 times 3 seconds

Figure 2. Comparison of plants condition from each experimental groups of duckweed (*Lemna minor*) a) control group, b) plants irradiated with laser diode (660 nm), irradiation time: 3 times 3 seconds, c) plants irradiated with argon laser (514 nm), irradiation time: 3 times 3 seconds

of *Phelum pratense* and plants growth in soil contaminated with Cd. Cadmium significantly impairs metabolic processes in plants (photosynthesis, transpiration, metabolism of nitrogen compounds, changes in the permeability of cell membranes and DNA structures).

Differences were found in the germination of plant and the condition of plants from each experimental group. The average length of coleoptile, after 14 days of growth, was 3.4 cm for the plants in the control group, 3.2 cm for the plant irradiated with laser diode (473 nm) and 3.7 cm for plant irradiated with impulse laser (670 nm) (Figure 3).

On the other hand average root length, measured after 14 days of growth, was 2.1 cm for the plants in the control group, 1.6 cm for the plant irradiated with laser diode (473 nm) and 1.8 cm for plant irradiated with impulse laser (670 nm) (Figure 4).

Assessment of growth of coleoptile and roots in the experimental groups of *Phelum pratense* requires repeating with regard to extending the irradiation time to increase the differences in growth of experimental plants.

The germination capacity, calculated after 10 days of duration of the experiment, indicated a beneficial effect the red laser diode (670 nm) for germination capacity (Figure 5), as well as shorten the time of germination (calculated by Pieper) as a result of irradiation of the material with blue laser diode (473 nm) (Figure 6).

The results of preliminary observations showed the impact of coherent light on the germination of plants: increase of germination and shortening of time of germination, which may be crucial for efficiency of soil reclamation. The described experiment requires a continuation, which will allow to confirm the assumed thesis.

Positive effects of laser light stimulation on the morphological and physiological features at the molecular level of other species (kidney bean, cabbage, grass pea, soybean, spring durum wheat, sugar beet) were also observed [Kacharava et al. 2009, Khalifa, El Ghandoor 2011, Prośba-

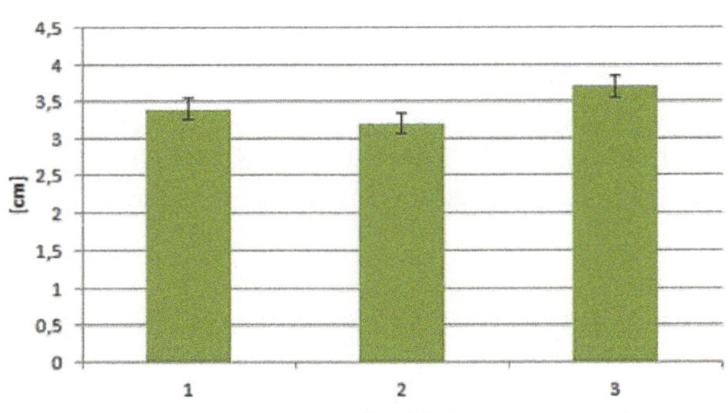

Figure 3. Comparison of growth of coleoptile [cm] *Phelum pratense* in experimental groups: 1 – control group, 2 – plants irradiated with laser diode (473 nm), 3 – plants irradiated with impuls laser (670 nm) (on the graph was marked bars bugs with the standard error)

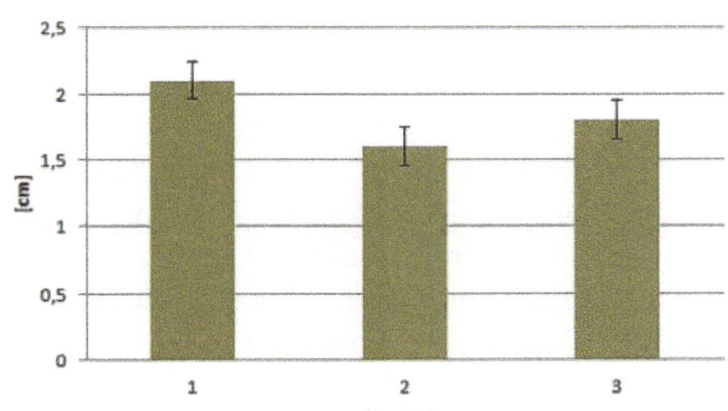

Figure 4. Comparison of growth of roots [cm] *Phelum pratense* in experimental groups: 1 – control group, 2 – plants irradiated with laser diode (473 nm), 3 – plants irradiated with impuls laser (670 nm) (on the graph was marked bars bugs with the standard error)

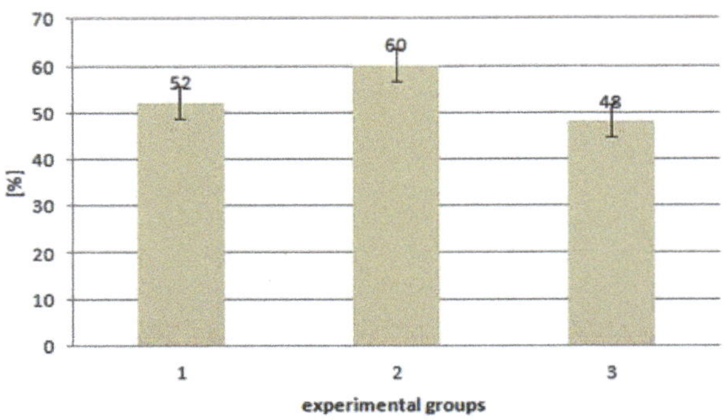

Figure 5. Comparison of ability of germination of *Phelum pratense* in experimental groups: 1 – control group, 2 – plants irradiated with laser diode (473 nm), 3 – plants irradiated with impuls laser (670 nm) (on the graph was marked bars bugs with the standard error)

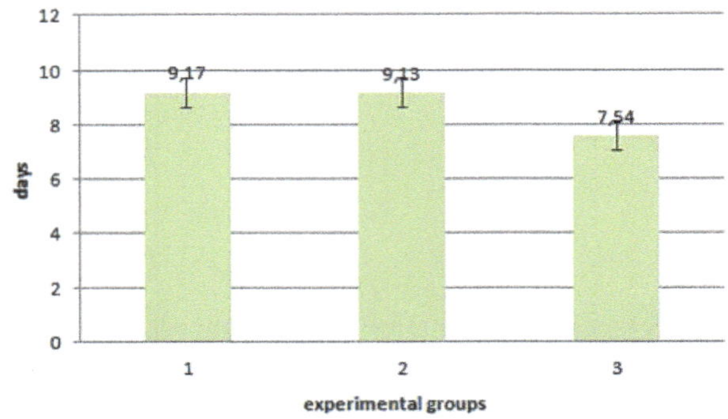

Figure 6. Comparison of germination rate of *Phelum pratense* in experimental groups: 1 – control group, 2 – plants irradiated with laser diode (473 nm), 3 – plants irradiated with impuls laser (670 nm) (on the graph was marked bars bugs with the standard error)

Białczyk et al. 2013, Ritambhara, Girjesh 2013, Zare et al. 2014]. Irradiation of sugar beet seeds resulted in significantly bigger length of the hypocotyl and root of seedlings, in a higher dry matter, chlorophylls and carotenoids content in leaves and roots during vegetation than in non-stimulated seeds, as well as favourably modified the chlorophyll *a* to *b* ratio. It also significantly increased yield and the saccharose content, and reduced sodium and potassium content [Prośba-Białczyk et al. 2013]. Changes of the antioxidants content in leaves of kidney bean, cabbage and beet was observed as well [Kacharava et al. 2009]. Biostimulation may be used for enhancing productivity of the *Lathyrus sativus* L. plants [Ritambhara, Girjesh 2013]. Pre-sowing laser biostimulation of seeds caused increase crops of maize by 10–20%, spring wheat by 20–30%, spring barley by 20-25% and sugar beets by 10–35% [Koper 1994]. The additional effects are stronger plant seedlings, higher resistance to cold, shortening the vegetation period of corn by about 10 days, earlier plant maturation, the better plants quality and possible diminishing of nitrogen fertilizers application without essential crops decrease, as well as changes in protein banding patterns and large subunit of Rubis CO [Koper 1994, Khalifa, El Ghandoor 2011]. Laser irradiation of spring durum wheat (*Triticum turgidum* L.) under salinity stress induced considerable biological changes in plant metabolism e.g. significantly increases free proline accumulation in leaves and salt tolerance [Zare et al. 2014], and may induce an increase of plants resistance to environmental pollutions and unfavorable environmental factors, possibility of higher efficient of sewage treatment and soil reclamation processes and increase of bioremediation abilities as well [Jakubiak, Gdowska 2013]. The results of the discussed experiments allow to conclude that it is possible to increase plant resistance to unfavorable environmental conditions e.g. a low temperature. Irradiation of plants with a coherent light source may significantly extended the growing season of plants (like *Lemna minor, Iris pseudoacorus*) used in hydrobotanical treatment plants [Śliwka 2005, Śliwka, Jakubiak 2007].

The observed effect of laser stimulation on the ability of germination and germinate rate of *Phelum pratense*, may contribute to an increase of biomass growth, and thus, the efficiency of biological remediation of soils contaminated with heavy metals, by increasing the resistance of plants to pollutants that cause phototoxic effects.

Researches carried out by Dobrowolski et al. [1995] and Zielinska-Loek [2003], related to the assessment of the impact of coherent light on different species of willows (*Salix viminalis, Salix acutifolia* and *Salix varieties* Rapp), showed a significant effect of stimulation on accelerate the growth and increase of plants resistance cultivated on lands belonging to ZGH – Mining and Metallurgical Factory "Bolesław" in Bukowno, Cement Plant "Chełm" and in the areas of express roads.

CONCLUSION

The implementation of laser stimulation of plants used in phytoremediation technologies (reclamation, wastewater treatment plants), should significantly contribute to increasing the efficiency of biological reclamation and revitalization of degraded areas.

Acknowledgments

The research was prepared within the statutory research for the Faculty of Mining and Geoengineering AGH-UST, Department of Environmental Engineering and Mineral Processing No. 11.11.100.482 (2013).

REFERENCES

1. Aladjadjiyan A. 2007. The use of physical methods for plant growing stimulation in Bulgaria. J. Cent. Eur. Agric., 8(3): 369–380.
2. Bryszewska M., Leyko W. 1997. Biofizyka dla biologów. Wyd. PWN Warszawa.
3. Ćwintal M. 2010. Wpływ zapraw nasiennych i stymulacji laserowej nasion na wschody oraz strukturę łanu i plonowanie koniczyny czerwonej w roku siewu. Ann. UMCS, Sec. EE, LXV(4): 84–93.
4. Dobrowolski J.W. 2002. Zastosowanie biostymulacji laserowej w ekoinżynierii i ekorozwoju. Inż. Ekol., 6: 194–196.
5. Dobrowolski J.W., Różanowski B. 1998. The influence of laser light on accumulation of selected macro, trace and ultra trace elements by some plants. Microelements and trace elements. [In:] Mengen- und Spurenelemente, Arbeitstagung 18, Verlag Harald Schubert, Leipzig, 147–156.
6. Dobrowolski J.W., Różanowski B., Zielińska A., Budzyński M., Walczak P. 1995. Próby zastosowania biostymulacji laserowej w celu przyspieszenia wzrostu niektórych gatunków roślin i rekultywacji terenów silnie skażonych. II Kraj. Konf. Nauk. Las-Drewno-Ekologia, 20–22 czerwca, PAN, Poznań, 25–29.
7. Dobrowolski J.W., Śliwka M., Mazur R. 2012. Laser biotechnology for more efficient bioremediation, protection of aquatic ecosystems and reclamation of contaminated areas. J. Chem. Technol. Biot. B, 87(9): 1354–1359.
8. Drozd D., Szajsner H. 2001. Promienie lasera jako czynnik fizyczny stymulujący wartość użytkową nasion. Acta Agrophys., 58: 71–79.
9. Drozd D., Szajsner H., Bielecki K. 2003. Wpływ światła lasera na aktywność alfa-amylazy w ziarniakach różnych genotypów pszenżyta. Biul. IHAR, 226/227(1): 177–180.
10. Friedman H., Lubart R. 1993. Nonlinear photostimulation: The mechanism of visible and infrared laser-induced stimulation and reduction of neural excitability and growth. Laser Therapy, 5(1): 30–42.
11. Góral S. 2001. Roślinność zielna w ochronie i rekultywacji gruntów. Inż. Ekol., 3: 161–178.
12. Gosh S., Bupp S. 1992. Stimulation of biological uptake of heavy metals. Water Sci. Technol., 26: 227–236.
13. Inyushin W.M. 1977. Laser technology in agricultural services. An attempt of using a new optic-quantum generator. Nowe Rolnictwo, 21/22: 21–26 (in Polish).
14. Inyushin W.M., Iliasov G.U., Fedorowa N.N. 1981. Laser light and crop. Kainar Publication, Ałma-Ata. 1981, 210: 165–168 (in Russian).
15. Jakubiak M. 2012. Degraded lands biological reclamation using modern methods of stimulating the growth of selected varieties of willows (*Salix* sp.). Pol. J. Environ. Stud., 21(5A): 129–133.
16. Jakubiak M., Gdowska K. 2013. Innovative virommental technology applications of laser light stimulation. Energetika *i* awtomatika, 3: 14–21.
17. Kabata-Pendias A., Pendias H. 1999. Biogeochemia pierwiastków śladowych. Wyd. PWN, Warszawa.
18. Kacharava N., Chanishvili S., Badridze G., Chkhubianishvili E., Janukashvili N. 2009. Effect of seed irradiation on the content of antioxidants in leaves of kidney bean, cabbage and beet cultivars. Australian J. Crop Sci., 3(3): 137–145.
19. Karlander E.P., Krauss R.W. 1968. The laser as a light source for the photosynthesis and growth of *Chlorella vannielii*. Biochim. Biophys. Acta, 153(1): 312–314.
20. Khalifa N.S., El Ghandoor H. 2011. Investigate the effect of Nd-Yag Laser Beam on soybean (*Glycine max*) leaves at the protein level. Intern. J. Biology, 3(2): 135–144.
21. Kopcewicz J., Lewak S. 2007. Fizjologa roślin. PWN, Warszawa.
22. Koper R. 1994. Pre-sowing laser biostymulation of seeds of cultivated plants and its results in agrotechnics. Int. Agrophys., 8(4): 593–596.
23. Prośba-Białczyk U., Szajsner H., Grzyś E., Demczuk A., Sacała E., Bąk K. 2013. Effect of seed stimulation on germination and sugar beet yield. Int. Agrophys., 27: 195–201.
24. Ritambhara S., Girjesh K. 2013. Biostimulating effect of laser beam on the cytomorphological aspects of *Lathyrus sativus* L. Annals Plant Sci., 02(05): 141–148.
25. Rozporządzenie Ministra Środowiska z dnia 9 września 2002 r. w sprawie standardów jakości gleby oraz standardów jakości ziemi, które określa dopuszczalne wartości stężeń zanieczyszczeń w glebie lub ziemi. Dz.U. 2002, Nr 165, poz. 1359.
26. Śliwka M. 2005. Wykorzystanie biostymulacji laserowej roślin do zwiększenia przyrostu ich biomasy oraz zdolności bioremediacyjnych. [In:] Obieg pierwiastków w przyrodzie, Monografia, 3: 637–640.
27. Śliwka M. 2007. Zastosowanie stymulacji laserowej wybranych gatunków hydrofitów do zwiększenia ich zdolności bioremediacyjnych. Roz-

prawa doktorska, Wydz. GGiIŚ, AGH Kraków, 242 p. (online: http://winntbg.bg.agh.edu.pl/rozprawy/9951/full9951.pdf)

28. Śliwka M., Jakubiak M. 2007. Wpływ stymulacji laserowej na zwiększenie przyrostu biomasy oraz zdolności bioremediacyjnych roślin wykorzystywanych w hydrofitowych oczyszczalniach ścieków. Ochrona Środowiska i Zasobów Naturalnych, 33: 103–107.

29. Śliwka M., Jakubiak M. 2009. The application of the innovative biotechnology in hydrobotanical wastewater treatment plants. Pol. J. Environ. Stud., 18(3A): 445–449.

30. Śliwka M., Jakubiak M. 2010. Application of laser biotechnology for more efficient phytoremediation of biogenic elements. Ecol. Chem. Eng. A, 17(2–3): 297–303.

31. Śliwka M., Jakubiak M. 2010. Application of laser stimulation of some hydrophytes species for more efficient biogenic elements phytoremediation. Proceedings of Ecopole, 4(1): 205–211.

32. Szajsner H., Drozd D., Błażewicz J. 2008. Wpływ stymulacji laserowej nasion na kształtowanie cech morfologicznych siewek i wielkość siły diastatycznej form żyta i pszenżyta. Biuletyn Instytutu Hodowli i Aklimatyzacji Roślin, 250: 177–187.

33. Wilczek M., Ćwintal M. 2009. Ocena możliwości poprawy wartości siewnej nasion koniczyny czerwonej poprzez zastosowanie przedsiewnej stymulacji laserowej. Acta Agrophys., 14(1): 221–231.

34. Wilczek M., Ćwintal M., Kornas-Czuczwar B., Koper R. 2006. Wpływ laserowej stymulacji nasion na plonowanie di- i tetraploidalnej koniczyny czerwonej w roku siewu. Acta Agrophys., 8(2): 527–536.

35. Zare N., Sadat Noori S.A., Kholgh Sima N.A.K., Mortazavian S.M.M. 2014. Effect of laser priming on accumulation of free proline in spring durum wheat (*Triticum turgidum* L.) under salinity stress. Int. Trans. J. Eng. Manag. Sci. Tech. 5(2): 119-130.

36. Zielińska-Loek A. 2002. Ocena możliwości wykorzystania fotostymulacji laserowej roślin do proekologicznego zagospodarowania rejonów dróg. Praca doktorska, Wydz. GGiIŚ, AGH Kraków, 173 p.

15

LEECH COMMUNITIES IN THE UPPER REACHES OF THE STEPNICA RIVER (NW POLAND)

Małgorzata Raczyńska[1]**, Juliusz C. Chojnacki**[1]**, Monika Hałupka**[1]

[1] Department of Marine Ecology and Environmental Protection, Faculty of Food Sciences and Fisheries, West Pomeranian Technological University in Szczecin, Kazimierza Królewicza 4, 71-550 Szczecin, Poland, e-mail: malgorzata.raczynska@zut.edu.pl; juliusz.chojnacki@zut.edu.pl

ABSTRACT

The paper describes taxonomic composition of leeches (Hirudinea) in the strongly anthropogenically altered upper reaches of the Stepnica River in Western Pomerania. The numbers of respective species of the Hirudinea were analyzed in the context of such environmental factors as: instantaneous flow, dissolved oxygen content, and calcium content. It was noted that most leech species inhabited the headstream, and factors deterrent to their occurrence included the presence of hydrogen sulfate and siltation.

Keywords: taxonomic composition of leeches, instantaneous flow, hydro-chemical indicators.

INTRODUCTION

According to hydrobiologists, the occurrence of leeches may be a good indicator of biodiversity and cleanliness of aquatic environment. However, they are endangered by rapidly progressing industrialization and excessive exploitation of natural resources. Continuous deterioration of the quality of inland waters results in degradation of leech habitats. It also affects other invertebrates which constitute nutritional basis for leeches, as well as molluscs and fish, which are hosts for parasitic species [Bielecki et al. 2011a]. Common leech species, being highly tolerant organisms, typically occur in large numbers in littoral zone of lakes, ponds and rivers. Furthermore, those predators and parasites, very important from the ecological point of view, are also abundant in urban freshwater habitats, even those which are considerably degraded and strongly altered by man. Some species within the group are highly specialized ecologically [Koperski, 2010].

Leeches are among invertebrates which have been the focus of scientific attention for a very long time [Stańczykowska, 1986]. The first studies on the Hirudinea, varying in intensity, were initiated in various regions of Poland at the break of 19th and 20th centuries [Bielecki et al. 2011a, b]. Presently, our knowledge about this group of invertebrates is still insufficient. Every new study expands the available knowledge and may prove useful in estimating various types of anthropogenic threats to particularly rare and highly specialized leech species.

The present study focuses on the quantitative and qualitative structure of leech communities in the Stepnica River, a strongly anthropologically altered river in Western Pomerania, in the context of environmental conditions. The study supplements the already available data on taxonomic composition of leeches occurring in water courses of Western Pomerania.

MATERIAL AND METHODS

The Stepnica River is the main water course of the Maszewo district and a right-bank tributary of the Gowienica River. It is 34.2 km long. Its headwaters are located in the town of Wisławie and it joins the Gowienica in the town of Bodzęcin. Human activities are the main source of pollution of this water course, which is especially striking in its upper reaches, where breeding ponds and poultry processing factories significantly contribute to the deterioration of the river water, as reported by Raczyńska et al. [2013].

Results presented in this study are based on benthic fauna samples collected with a grabber from four research sites (Table 1) during the period from 21 December 2004 until 12 December 2006, at monthly intervals. Simultaneously, hydrological studies of the water course were conducted: width, depth and instantaneous flow were measured, the latter by the flow method (Table 2).

Biological material was transported to a laboratory, sifted through sieves with 0.45 mm mesh size and transferred to laboratory trays. Leech specimens were singled out and preserved according to the procedure recommended by Pawłowski [1936]. The remaining benthic fauna material was transferred to containers and preserved in 75% alcohol. Qualitative identification of the Hirudinea down to the level of species was conducted according to Pawłowski's keys [1936, 1968].

In order to discern the relationships between environment quality and leech communities in the Stepnica River, the study used the results of detailed analyses providing average monthly values of selected hydro-chemical indicators, i.e. dissolved oxygen content and calcium content, from the recent study by Raczyńska et al. [2013].

RESULTS

In the years 2004–2006, 60 specimens of the Hirudinea were found in the analyzed material from the Stepnica River. They belonged to 3 families:

- family Erpobdellidae, represented by 2 species: *Erpobdella nigricollis* (Brandes) and *Erpobdella octoculata* (Linnaeus),
- family Glossiphoniidae, represented by 2 species: *Glossiphonia complanta* (Linnaeus) and *Helobdella stagnalis* (Linnaeus)
- family Haemopidae, represented by 1 species: *Haemopis sanguisuga* (Linnaeus).

The most abundant were the species from the family Erpobdellidae (37 specimens), followed by the species from the family Glossiphonidae (21 specimens). Only 2 specimens representing the family Haemopidae were found during the whole research period. The highest number of species representing the family Hirudinea were found in research sites 1 and 2: 4 and 5 species respectively. In each of the remaining sites only one species belonging to the family Haemopidae was found.

Table 1. The characteristics of research sites on the Stepnica River (after Hałupka [2013])

Site	Coordinates	Cover the bottom [%]	Overshadowing of the riverbed [%]
1 - Wisławie	53°31'8.2"N 15°03'58.01"E	60% organic substrate 40% silt	100
2 - Maszewo	53°30'0.46"N 15°03'52.11"E	5% organic substrate 70% sand 25% silt	0
3 - Radzanek	53°30'9.23"N 15°01'21.17"E	70% organic substrate 30% silt	0
4 - Maciejewo	53°33'27.86"N 15°0'14.24"E	50 organic substrate 10% sand 40% silt	less than 25

Table 2. Results of hydrological measurements conducted on the Stepnica River in the years 2004–2006

Site	Width [m]			Average depth during the study [m]			Other information
	min	max	$\bar{x}$	min	max	$\bar{x}$	
1 - Wisławie	0.30	1.20	0.71	0.05	0.65	0.25	• absence of water (28.09.2005, 09.08.2006, 03.10.2006)
2 - Maszewo	0.10	2.00	1.00	0.05	0.60	0.25	• riverbed virtually 100% overgrown with reeds and sweet flag (11.07.2006, 09.08.2006, 03.10.2006)
3 - Radzanek	0.30	4.00	1.85	0.20	1.50	0.65	• frozen water, mid-current with ca. 5 cm ice cover (20.12.2005) • silted water with H_2S, riverbed virtually 100% overgrown with reeds and sweet flag (11.07.2006, 09.08.2006, 03.10.2006) • water outside the riverbed (07.11.2006)
4 - Maciejewo	0.40	4.00	1.53	0.03	1.50	0.57	• silted water with H_2S (20.12.2005, 31.05.2006, 11.07.2006, 09.08.2006, 06.09.2006, 03.10.2006) • white "mould" layer on the bottom (03.10.2006)

Erpobdella octoculata was a dominant species. In the research sites 1 and 2 it constituted 54.5% and 63.6% of the total leech abundance, respectively. The remaining species found in those research sites were adominants. *Glossiphonia complanata* was an absolute dominant in the research site 3, and *Helobdella stagnalis* was an absolute dominant in the research site 4, due to the fact that in the course of the research only one specimen of each of those species was found (Table 3).

As for seasonal leech abundance, the highest abundance was noted in spring and in summer (March – August) 2005 and 2006, whereas in winter the leeches occurred only incidentally (Tables 4–6). The analysis of leech abundance distribution in respective research sites did not reveal any repetitive pattern of occurrence in any particular months or years (Tables 4–6). The only exception was *Erpobdella octocullata*, which occurred in March and June in both 2005 and 2006 in the research site 2, whereas in the research site 1 it occurred in April in both 2005 and 2006 (Table 4).

DISCUSSION

The studied stretch of the Stepnica River was characterized by such specific hydrological conditions as shallowness and low flow velocity (Table 2, Table 6), which affected the quality of the river water and thus determined the formation of specific community structures of macrobenthic fauna, including the Hirudinea. Additional factors observed in the course of the study which might influence the studied leech communities included: periodical absence of water in the research site 1; vegetation overgrowing the riverbed and reducing water flow as well as causing organic matter accumulation in sites 2 and 3; siltation and the presence of hydrogen sulfate in sites 3 and 4 (Table 2).

Leech species occurring in the Stepnica River were species characteristic for lotic waters and highly immune to pollution. One of them was *Erpobdella octoculata*, which not only tolerated strongly eutrophicated waters but could also temporarily survive in anaerobic conditions [Bielecki et al. 2011b]. The research conducted on the Stepnica River did not reveal a strong relationship between the oxygen content and abundance of *E. octoculata*. The species occurred regardless of whether the oxygen content amounted to 16.6 or 3.4 $mgO_2 \cdot dm^{-3}$. Furthermore, 3 specimens were found in the first case, and 5 in the second, which might indicate that the species was highly tolerant of water oxygenation (Table 4). The occurrence of *E. octoculata* solely in sites 1 and 2 (Table 4), where the amount of total suspension and organic matter indicated low (lower than good) water quality according to Raczyńska et al. [2013], might be due to strong immunity of the species to pollution. But why did the species not occur in the other sites, where values of the above physico-chemical indicators were similar [Raczyńska et al. 2013]? Perhaps the siltation present practically throughout the study in sites 3 and 4, and temporary occurrence of hydrogen sulfate in the two sites (Table 2), were the reasons.

Apart from *E. octoculata*, another leech species characterized by a high tolerance of environmental pollution was *Helobdella stagnalis*, uncommon in slightly eutrophicated waters [Bielecki et al. 2011b], and often found in polluted lowland waters [Pawłowski, 1936]. However, the species had been recorded previously in clean post-glacial lakes in Myślibórz Lakeland [Agapow 1982, Agapow & Nadobnik 2006]. In the Stepnica River the species occurred in research stations 1 and 2, and one time in September 2006 it was also recorded in site 4 (Table 5). As has been mentioned earlier, sites 1 and 2, where it was typical for *H. stagnalis* to occur, were characterized by polluted

Table 3. The number of specimens of the respective leech species in the respective research sites on the Stepnica River in the years 2004–2006

Family	Species	Site 1 Wisławie	Site 2 Maszewo	Site 3 Radzanek	Site 4 Maciejewo	Total in the river
Glossiphonidae	*Glossiphonia complanta*	1	6	4	0	11
	Helobdella stagnalis	2	7	0	1	10
Erpobdellidae	*Erpobdella nigricollis*	2	1	0	0	3
	Erpobdella octoculata	6	28	0	0	34
Haemopidae	*Haemopis sanguisuga*	0	2	0	0	2
Number of individuals at the sampling site		11	44	4	1	60
Number of species at the sampling site		4	5	1	1	5

Table 4. The annual abundance of *Erpobdella octoculata* (ind.) and dissolved oxygen content [$mgO_2 \cdot dm^{-3}$] in the respective research sites on the Stepnica River (grey colour indicates oxygen content values coinciding with the occurrence of *E. octoculata*)

Date	*Erpobdella octoculata* [os.]				Dissolved oxygen [mg $O_2 \cdot dm^{-3}$]			
	Site 1	Site 2	Site 3	Site 4	Site 1	Site 2	Site 3	Site 4
21.12.04					7.5	12.2	7.5	12.3
22.03.05		1			16.5	11.5	10.2	10.2
11.04.05	3	3			7.2	16.6	4.6	16.5
27.05.05					3.2	7.2	2.4	7.2
21.06.05	2	4			2.1	8.5	1.3	9.0
20.07.05					1.4	9.1	0.6	2.2
10.08.05					1.0	8.8	0.2	0.0
28.09.05					absence of water	9,0	1,8	2,4
11.11.05					2.7	7.0	5.1	9.8
20.12.05					7.5	11.7	9.9	12.2
22.02.06					7.0	13.0	7.5	6.6
29.03.06		2			5.6	8.3	6.9	6.7
25.04.06	1				4.2	7.4	3.4	10.4
31.05.06		6			3.8	5.4	1.4	2.9
14.06.06		4			6.9	4.5	2.6	2.7
11.07.06		5			2.9	3.4	2.7	2.1
09.08.06					absence of water	5.3	0.8	H_2S
06.09.06		1			0.6	5.1	1.4	0.0
03.10.06					absence of water	5.4	1.6	0.0
07.11.06					3.0	5.4	4.5	7.4
12.12.06		2			6.9	5.6	8.0	5.8

Table 5. The annual abundance of *Glossiphonia complanata* and *Helobdella stagnalis* (ind.) and calcium content [$mgCa \cdot dm^{-3}$] in the respective research sites on the Stepnica River (grey colour indicates calcium content values coinciding with the occurrence of *G. complanata*)

Date	*Glossiphonia complanata* [ind.]				*Helobdella stagnalis* [ind.]				Calcium [mg $Ca \cdot dm^{-3}$]			
	Site 1	Site 2	Site 3	Site 4	Site 1	Site 2	Site 3	Site 4	Site 1	Site 2	Site 3	Site 4
21.12.04									38	38	36	26
22.03.05									62	77	59	76
11.04.05		2			1				95	100	91	81
27.05.05									101	104	81	55
21.06.05	1	2							105	94	83	66
20.07.05									107	86	71	67
10.08.05						5			107	94	79	81
28.09.05									absence of water	102	93	87
11.11.05						1			122	111	89	71
20.12.05			1			1			124	123	101	78
22.02.06									94	96	88	81
29.03.06		1							99	106	93	91
25.04.06									108	112	92	75
31.05.06					1				108	92	96	81
14.06.06									117	106	85	56
11.07.06									81	100	87	65
09.08.06									absence of water	55	40	41
06.09.06								1	100	56	43	35
03.10.06		1							absence of water	60	44	48
07.11.06									66	63	42	46
12.12.06			3						63	61	46	38

waters, whereas the absence of the species from site 4 – analogically to the absence of *E. octoculata* – might be due to siltation and the presence of hydrogen sulfate (Table 2).

A species competing with *Helobdella stagnalis* (L.) for food and habitat was *Glossiphonia complanata* [Wrona et al. 1981], which preferred waters with a high calcium content (http://www.naturespot.org.uk/species/glossiphonia-complanata). That leech species occurred only in sites 1 and 2, where higher calcium content values were noted in respective months in comparison to sites 3 and 4, and where the average calcium content amounted to 100 $mgCa{\cdot}dm^{-3}$ (except October and December 2006, when it was equal to 60 and 61 $mgCa{\cdot}dm^{-3}$ respectively) (Table 5). The occurrence of *G. complanata* solely in sites 1 and 2 confirmed the data presented by Pawłowski [1936], according to whom a typical habitat of that species was a small water course overgrown with vegetation, characterized by an organic substrate with high calcium content. A competition for resources between *G. complanata* and *Helobdella stagnalis* on the Stepnica River was revealed by the analysis of their distribution, since throughout the study the two species never co-occurred, even though *H. stagnalis* also preferred sites located in the upper reaches of the studied river (Table 5). As in the cases discussed above, the occurrence of the two species in sites 3 and 4 was reduced by the high content of silt and hydrogen sulfate (Table 2).

The fourth leech species encountered during the studies on the Stepnica River, namely *Erpobdella nigricollis*, had been frequently recorded in river stretches with slow-flowing or even stagnant waters [Bielecki et al. 2011b] and associated with habitats strongly overgrown with vegetation [Agapow 1986, Kołodziejczyk & Koperski 2000]. Furthermore, similarly to the species discussed previously, it was characterized by a high tolerance of polluted waters [Agapow et al. 2008]. The preference of this species for habitats with stagnant water was confirmed during the studies on the Stepnica River: *Erpobdella nigricollis* occurred only in sites 1 and 2, three times during the absence of the flow (Table 6).

Table 6. The annual abundance of *Erpobdella nigricollis* [ind.] instantaneous flow [m/s] in the respective research sites on the Stepnica River (grey colour indicates instantaneous flow values coinciding with the occurrence of *E. nigricollis*)

Date	*Erpobdella nigricollis* [ind.]				Instantaneous flow [m/s]			
	Site 1	Site 2	Site 3	Site 4	Site 1	Site 2	Site 3	Site 4
21.12.04					0.00	0.20	0.25	0.20
22.03.05					0.10	0.38	0.53	0.19
11.04.05					0.06	0.23	0.03	0.24
27.05.05					0.00	0.29	0.33	0.45
21.06.05					0.00	0.12	0.24	0.30
20.07.05					0.00	0.00	0.00	0.21
10.08.05					0.00	0.05	0.00	0.24
28.09.05					absence of water	0.00	0.04	0.20
11.11.05	1				0.00	0.00	0.20	0.31
20.12.05					0.00	0.14	0.13	0.28
22.02.06					0.08	0.37	0.16	0.38
29.03.06					0.19	0.34	0.21	0.21
25.04.06	1				0.00	0.14	0.16	0.16
31.05.06					0.00	0.17	0.00	0.20
14.06.06					0.00	0.00	0.09	0.16
11.07.06					0.00	0.18	0.00	0.14
09.08.06		1			absence of water	0.00	0.00	0.07
06.09.06					0.00	0.00	0.00	0.16
03.10.06					absence of water	0.00	0.00	0.00
07.11.06					0.00	0.00	0.27	0.19
12.12.06					0.06	0.19	0.13	0.15
min					0.00	0.00	0.00	0.00
max					0.19	0.38	0.53	0.45
$\bar{x}$					0.03	0.14	0.14	0.21

The last of the encountered species, *Haemopis sanguisuga*, was found only in site 2, and throughout the study was represented by only two specimens (Table 3), which was somewhat surprising, as the species was believed to prefer water courses with a high silt content and thus its presence was expected either in site 3 or 4. A location of this species in Western Pomerania had been found by Bielecki et al. [2008] in Słowiński National Park, in the Pustynka River characterized by strong siltation.

CONCLUSIONS

The fauna of the Hirudinea encountered in the studied stretch of the Stepnica River in the years 2004–2006 was neither diversified nor very abundant. Only 5 species representing three families were recorded. The most abundant and the most common species included *Erpobdella octoculata* and *Helobdella stagnalis*, followed by the less abundant *G. complanata,* whereas the remaining species, *E. nigricollis* and *Haemopis sanguisuga* occurred only sporadically. All those species were characterized by a high immunity to pollution; however, their distribution in the respective sites indicated that they were not tolerant of siltation and the presence of hydrogen sulfate.

Acknowledgments

The authors thank to Professor Lucjan Agapow for consultation and assistance in the determination of leeches.

REFERENCES

1. Agapow L. 1982. Conditions of life and occurrence of leeches (Hirudinea) in the Myślibórz Lake district. Monogr. AWF Poznań 150, pp. 164 (in Polish).
2. Agapow L. 1986. Rzadkie gatunki pijawek (Hirudinea) projektowanego Drawieńskiego Parku Narodowego. Przegl. Zool., 30, 3, 293-297.
3. Agapow L., Nadobnik J., 2006. Fauna of leeches (Hirudinea) in postglacjal tarns in the Dobiegniewskie Lakeland. Teka Kom., Ochr., Kszt., Środ., Przyr., 3, 13-18.
4. Agapow L., Korościński B., Nadobnik J., 2008. Preliminary studies on the fauna of Hirudinea and Mollusca in Lake Bulikowskie. Limnological Papers, 3(1), 9-18.
5. Bielecki A., Palińska K., Marczak D. 2008. Leeches (Hirudinida) of Słowiński National Park. Parki nar. Rez. Przyr., 27(4), 35-41.
6. Bielecki A., Cichocka J.M., Jeleń I., Ropelewska E., Adamiak-Brud Ż., Biedunkiewicz A., JaninDziekońska-Rynko J. 2011a. *Batracobdelloides moogi* Nesemann et Csányi, 1995 (Hirudinida: Glossiphoniidae): Morphometry and structure of the alimentary tract and reproductive system. Biologia, 66, 5, 848-855.
7. Bielecki A., Świątek P., Cichocka J.M., Ropelewska E., Jeleń I., Adamiak-Brud Ż., 2011b. Pijawki (Hirudinida) wód powierzchniowych Olsztyna. Forum Faunistyczne 1(1), 12-34.
8. Hałupka M., 2013. Hirudinea i Mollusca – potencjalna baza pokarmowa dla ryb nizinnej rzeki Stepnica, woj. zachodniopomorskie w latach 2004-2006. Rozprawa doktorska, ZUT w Szczecinie.
9. http://www.naturespot.org.uk/species/glossiphonia-complanata)
10. Kołodziejczyk A., Koperski P. 2000. Bezkręgowce słodkowodne Polski. Klucz do oznaczania oraz podstawy biologii i ekologii makrofauny. Wyd. Uniwersytetu Warszawskiego, Warszawa, 1-250.
11. Koperski P. 2010. Urban environments as habitats for rare aquatic species: The case of leeches (Euhirudinea, Clitellata) in Warsaw freshwaters. *Limnologica*, 40, 233-240.
12. Pawłowski L.K., 1936. Fauna słodkowodna Polski. Opracowanie zbiorowe pod redakcją T. Jaczewskiego i T. Wolskiego. Zeszyt 26. Pijawki, 1-176.
13. Pawłowski L.K. 1968. Pijawki. Hirudinea. Katalog fauny Polski XI. PWN Warszawa, 3, 1-94.
14. Raczyńska M., Machula S., Grzeszczyk-Kowalska A., 2013. Stan ekologiczny rzeki Stepnicy (Pomorze Zachodnie). Inżynieria ekologiczna (Ecological Engineering) – Kształtowanie i Ochrona Środowiska, 35, 46-59.
15. Stańczykowska A. 1986. Zwierzęta bezkręgowe naszych wód. Wyd. Szkolne i Pedagog., Warszawa, 1-317.
16. Wrona F.J., Davies D., Ronald W., Linton L., Wilkialis J., 1981. Competition and coexistance between *Glossiphonia complanata* and *Helobdella stagnalis* (Glossiphonia: Hirudinoidea). Oecologia, 48(1), 133-137.

16

CLASSIFICATION OF PRECIPITATION INTENSITY DURING VEGETATION SEASON IN CENTRAL-EASTERN POLAND (1971–2005)

Elzbieta Radzka[1]

[1] Department of Agrometeorology and Land Reclamation, University of Natural Sciences and Humanities, B. Prusa 14, 08-110 Siedlce, Poland, e-mail: elzbieta.radzka@uph.edu.pl

ABSTRACT

The data used in this paper come from nine IMGW stations from central-eastern region of Poland and they were registered between the years 1971–2005. On the basis of daily precipitation sums analysis during vegetation season, the number of days with precipitation was determined in particular classes. In central-eastern Poland during vegetation season days with very little and little precipitation were noted most frequently. Very little precipitation occurred most frequently in April; little, moderate and moderately strong precipitation - in June, and strong precipitation - in July. Whereas very strong precipitation was noted most frequently in June and July. Days with very little and little precipitation were noted most frequently in northern and north-eastern part of the area examined. While the number of days with moderate, moderately strong and strong precipitation showed little spatial diversity. Very strong precipitation was noted the most frequently in the area of Legionowo and Ostroleka. The number of days with very little precipitation during vegetation season was decreasing in subsequent years of the study. Statistically significant negative directional coefficients of the trend were noted most frequently in April. The number of days with very strong precipitation was also decreasing significantly in most of the stations.

Keywords: number of days with precipitation, vegetation season, changes tendencies, central-eastern Poland.

INTRODUCTION

In recent years large variability of the precipitation conditions can be observed, therefore, many studies relate to the characteristics of this meteorological element [Banaszkiewicz et al. 2004, 2007, Hutorowicz et al. 1996, Nowicka and Grabowska 1989, Szwejkowski et al. 2002, Scigalska and Labuz 2009]. Longer periods of precipitation deficiency are the cause of droughts, while even short duration of high intensity precipitation may result in soil erosion [Zyromski 2002] and cause floods. According to Kirschenstein [2002a, b, 2004] on the basis of long-term observation series the variation of a specific meteorological element can be evaluated. The aim of this study was to determinate the number of days with different intensity of precipitation in central-eastern Poland during the multiyear 1971–2005.

MATERIAL AND METHODS

Data used in this paper refer to daily precipitation sums and they were registered in nine IMGW stations from central-eastern region of Poland in the years 1971–2005 (Table 1).

During vegetation season the number of days with precipitation was determined in particular classes according to Olechnowicz-Bobrowska criterion [1970]:

Precipitation classes (mm)	**Days with precipitation:**
0,1–1,0	very little
1,1–5,0	little
5,1–10,0	moderate
10,1–20,0	moderately strong
20,1–30,0	strong
>30,0	very strong

Table 1. Geographic coordinates of synoptic and climatic IMGW stations in central-eastern Poland

Station	Geographic coordinates		H_s
	<P°	1°	
Ostroleka	53° 05'	21° 34'	95
Bialowieza	52° 42'	23° 51'	164
Wlodawa	51°33'	23° 32'	163
Szepietowo	52° 51'	22°33'	150
Legionowo	52° 24'	20° 58'	93
Biala Podlaska	52° 02'	23° 05'	133
Sobieszyn	51° 37'	22° 09'	135
Pultusk	52° 44'	21° 06'	95
Siedlce	52° 11'	22° 16'	146

Explanations: cp° – geographic latitude, *V* – geographic longitude, H_s – elevation above sea level.

The direction and significance of changes tendencies of the number of days with precipitation in particular classes was determined on the basis of linear trend equations. The significance of directional coefficient of the trend was estimated with t-Student test where significance level $\alpha = 0,05$.

RESULTS AND DISCUSSION

In central-eastern Poland during the vegetation season (1971–2005) the most frequently were noted days with very little precipitation and little precipitation, 20,73 days and 26,97 days respectively (Table 2). During the analyzed period among all the precipitation classes the least numerous were the days with strong precipitation (2,35 days). While there were almost two times more (4,16) days with very strong precipitation (>30 mm). Days with moderate and moderately strong precipitation during vegetation season were noted 11,24 and 7,15 respectively. Very little precipitation occurred most frequently in April, little, moderate and moderately strong precipitation – in June, and strong – in July. Whereas very strong precipitation (>30 mm) was noted the most frequently in June and July. Days with observed precipitation (>0,l mm) constituted about 40% of all the days of vegetation season (IV–IX).

The analysis of spatial distribution of days with precipitation showed that days with very little and little precipitation were noted most frequently in northern and north-eastern part of the area examined (Figure 1). While the number of days with moderate, moderately strong and strong precipitation showed little spatial variety. The number of the days with moderate precipitation during vegetation season ranged from 10,62 (Biala Podlaska) to 11,81 (Bialowieza), with moderately strong precipitation from 6,50 (Siedlce) to 7,79 (Bialowieza), with strong precipitation from 1,78 (Pultusk) to 2,9 (Legionowo). Very strong precipitation was noted most frequently in the area of Legionowo (5,06 days) and Ostrolejca (5,0 days). While the smallest number of days with very strong precipitation during vegetation season occurred in Siedlce (2,67 days). Characterizing precipitation conditions on meteorological station Puczniew between the years 1972–2002 Baryla et al. [2012] concluded that an important characteristics was the probability of high daily precipitation (assumed > 10,1 mm) occurrence. Among that 65% constituted precipitation from the range of 10,1–20,0 mm, occurring throughout the year. Maxima > 50 mm, which may cause material loss in agriculture, constituted barely 2,0% of all the cases and occurred in the period of time between May and August.

On the basis of the analysis of directional coefficients of the trend of the number of days with precipitation in central-eastern Poland, it can be stated that the number of days with very little precipitation during vegetation season decreased in subsequent years of the study (Table 3). Sta-

Table 2. Average number of days with precipitation during vegetation season in central-eastern Poland between the years 1971–2005

Month/Class	0,1–1,0 mm	1,1–5,0 mm	5,l–10,0 mm	10,1–20 mm	20,1–3 0 mm	>30 mm
IV	4,26	4,76	1,51	0,74	0,28	0,40
V	3,37	4,48	1,75	1,01	0,36	0,59
VI	3,21	4,98	2,29	1,51	0,38	0,87
VII	3,30	4,44	2,06	1,48	0,55	0,87
VIII	2,95	3,72	1,79	1,26	0,45	0,77
IX	3,66	4,59	1,84	1,14	0,33	0,66
Sum IV–IX	20,73	26,97	11,24	7,15	2,35	4,16

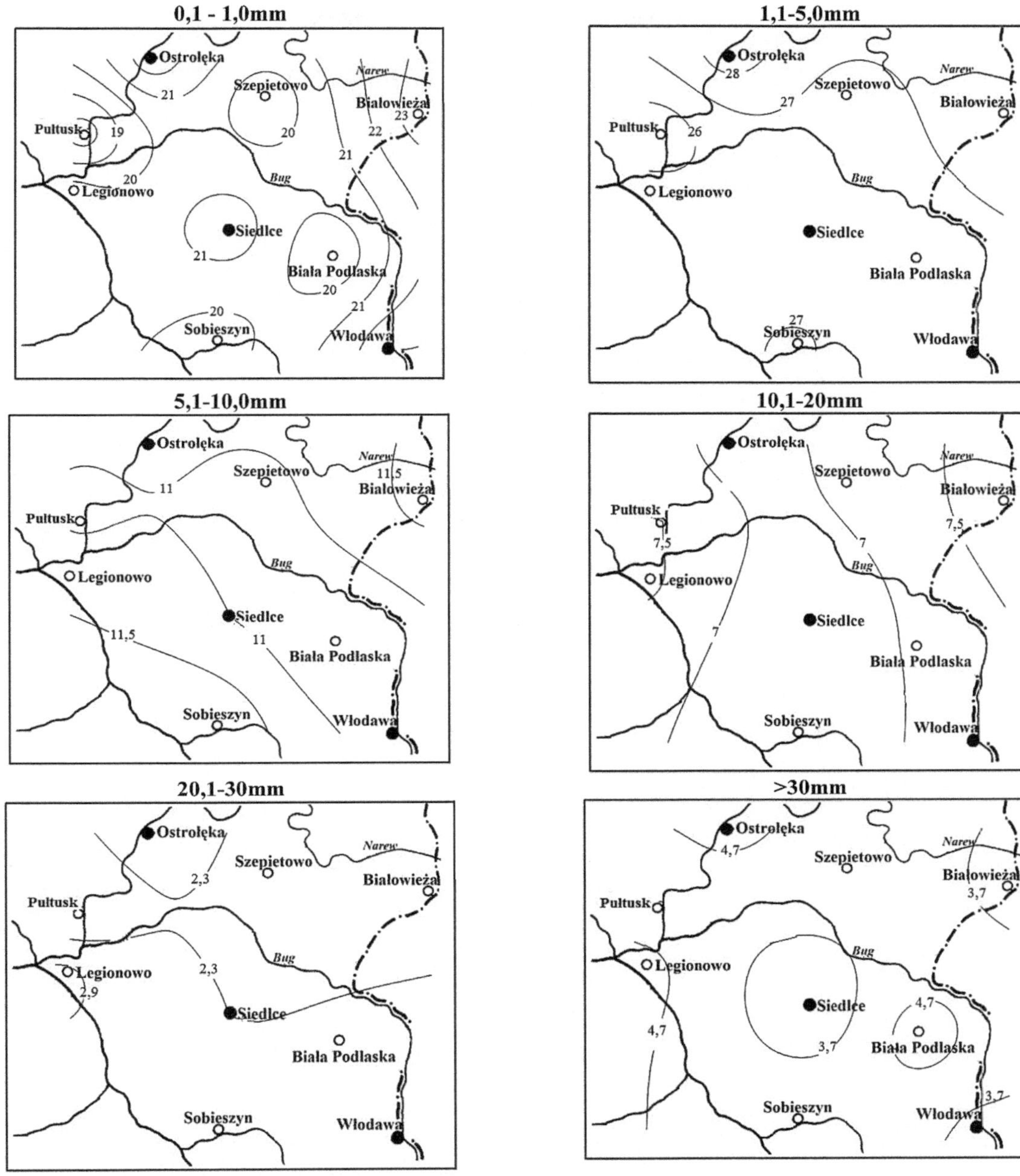

Figure 1. Spatial distribution of the number of days with precipitation during vegetation season (IV–IX) in central-eastern Poland between the years 1971–2005

tistically significant negative directional coefficients of the trend were noted most frequently in April. While in station Pultusk they were significant in all of the vegetation season's months, except for July. The number of days with moderate precipitation increased significantly about 0,5 day in 10 years during vegetation season, especially in July. While the number of days with very strong precipitation during vegetation season was decreasing significantly in most of the stations, however, directional coefficients of the trend values were low (from -0,11* to -0,042*). In stations Bialowieza and Ostrolejca there were no significant changes in the number of days with very strong precipitation, whereas in stations Wlodawa nad Legionowo significances occurred only in June. Uscka-Kowalkowska and Kejna [2009] characterizing variability of thermal-precipitation conditions in Koniczynka (Chelmihskie Lakeland) in the period 1994–2007 concluded that in case of precipitation not only their sum is important but also the temporal distribution. On average 169,9 days with precipitation (> 0,1 mm) per year occurred in Koniczynka in the period 1994–2007 . Their number was highly variable year by year. The least number of days

Table 3. Directional coefficients of the trend of the number of days with precipitation during vegetation season (IV–IX) in central-eastern Poland between the years 1971–2005

	0,1–1,0 mm	1,1–5,0 mm	5,1–10,0 mm	10,1–20,0 mm	20,1–30,0 mm	>30,0 mm
Białowieża						
IV	**-0,106***	-0,035	**-0,040***	-0,013	0,005	0,007
V	-0,060	-0,032	0,016	0,007	0,001	0,010
VI	-0,017	-0,036	-0,007	0,004	**-0,017***	-0,005
VII	-0,025	-0,015	-0,021	-0,012	-0,006	0,000
VIII	**-0,074***	0,082*	0,002	-0,003	**-0,021***	0,004
IX	**-0,058***	-0,010	-0,033	0,001	-0,005	0,012
Ostrołęka						
IV	0,003	-0,020	0,009	0,013	0,010	0,011
V	-0,063	-0,024	0,035	-0,024	-0,016	-0,011
VI	-0,068	0,012	-0,049	-0,005	-0,006	-0,025
VII	-0,020	**-0,106***	0,038	-0,034	-0,021	-0,009
VIII	-0,022	-0,057	-0,005	0,005	-0,021	-0,014
IX	-0,007	-0,046	-0,035	-0,002	-0,004	-0,003
Siedlce						
IV	-0,049	0,029	0,001	-0,010	0,001	**-0,019***
V	-0,018	0,027	-0,009	-0,008	-0,005	**-0,029***
VI	0,009	0,018	0,001	-0,025	0,002	-0,043
VII	**-0,063***	-0,002	**0,060***	-0,017	-0,017	**-0,036***
VIII	-0,042	0,025	-0,010	-0,027	0,011	**-0,038***
IX	-0,015	-0,041	0,046	0,001	**-0,026***	**-0,027***
Włodawa						
IV	**-0,064***	0,060	0,000	0,012	0,013	-0,003
V	0,014	0,014	-0,006	0,013	0,011	-0,015
VI	**0,014***	0,008	-0,008	-0,045	0,018	**-0,027***
VII	-0,031	-0,034	0,027	-0,009	-0,012	-0,024
VIII	0,022	0,015	0,010	0,015	0,014	-0,018
IX	0,011	-0,042	0,040	**0,038***	-0,005	-0,013
Legionowo						
IV	-0,024	0,022	-0,008	**0,037***	0,011	-0,010
V	0,034	-0,032	0,020	0,018	0,019	-0,006
VI	0,013	0,032	-0,017	-0,031	0,012	**-0,027***
VII	-0,020	-0,031	**0,057***	0,008	-0,010	-0,017
VIII	0,030	0,038	0,002	**0,043***	0,003	-0,020
IX	0,044	-0,039	-0,001	0,009	-0,006	-0,004
Pułtusk						
IV	**-0,123***	0,011	-0,019	-0,001	-0,005	**-0,017***
V	**-0,055***	0,029	0,015	-0,011	-0,002	**-0,029***
VI	**-0,074***	0,045	-0,011	-0,005	-0,007	**-0,035***
VII	-0,043	-0,062	0,035	0,003	-0,005	**-0,032***
VIII	**-0,074***	0,038	0,011	-0,024	-0,014	**-0,038***
IX	**-0,077***	-0,023	-0,018	**-0,003***	0,001	**-0,030***
Biała Podlaska						
IV	**0,071***	0,040	**0,035***	-0,011	0,008	-0,013
V	0,000	0,025	0,011	-0,008	-0,012	**-0,028***
VI	-0,010	0,035	-0,001	-0,011	-0,001	**-0,039***
VII	-0,017	-0,088	0,020	-0,013	-0,024	**-0,027***
VIII	0,061	-0,063	0,034	0,036	-0,001	**-0,024***
IX	0,068	-0,041	**0,076***	-0,031	0,000	-0,009
Sobieszyn						
IV	-0,006	0,020	-0,018	**-0,020***	0,013	**-0,022***
V	-0,003	-0,055	-0,026	0,008	-0,003	**-0,031***
VI	0,036	0,027	-0,005	-0,033	0,013	**-0,042***
VII	-0,043	0,004	**0,068***	0,027	0,011	**-0,040***
VIII	-0,017	0,004	-0,003	0,019	-0,014	**-0,033***
IX	-0,007	0,000	-0,033	0,017	-0,010	-0,020
Szepietowo						
IV	**-0,065***	0,039	-0,022	-0,020	0,004	**-0,011***
V	-0,034	0,003	0,014	-0,006	-0,004	**-0,026***
VI	-0,044	**0,113***	-0,029	-0,025	-0,038	**-0,038***
VII	-0,010	-0,068	**0,041***	-0,025	0,004	**-0,027***
VIII	-0,042	0,034	-0,010	-0,025	-0,003	**-0,029***
IX	-0,037	-0,011	-0,002	-0,007	0,010	**-0,019***

with precipitation took place in the year 2003 (145 days), while the highest number – in 2004 (199 days). The number of days with precipitation was characterized by slow increase which, however, was not statistically significant. Analyzing pluvial conditions in Zielonka Forest in the years 1987–2008 Grajewski [2011] noticed that the frequency of days with precipitation from the range of 1,0–9,9 mm shows relative stability, confirmed by a linear trend. Whereas the number of precipitation from the other ranges presents visible downward tendency, the most clearly pointed out in case of precipitation < 1,0 mm. Directional variability of this value of precipitation was not only the highest but also statistically significant ($p < 0,05$).

CONCLUSIONS

1. In central-eastern Poland between the years 1971–2005 days with precipitation (>0,1 mm) constituted about 40% of all the vegetation season's days (IV–IX). Days with very little and little precipitation were noted the most frequently, while days with strong and very strong precipitation – the most rarely.
2. Very little precipitation was noted most frequently in April; little, moderate and moderately strong precipitation – in June, and strong – in July. Very strong (>30,0mm) precipitation occurred most frequently at the beginning of summer (June, July).
3. On the basis of analysis of spatial distribution of the days with precipitation, it was noticed that days with least intense precipitation (till 5,00 mm) occurred most frequently in northern and north-eastern part of the area examined, while days with very strong precipitation – in western and north-western part.
4. During vegetation season the most frequently was noted statistically significant decrease of the number of days with very strong and very little precipitation (especially in April). While the number of days with moderate precipitation showed a rising tendency.

REFERENCES

1. Banaszkiewicz B., Grabowska K., Suchecki S. 2007. Charakterystyka warunków opadowych Pojezierza Suwalskiego w latach 1971-2000. Prz. Nauk. Inż. Kszt. Środ. 3(37), 23-32.
2. Banaszkiewicz B., Grabowska K., Szwejkowski Z. 2004. Charakterystyka opadów atmosferycznych na terenie województwa warmińsko-mazurskiego w latach 2000-2002. Acta Agrophysica, 3(1), 5-11.
3. Baryła A., Karczmarczyk A., Frąk M. 2012.Charakterystyka warunkow opadowych na stacji meteorologicznej Puczniew w latach 1972–2002. Infrastruktura i Ekologia Terenów Wiejskich, 3/III, 17-26.
4. Grajewski S. 2011. Warunki pluwialne w Puszczy Zielonka w latach 1987-2008. Nauka Przyroda Technologie, 5, 6, 1-12.
5. Hutorowicz H., Grabowska K., Nowicka A. 1996. Charakterystyka warunków klimatycznych Pojezierza Mazurskiego. Zesz. Probl. Post. Nauk Roln., 431, 21-29.
6. Kirschenstein M. 2002a. Cyrkulacyjne uwarunkowania opadów w północno-zachodniej Polsce. Przegląd Geofizyczny, 47, 1-2.
7. Kirschenstein M. 2002b. Opady a typy cyrkulacji atmosferycznej w północno-zachodniej Polsce. Badania Fizjograficzne nad Polską Zachodnią, Seria A – Geografia Fizyczna, 53, 57-82.
8. Kirschenstein M. 2004. Rola cyrkulacji atmosferycznej w kształtowaniu opadów w północno-zachodniej Polsce. Wyd. Pomorskiej Akademii Pedagogicznej, Słupsk. pp. 193.
9. Nowicka A., Grabowska K. 1989. Charakterystyka ważniejszych elementów klimatu Pojezierza Warmińsko-Mazurskiego, IV. Opady atmosferyczne. Acta Acad. Agricult. Techn. Olst. Agricultura, 59, 105-113,
10. Olechnowicz-Bobrowska B. 1970. Częstość dni z opadem w Polsce. Prace Geogr. IG PAN, 86, 1-75.
11. Uscka-Kowalkowska J., Kejna M. 2009.Zmienność warunków termiczno-opadowych w Koniczynce (Pojezierze Chelmińskie) w okresie 1994–2007. Acta Agrophysica, 14(1), 203-219.
12. Szwejkowski Z., Nowicka A., Dragańska E. 2002. Klimat Pojezierza Mazurskiego, Cz. I, temperatura i opady atmosferyczne w okresie 45-lecia 1951-1995, Fragmenta Agronomica, XVIX, 2(74), 285-295.
13. Ścigalska B., Łabuz B. 2009. Analiza zmienności warunków pluwiotermicznych od kwietnia do lipca w okolicach Krakowa (1961–1990). Acta Agrophysica, 13(2), 505-521.
14. Żyromski A. 2002. Zmienność wysokich sum dobowych opadow atmosferycznych wiosną i jesienią w Polsce w okresie 1961–1980. Zesz. Probl. Post. Nauk Rol. 487, 389-397.

CHANGES IN HYDROGEN ION EXPONENT OF SEWAGE SLUDGE IN THE PROCESS OF AUTOTHERMAL THERMOPHILIC AEROBIC DIGESTION

Izabela Bartkowska[1]

[1] Chair of Environment Engineering Systems, Technical University in Białystok, Wiejska 45A, 15-351 Białystok, Poland, e-mail: izabela.bartkowska@neostrada.pl

ABSTRACT

This paper presents a study on the evaluation of digested sewage sludge during the process of Autothermal Thermophilic Aerobic Digestion (ATAD), taking place as a three-tank process at a municipal waste water treatment plant in Luban, Poland. ATAD installation was manufactured by FUCHS Enprotec GmbH Mayen-Deutschland. Over the period from June 2006 to February 2010 sewage sludge digested by tha ATAD-Process was examined. The hydrogen ion exponent was measured in every tank. The results obtained indicated changes in the composition of the digesting sludge at successive stages of the process. Over the study period the ATAD-installation was in both a two- and a three-stage process. pH of sludge under study during the process of the thermophilic stabilisation changes and its value grows significantly, with the installation working in a two-stage arrangement from 6,63 to 7,99, and when the installation was operated as a three-stage system from 6,60 to 8,14. The results collected were subject to the statistical analysis. The paper presents conclusions drawn from the study and own experience.

Keywords: activated sludge, hydrogen ion exponent, sewage sludge, sludge processing, thermophilic aerobic digestion.

INTRODUCTION

The least-expensive method of disposal or reuse of sewage sludge is to return it to the natural environment [Woźniak et al. 2004]. However, the sludge can be naturally utilised only when stabilised and without metals [Skowron et al. 2011]. Appropriate processing of sewage sludge is required by legal regulations. Initial and excessive precipitate generated at every waste water treatment plant, after chemical precipitation and addition of organic carbon from external source with this carbon required for some processes, e.g. to remove nitrogen compounds (denitrification), require engineering the installation process to meet expectations, depending on their quantity and quality. This qualifies the ATAD process as useful.

CHARACTERISTICS OF THE PROCESS

The ATAD process allows to stabilise sewage sludge and to achieve pathogenic reduction in one installation. As a result, sewage sludge turns into Class A Biosolids and is required to meet fewer restrictions placed on its use [Augustin et al. 2007].

Biological stabilization of sludge is based on the reduction of organic substances contained in sewage sludge. In the ATAD-process the contents of these substances is reduced by microorganisms. Aerobic transformation is an exothermic process, i.e. a lot of heat is released. The final products of this process are simple substances like H_20 and CO_2 [Bartkowska 2005]. From a chemical point of view this chemical reaction can have a form of simplified equations basing on a hypothetical matter of an organic cell $C_5H_7NO_2$, representing an active sludge [Wersocki, Hupka 2006, Zupančič, Roš 2008]:

$$C_5H_7NO_2 + 5O_2 \rightarrow 5CO_2 + 2H_2O + NH_3 + energy \quad (1)$$

$$C_5H_7NO_2 + 7O_2 \rightarrow 5CO_2 + 3H_2O + H^+ + energy \quad (2)$$

Good retention of heat released during the decomposition produces high working temperature (>50 °C), followed by high degree of organic material decomposition as well as destruction of pathogens. This process requires sludge pre-thickening to more than 4% D.S., and therefore, higher elementary organic matter content can be achieved, which cannot be less than 40.0 g/l expressed by the COD value [Bartkowska et al. 2005a, b].

If a proper oxygen amount is supplied at a temperature from 55 °C, the process develops automatically. For most plants this process is limited to 55–60 °C, with possible recovery of heat. High temperature along with smaller dimensions of a reactor (the design dwell time is several days) allow a substantial reduction of the contents of organic substances down to 38–50% of dry mass by volume [Layden 2007] and a safe sludge; i.e. safe in terms of the contents of pathogenic factors [Nosrati 2007].

The ATAD installation consists of reactors working in series. The system is batch-supplied once a day after which reactors are isolated and closed in order to minimise the heat loss. The process is effective thanks to aeration, which causes the foam to generate fast. The foam acting as insulation ensures a better utilisation of oxygen and an increase in the biologic activity. The use of foam disintegrators allows to control the formation, the thickness and to some extent the density of the foam layer [Bartkowska, Dzienis 2007].

In the first stage of the installation the temperature keeps within the lower thermophilic distribution: (40–50 °C). The disinfection maximum is reached in the last stage with the temperature between 50 and 60 °C. The daily dump of neutralised sewage sludge takes place from the last stage. With the successive dump completed the raw sludge is fed to the first stage while partly transformed sediment is moved to the next reactor. As the sludge is moved from the first reactor to the next one, a small temperature drop occurs. After being fed the reactors remain isolated for 23 hours and the thermophilic decomposition occurs [Bartkowska 2005, Bartkowska, Dzienis 2007]. To limit the temperature growth in the last reactor a heat exchanger is installed.

The ATAD installation under study comprises of three reactors in a form of closed steel tanks protected by anti-corrosion coats, thermally insulated, equipped with control instrumentation, aeration and foam disintegrating equipment and equipment for photo catalytic oxidation of odours (PCO), which are generated during the process. The installation was designed in order to process the formation of initial sludge and excessive sludge to the amount of 58 m³/day, thickened to about 5% of dry mass and containing some 75% of organic substances. The diameter of every single tank is 8.57 m. Every tank is filled up to the height of 3.0 m. The working capacity of single tank is therefore 173.0 m³, with the dwelling time of sludge inside it if some 9 days.

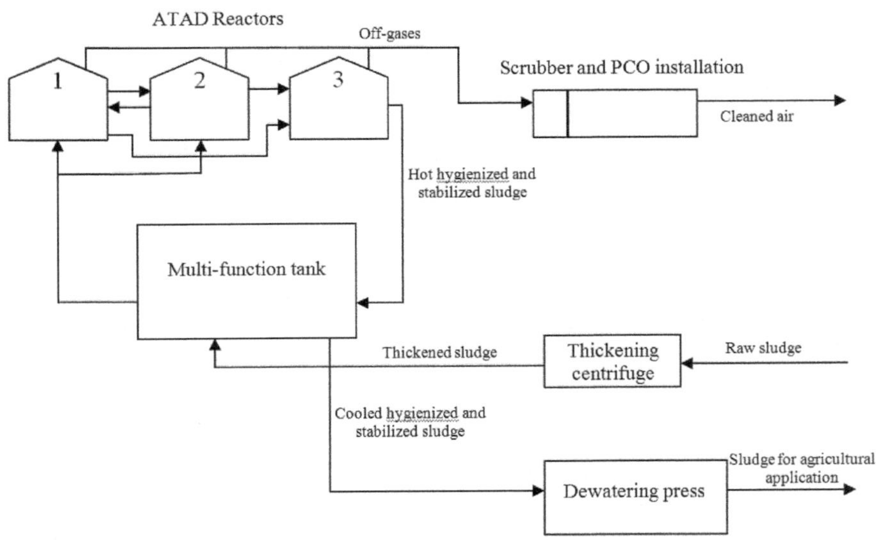

Figure 1. Simple diagram of sludge complex in Lubań wastewater treatment plant

With these three reactors the installation can be run in two or three working stages. In a two-stage system reactors No. 1 and 2 run simultaneously as the first stage and No. 3 as the second stage. In a three-stage arrangement all reactors work in series as successive stages. The reaction time is 23 hours and the emptying and charging time is 1 hour for both systems. The scheme of sedimentation node of the plant is shown in Figure 1.

METHODOLOGY, RESEARCH RESULTS AND DISCUSSION

The research was conducted at an existing object from June 2006 to February 2010. The goal of this study was to evaluate the appropriateness of selected process parameters for the estimate of the process of the Autothermal Thermophilic Aerobic Digestion for sewage sludge stabilisation, with a special focus on an easy method of measuring pH. The pH was examined in the raw, thickened sludge and in successive reactors of the installation. The research was carried out during operation of both a two-stage system and a three-stage system.

The pH was measured by a potentiometer method using an ion-selective electrode according to the PN-90/C-04540.01 standard.

All gathered results were subject to statistical analysis. The summary of the data set in order to draw basic conclusions and generalisation was conducted by a method of the descriptive stats. Table 1 presents the results of the statistical analysis.

Regarding the ATAD process as the oxidation of organic matter from dead microorganisms by thermophilic bacteria we can observe changes occurring there, which are visible in the analysis of the processed sludge pH. The pH of sludge under study during the process of thermophilic stabilisation changes and its value grows significantly.

With the installation working in a two-stage arrangement pH of sludge, thickened prior to the ATAD process, and was on the average of 6.63 pH and after the first stage of the installation it rose to 7.67 pH after it left tank No. 1 and to 7,62 after it left tank No. 2 The pH for stabilised and disinfected sludge upon leaving tank No. 3 was at average value of 7.99. The value of the tested parameter in thickened sludge was ranging from the minimum of 6.41 pH to the maximum of 7.05 pH. After the first stage of the installation pH increased from 6.98 to 8.60 (tank No. 1) and from 6.89 to 8,52 (tank No. 2). The pH of digesting sludge was characterised by the values within the range of 7.14 to 8.76 . The lowest diversification in the pH results was characteristic for the thickened sludge, for which the value of the standard deviation was 0.14 and the variance was 0.02. After the first process stage the differences in results were higher. For sludge from tank No. 1 the standard deviation was 0.45 and the variance was 0.20. Results for sludge from tank No. 2 show the standard deviation of 0.43 and the variance

Table 1. Values of selected distribution units

Distribution units	Sediment thickener		Reactor No. 1		Reactor No. 2		Reactor No. 3	
	Sludge pH during operation of the installation in a 2 or 3-stage arrangement		Sludge pH during operation of the installation in a 2 or 3-stage arrangement		Sludge pH during operation of the installation in a 2 or 3-stage arrangement		Sludge pH during operation of the installation in a 2 or 3-stage arrangement	
	2-stage	3-stage	2-stage	3-stage	2-stage	3-stage	2-stage	3-stage
Minimum value	6.41	6.23	6.98	7.07	6.89	7.48	7.14	7.72
Maximum value	7.05	7.08	8.60	7.93	8.52	8.34	8.76	8.91
Arithmetic mean	6.63	6.60	7.67	7.50	7.62	7.84	7.99	8.14
Median	6.62	6.57	7.58	7.49	7.63	7.81	7.92	8.09
Mode	6.62	multiple	7.68	multiple	7.53	multiple	7.71	multiple
Standard deviation	0.14	0.22	0.45	0.20	0.43	0.20	0.37	0.27
Variance	0.02	0.05	0.20	0.04	0.19	0.04	0.14	0.07
Range	0.64	0.85	1.62	0.86	1.63	0.86	1.62	1.19
Skewness coefficient	0.78	0.11	0.42	0.02	-0.01	0.85	0.18	0.92
Curtosis	0.95	-0.87	-0.77	-0.05	-0.78	0.56	-0.16	0.53
Percentile 10%	6.46	6.32	7.05	7.23	6.99	7.62	7.62	7.90
Percentile 90%	6.79	6.88	8.37	7.76	8.18	8.17	8.54	8.52

of 0.19. The pH of sludge after the ATAD process shows the standard deviation of 0.37 and the variance of 0.14. The range of pH values of the sludge processed by the ATAD process are shown in Figure 2. The diagram demonstrates that during the ATAD due to occurring biochemical processes pH increases gradually.

For thickened sludge the expected value of pH below and above the same number of observations is 6.62. The median value for sludge after the first ATAD stage in tank No. 1 was 7.58 pH and respectively 7.63 in tank No. 2 and for sludge after the entire process it was 7.92 pH. The most often occurring value of the sediment pH at individual sets determines the mode, which for thickened sludge amounts is 6.62 for sludge from tank No 1 is 7.68 and from tank No. 2 amounts to 7.53. The most often measured value in stabilised sediments was 7.71 pH. The results of the sediment pH are distributions of a right hand symmetry. The largest concentration around the average value is characteristic for the thickened sludge. With a 90 % probability we can also state that the pH value for the thickened sludge will be 6.79 or less. After the first stage of the installation 90% of results show values below 8.37 pH in tank No. 1 and similarly 8.18 pH in tank No. 2, where in the stabilised sludge these values will be lower than 8.54 pH.

Over the period under study the growth of the pH for sludge due to the autothermal thermophilic stabilisation was on the average 1,36, however, for 10% of the results it was 1.90 to 2.23 and in case of 90% of observations the growth was above 0.98. A significant increase was observed after the first stage reactors. After tank No. 1 the pH increased on the average by 1.04 units and after tank No. 2 by 0.99 units. After tank No. 1 the pH increased on the average by 1.04 unit and after tank No. 2 by 0.99 unit.

After reactor No. 3 the pH value grew by 0.32 units on average against reactor 1 and by 0.37, comparing to reactor 2.

When the installation was operated as a three-stage system the characteristic average value of pH for the thickened sludge was 6.60 pH, but ranging between the low of 6.23 and the high of 7.08 pH. After the first stage of the ATAD the pH rose with its mean value being 7.50 and ranging from 7.07 to 7.93. Higher values were recorded in sludge after the second stage of the installation being on the average 7.84 pH and ranging between 7.48 and 8.34 pH. For the stabilised sludge with the process completed the pH value ranged between 7.72 and 8.91, with the calculated average value being 8.14. The largest differentiation of the pH results was found out for the stabilised sludge. The ranges of the pH in sewage sludge under study are shown in Figure 3.

The presented diagram confirms a constant growth of pH for the observed sludge after successive stages of the installation. The median as a consistent and asymptotically unbiased estimator of an expected value shows a growing tendency of the analysed feature, the value of which was successively: for the thickened sludge 6.57 pH after the first installation stage and 7.82 pH after the second stage and 8.09 pH in the stabilised sludge. The results of the sludge pH are distributions of a right hand symmetry. The curve of size of the analysed feature is in case of the thickened sludge more flat than with a normal distribution, however, as the process advances after successive stage it slims. For the analysed results, 90%

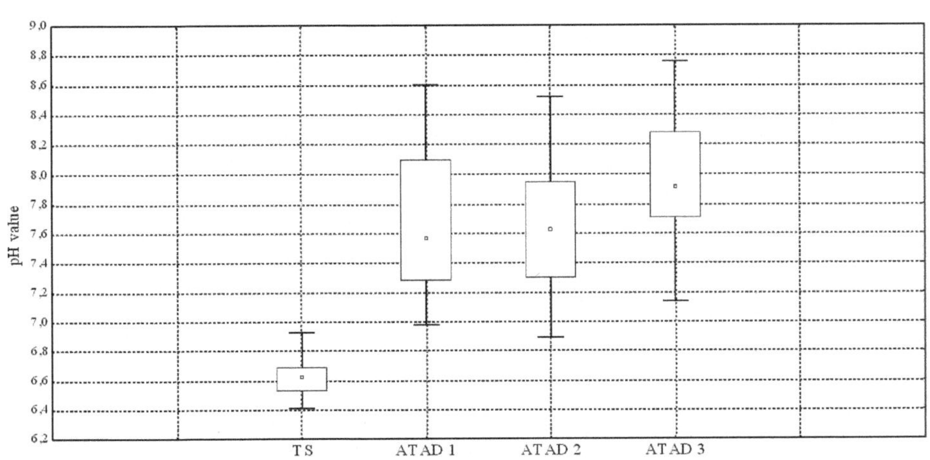

Figure 2. The ranges of the pH value for sludge measured during operation of a two-stage installation

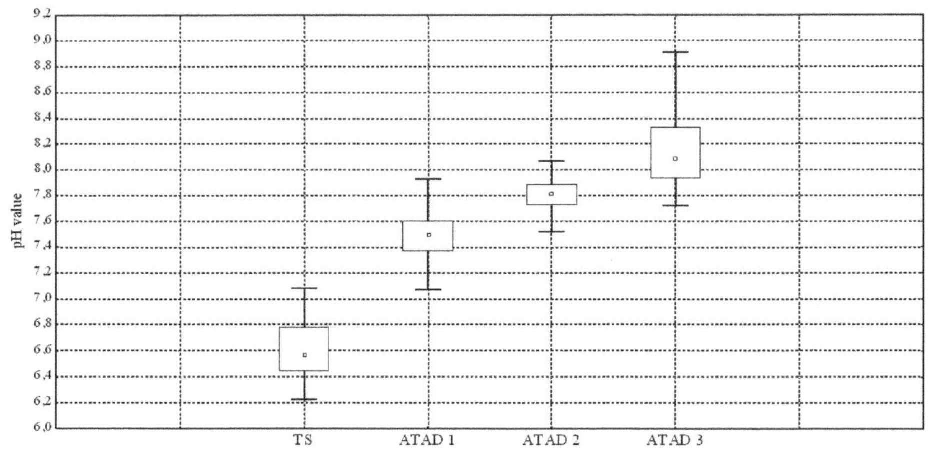

Figure 3. The ranges of pH value for sludge under study during operation of a three-stage installation

of these showed for the thickened sludge values lower than 6.68 pH, after the first stage under 7.76 pH and after the second stage under 8.17 pH. For sludge after the ATAD process 90% of determinations were below the pH value of 8.52.

During the study a growth of the pH value due to the deposits stabilisation was observed – on average by 1.54, however, for 90% of all the analysed results it was above 1.10. The greatest increase in the sludge pH occurred after the first ATAD stage – by 0.90 unit on average. After the second installation stage this value was 0.34 pH on average and after the third stage by 0.30 pH.

Taking into account the chemistry of the Autothermal Thermophilic Aerobic Digestion of sewage sludge certain growth of pH seems to be understandable. The generation of ammonia in the process with it being carried away along with generated gases, however, still present in the liquid and creating a weak base as well as retarding the nitrification process – causing pH decrease – as caused by thermophilic temperatures, may explain this phenomenon. Changes of pH in sludge caused by the Autothermal Thermophilic Aerobic Digestion for their stabilisation according to available bibliography point to ranges similar to those received. A most common growth of pH starts from 6.3 for the raw sludge and goes up to 9.1 for the stabilised sludge [Nosrati 2007, Piterina et al. 2010]. Also narrower ranges of changes are quoted from 7.0 pH for sludge prior to ATAD up to 8.2 pH after the process [Movahedian et al. 2005, Zhelev et al 2008]. Absolute sediment pH growth values due to the ATAD amount usually to 1–2 pH units [Kelly 2007, Zupančič, Roš 2008].

CONCLUSIONS

The Autothermal Thermophilic Aerobic Digestion is a process, which ensures a full sanitation of the sewage sludge and its effective stabilisation [Bartkowska et al. 2005a]. A full automatic control of the process ensures receiving sludge devoid of bacterial contaminations and as being a precious fertiliser a one which can be successfully returned to the natural circulation. The process is characteristic due to its very stable run of biochemical reactions with a short design dwelling time of sludge, therefore, the installation requires only a small area, thus resulting in low investment costs.

The research of the initial operation period of the ATAD-plant carried out at the subject plant confirms that the ATAD-process can be an effective way to stabilise and sanitize sewage sludge. Determination of the sediment pH allows a quick, easy and clear confirmation of the correctness of occurring biochemical processes. Results recorded over the period under discussion and rich experience from observations justify the following conclusions:

- The examination of the sediment pH allow an evaluation of the operation of an installation for the autothermal stabilisation, it confirms the direction of biochemical processes, this direction is additionally confirmed by the results of other parameters, therefore, it can serve as one of process indicators.
- Definitely larger increases in the pH value are observed at the first stage ATAD, where one can deem most of the biochemical processes. So one may consider a greater cubage of the

first installation stage, which will improve the energy balance of the process. The sludge in the first stage reactor will heat up quicker reaching a higher temperature. An increased capacity will also allow supplying more air according to the demand.

- A greater relative increase in the pH value was observed during a three-stage operation of the ATAD-installation. Therefore, we can suppose that in this operation cycle, processes occur more intensively.
- Results gathered during the research are, however, not clear to the extent they might point to a selection of a three-stage system. We must remember that a two-stage operation guarantees greater reliability.
- The presented technology enables to prepare sewage sludge to return it back to the natural cycle.

The differences observed during the research are, however, not so distinct to indicate a more advantageous manner to realise the process. This can be due to a short time of observations or can result from a different composition of sludge supplied to reactors in both cases. Another problem is that the operation regime of a centrifuge for initial compacting of sludge is difficult to maintain. Practical reasons favour a two-stage stabilisation, which with three tanks operating ensures a continuous operation of the installation, even with one tank failed.

REFERENCES

1. Augustin O., Bartkowska I., Dzienis L. 2007. Moving Forward, Wasterwater Biosolids Sustainability: technical, managerial and public synergy, conference proceedings: Efficiency of wastewater sludge disinfection by autoheated thermophilic aerobic digestion (ATAD). In IWA Specialist Conference, pp.1037-1043. Moncton, Canada.
2. Bartkowska I. 2005. Operation's initial period of the autothermal thermophilic sludge instalation (ATAD) in Giżycko waste water treatment plant. Pol. J. Natur. Sci., 18(1), 105-116.
3. Bartkowska I., Augustin O., Dzienis L. 2005a. Skuteczność higienizacji osadów ściekowych w procesie autotermicznej tlenowej stabilizacji. Monografie Polska Akademia Nauk; Komitet Inżynierii Środowiska, 30, 763-774.
4. Bartkowska I., Augustin O., Dzienis L. 2005b. Zastosowanie autotermicznej tlenowej stabilizacji do unieszkodliwiania osadów z oczyszczalni ścieków przemysłu papierniczego. Monografie Polska Akademia Nauk; Komitet Inżynierii Środowiska, 32, 1025-1033.
5. Bartkowska I., Dzienis L. 2007. Technical and economic aspects of autothermal thermophilic aerobic digestion exemplified by sewage treatment plant in Giżycko. Env. Prot. Eng., 33(2), 17-25.
6. Kelly H.G. 2007. Emerging processes in biosolids treatment. J. Environ. Eng. Sci., 5, 175-186.
7. Layden N.M. 2007. An evaluation of autothermal thermophilic aerobic digestion (ATAD) of municipal sludge in Ireland. J. Environ. Eng. Sci., 6, 19-29.
8. Movahedian A.H., Bina B., Moeinian K. 2005. Effects of aeration rate and detention time on thermophilic aerobic digestion of mixed sludge and its dewaterability. Int. J. Environ. Sci. Tech., 2(2), 105-111.
9. Nosrati M., Sreekrishnan T.R., Mukhopadhyay S.N. 2007. Energy Audit, Solids Reduction, and Pathogen Inactivation in Secondary Sludges during Batch Thermophilic Aerobic Digestion Process. Journal of Environmental Engineering, 133(5), 477-484.
10. Piterina A.V., Bartlett J., Pembroke T.J. 2010. Evaluation of the Removal of Indicator Bacteria from Domestic Sludge Processed by Autothermal Thermophilic Aerobic Digestion (ATAD). Int. J. Environ. Res. Public Health., 7(9), 3422-3441.
11. Skowron K., Olszewska H., Pawłowska A. 2011. Ocena wpływu procesu kompostowania na zmiany liczebności populacji grzybów drożdżoidalnych i pleśniowych w osadach ściekowych. Rocznik Ochrona Środowiska, 13, 1485-1506.
12. Wersocki S., Hupka J. 2006. Autotermiczna higienizacja tlenowa osadów nadmiernych. Inżynieria ekologiczna, 14, 63-69.
13. Woźniak M., Żygadło M., Latosińska J. 2004. Ocena chemicznej stabilności osadów ściekowych składowanych w naturalnych warunkach środowiskowych. Ochrona Środowiska, 26(1), 25-31.
14. Zhelev T., Jamniczky-Kaszas D., Brzyszcz B., Trévarain M., Kovacs R., Vaklieva-Bancheva N. 2008. Energy efficiency improvement of wastewater treatment processes using process integration techniques. Report prepared for the Environmental Protection Agency by University of Limerick, 31.
15. Zupančič G.D., Roš M. 2008. Aerobic and two-stage anaerobic–aerobic sludge digestion with pure oxygen and air aeration. Bioresource Technology, 99, 100–109.

EFFECT OF GROWTH REGULATORS ON SELECTED MORPHOLOGICAL FEATURES OF YELLOW PINE

Kazimierz Jankowski[1], Jacek Sosnowski[1], Agnieszka Wilk[1], Elżbieta Malinowska[1], Beata Wiśniewska-Kadżajan[1]

[1] Department of Grassland and Green Areas Creation, Siedlce University of Natural Sciences and Humanities, B. Prusa 14, 08-110 Siedlce, Poland, e-mail: laki@uph.edu.pl

ABSTRACT

The aim of study was to compare the effects of two growth regulators on selected morphological features of yellow pine. In April 2012, 9 plants of yellow pine Lawson variety were planted to the soil. In April 15, 2013 a research agent in a form of following growth regulators: Asahi SL and Moddus 250-EC was introduced. There were the following experimental objects: C – control (only water spraying), As – Asahi spraying, Md – Moddus spraying. The tested features were: increase in plant height (cm), the increase in needle length (cm) of yellow pine. Measurements of increases in length of needles were made in each case on the same specially marked parts of the plants and performed on 27th of each month from beginning of May to ending of September. Among the used of growth regulators the greatest influence on the height increase of the tested trees as well as on the needle length of yellow pine more affected regulator Moddus. The largest total increase in the needles length of yellow pine was on the object with Modus and Asahi, and on control object it was significantly lower than on objects with growth regulators.

Keywords: biostymulants, Moddus 250-EC, Asahi SL, yellow pine.

INTRODUCTION

Stimulators are commonly referred to as biostimulants [Smoleń and Szura, 2008]. According to Jankowski and Dubis [2008] biostimulators are biologically active substances in its composition containing hormones, enzymes, proteins, amino acids, trace elements and other compounds which are typically used in small doses, activate metabolism, mainly in the direction of growth and development of the whole plant. Their main task is to participate in the regulation of life processes at a cellular level, organ and whole organism. Effect of bio-stimulators for plants does not result from direct participation in the regulation of life processes, but from the impact on the metabolism in a broad sense. They stimulate the synthesis of natural hormones, sometimes increase their activity, improve the intake of minerals from the ground and induce root growth.

According to Przybysz et al. [2010] first preparation stimulates of plant growth and development introduced widely into orchard practice was Asahi SL.

In Poland intensive research [Kozak et al. 2008] has been conducted for a few years on the mechanism of interaction of biostimulator Asahi SL on growth, development and yield of field crops, vegetables and fruits.

Another growth regulator is Moddus 250-EC, whose active ingredient is trinexapac ethyle – 250 g/l (25.5%). This preparation has been studied practically only in crops, such as winter and spring wheat, spring and winter barley, rye, winter triticale and oat [Gruszczyk and Berbeć 2004; Michalski et al. 2008; Kozak et al. 2008; Budzyński et al. 200:, Malarz et al. 2008], grown in intensive technologies with high levels of nitrogen fertilization on grasses lawn [Jankowski et al. 2012; Czeluścinski et al.2008]. In the literature, there is very little data on the effects of growth regulators on ornamental trees. If they are connected with leafy trees [Adamiak and Hetman 2007; Bąbelewski and Dębicz 2006; Górnik and Grzesik 2005; Hetman and Adamiak 2003 b] there is no such data in relation to conifers.

Therefore, the aim of the study was to compare the effects of these two growth regulators on selected morphological features of yellow pine.

MATERIALS AND METHODS

The experiment was established at forest school area in Ceranów. In April 12, 2012 9 plants of yellow pine (Pinus ponderosa Dougl. Ex C.) Lawson variety were planted to the soil (soil quality class 5). In first year the plants were only cultivated and infested between rows. In April 15, 2013 a research agent in the form of following growth regulators was introduced: Asahi SL and Moddus 250-EC, at a dose of 100 ml of preparate in 100 l of water.

There were the following experimental objects: C – control (only water spraying), As – Asahi spraying, Md – Moddus spraying. The tested features were: increase in plant height (cm), increase in needle length (cm) of yellow pine. Measurements of increases in length of needles were made in each case on the same specially marked parts of the plants and performed on 27th of each month from beginning of May to the end of September.

The results were evaluated statistically by performing variance analysis. Medium differentiation was verified by Tukey's test at a significance level of $p \leq 0.05$.

Meteorological data for the study area were obtained from the Hydrological and Meteorological Station in Siedlce. However, in order to determine the temporal variability of meteorological elements and their influence on plant growth, the hydrothermal coefficient of Sielianinov was evaluated.

From the data presented in Table 1 resulted that the most favourable distribution and magnitude of rainfall, with optimal air temperatures in the period of plants growing were in May and September. For those months the highest values of coefficient (K>2), were recorded which indicates the optimal relation to the value of the air temperature and precipitation. It should also be noted that in August was aqueous deficit what was transformed into strong drought (K = 0.25).

RESULTS AND DISCUSSION

Increase of pine height (Figure 1) was varied and depend both on the applied biopreparation and the month in which this measurement was made. In regard to the nature of used in the research of growth substance, it was found that the largest increases in pine height (19 cm) was obtained in June, as a result of Asahi formulation. In the successive months of research until September a systematic decrease in the height growth of trees was observed. It is worth noting that in all the months higher increases in plant height were obtained for objects with regulators than for the control object. This relationship has also been confirmed in the assessment of total growth of yellow pine (Figure 2). The highest total growth was achieved on the object with Moddus (42.1 cm) and only slightly less on the object with Asahi (41.0 cm). The differences between these objects were not statistically significant. The total regrowth of yellow pine from objects with regulators differed significantly in compare to the control object (31.0 cm).

These results are confirmed in Hetman and Adamiak [2003 a,b] studies, where the treatment of plants for root-stock with Asahi SL and Titanite preparations had a positive impact on their quality .

Another estimated parameter was the increase in the length of the needles (Table 2). These studies showed different response of needle length to the applied of growth regulators. The largest increase in the needle length was achieved in May after applying of Moddus preparation (1.10 cm). A slightly smaller increase reached the needles after application of Asahi (0.80 cm). The differences between these objects were not significant statistically. In turn, on the control object the increase of the needles length was the lowest (0.30) and differed significantly to the other objects. Also in Bąbelewski [2008] studies was found beneficial effects, eg in the production of bicolor acidantara var Murielska and root-stocks of rose multiflorum [Kocira, Laskowska 2005].

Table 1. The value of hydrothermic coefficient of Sielianinov (K) in each month of the growing season

Year	Month						
	IV	V	VI	VII	VIII	IX	X
2013	1.60	2.20	1.80	1.50	0.25	2.70	1.22
K<0.5 – strong drought; 0.51–0.69 – drought; 0.70–0.99 – poor drought; K> 1 – no drought							

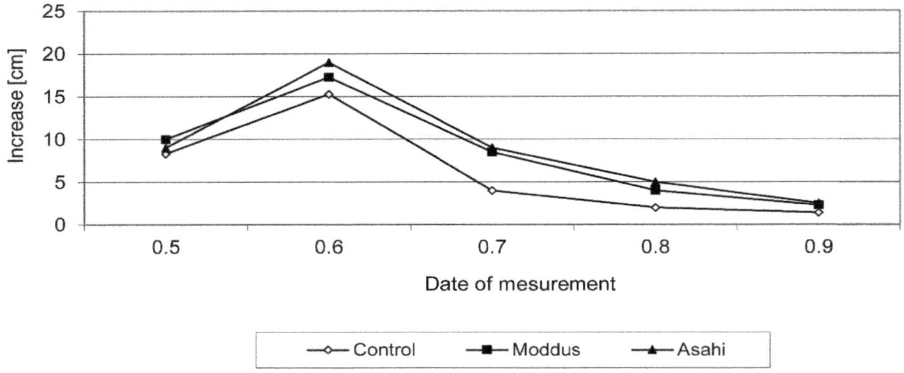

Figure 1. Increase in the height of yellow pine in particular time measurement terms depending on growth regulator

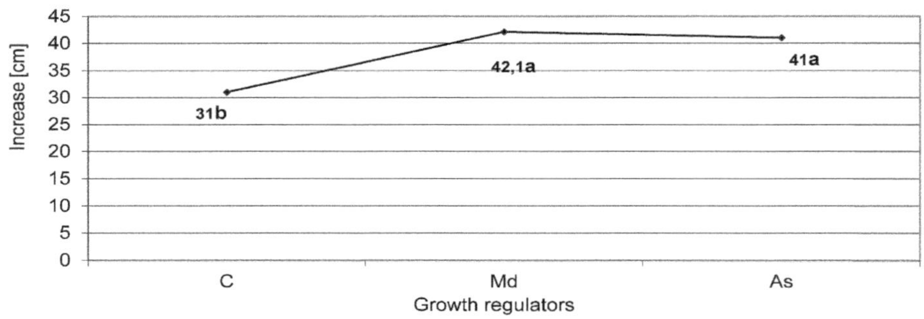

Figure 2. Annual growth in the height of yellow pine, depending on the growth regulator

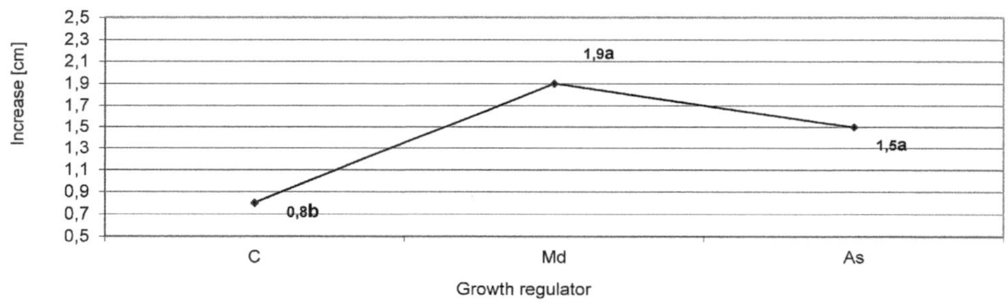

Figure 3. The total increase in the needles length of yellow pine, depending on the growth regulator

Table 2. Increase in the needle length of yellow pine in particular measurement terms depending on growth regulator

Regulator	Date of masarement (month)					Mean
	0.5	0.6	0.7	0.8	0.9	
C	0.30^{Ba}	0.10^{Bb}	0.20^{Bab}	0.10^{Ab}	0.10^{Ab}	0.16^{B}
Md	1.10^{Aa}	0.20^{Ab}	0.40^{Aab}	0.10^{Ab}	0.10^{Ab}	0.38^{A}
As	0.80^{Aa}	0.20^{Ab}	0.30^{ABb}	0.10^{Ab}	0.10^{Ab}	0.30^{A}
Mean	0.73^{a}	0.16^{b}	0.30^{b}	0.10^{b}	0.10^{b}	–

Mean in columns estimated with the same big litters don't differ significantly
Mean in lines estimated with the same small letters don't differ significantly

Asahi SL at concentrations of 0,1% to 0,6% increased the diameter and length of the root colour and the above ground dry matter and the root system [Hetman and Adamiak 2003 b]. These studies showed that to July the effects of growth regulators applied were visible, in the form of different length needles increments. However, in August and September the increase of needle length was 0.10 cm and did not differ between the objects.

The effect of applied growth regulators can be seen in the assessment of the total increase in the needles length of yellow pine (Figure 3) in relation to the growing season (May - September). The largest total increase in the length of needles were obtained on the object with Moddus (1.9 cm) and slightly less on the object with Asahi (1.5 cm). The differences between these objects were not significant statistically. Significantly smaller increase reached the needles on the control object (0.8 cm).

In a study conducted by Hetman and Adamiak [2005] was found that in the rose bushes treated with Asahi SL preparation was obtained from 7 to 12% more shoots of first row than in the control shrubs. Foliar application of preparations with properties stimulating growth and development is a fast way to stimulate plants for proper growth and development.

CONCLUSIONS

1. Among the used growth regulators the greatest influence on the increase height of the tested trees as well as on the needle length of yellow pine were more affected by Moddus regulator.
2. From the vegetable period the largest increase of yellow pine was recorded in June.
3. The largest total increase in the needles length of yellow pine was recorded for the object with Moddus and Asahi, and on control object was significantly lower than on the objects with growth regulators.

REFERENCES

1. Adamiak J., Hetman J. 2007. Wpływ następczy preparatów Asahi Sl i tytanitu stosowanego w uprawie Rosa multiflora Thunb. na jakość okulantów róż odmiany Flamingo. Rocz. AR Poznań. CCCLXXXIII, Ogrodn. 41, 5–10.
2. Bąbelewski P. 2008. Wpływ preparatu Ashai SL na jakość ukorzenionych sadzonek wybranych krzewów ozdobnych z rodzaju żywotnik (Tuja sp.). Zesz. Prob. Post. Nauk. Roln. 525, 33–38
3. Bąbelewski P., Dębicz R. 2006. Wpływ preparatu Asahi SL na wzrost i jakość rozsady torenii ogrodowej (Torenia Fournieri Linden) i fuksji mieszańcowej (Fuchsia Hybryda). Zesz. Nauk UP Wrocław, Rolnictwo LXXXIX 546, 37–41.
4. Budzyński W., Dubis B., Jankowski A., 2008. Response of Winter oilsed rape to the biostymulator Asahi SL in spring. Monographs series. Biostimulators in modern agriculture, Field Crop. Wieś Jutra, 47–55.
5. Czeluściński W., Jankowska J., Ciepiela G. A., Jankowski K. 2008. Wpływ regulatora wzrostu Moddus 250 SC na tempo odrostu muraw trawnikowych, w zależności od rodzaju zastosowanych nawozów. Biostymulatory w nowoczesnej uprawie roślin. Publikator Wieś Jutra Sp. z o.o.
6. Górnik K., Grzesik M. 2005. China aster plant growth, seed field and quality as influence by Asahi SL treatment. Folia Horticulturae. Ann. 17/2 119–127.
7. Gruszczyk M., Berbeć S. 2004. Porównanie wpływu wybranych preparatów stosowanych dolistnie na plony i jakość surowca złocienia maruny (*Chrysanthemum parthenium* L.). Annales UMCS. Sec. E, Agricultura 59(2), 755–759.
8. Hetman J., Adamiak J. 2003 a. Wpływ tytanitu na jakość podkładki róży wielokwiatowej (Rosa multiflora Thunb.). Acta Agroph. 85, 251–256.
9. Hetman J., Adamiak J. 2003 b. Wpływ Ashai SL na jakość podkładki róży wielokwiatowej (Rosa multiflora Thunb.). Zesz. Prob.. Post. Nauk. Roln. 491, 61–65.
10. Hetman J., Adamiak J. 2005. Wpływ przemiennego dolistnego stosowania Asahi SL i tytanitu na jakość róży wielokwiatowej (Rosa multiflora Thunb.). Zesz. Prob. Post. Nauk. Roln. 504 , 399–406.
11. Jankowski K., Czeluściński W., Jankowska J., Kolczarek R., Sosnowski J. 2012. The influence of the growth regulator trinexapac-ethyl on the regrowth rate of lawn grasess. Acta Sci. Pol., Hortorum Cultus 11(4), 67–76.
12. Jankowski K., Dubis B., 2008. Biostymulatory w polowej produkcji roślinnej, Biostymulatory w nowoczesnej uprawie roślin. Publikator Wieś Jutra.
13. Kocira A., Laskowska H. 2005. Plonowania Acidanthery bicolor var. Murielae Perry w zależności od stężenia i formy aplikacji Ashai SL. W, Ogólnopolska konferencja „Postęp w produkcji roślin ozdobnych" Skierniewice, 167–168.
14. Kozak M., Malarz W., Serafin-Andrzejewska M., Kotecki A. 2008. The effect of sowing rate and Asahi SL, biostimulator on soybean sowing and yield . Monographs series: Biostimulators in modern agriculture, Field Crops. Wieś Jutra, 77–84.
15. Malarz W., Kozak M., Kotecki A. 2008. The use of Asahi SL biostimulator in spring rape growing. Monographs series: Biostimulators in modern agriculture, Field Crops. Wieś Jutra, 25–32.
16. Michalski T., Bartos-Sychała M., Maciejewski T., Jarosz A. 2008. Effect of biostimulator Asahi SL on cropping of maize grown for grain. Monographs series: Biostimulators in modern agriculture, Field Crops. Wieś Jutra, 66–76.
17. Przybysz A., Wrochna M., Słowiński A., Gawrońska H. 2010. Stimulatory effect of Asahi SL on selected plant species. Acta Scien. Polo. Hortorum Cultus 9(2) 53–64.
18. Smoleń S., Szura A.2008. Regulatory wzrostu i stymulatory odporności. Hasło ogrodnicze 4.

EXPANSION OF *DIDYMOSPHENIA GEMINATA* (LYNGBE) M. SCHMIDT (BACILLARIOPHYCEAE) IN RUNNING WATERS IN S-E POLAND: NEW RECORDS IN THE PODKARPACIE REGION

Teresa Noga[1], Jadwiga Stanek-Tarkowska[2], Anita Pajączek[2], Łukasz Peszek[3], Natalia Kochman[3]

[1] Department of Biological Foundations of Agriculture and Environmental Education, Faculty of Biology and Agriculture, University of Rzeszów, Ćwiklińskiej 2, 35–601 Rzeszów, Poland, e-mail: teresa.noga@interia.pl

[2] Department of Soil Studies, Environmental Chemistry and Hydrology, Faculty of Biology and Agriculture, University of Rzeszów, Ćwiklińskiej 2, 35–601 Rzeszów, Poland, e-mail: jagodastanek@wp.pl; chuanita66@poczta.fm

[3] International Doctoral Studies in Natural Sciences, Faculty of Biology and Agriculture, University of Rzeszów, Ćwiklińskiej 2, 35–601 Rzeszów, Poland, e-mail: lukaspeszek@gmail.com; kochman_natalia@wp.pl

ABSTRACT

Nowadays *Didymosphenia geminata* is one of the most frequently studied diatoms in the world, due to its massive development in most continents. In Poland, up to the 1990's only a small number of specimens were recorded, from streams in the Tatra Mountains and the Dunajec River. *D. geminata* was found in 2007–2008 in the middle and upper sections of the Wisłok River, while in 2009 and 2011 it was also found in the Wisłoka, Ropa, Bielcza and San. The cells belonged to one morphotype "geminata" and were characterized by high variability of size. It develops in mesotrophic, well-oxygenated, upper parts of flowing waters with rapid currents. It was most abundant below reservoirs on the San, Wisłok and Wisłoka.

Keywords: *Didymosphenia geminata*, running waters, ecology.

INTRODUCTION

Until the 1990's *Didymosphenia geminata* (Lyngbe) M. Schmidt was a species known mainly from the boreal and alpine regions of Europe, Asia (Himalayas), and North America. In Central Europe, until recently, it was very rare and found only in cold, oligotrophic waters, with moderate or high electrolyte content, in the Alps [Krammer, Lange-Bertalot 1986]. The above features confirm the presence of this diatom in unpolluted waters in Scandinavia [Cleve-Euler 1955]. It has been reported in Finland [Kawecka, Eloranta 1987], Sweden [Johansson 1980] and also in Norway, where it was found, amongst other places, in the river Glåma, which is considered to be its natural environment [Skulberg, Lillehammer 1984].

D. geminata was recorded in Poland in the 1960s, but only in oligotrophic streams in the Tatra Mts. [Siemińska 1964] and some other places, e.g. in the Białka River [Kawecka 1965]. In the 1990's this species started to spread to other Carpathian rivers: San, Soła, Skawa, and Raba [WIOŚ 1996, 2000]. Massive development, in a form of mucilaginous colonies, of *D. geminata* was observed below the Solina and Myczkowce reservoirs [Kawecka, Sanecki 2003]. At that time, it was also found in the seston of the Vistula [Bucka 2000, Kasza, Galas 2000]. In 2002 it was reported in the Czarna Orawa River and its tributaries in the Orawa region [Noga 2003]. The species is more and more frequently found in the running waters of SE Poland [Kawecka, Sanecki 2003, Mrozińska-Broda, Czerwik-Marcinkowska 2004, Mrozińska-Broda et al. 2006, Noga et al. 2012].

D. geminata is currently one of the most frequently studied diatom species, and it is spreading, often in large numbers, not only in Central

Europe but also on other continents [Bhatt et al. 2008, Blanco, Ector 2009, Whitton et al. 2009].

The aims of this study were: to investigate the distribution of *Didymosphenia geminata* in the Wisłok River and other rivers and streams in the Podkarpacie region, to make a taxonomic analysis of the population, and to determine its living conditions.

STUDY AREA

The Wisłok River is the longest left-bank tributary of the San, 228.5 km long, with 3516 km^2 of catchment area. The river rises at Mount Kanasiówka (823 m a.s.l.) in the Beskid Niski Mts., near the border between Slovakia and Poland. The catchment of the Wisłok is characterized by a varied landscape, as the river flows through seven geographic mesoregions distinguished within the western part of the Polish Carpathians: starting from the Beskid Niski Mts., through the Bukowskie Foothills (Pogórze Bukowskie) and the Jasielsko-Krośnieńska Basin (Kotlina Jasielsko-Krośnieńska), then forming a border between the Strzyżowskie Foothills (Pogórze Strzyżowskie) and the Dynowskie Foothills (Pogórze Dynowskie), while in its lower section the river crosses the Rzeszowskie Foothills (Podgórze Rzeszowskie) and finally meets the San River in the Podkarpacie Proglacial Valley (Pradolina Podkarpacka) at the altitude of 182 m a.s.l. In its upper section, the Wisłok flows through wooded, mountainous areas (relatively undisturbed by human activity), whereas in the middle and lower sections, it drains farmlands and industrial areas [Kondracki 2001, Dąbal 2005].

The Wisłok River and its major tributaries: Morwawa, Pielnica, Stobnica, Strug and Mleczka belong to the catchment of the San River, which is the left tributary of the Vistula. The Wisłok is a relatively shallow river (especially in its upper course), and it has a rocky bottom, overgrown in some places by aquatic plants. Mleczka River is of different character, its bottom in the middle and lower reaches is silted and the current is very slow. A detailed description of the sampling sites on the Wisłok River is shown in Table 1.

Because of its varied landscape and high conservation value, much of the catchment area is protected by law within the Jaśliski Landscape Park and the Czarnorzecko-Strzyżowski Landscape Park.

The upper section of the river ends in the Besko Reservoir, at the concrete dam in Sieniawa (Figure 1, site 3). The Rzeszów Reservoir (Zbiornik Rzeszowski) near the city of Rzeszów was created to ensure water intake for the city, to protect it against floods, and for recreation. The waters of the Wisłok River are used for industrial and communal purposes. The river receives urban sewage, industrial

Table 1. Morphological and physico-chemical profile of Wisłok River on chosen sites (2007–2008)

Site	1. Wisłok Wielki	2. Rudawka Rymanowska	3. Besko	4. Żarnowa
Section of river	upper			middle
Insolation	average	large	large	large
Bottom	stony, inshore muddy	stony, inshore muddy	stony, overgrown with algae	stony, overgrown with algae and plants
Width [m]	2–4	20–25	20–22	22–25
Depth [m]	0.1–0.2	0.1– 0.5	0.3– 0.5	0.2–1.0
Temperature [°C]	4.8–19.7	5.3–27	4.2–26	3.8–24.2
pH	7.3–8.1	7.4–8.4	7.5–8.2	8.0–8.4
Conductivity [µS cm^{-1}]	212–327	165–410	299–322	460–500
Oxygen dissolve [mg O_2 L^{-1}]	–	7.4–12.4	8–10.5	7.85–11.18
BOD_5 [mg O_2 L^{-1}]	0.9–1.5	0.7–2.1	0.9–2.3	1.1–2.3
COD_{Cr} [mg O_2 L^{-1}]	<10–12.6	<10–22.1	10.6–13.4	13.7–28.7
PO_4 – P [mg L^{-1}]	< 0.05–0.058	< 0.05	< 0.05	0.099–0.392
Total P [mg L^{-1}]	<0.015–0.055	< 0.05	< 0.05	0.068–0.172
Ca [mg L^{-1}]	35–54	50–62.7	47.7–50	69–76
N_{NH4} – N [mg L^{-1}]	<0.1–0.366	0.076–0.106	0.093–0.14	<0.163–0.7
N_{NO2} – N [mg L^{-1}]	0.0031–0.0118	< 0. 003–0.0045	0.006–0.015	0.0169–0.0426
N_{NO3} – N [mg L^{-1}]	<0.2–0.711	0.124–0.928	0.532–1.23	1.31–2.02

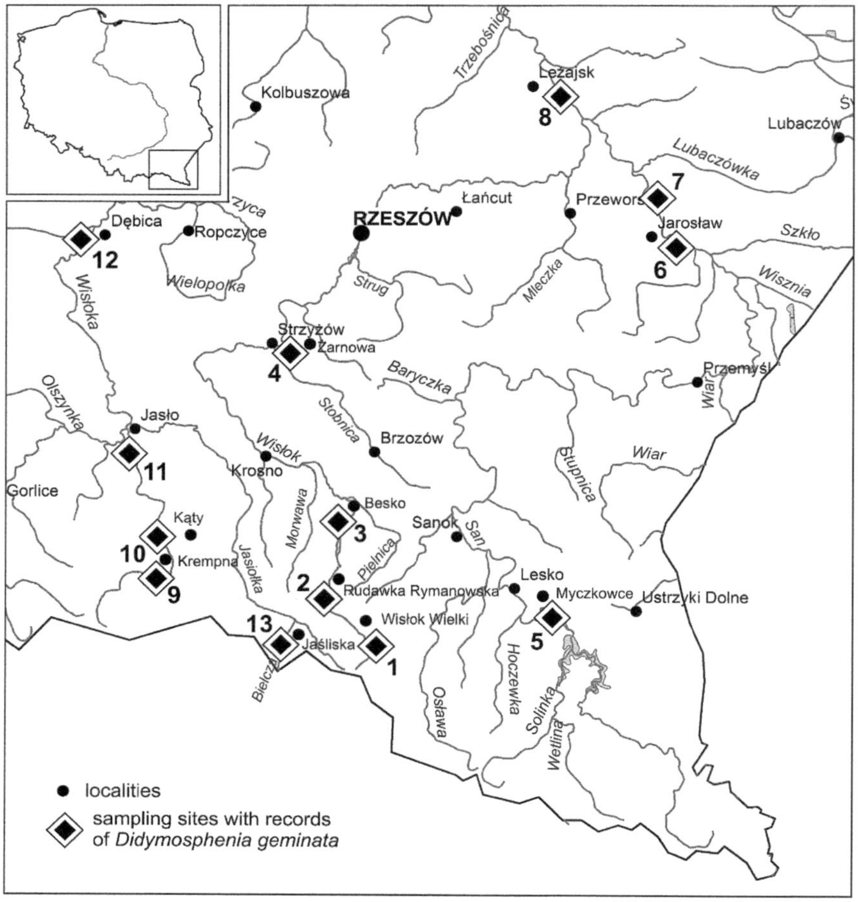

Figure 1. The distribution of sites with records of *Didymosphenia geminata* in Wisłok River and others rivers in the Subcarpatian region

waste, and wastewater from rural areas, discharged directly or indirectly by way of tributaries.

The catchment of the Wisłok is characterized by many large sources of pollution, both communal and industrial. As a result, the Wisłok is distinguished by higher concentrations of organic pollution and nutrients, when compared with other large rivers of the Podkarpacie Region [WIOŚ 2009].

The Wisłoka River is a right bank tributary of the Vistula River, flowing through the area of the Beskid Niski, Jasielskie Foothills (Pogórze Jasielskie), Jasielsko-Krośnieńska Valley (Kotlina Jasielsko-Krośnieńska), Strzyżowskie Foothills (Pogórze Strzyżowskie), Lower Wisłoka Valley (Dolina Dolnej Wisłoki) and Nadwiślańska Lowland. Wisłoka River is 164 km long, with a catchment area of 4110 km^2. It drains half of the area of the Magura National Park, and its sources are near the Dębi Wierch Peak. A small dam was built in 1972 in Krempna on the Wisłoka River, forming a reservoir with the area of 3.2 ha. The initial capacity of the reservoir was 112 000 m^3 (Figure 1, site 9), which at present is over 50% full [Bednarczyk, Michalec 1997]. In its upper reaches the Wisłoka River is a mountainous river, shallow, with a rocky bottom, with a predominance of small and medium-sized stones in the riverbed. All sampling sites were characterized by high insolation and a rapid current (Figure 1, sites 9–12).

The Bielcza River is a left-bank tributary of the Jasiołka (a right-bank tributary of the Wisłok River). On the sampling site (Figure 1, site 13), it was about 3–4 m wide and 30 cm deep. It was characterized by a rocky bottom with a predominance of small and medium-sized stones and a rapid current. The site was well insolated.

The San River is the largest river of the Carpathian Region, it is 443 km long with a 16 861 km^2 catchment area. It flows from Ukraine, from the Pinaszkowski Slope, at the altitude of 961 m a.s.l. [Michałowicz-Kubal, Kubal 2001]. The first sampling site was under the dam in Myczkowce (Figure 1, site 5). The river was about 4–5 m wide and 30–60 cm deep, flowing over quite a wide riverbed, in some places growing clumps of grass vegetation. The river bottom was rocky, with small amounts of mud on the shore and well insolated. The San River (Figure 1, sites 6–8) was about 80 m wide and over 2 m deep in its middle reaches on the other sites. The sites were well insolated, the bottom in the middle was stony, with a predominance of silt and sand at the shore.

MATERIALS AND METHODS

Material for this study was collected from the summer of 2007 to the spring of 2008, in four seasons (07.2007, 10.2007, 02.2008, 05.2008), in the Wisłok River and its major tributaries. In 2009 and 2011 (from May to October, San – 05.2010, Bielcza – 07.2010, Wisłoka – 09.2019 – upper, middle section, and 05.2011 – site in Dębica city), single samples were taken from various sites in the Podkarpacie region, to check if the species was also present in other rivers and streams.

Samples were taken at regular intervals along the Wisłok in the lower, middle and upper parts of its course. In other rivers the sampling came from different places, mainly below major towns and villages along the rivers. The map shows only the localities where the occurrence of *D. geminata* was confirmed (Figure 1).

Samples were collected from stones, sediments, and aquatic plants. All the samples were preserved in 4% formalin, with application of methods used in research on river algae by Kawecka [1980, 2012]. To obtain clean diatom valves, part of each sample was macerated in a mixture of sulphuric acid and potassium dichromate at a ratio of 3:1. Next, the material was washed in a centrifuge (2500 rpm). Finally, the valves were embedded in the synthetic resin Pleurax (refractive index 1.75).

The diatoms were identified under a light microscope (Nikon ECLIPSE 80i), using the keys: Krammer, Lange-Bertalot [1986, 1988, 1991a,b], Lange-Bertalot [1993, 2001], and Krammer [2000].

The abundance of species was estimated by counting the individuals of all diatom species in a randomly selected microscopic field of view, up to a total of 400 cells. The species whose contributions exceeded 5% of the total were regarded as the most common.

RESULTS

The sites where *Didymosphenia geminata* was found in the Wisłok River had alkaline water (pH: 7.3–8.4), and conductivity ranging from 165 to 500 μS×cm^{-1} (Table 1). The water had a mainly mesotrophic character, also eutrophic in its middle and lower courses, whereas in its upper course it was oligotrophic or ranging between oligo- and mesotrophic (Table 1).

1. Distribution of *Didymosphenia geminata* in the Wisłok River and its tributaries

The studies conducted in 2007–2008 showed the presence of *D. geminata* in the upper and middle course of Wisłok River (Figure 1, sites 1–4). It was not found in the studied tributaries (Morwawa, Pielnica, Stobnica, Strug, Mleczka). *D. geminata* developed on rocks in rapid currents. Single deciduous trees were growing along the river. Site 1 was slightly shaded, others were in full sun (Table 1).

In the Wisłok, the species was represented by single individuals, and while it was more abundant at site 3, below the Besko Reservoir (Figure 1), it was still not a dominant species there. At site 4, located in Żarnowa, only a few cells were found in samples, and only in winter (02.2008) and spring (05.2008). At the sites where this species was recorded, the most common diatom species were: *Achnanthidium pyrenaicum, A. minutissimum* var. *minutissimum, Navicula lanceolata, Diatoma moniliformis, Fragilaria vaucheriae* and *Encyonema minutum* (Table 2).

2. Distribution of *Didymosphenia geminata* in other Carpathian rivers

From spring to autumn 2010, the species was found also in the San, Wisłoka, and Bielcza (Figure 1).

Catchment of the river Wisłoka

In the Wisłoka, this species was recorded on four sites (Figure 1, sites 9–12). The diatom was very abundant on site 9, in the village of Myscowa, in September (Plate 1: 7). Mats were observed only on the rocks where the river current was rapid. It was less numerous on two sites (10 and 11), where it formed mucilaginous stalks. It was recorded on site 4 in Dębica, as single cells. All the localities were fully insolated for most of the day, with only single trees scattered along the edges (water temperature 19–20°C, pH 7.5–7.8, conductivity 305–320 μS×cm^{-1}).

D. geminata was also found in the Bielcza stream (a small left-bank tributary of the Jasiołka River, site 13). It developed in large numbers in July and formed gelatinous stalks.

Table 2. The most frequent diatom species coexisting with *Didymosphaenia geminata* and them proportional share on individual sites (2007–2008): 1: <5%; 2: 5–10%; 3: 10–30%; 4: 30–50%

Taxa \ Site	1. Wisłok Wielki	2. Rudawka Rymanowska	3. Besko	4. Żarnowa
Achnanthidium pyrenaicum (Hust.) Kobayasi	3	2	3	1
A. minutissimum (Kütz.) Czarnecki var. *minutissimum*	3	3	4	1
Amphora pediculus (Kütz.) Grunow	1	1	1	3
Cocconeis pediculus Ehrenb.	2	1	2	1
C. placentula var. lineata (Ehrenb.) Van Heurck	1	1	1	2
Cymbella excisa Kütz.	1	2	1	1
Diatoma moniliformis Kütz.	2	2	1	1
Didymosphenia geminata (Lyngbe) M. Schmidt	1	1	1	1
Encyonopsis microcephala (Grunow) Krammer	1	2	1	1
Encyonema minutum Hilse	1	2	2	1
Fragilaria vaucheriae (Kütz.) Petersen	1	2	1	1
Navicula lanceolata (Ag.) Ehrenb.	2	1	1	3
N.reichardttiana Lange-Bert.	1	1	1	2

Catchment of the river San

D. geminata was found in the San, along almost the whole length of the river. Single cells were recorded as far downstream as near the town of Leżajsk. However, the best conditions for it were to be found in the upper section (05.2010), at a site below the dam at Myczkowce (water temperature 8°C, pH 7.5, conductivity 165 $\mu S \times cm^{-1}$), where numerous individuals were observed and the cells formed mucilaginous stalks (Plate 1: 5–6). No mucilaginous mats were observed on pebbles, unlike in the upper section of the Wisłoka.

3. Taxonomy of *Didymosphenia geminata* (Lyngbe) M. Schmidt

Morphological variables of 100 randomly selected cells of *D. geminata* were compared from all sampling sites in the Wisłok River (Plate 1). Their cells were 47.0–107.5 μm long and 20.2–30.0 μm wide. The number of striae per 10 μm ranged from 9 to 13, while the number of stigmata in the central area, from 1 to 3.

With respect to morphology, the individuals from the Wisłok River are slightly different from the diagnosis of this species found in the identification keys (length: 48–140 μm; width: 25–45 μm; number of striae per 10 μm: 8–10, according to Krammer, Lange-Bertalot 1986 and Hofmann et al. 2011), single cells were observed that were shorter and narrower (Plate 1: 1).

DISCUSSION

The Podkarpacie region is still poorly studied with respect to phycology. The diversity of diatoms has been monitored for a few years, mostly in the catchment of the Wisłok River, including observations on the distribution of *Didymosphenia geminata* [Noga 2012, Pajączek et al. 2012]. Single studies were also carried out in the upper section of the San, River, because of the high abundance of the diatom *D. geminata* there [Kawecka, Sanecki 2003]. In 2010–2011 *D. geminata* was found in the rivers Ropa and Biała Tarnowska, over their entire length. On most sampling sites it was mainly recorded in a form of single specimens, on stones. Larger populations were observed only in the Ropa River below the "Klimkówka" dam reservoir, during the production of gelatinous stalks in the spring and autumn. In autumn 2010, a brown mat with a thickness of 2 cm was also observed on stones, covering the bottom of the river by about 40% [Noga et al. 2012]. The present study provides the first data on the occurrence of *D. geminata* in the Wisłok and other Carpathian rivers (except for the Wisłoka, Bielcza and San – and with the exception of the upper section).

Until the 1990's *Didymosphenia geminata* had been regarded as a northern alpine species [Krammer, Lange-Bertalot 1986] and an indicator of xenosaprobic, unpolluted waters [Sladeček 1986]. *D. geminata* was rare in Poland until recently, represented by a small number of individuals, mostly in streams in the Tatra Mountains and the Dunajec River [Siemińska 1964]. How-

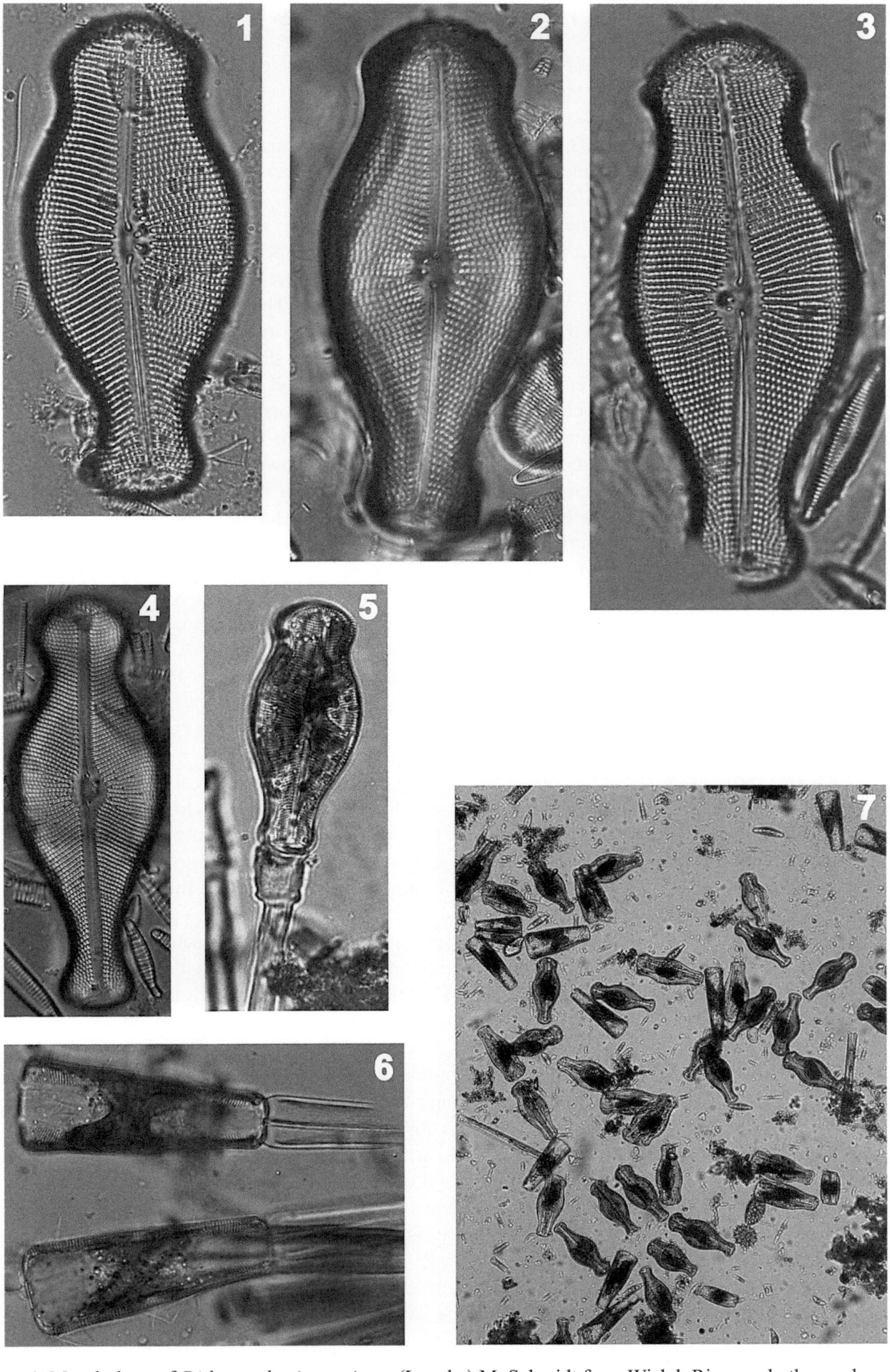

Plate 1. Morphology of *Didymosphenia geminata* (Lyngbe) M. Schmidt from Wisłok River and others subcarpatian rivers; 1–3 (1000×), 4 (400×), 5–6 (200×), 7 (100×)

ever, in recent years it has formed large populations in some rivers and streams of Europe, North America, and New Zealand [Biggs et al. 2006, Blanco, Ector 2009, Whitton et al. 2009]. It is also more and more frequent in the running waters of southern Poland [Kawecka, Sanecki 2003, Noga 2003, Mrozińska-Broda, Czerwik-Marcinkowska 2004, Mrozińska et al. 2006, Noga 2012, Noga et al. 2012]. Currently, it also occurs in the rivers of the Gorce Mountains: Ochotnica, Forendówka and Jaszcze [Noga, unpubl. data].

D. geminata cells identified in the Wisłok River mostly match the dimensions given in the keys [Krammer, Lange-Bertalot 1986, Hofmann et al. 2011]. Single specimens have different dimensions (they were smaller and narrower). One morphotype ("Morphotype geminata") appeared in all the analyzed samples. Metzeltin and Lange Bertalot [1995] noted "Morphotype geminata" from different parts of the northern hemisphere and showed large differences in the size of cells. The smallest specimens of *D. geminata* are recorded in Norway (48–97×26–35 μm) and Siberia (60–110×25–39 μm). In the Wisłok River single cells were observed, which were slightly smaller (47–107.5 μm) and narrower (20.2–30 μm) – Plate 1: 1.

In the Wisłok River *D. geminata* was the most abundant on site 3, below the Besko Reservoir, in mesotrophic waters (Table 1). Nevertheless, it did not form water blooms and in none of the studied seasons was it a dominant species. Among the dominant species on that site, the most abundant were *Achnanthidium pyrenaicum*, *A. minutissimum* var. *minutissimum*, *Cocconeis pediculus* and *Encyonema minutum* (Table 2). On other sites on the Wisłok River *D. geminata* was mostly represented by single individuals during the whole study period. It was not detected in any of the tributaries of the Wisłok (Morwawa, Pielnica, Stobnica, Strug, Mleczka). Most of them are similar to lowland rivers and streams, with a slower water flow or even high turbidity (the Mleczka River).

The materials collected in 2009 from rivers and streams in the Podkarpacie Region showed that *D. geminata* is present also in the catchments of the Wisłoka and San (Figure 1). A very high abundance was observed in September in the upper section of the Wisłoka below the dam in Krempna, where mucilaginous mats (about 1 cm thick) were visible on stones. It was also abundant in May in the upper section of the San, below the Myczkowce Reservoir. In the river, on stones, no mats were noticeable, but under a microscope numerous and long mucilaginous stalks were visible.

The above data on site conditions suggest that a low water level, full light, and well-oxygenated water are favourable for this diatom species. Chemical analyses of water (Table 1) in the Wisłok River showed that it contains high concentrations of nutrients, and thus the river is mesotrophic or even eutrophic. This is also confirmed by the dominant diatom species, most frequently *Encyonema minutum* and *Cocconeis pediculus*. They are α-β-mesosaprobic species, found also in eutrophic waters at pH ≥ 7 [Van Dam et al. 1994]. A frequent dominant, particularly in winter and spring, was *Navicula lanceolata*, a cosmopolitan diatom, with a wide ecological spectrum: from springs and brackish waters. It prefers lower temperatures and forms large populations in the colder part of the year [Krammer, Lange-Bertalot 1986]. The species is eutrophic, α-mesosaprobous and alkaliphilous [Van Dam et al. 1994]. Among the dominants *Achnanthidium minutissimum* var. *minutisimum* was the most abundant. It is a cosmopolitan species, found in a wide range of pH, from 4.3 to 9.2 [Krammer, Lange-Bertalot 1991b], preferring well-oxygenated waters. *A. pyrenaicum* was also abundant in mesotrophic waters, rich in calcium, pH>7 [Van Dam et al. 1994].

D. geminata grows when there are no great floods, in full sun, with a pH close to neutral, and in the presence of specific nutrient chemistry. It is frequent, but not always, in waters where the N:P ratio is high for much of the year. However, the key factor is the ratio of organic to inorganic phosphate. *D. geminata* probably prefers low temperatures and full sun in the spring season [Whitton et al. 2009]. In such conditions it was also observed in the San, at Myczkowce.

Its mats were formed in later seasons, if the dominant form of phosphorus changed from organic to inorganic [Whitton et al. 2005]. In the Wisłoka River, mats were observed in September, i.e. at the end of summer, only on stones and in rapid currents. Mucilageous overgrowth was not observed on stones in places where the current was slow. Our results show that in the Podkarpacie Region *D. geminata* is always the most abundant below reservoirs, located on the San, Wisłok and Wisłoka. It was the most frequent on the Ropa River, also below the "Klimkówka" dam reservoir [Noga et al. 2012]. These sites were always fully insolated.

D. geminata grows in similar conditions in the San River, where it forms very large populations, covering up to 100% of the river bed [Kawecka, Sanecki 2003], and also in the Czarna Orawa River and its tributaries [Noga 2003].

The spreading of *D. geminata* in recent years in Europe, including the southern part of Poland and in other continents, indicates that this species could be expansive. Its required environmental conditions (alkaline pH, moderate values of electrolytic conductivity, well oxygenated, rapid current and mesotrophic character of water) indicate that this species has a greater adaptive potential than believed so far. Our preliminary research in the Podkarpacie Region suggests that *D. geminata* is probably also present in many other rivers and streams in that region, particularly in their upper sections, and its abundance varies between seasons. Further research is needed to determine a more detailed number of sites where the species is present, and the seasons and physicochemical conditions are favourable for its growth.

Acknowledgments

This study was supported partly by the Polish Ministry of Education and Science (grant No. N304 0143 2/0936).

REFERENCES

1. Bednarczyk T., Michalec B. 1997. Wpływ zamulania małych zbiorników karpackich na ich eksploatację. [In:] Współczesne problemy inżynierii wodnej. II Konferencja Naukowa, Wisła 23–25.04.1997. Politechnika Krakowska.
2. Bhatt J.P., Bhaskar A., Pandit M.K. 2008. Biology, distribution and ecology of *Didymosphenia geminata* (Lyngbye) Schmidt an abundant diatom from the Indian Himalayan rivers. Aquatic Ecol. 42: 347–353.
3. Biggs B.J, Kilroy C., Vieglais C.C. 2006. A New Zeland science response to help manage *Didymosphenia geminata* – an unwanted diatom invader of freshwaters, p. 29 [In:] Current Knowledge of *Didymosphenia geminata*: Developing a Research and Management Response. Federation of Fly Fishers and EPA region 8, held in association with Western Division American Fisheries Society Annual Meeting, May 15–16, Montana State University, USA.
4. Blanco S., Ector L. 2009. Distribution, ecology and nuisance effects of the freshwater invasive diatom *Didymosphenia geminata* (Lyngbye) M. Schmidt: a literature review. Nova Hedwigia 88(3–4): 347–422.
5. Bucka H. 2000. Upper Vistula River: response of aquatic communities to pollution and impoudment. VI. Net phytoseston communities. Pol. J. Ecol. 50: 167–189.
6. Cleve-Euler A. 1955. Die Diatomeen von Schweden und Finland. Handl. K. Svenska Vetensk Akad., Fjärde Ser. 5(4): 1–232.
7. Dąbal L. 2005. Wisłok – niedoceniany ekosystem. Dzikie Życie 9: 135.
8. Hofmann G., Werum M., Lange-Bertalot H. 2011. Diatomeen im Süβwasser – Benthos vom Mitteleuropa. Bestimmungsflora Kieselalgen für die ökologische Praxis. Über 700 der häufigsten Arten und ihre Ökologie. [In:] H. Lange-Bertalot (ed.), pp. 908. A.R.G. Gantner Verlag K.G.
9. Johansson C. 1980. Attached algae vegetation in some streams from the Norssag area South Greenland. Acta Phytogeogr. Svecica 68: 89–96.
10. Kasza H., Galas J. 2000. Upper Vistula River: response of aquatic communities to pollution and impoudment. II. Chemical composition of water and sediment. Pol. J. Ecol. 50: 123–135.
11. Kawecka B. 1965. Communities of benthic algae in the River Białka and its Tatra tributaries The Rybi Potok and Roztoka. Komitet Zagosp. Ziem Górskich PAN 11: 113–129.
12. Kawecka B. 1980. Sessile algae in European mountains streams. 1. The ecological characteristics of communities. Acta Hydrobiol. 22: 361–420.
13. Kawecka B. 2012. Diatom diversity in streams of the Tatra National Park (Poland) as indicator of environmental conditions. W. Szafer Institute of Botany Polish Academy of Sciences, Kraków.
14. Kawecka B., Eloranta P. 1987. Communities of sessile algae in some small streams of Central Finnland. Comparision of the algae of the high mountains of Europe and those of its northen regions. Acta Hydrobiol. 29: 403–415.
15. Kawecka B., Sanecki J. 2003. *Didymosphenia geminata* in running waters of southern Poland – symptoms of change in water quality? Hydrobiologia 495: 193–201.
16. Kondracki J. 2001. Geografia regionalna Polski. PWN, Warszawa,
17. Krammer K. 2000. The genus *Pinnularia.* In: H. Lange-Bertalot (ed.), Diatoms of Europe. 1, A.R.G. Ganter Verlag K.G., Vaduz.
18. Krammer K., Lange-Bertalot H. 1986. Bacillariophyceae. 1. Naviculaceae. In: H. Ettl, J. Gerloff, H. Heyning, D. Mollenhauer (eds), Süsswasserflora von Mitteleuropa. 2(1), G. Fischer Verlag, Stuttgart – New York.
19. Krammer K., Lange-Bertalot H. 1988. Bacillariophyceae. 2. Bacillariaceae, Epithemiaceae, Surirellaceae. [In:] H. Ettl, J. Gerloff, H. Heyning, D. Mollenhauer (eds.), Süsswasserflora von Mitteleuropa. 2(2), G. Fischer Verlag, Stuttgart–New York.

20. Krammer K., Lange-Bertalot H. 1991a. Bacillariophyceae. 3. Centrales, Fragilariaceae, Eunotiaceae. [In:] H. Ettl, J. Gerloff, H. Heyning, D. Mollenhauer (eds), Süsswasserflora von Mitteleuropa. 2(3), G. Fischer Verlag, Stuttgart – Jena.
21. Krammer K., Lange-Bertalot H. 1991b. Bacillariophyceae. 4. Achnanthaceae, Kritische Ergänzungen zu *Navicula* (Lineolate) und *Gomphonema* Gesamtliteraturverzeichnis. [In:] H. Ettl, J. Gerloff, H. Heyning, D. Mollenhauer (eds.), *Süsswasserflora von Mitteleuropa.* 2(4), G. Fischer Verlag, Stuttgart – Jena.
22. Lange-Bertalot H. 1993. 85 new taxa and much more than 100 taxonomic clarifications supplementary to Süsswasserflora von Mitteleuropa. 2(1–4), Bibliotheca Diatomologica 27. J. Cramer, Berlin – Stuttgart.
23. Lange-Bertalot H. 2001. Diatoms of the European inland waters and comparable habitats. *Navicula* sensu stricto, 10 Genera Separated from *Navicula* sensu lato, *Frustulia.* [In:] H. Lange-Bertalot (ed.), Diatoms of Europe. 2, A.R.G. Gartner Verlag, K.G., Vaduz.
24. Metzeltin D., Lange-Bertalot H. 1995. Kritische Wertung der Taxa *Didymosphenia* (Bacillariophyce). Nova Hedwigia 60: 381–405.
25. Michałowicz-Kubal M., Kubal G. 2001. Powiat Jarosławski. Oficyna Wydawnicza „ALPA", Krosno (in Polish).
26. Mrozińska-Broda T., Czerwik-Marcinkowska J. 2004. Eucaryotic algae and cyanobacteria in the River Dunajec upstream and downstream from the New dam reservoirs in Czorsztyn and Sromowce and their use for monitoring. Oceanological and Hydrobiological Studies 33(3): 83–97.
27. Mrozińska-Broda T., Czerwik-Marcinkowska J., Gradziński M. 2006. A new species of *Didymosphenia* (Bacillariophyceae) from the Western Carpathian Mountains of Poland and Slovakia. Nova Hedvigia 83: 499–510.
28. Noga T. 2003. Dispersion of *Didymosphenia geminata* in the flowing waters of Southern Poland – new sites of species occurrence in the Orawska Watershed and the Orawska Basin. Oceanological and Hydrobiological Studies 32(4): 159–170.
29. Noga T. 2012. Diversity of diatom communities in the Wisłok River (SE Poland). [In:] K. Wołowski, I. Kaczmarska, J.M. Ehrman, A.Z. Wojtal (eds), Phycological Reports: Current advances in algal taxonomy and its applications: phylogenetic, ecological and applied perspective. pp. 109–128. Institute of Botany Polish Academy of Sciences, Krakow.
30. Noga T., Stanek-Tarkowska J., Irlik E., Soliwoda K., Peszek Ł. 2012. Nowe stanowiska *Didymosphenia geminata* w Ropie i Białej Tarnowskiej (Polska południowa). Inżynieria Ekologiczna 30: 257–265 (in Polish with English summary).
31. Pajączek A., Musiałek M., Pelczar J., Noga T. 2012. Diversity of diatoms in the Mleczka River, Morwawa River and Różanka Stream (tributaries of the Wisłok River, SE Poland), with particular reference to threatened species. [In:] Phycological Reports: Current advances in algal taxonomy and its applications: phylogenetic, ecological and applied perspective, K. Wołowski, I. Kaczmarska, J.M. Ehrman, A.Z. Wojtal (eds). Institute of Botany Polish Academy of Sciences, Krakow, 129–152.
32. Siemińska J. 1964. Bacillariophyceae – Okrzemki. In: K. Starmach (ed.), Flora Słodkowodna Polski. 6. PWN, Warszawa.
33. Skulberg O.M.A., Lillehammer A. 1984. Glåma. In: B.A. Whitton (ed.), Ecology of European Rivers, pp. 469–498. Blackwell Sci. Publ., Oxford-Melbourne.
34. Sladeček V. 1986. Diatoms as indicators of organic pollution. Acta Hydrochim. Hydrobiol. 14(5): 555–566.
35. Van Dam H., Martens A., Sinkeldam J. 1994. A coded checklist and ecological indicator Values of freshwater diatoms from the Netherlands. Netherlands J. Aquatic Ecol. 28: 117–133.
36. Whitton B.A., Al-Shehri A.M., Ellwood N.T., Turner B.L. 2005. Ecological aspects of phosphatase activity in cyanobacteria, eukaryotic algae and bryophytes, [In:] B.L. Turner, E. Frossard, D.S. Baldwin (eds), Organic Phosphorus in the Environment, pp. 205–241. CAB International, Wallingford, UK.
37. Whitton B.A., Ellwood N.T., Kawecka B. 2009. Biology of the freshwater diatom *Didymosphenia*: a review. Hydrobiol. 630: 1–37.
38. WIOŚ – Jasło. 1996. Raport o stanie środowiska w województwie podkarpackim, (The state of the environment, report in Subcarphatian Voivodeship).
39. WIOŚ – Kraków. 2000. Raport o stanie środowiska w województwie krakowskim, (The state of the environment, report in Cracow Voivodeship).
40. WIOŚ – Rzeszów. 2009. Raport o stanie środowiska w województwie podkarpackim, (The state of the environment, report in Subcarphatian Voivodeship).

MULTI-TRAIT EVALUATION OF VALUE FOR CULTIVATION AND USE OF EARLY MATURING EDIBLE POTATO CULTIVARS REGISTERED IN POLAND

Katarzyna Rymuza[1]

[1] Department of Quantitative Methods and Spatial Management, University of Natural Sciences and Humanities in Siedlce, Prusa 14, 08-110 Siedlce, Poland, e-mail: katarzyna.rymuza@uph.edu.pl

ABSTRACT

The work presents an analysis of diversity and comparison of value for cultivation and use of early maturing potato cultivars registered with the Polish National Register of Cultivars. The comparison was based on 17 yield and appearance traits and quality attributes of tubers as well as their resistance to diseases. The analysis employed the following multi-dimensional statistical methods: principal component analysis and cluster analysis. The principal component analysis revealed that over 70% of the total variation was associated with the first 6 principal components. Cluster analysis yielded 4 groups of genotypes. The first group consisted of the cultivars which produced tubers with the most shallow eyes, the best flavour and the least severe darkening of raw flesh. The cultivars in the second group produced high yields and were low in starch, dry matter and glycoalkaloids. The tubers of cultivars which were classified into the third group had the highest starch, dry matter and vitamin C contents. However, they produced the lowest yields and were quite susceptible to most diseases. The fourth group was made up of high-yielding cultivars which tended to accumulate glycoalkaloids but were most resistant to hollow heart in tubers.

Keywords: edible potato, potato cultivar, principal component analysis, cluster analysis

INTRODUCTION

Potato (*Solanum tuberosum*) is one of the major crop plants cultivated worldwide due to its versatility as it is consumed fresh, in processed forms and also used for industrial processing purposes [Oruna-Concha et al. 2002, Zgórska and Grudzińska 2012]. The plant's popularity is obvious when one considers the number of potato cultivars grown worldwide (over 4000) [Hamouz et al. 2010]. The most valuable cultivars produce high and stable yields of tubers which have attractive appearance and are of good quality. Apart from sensory traits, potato quality is also influenced by dry matter, starch and vitamin C contents [Hamouz et al. 2006, Murnice et al. 2011, Arvanitoyannis et al. 2012]. Moreover, good quality tubers should contain no or low levels of harmful substances, glycoalkaloids in particular [Trawczyński and Wierzbicka 2011].

In Poland, like in other EU countries, cultivation of early maturing potatoes is gaining popularity [Dwořák et al. 2006, Hamouz at.al. 2006, Wierzbicka 2012. As a result, the number of such genotypes registered is also increasing Rykaczewska 2007]. Cultivars registered (in variety-testing experiments and post-registration testing) are evaluated with respect to many traits which are then used to prepare a cultivar description. A great number of traits makes it difficult to select a suitable cultivar. When this is the case, multi-dimensional statistical analyses may be employed and they are increasingly popular in natural sciences [Madlaina et al. 2008, Grużewska et al. 2009, Rakonjac et al. 2010, Mohamed et al. 2011, Rymuza et al. 2012]. The methods make it possible to evaluate variation of objects with respect to many traits and to separate them into groups which share similar characteristics. Principal component analysis (PCA) and cluster analysis are most frequently used to this end [Veronesi and Falcinelli 1988, Charmet et al. 1994, Ahmadizadeh and Felenji 2011, Arvanitoyannis et al. 2012], because, according to Pluta et al. [2012], they are most effective.

Bearing the above in mind, an attempt was made to evaluate the multi-trait diversity of value for cultivation and use of early maturing edible potato cultivars registered in Poland in 2012 and classify them, by means of cluster analysis, into groups sharing similar characteristics.

MATERIALS AND METHODS

The analysis was based on the data describing potato cultivars registered in the Register of Cultivars (National List). Research centers all over Poland conduct studies to assess the registered cultivars in terms of characteristics associated with agrotechnology and use. The results of these studies are further analysed synthetically to produce a final description of potato cultivars. The data were also used in the present work.

The following groups of traits describing a cultivar's value for cultivation and use were considered in this analysis:

- tuber yield (t/ha);
- tuber morphological characteristics: tuber size (9-point scale), tuber shape regularity (9-point scale where 9 – perfect regularity, 1 – deformed), eye depth (9-point scale where 9 – very shallow eyes, 1 – very deep eyes);
- usable and technological value: starch content (%), dry matter content (%); vitamin C content (mg%) glycoalkaloid content (mg×kg^{-1}), flavour (9-point scale where 9 – delicious flavour, 1 – bad flavour, tuber unsuitable for consumption), darkening of raw and cooked (by boiling) tuber flesh (9-point scale where 9 – no darkening, a 1 – the most severe darkening), resistance to mechanical injury;
- plant resistance to diseases and pests: resistance to potato virus Y (PVY) (9-point scale where 9 – very high resistance, 1 – very low resistance), resistance to potato leafroll virus (PLRV) (9-point scale where 9 – very high resistance, 1 – very low resistance), resistance to potato blight (9-point scale where 9 – very high resistance, 1 – very low resistance), resistance to hollow heart (5-point scale where 1 – very low resistance, 5 – very high resistance), and resistance to internal rust spot (5-point scale where 1 – very low resistance, 5 – very high resistance).

The analysis was based on two multi-dimensional statistical analyses: principal component analysis and cluster analysis [Badenes et al. 2000, Lotti et al. 2008]. As the variables used in the analysis had different units, they were standardised. Principal component analysis (PCA) was used to determine multi-dimensional associations between the traits studied. PCA made it possible to reduce a relatively high number of diagnostic variables to only several formal variables called principal components [Mohammadi and Prasanna 2003, Kreutzmann et al. 2011, Arvanitoyannis et al. 2012]. The number of components for consideration was selected based on the Kaiser's criterion according to which only the variables with the value of more than 1 are analysed [Iezzoni and Pritts 1991, Lotti et al. 2008]. Such components carry only the most significant information which reflects variation of the objects so they were used in the second part of the analysis – cluster analysis. The purpose of this analysis is to group objects which are similar with respect to many characteristics (variables) [Mądry et al. 2010, Pluta et al. 2012]. Euclidean distance was used as a measure of distance between objects and Ward's procedure as an agglomeration method [Nikolic et al., 2010]. The intersection point was determined applying the Mojena's rule [Rymuza and Bombik 2012].

All the calculations were performed using Statistica 10.0 PL.

RESULTS AND DISCUSSION

Following works by Arslanoglu et al. [2011 and Haydar [2007], principal component analysis and cluster analysis were used in order to examine multi-dimensional relationships between characteristics describing *Solanum tuberosum* L genotypes. The methods yielded groups of cultivars with similar parameters reflecting potato value for cultivation and use.

Principal component analysis demonstrated that early maturing potato quality was associated with the first six components: PC1, PC2, PC3, PC4, PC5 and PC6. The components explained over 78.34% of the total variance – that is overall multi-dimensional variation of traits (Table 1).

The first principal component, explaining over 21% of total variation, was most closely correlated with dry matter content (r = 0.858) and starch content (r = 0.754). Similar relations were

Table 1. Factor loads, eigenvalues and proportion of the total variance in potato cultivars, as explained by the first five principal components

Traits	PC 1	PC 2	PC 3	PC 4	PC 5	PC 6
Tuber field	-0.172	-0.095	0.016	0.157	-0.827	-0.062
Tuber size	-0.154	0.232	-0.107	-0.071	-0.817	-0.037
Tuber shape regularity	0.364	0.336	0.663	-0.240	0.074	0.086
Eye depth	0.008	0.302	0.772	-0.030	0.430	-0.091
Delicious flavour	-0.550	0.045	0.209	0.711	-0.076	0.045
Darkening of raw tuber flesh	0.063	0.739	0.369	0.187	-0.296	-0.240
Darkening of cooked tuber flesh	-0.532	-0.358	0.453	0.004	0.140	0.437
Starch content	0.754	-0.058	-0.041	0.061	0.484	0.102
Dry matter content	0.858	0.038	0.210	-0.129	0.064	-0.194
Vitamin C content	0.164	-0.008	0.060	-0.056	-0.059	-0.890
Glycoalkaloid content	0.302	-0.731	0.176	0.393	-0.126	0.045
Resistance to internal rust spot	0.105	-0.010	-0.021	0.895	-0.074	-0.057
Resistance to hollow heart	0.200	-0.069	-0.317	0.435	-0.533	0.322
Resistance to mechanical injury	-0.306	0.082	-0.368	0.666	0.072	0.183
Resistance to potato virus Y (PVY)	0.045	0.096	-0.766	-0.092	0.199	0.068
Resistance to potato leafroll virus (PLRV)	-0.241	0.491	0.041	0.082	0.189	-0.510
Resistance to potato blight	-0.234	-0.814	-0.163	-0.117	0.020	-0.125
Eigenvalue principal components	3.66	2.74	2.16	1.87	1.67	1.22
Explained proportion of tatal variance(%)	21.52	16.13	12.69	11.01	9.8	7.19
Comulative proportion of total vaiance %	21.52	37.65	50.34	61.35	71.15	78.34

reported by Ahmadizadeh and Felenja [2011]. The values of correlation coefficients indicated that cultivars rich in starch and dry matter have got worse flavour and tend to darken after cooking (Table 2). Also, Thybo et al. [2000], Thygensen et al. [2001] and Kreutzmann et al. [2011] reported similar finding and concluded that dry matter and starch contents are the main factors determining the culinary value of potato tubers. Moreover, Murnice et al. [2011] as well as Zgórska and Grudzińska [2012] have pointed to the relationship of tuber texture and flavour with starch and dry matter content.

The second principal component, strongly associated with plant resistance to potato blight (r = -0.814), after-cooking darkening (r = 0.739) and glycoalkaloid content (-0.731 mg), explained over 16% of total variation. Potato tubers whose raw flesh tended to darken were also less resistant to potato blight and had higher glycoalkaloid contents. Negative correlation of raw tuber darkening and resistance to potato blight with PC2 can be explained by the fact that pathogen-resistant tubers contain more phenolic compounds, in particular chlorogenic acid and tyrosine [Styszko, 2001, 2004].

The third principal component is mainly the carrier of information about tuber morphological characteristics (eye depth and regularity of shape) which are very important from consumer's point of view [Styszko et al., 2003]. What is more, the component is quite strongly correlated with resistance to potato virus Y (-0.766). Cultivars which have tubers more regular in shape are usually more shallow-eyed but less resistant to PVY.

The fourth principal component, accounting for 11% of total variation, was predominantly correlated with resistance to internal rust spot (r = 0.895), flavour (r = 0.711) and resistance to mechanical injury (r = 0.666). Potato tubers with better flavour are more susceptible to internal rust spot and mechanical injury. PC5 explained only under 10% of the total variation of genotypes studied and is strongly correlated with yield (r = 0.827) and tuber size (r = 0.817). The last component, which explained only 7% variability, carried information about vitamin C content (r = -0.890) and resistance to PVY (r = -0.510).

The cluster analysis, based on the above-discussed principal components, yielded a dendrogram presented in Figure 1.

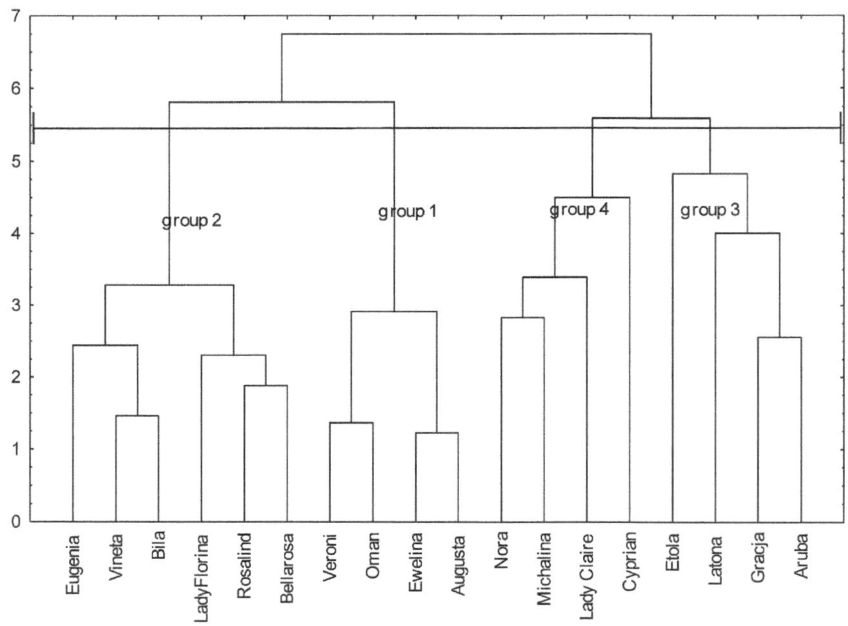

Figure 1. Dendrogram for the 18 cultivars obtained for the first six principal components

The cultivars were divided into 4 groups with different properties. Mean values of traits for each group are compiled in Table 2. Figure 2 shows spatial variation of the cultivars as influenced by the first three components, which explained over 50% of the total variation. The first group was made up of two German cultivars (Augusta, Ewelina), one Polish cultivar (Oman) and one Dutch cultivar (Veroni) (Table 3). The cultivars were slightly (positively or negatively) associated with PC1, PC2 and PC3. They had the most shallow eyes (7.55, on average), the best flavour (7.20 on average) and their raw flesh darkened least (7.95), which indicates that they are the best cultivars for consumption when fresh and for processing [Tarant, 2005, Chotkowski and Rembeza, 2005]. Other characteristics of this cultivar included the highest resistance of tubers to internal rust spot (4.75) and mechanical injury (7.0), and the highest susceptibility to potato virus Y (4.35).

Table 2. Means of the observed traits for the potato cultivars for the four groups distinguished by the cluster analysis

	Group 1	Group 2	Group 3	Group 4
Tuber field	37.10	38.67	35.95	41.13
Tuber size	7.75	8.42	7.63	8.25
Tuber shape regularity	7.30	7.22	7.33	7.15
Eye depth	7.55	7.18	7.15	7.03
Delicious flavour	7.20	7.02	6.55	6.95
Darkening of raw tuber flesh	7.95	7.93	7.03	6.73
Darkening of cooked tuber flesh	8.43	8.17	7.88	8.68
starch content	12.95	12.58	14.28	13.15
Dry matter content	20.60	19.78	22.58	20.48
Vitamin C content	20.80	18.78	22.20	16.78
Glycoalkaloid content	74.25	38.17	71.50	90.25
Resistance to internal rust spot	4.75	3.83	3.50	3.75
Resistance to hollow heart	3.75	4.17	3.75	4.50
Resistance to mechanical injury	7.00	6.83	5.00	5.75
Resistance to potato virus Y (PVY)	4.35	6.42	6.38	5.50
Resistance to potato leafroll virus (PLRV)	6.50	6.58	5.88	4.50
Resistance to potato blight	3.00	2.67	3.50	3.50

Table 3. List of the potato cultivars, their geographic origins and their assigment to homogenous group, as by cluster analysis

Cultivar	Year	Country of origin	Group number	Cultivar	Year	Country of orgin	Group number
Aruba	2007	Poland	3	Lady Claire	2001	Holand	4
Augusta	2003	German	1	LadyFlorina	2004	Holand	2
Bellarosa	2006	German	2	Latona	1997	Holand	3
Bila	1994	Poland	2	Michalina	2010	Poland	4
Cyprian	2007	Poland	4	Nora	2003	Greman	4
Etola	2009	Poland	3	Oman	2005	Poland	1
Eugenia	2006	Poland	2	Rosalind	2001	German	2
Ewelina	2006	German	1	Veroni	2005	Holand	1
Gracja	2002	Poland	3	Vineta	1999	German	2

The second group consisted of cultivars which were only slightly correlated with PC1 and PC2, and moderately and positively associated with the third principal component. The group included three German cultivars: Bellarosa, Rosalind and Vinieta, two Polish cultivars: Bila and Eugenia, and Lady Florina – a Dutch cultivar. They are high-yielding varieties (38.67 t/ha) which produce large tubers (8.42) with low starch, dry matter and glycoalkaloid contents (12.58%, 19.78% and 38.17mg/kg, respectively). It is assumed that the levels of glycoalkaloids in tubers should not exceed 200 mg/kg [Friedman and Mc Donald 1997 Hall 1992]. A low tendency for these cultivars to accumulate TGA indicates that it is unlikely that dangerous concentrations of glycoalkaloids could be exceeded in potato under field conditions of production [Trawczyński and Wierzbicka 2011]. The cultivars in this group are quite resistant to potato virus Y, potato leafroll virus and mechanical injury but, at the same time, susceptible to potato blight (2.67) (table 2,3, fig.1)

As shown in Figure 2, most cultivars in the third group are highly and negatively correlated with PC1 and positively correlated with PC2. Tubers of these cultivars are quite tasty 6.55), tend to darken after cooking (7.88) and accumulate most starch (14.28%), dry matter (22.58%) and vitamin C (22.20 mg%). However, their yields are the lowest and they are most susceptible to internal rust spot, hollow heart and mechanic injury.

Cultivars in the fourth group were positively correlated with PC1 and PC2 and negatively associated with PC3. The group was made up of the following cultivars: Nora, Michalina, Lady Clarine and Cyprian. They produced the highest yields (41.13 t/ha) of tubers with the least after-cooking discolouration (8.68) and low resistance to potato

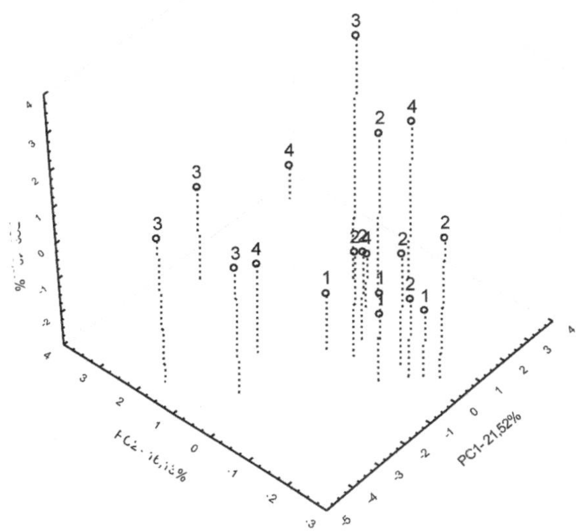

1, 2, 3, 4 – espective signs denote cultivars assigned to particular groups identified by cluster analysis

Figure 2. Dystrybution of 18 potato cultivars terms of 17 traits in the system of the three components

leafroll virus (4,50) which tended to accumulate most glycoalkaloids (90.25) and were most resistant to hollow heart (4.50). The considerably high yield potential is clear a evidence of the fact that breeding progress makes it possible to create genotypes which give stable crops under favourable conditions irrespective of geographical location: Michalina and Cyprian are Polish cultivars, Lady Clarine is a Dutch cultivar and Nora is a German cultivar (Table 2, 3 and Figure 1, 2).

CONCLUSION

The analytical tools presented here are useful when one wants to conduct a complex and detailed analysis of trait diversity among cultivars.

These tool make it possible to classify genotypes simultaneously in terms of many characteristics and describe them (typology). Cultivar classification makes it easier to choose the best cultivars depending on their use, e.g.: to produce chips, crisps or to grow organic potatoes.

REFERENCES

1. Ahmadizadeh M., Felenji H., 2011. Evaluating diversity among potato cultivars using agro-morphological and yield components in fall cultivation of Jiroft area. American-Eurasin Journal of Agricultural and Environmental Science, 11(5), 655–662.
2. Arslanoglu F., Aytac S., Karaca-Oner E., 2011. Morphological characterization of the local potato (Solanum tuberosum L.) genotypes collected from Eastern Black Sea region of Turkey. African Journal of Biotechnology, 10(6), 922–932.
3. Arvanitoyannis I.S., Mavromatis A.G., Vaitsi O., Korkovelos A., Golia E., 2012. Effect of genotype and geographical origin on potato properties (physical and sensory) for authenticity purposes. Journal of Agricultural Science, 4(4), 63–74.
4. Badenes M.L., Martinez-Calvo J., Llacer G., 1998. Analysis of apricot germplasm from the European ecogeogaphical group. Euphytica 102, 93–99.
5. Charmet G., Balfourier F., Monestiez P., 1994. Hierarchical clustering of perennial ryegrass populations with geographic contiguity constraint. Theoretical and Applied Genetic, 88, 42–48.
6. Chotkowski J., Rembeza J., 2005. Consumer Preferences and Changes in Potato Consumption. Roczniki Naukowe Stowarzyszenia Ekonomistów Rolnictwa i Agrobiznesu 7(8), 42–49 (in Polish).
7. Dwořák P., Homouz K., Jůzl M., Erhatova D., 2006. Influence of row covering with non-woven textile on tubers quality in early potatoes. Zeszyty Problemowe Postępów Nauk Rolniczych, 511, 225–231.
8. Friedman M., McDonald G.M., 1997. Potato glycoalkaloids: chemistry, analysis, safety and plant physiology. Critical Reviews in Plant Sciences, 16, 55–132.
9. Grużewska A., Biesiada-Drzazga B., Markowska M., 2009. An application of multidimensional comparative analysis in research on broiler chicken commercial lines. Archiv für Geflugelkunde, 73, 132–138.
10. Hall R.L., 1992. Toxicological burdens and the shifting burden of toxicology. Food Technology, 46, 109–114.
11. Hamouz K., Lachman J., Dvořák P., Trnková E., 2006. Influence of non-woven fleece on the yield formation of early potatoes. Plant, Soil and Environment, 52 (7), 289–294.
12. Hamouz K., Lachman J., Hajtmánkoa K., Pazderul M., Ćižek M, Dvořak P., 2010. Effect of natural and growing conditions on the content of phenolics in potatoes with different flesh colour. Plant, Soil and Environment, 56, 368–374.
13. Haydar A., Ahmed A.B., Hannan M.M., Razvy M.A., 2007. Analysis of genetic diversity in some potato varieties grown in Bangladesh. Middle East Journal of Scientific Research, 2(3-4), 143–145.
14. Iezzoni A.F., Pritts M.P., 1991. Application of principal components analysis to horticultural research. Horticulturae Science, 26(4), 334–338.
15. Kreutzmann S., Bassompierre M., Kistrup A., Buch L., Engelsen S.B., 2011. Exploratory study of potato cultivar differences in sensory and hedonistic applicability tests. Potato Research, 54, 13–28.
16. Lotti C., Marcotrigiano A.R., De Giovanni C., Resta P., Ricciardi A., Zonno V., Fanizza G., Ricciardi L., 2008. Univariate and multivariate analysis performed on bio-agronomical traits of Cucumis melo L. germplasm. Euphytica, 55, 511–522.
17. Madlaina K., Schmid P., Kölliker R., Boller B., 2008. Value of permanent grassland habitats as reservoirs of Festuca pratensis Huds. and Lolium multiflorum Lam. populations for breeding and conservation. Euphytica, 164, 239–253.
18. Mądry W., Pluta S., Sieczko L., Studnicki M., 2010. Phenotypic diversity in a sample of blackcurrant (Ribes nigrum L.) cultivars maintained in the Fruit Breeding Department at the Research Institute of Pomology and Floriculture in Skierniewice, Poland. Journal of Fruit and Ornamental Plant Research, 18, 23–37.
19. Mohamed V.O., Zein E., Fouteye M., Taleb K., Trifi M., Ali O., 2011. Use of multivariate analysis to assess phenotypic diversity of date palm (Phoenix dactylifera L.) cultivars. Scientia Horticulturae, 127, 367–371.
20. Mohammadi S.A., Prasanna B.M., 2003. Analysis of genetic diversity in crop plants – Salient statistical tools and considerations. Crop Science, 43, 1235–1248.
21. Murnice I., Karklina D., Galoburda R., Santare D., Skrabule I., Costa H.S., 2011. Nutritional composition of freshly harvested and stored Latvian potato varieties depending on traditional cooking methods. Journal of Food Composition and Analysis, 24, 699–710.
22. Nikolic D., Rakonjac V., Milatovic D., Fotiric M., 2010. Multivariate analysis of vineyard peach [Prunus persica (L.) Batsch.] germplasm collection. Euphytica, 171, 227–234.
23. Oruna-Concha M.J., Duckham S.C., Ames J.M., 2002. Comparison of the volatile components of eight cultivars of potato after microwave baking. Lebensmittel-Wissenschaft und Technologie, 35(1), 80–86.

24. Pluta S., Mądry W., Sieczko L., 2012. Phenotypic diversity for agronomic traits in a collection of blackcurrant (Ribes nigrum L.) cultivars evaluated in Poland. Scientia Horticulturae, 145, 136–144.
25. Rakonjac V., Fotiric A.M., Nikolic D., Milatowic D., Colic S., 2010. Morphological characterization of Oblacinska sour cherry cultivars in a germplasm bank at Portugal. Genetic Resources and Crop Evolution, 55, 593–601.
26. Rykaczewska K., 2007. Productivity of early lock and foreign potato cultivars. Zeszyty Problemowe Postępów Nauk Rolniczych, 517, 629–638 (in Polish).
27. Rymuza K., Bombik A., 2012. Comparison of edible potato cultivars by means of some methods of multidimensional analysis. Acta Scientiarum Polonorum Agricultura, 11(1), 67–77.
28. Rymuza K., Turska E., Wielogórska G., Bombik A., 2012. Use of principal component analysis for the assessment of spring wheat characteristics. Acta Scientiarum Polonorum Agricultura, 11(1), 79–90.
29. Styszko L., 2004. Breeding of starch potato cultivars with high utility parameters. Biuletyn IHAR, 232, 275–283 (in Polish).
30. Styszko L., Modzelewski T., Kamasa J., Majewski A. 2001. Relationship between morphological characteristics of potato tubers and starch content and starch yield. Zeszyty Naukowe AR Wocław, 415(80), 283–293 (in Polish).
31. Styszko L., Modzelewski T., Kamasa J., Majewski A., 2003. The relationship between a type of cooking quality (culinary) and morphological features of potato tubers. Biul IHAR, 226/227/2, 533–538.
32. Tarant Sz., 2005. Taste as a main criterion of food products in consumers' evaluation. Roczniki Naukowe Stowarzyszenia Ekonomiki Rolnictwa i Agrobiznesu 7(8), 267–270 (in Polish).
33. Thybo AK., Bechmann IE., Martens M., Engelsen SB., 2000. Prediction of sensory texture of cooked potatoes using uniaxial compression, near infrared spectroscopy and low field 1H NMR spectroscopy. Journal of Food Science and Technology, 33(2), 103–111.
34. Thygesen LG., Thybo AK., Engelsen SB., 2001. Prediction of sensory texture quality of cooked potatoes from low-field 1H NMR of raw potatoes. The role of chemical constituents. Journal of Food Science and Technology, 34(7), 469–477.
35. Trawczyński C., Wierzbicka A., 2011. The reaction of new potato cultivars to nitrogen fertilization. Bul. IHAR, 259, 193–201 (in Polish).
36. Veronesi F., Falcinelli M., 1988. Evaluation of an Italian germplasm collection of Festuca arundinacea Schreb. through a multivariate analysis. Euphytica, 28, 211–220.
37. Wierzbicka A., 2012. Effect of variety, harvest date and nitrogen fertilization on the content of starch and dry matter in early potatoes tubers. Fragmenta Agronomica, 29(2), 134–142 (in Polish).
38. Zgórska K., Grudzińska M., 2012. Changes in selected quality parameters of potato tubers during storage. Acta Agrophysica, 19(1), 203–214 (in Polish).

EFFECTS OF DIFFERENT KINDS OF SEWAGE SLUDGE AMENDMENT ON URBAN LAWN GRASSES

Elżbieta Wołejko[1], Andrzej Butarewicz[1], Urszula Wydro[1], Tadeusz Łoboda[1]

[1] Division of Sanitary Biology and Biotechnology, Faculty of Civil and Environmental Engineering, Bialystok University of Technology, 45d Wiejska Str., 15–351 Białystok, Poland, e-mail: e.wolejko@pb.edu.pl

ABSTRACT

The paper presents the results of research on the use of sewage sludge from the municipal sewage treatment plant in Białystok for fertilization of urban lawns. To fertilize the ground two types of deposits were used: stabilized sludge after the dewatering process in the press and granular sludge after drying in the sludge dryer. Two doses of sludge were used – 5 and 10 kg/m^2. Effective microorganisms (EM) were added to the test plots. Both doses of the applied sewage sludge contributed significantly to the growth of grass biomass. Aboveground biomass of the plants was significantly correlated with the concentration of Cd, Cu, Pb and Zn in the surface layer of the soil. The addition of EM to soil containing the dewatered sludge from the press reduced significantly the content of heavy metals in the soil, as opposed to the sludge in a form of granules with addition of EM, which contributed to the increase of their concentration in the soil.

Keywords: sewage sludge, soil, heavy metals, effective microorganisms.

INTRODUCTION

Sewage sludge is an essential product of wastewater treatment processes. Sludge extracted from wastewater is a complex organic-mineral substance, composed mainly of carbon, hydrogen, oxygen, nitrogen, sulfur, and chlorine [Wilk, Wolańczyk 2008].

Most of the sludge produced in municipal sewage treatment plants in Poland should be designed for natural use, including use in agriculture. The only obstacle to use sludge in agriculture on a large scale may be the presence of heavy metals and pathogenic organisms in it. Therefore, the preferred direction of the management of sediment should be the use of such processes, which would allow for the inactivation of pathogenic microorganisms and heavy metals [Butarewicz 2013]. If this condition is met, settlements are an excellent product to be used in nature, which makes it possible to recycle valuable components, such as organic matter, nitrogen, phosphorus and other nutrients for plants. Fertilization with stabilized sludge helps to increase humus in soil, as well as to improve its quality. Fertilization deposits may help to increase the yield of various crops, such as cereals, vegetables, trees and shrubs, thus influencing the landscape [Pathak et al. 2009]. However, the potential benefits of adding sewage sludge to soils are balanced by the risk of increasing the content of heavy metals in soil. The conditions for the application of sewage sludge into soil are regulated by EC Directive 86/278/EEC and are based on heavy metal content in both soil and sludge [Paz-Ferreiro et al. 2012].

The heavy metals in sewage sludge are present in dissolved, precipitated and co-precipitated forms with metal oxides or adsorbed on the particles of biological remains [Alvarez et al. 2002]. According to Weiner and Matthews [2003], metals can be in a form of oxides, hydroxides, sulfides, sulfates, phosphates, silicates, organic connections in a form of complexes with humic compounds and complex sugars. These changes affect the diversity of chemical forms of heavy metals accumulated in sewage sludge. However, some mobile forms can penetrate from the sediment to the soil environment [Alvarez et al. 2002; Weiner,

Matthews 2003; Skoczko, Piekutin 2014]. The intensity and magnitude of soil and plants contamination with heavy metals often depend not only on local factors. A major source of trace elements deposited on the surface of the soil and plants is the spread of pollutants over long distances [Kabata-Pendias, Pendias 1999].

Green city areas, including lawns, have various kinds of functions. On the one hand, well cared lawns enhance the aesthetic value of the city, on the other hand, they play a phytoremediation role, both for air and soil [Gawroński 2009]. For the effective phytoremediation plants should be characterized by high resistance to difficult environmental conditions, to have the ability to accumulate xenobiotics from the soil as well as to rapid growth and high dry mass production. Grasses are characterized by high phytoremediation potential [Singh et al. 2009]. According to Gawroński [2009], grasses in urban areas can fulfill the above requirements when they have good conditions for rapid growth and high dry mass production. For this purpose, sewage sludge may be used as a substitute for fertilizers, and at the same time effective microorganisms (EM) could be used to stimulate the growth of grasses. EM has a wide application in the field of agriculture, natural farming, livestock, gardening, composting and bioremediation. This consortium of microorganisms accelerates breakdown of proteins, sugars, fats and fibers, and thus, promotes rapid organic matter decomposition. EM works in two primary ways: competitive exclusion of harmful microorganisms and production of beneficial products such as enzymes, organic acids, amino acids, hormones and antioxidants that improve conditions of the environment [Khaliq et al. 2006].

The aim of this study was to examine the applicability of two different types of sewage sludge from the Municipal Wastewater Treatment Plant in Bialystok: dehydrated sludge from the press and dried in the form of granules to the turf lawns. At the same time the possibility of using EM to stimulate the decomposition of sludge and grass growth was also examined.

MATERIALS AND METHODS

The study was conducted on the campus of the Białystok University of Technology (Białystok, Wiejska Str. 45). The experiment was established in April 2012 in a randomized block design (on plots of 2 m^2). The plots were located in the green belt along the internal road. The surface of the research was divided into three blocks with an area of 20 m^2 each, constituting replicates. To fertilize the studied soil two different types of sewage sludge from the sewage treatment plant in Białystok were used: dewatered sludge after press and dried in the form of pellets. Prior to the use of sludge both types of sludge were tested according to the Regulation of the Minister of Environment of 13 July 2010 on municipal sewage sludge (Journal of Laws of 2010 No 137 item. 924). The analyses of *Salmonella* genus were carried out on EPA Method 1682: *Salmonella* in Sewage Sludge (Biosolids) by Modified Semisolid Rappaport-Vassiliadis (MSRV) Medium. The parasitological tests were carried out in accordance with Polish Standard PN-Z-19000–4 and recommended by US EPA.

Two doses of dehydrated sludge 5 and 10 kg/m^2 and two doses of the granular sludge, 1.5 and 3 kg/m^2, with or without 5 l EM per square meter were used on each experimental field. EM were from Greenland Technology EM, currently the only licensed manufacturer of EM in Poland. There were also experimental fields without sewage sludge (control).

In the experiment grass mixture Roadside from Barenbrug was used, which contains 32% of *Lolium perenne* cv. Barmedia, 5% of *Poa pratensis* cv. Baron, 52% of *Festuca rubra rubra* cv. Barustic, 5% of *Festuca rubra commutata* cv. Bardiva (BE) and 6% of *Festuca rubra commutata* cv. Bardiva (NL).

Due to adverse weather conditions the grass mixture was sown in an amount of 30 g per plot on 29 April 2012. On 4 June 2012, the first harvest was done. The second harvest was on 1 July 2012. In September 2012 samples of soil (0–20 cm) were collected. Heavy metals concentrations in soils fertilized with sewage sludge and in plant material were determined using Atomic Absorption Spectrometry AAS. The samples of soil were mineralized in temperature at about 450°C and remains were dissolved in aqua regia (3:1 mixture HCl and HNO_3) according to PN-ISO 11047:2001. The samples of mixtures of grasses were mineralized at the temperature of about 450°C and remains were dissolved in concentrated HNO_3 [Filipek et al. 2003].

Plants for the dry matter determination were taken in June, July and September 2012, then they were dried in an oven at 105 °C for 24 h, then at

75 °C until complete evaporation of water. Bioconcentration factors (BCF) for analyzed heavy metals were determined as quotients of average concentration of a given element in plants with relation to its average concentration in soil. The correlation between heavy metals concentration in above ground of plants and in soil fertilized with different doses of sewage sludge were calculated using Spearman's correlation factor r for $p \leq 0.05$ by using Statistica 10.0.

RESULTS AND DISCUSION

Białystok Wastewater Treatment Plant operates as an important element of the Polish Green Lungs. It is one of the ten largest sewage treatment plants in Poland. It was opened in 1994. Designed for the capacity of 176 500 m^3/day, it turned out to be receiving a much greater amount of actually sewage than it had been planned. Currently, the actual throughput is about 100 thousand m^3/d. It is a mechanical-biological treatment plant with sludge treatment and biogas exploitation. The treatment plant has a modern sanitary sludge treatment, which comprises a sludge dryer which produces granular sludge. These granules were tested and they are suitable for use on all types of soils. It is especially recommended for weak mineral soils with low organic matter content. It can be used in the cultivation of agricultural crops, lawns, squares of green, to fertilize outdoor trees, shrubs and ornamental plants. In the case of the use of dehydrated sludge after the press one must reckon with the possibility of depositing the chemicals in them, and especially dangerous organisms from a sanitary point of view. Some amount of impurities coming to the wastewater treatment plant is decomposed in the biological treatment processes, some importing impurities accumulate in sludge formed in the purifying processes and some are discharged with wastewater to the receiver. The content of toxic substances, heavy metals and biological contamination, depends on the amount and type of waste taken to the municipal sewer system. For the most part of sludge, the amount of these impurities is small and does not pose a serious risk to health and environment [Bernacka, Pawłowska 2000, Fukas-Płonka 2011].

Before the onset of the experiment both sewage sludge and soil from each combination were analyzed according to the Resolution of Ministry of Environment of July 13th, 2010 concerning municipal sewage sludge. The analyses were done by the Regional Chemical and Agricultural Station in Białystok (Table 1 and 2). Additionally, soil pH in 1M KCl were determined. Biological and parasitological indications were conducted in the Division of Sanitary Biology and Biotechnology, Białystok University of Technology. The results are shown in Tables 1 and 2. According to the cited Regulation, the tested samples from sewage sludge treatment plant in Białystok and averaged soil samples taken from the top layer of plots met all the criteria set out in the Regu-

Table 1. Selected physical and chemical properties of soils

Properties		The limits of heavy metals in the surface of light soil layer (0–25 cm) with the application of municipal sewage sludge for land reclamation for non-agricultural (Regulation 2010)
pH	7.65	–
Textural class	loamy sand	–
P_2O_5 [mg×100 g^{-1}D.M.]	14.43	–
Cd [mg/kg D.M.]	<0.50	3
Cu [mg/kg D.M.]	17.35	50
Pb [mg/kg D.M.]	25.2	50
Zn [mg/kg D.M.]	138.95	150

Table 2 Selected properties of municipal sewage sludge

Properties	Granular sludge	Dehydrated sludge
pH	7.6	7.4
Dry weight (%)	88	19.8
Organic matter (% D.W.)	59.2	58.4
Total P (% D.W.)	3.2	3.0
Total N (% D.W.)	4.6	4.4
Ammonium N (% D.W.)	0.32	0.2
Ca (% D.W.)	4.1	4.6
Mg (% D.W.)	0.7	0.6
Pb (mg×kg $D.W.^{-1}$)	21.9	20.7
Cd (mg×kg $D.W.^{-1}$)	1.2	<1.26
Cu (mg×kg $D.W.^{-1}$)	208	179
Zn (mg×kg $D.W.^{-1}$)	1109	1208
The number of viable helminth ova of *Ascaris sp., Trichris sp., Toxocara sp.*	Not detected	Not examinated
Bacteria of the genus *Salmonella* in 100 g of sludge	Not detected	Not examinated

lation. Dehydrated sludge from the press did not contain *Salmonella* and invasive nematode ova. In the case of granules, it was not necessary to do biological studies, because sewage sludge was dried at 130 °C.

Basing on the results, it was possible to use both types of deposits in the later part of the experiment and the establishment of the experimental plots.

Grassy areas in the passageways are usually established on degraded soils characterized by a mechanical malfunction and poor trophic conditions. Before the introduction of plants or at the stage of change of land use, soils often require the enrichment with organic substance to improve the physical and chemical properties, to provide a source of macro- and micronutrients, and to accelerate positive biological changes [Karczewska et al. 2012]. However, psychological barrier associated with the source of their origin, the specific structure of the sediment and usually a high content of certain heavy metals make it very difficult to find a buyer who could reasonably use sewage sludge [Dmochowski et al. 2011].

Heavy metals are not broken down by microorganisms in the biological treatment of wastewater. The soluble form is especially toxic, as it is generally found in plants and animals and can bioaccumulate in these organisms, as well as in humans. Moreover, heavy metals are involved in many chemical reactions. They can change forms from a soluble to insoluble state and from an absorbed to a free state [Zhu, Zang 2001]. The dynamics of changes in forms of metals is most intense in the top layer of the soil and it depends on diverse populations of microorganisms, content of organic matter, cation exchangeable capacity (CEC) and the biological interaction associated with microorganisms activity of rhizosphere and also on the plant itself [Adriano et al. 2004]. In our study, the content of metals was examined in plant, which grew on plots where different kinds and various doses of sewage sludge and sludge with EM were applied. There were significant correlations between concentrations of metals in the soil and the metal content in the aerial parts of lawn grasses. There was a negative correlation between the content of Zn in soil fertilized with sludge from the press and Cu in plants ($r = -0.82$), while for the plots fertilized with sludge granules there was also a correlation between Zn and Cu in the soil and in plants ($r = 0.88$). The correlation analysis was also done for press sludge and granulated sludge with the addition of EM. There were significant correlations between concentrations of metals in the soil and the metal content in plants. There was a positive correlation between high lead content in soil fertilized with sludge from the press and Cd in soil – for these elements at the level of correlation was $r = 0.94$ at $p < 0.05$.

The fertilization sludge from the press in the dose of 10 kg/m^2 was used, one observed an increased accumulation of the metals in above ground parts of plants about 70% in the plants with 5 l EM/m^2 compared with plots with sludge but without EM. In turn, the fertilization with granular sludge and 5 l EM/m^2 resulted in a lower content of heavy metals in above ground parts of plants, which shows that the sediment from the press plants have more phytoremediation properties than sludge granules. In addition, applying EM into the soil with the sludge from the press led to the strengthening of this effect (Table 3).

The differentiation of geochemical background in Poland is mainly related to variable chemical composition of the rocks from which they arise. Against this background a number of regional and local anomalies of geological and

Table 3. Contents of heavy metals and dry matter of grasses fertilized with different doses of sludge with or without EM

Dose of sewage sludge	Cd [mg/kg]		Cu [mg/kg]		Pb [mg/kg]		Zn [mg/kg]		Dry matter [g /m^2]	
	Dose of EM [l/m^2]									
	0	5	0	5	0	5	0	5	0	5
0 [kg/m^2]	0.25	0.16	12.1	9.4	5.67	2.80	79.92	83.18	367.5	755.0
Dehydrated sludge FROM THE PRESS										
5 [kg/m^2]	0.19	0.25	17.5	21.05	3.57	5.03	109.245	144.96	1277.0	1391.5
10 [kg/m^2]	0.24	0.35	21.25	30.7	5.16	6.80	134.575	171.03	1108.5	1496.5
Granular sludge										
1.5 [kg/m^2]	0.17	0.18	16.6	13.9	2.23	2.14	102.75	81.22	901.0	1327.0
3 [kg/m^2]	0.18	0.19	16.3	15.7	2.43	2.93	106.39	87.63	977.0	766.0

anthropogenic origin can be distinguished [Lis, Pasieczna 1995]. The sequence of heavy metal concentration (mg/kg) in the meadow soils in this research was: Zn>Cu>Pb>Cd. The content of cadmium in the soil on the test plots ranged from 0.26 to 0.67 mg/kg d.m., when the natural values of cadmium specific to Podlasie is Cd <0.5 ppm [Lis, Pasieczna 1995]. The addition of EM reduced the quantity of cadmium in the soil, which may be due to the properties of the sludge and its greater availability to plants. At 5 kg/m^2 sludge from the press and with addition of 5 l EM/m^2, the amount of Cd decreased from 0.47 to 0.32 mg/kg d.m., whereas at 10 kg/m^2 of sediment from the press and with the addition of 5 l EM/m^2 it decreased from 0.33 to 0.26 mg/kg d.m. The treatments applied in granular sludge with the addition of 5 l EM/m^2 resulted in an increase of cadmium in the soil. The granular sludge decomposed more slowly and metals contained in the sediment were gradually released into the soil, hence their less uptake by plants and higher content in the soil. Fertilizing plots with granular sludge in an amount of 1.5 kg/m^2 and a precipitate containing 5 l EM/m^2 increased the value respectively from 0.37 to 0.45 mg/kg dry matter, and with 3 kg/m^2 granular sludge with the addition of 5 l EM/m^2 respectively 0.31 to 0.41 mg/kg d.m. (Table 4).

The content of copper in the soils was very diverse (from 6.20 to 27.6 mg/kg d.m.). The average copper content was located within the natural values of this element in soils for Podlasie region in Poland (20–40 ppm) [Lis, Pasieczna 1995]. The content of zinc in soils varied between 47,60 and 110,76 mg/kg d.m. with the natural values set at 25–400 ppm Zn, but the content of lead on the test plots where sludge from the press and granular were applied and EM were added to both kinds, was below the natural range of 12.5–100 ppm Pb given by Lis and Pasieczna [1995] (Table 4).

The pH is an important factor determining the solubility of heavy metals in soil. It depends on e.g. ion balance – sorption and desorption processes of hydrogen cations and metal cations. Generally, heavy metals solubility is low in the neutral and alkaline range of pH and increases with decreasing pH [Kabata-Pendias, Pendias 1999]. Rosada's [2007] studies confirm that the soil pH determines the uptake of metals by plants and their mobility in the soil. There was no significant correlation between the content of the investigated metals in soil and plants and in soil pH

Trace metals are characterized by a very long period of persistence in the environment. Once entered, they practically remain in the environment and can cause bioaccumulation in the plants and can enter different biological metabolic pathways. Some heavy metals in trace amounts are essential to life but their excess has a negative impact on the health of plants, animals and humans [Bernacka, Pawłowska 2000]. In addition, the mobility of metals plays an important role in the rhizosphere compartment, which corresponds to the whole interface between roots and soil, where interactions between the soil microorganisms, roots and soil constituents take place [Kabata-Pendias, Pendias 1999, Lin et al. 2004]. The release of protons, exudates and metabolites into the rhizosphere by the roots and the microorganisms can modify the pH, depending on the available source of nitrogen nutrients or iron deficiency [Singh et al. 2009, Węglarzy 2007], thus modifying the fixation and mobility of heavy metals in soils [Lin et al. 2004]. The degree of bioaccumulation of a metal is the result of balance between the input rate and the rate of elimination. Bioaccumulation occurs primarily due to the inability to excrete necessary

Table 4. Effects of fertilizing with different doses of sewage sludge with or without EM on soil pH and contents of heavy metals in soil

Dose of sewage sludge	pH_{KCl}		Cd [mg/kg]		Cu [mg/kg]		Pb [mg/kg]		Zn [mg/kg]	
	Dose of EM [l/m^2]									
	0	5	0	5	0	5	0	5	0	5
0 [kg/m^2]	7.20	7.06	0.67	0.30	27.60	12.60	12.98	7.78	110.76	47.60
Dehydrated sludge from the press										
5 [kg/m^2]	7.22	7.06	0.47	0.32	17.35	8.15	10.86	8.53	80.94	48.02
10 [kg/m^2]	7.23	6.96	0.33	0.26	18.80	6.20	7.58	9.25	61.03	46.23
Granular sludge										
1.5 [kg/m^2]	7.03	7.18	0.37	0.45	12.75	13.20	8.45	12.25	64.34	63.38
3 [kg/m^2]	7.23	7.15	0.31	0.41	7.15	8.55	7.98	9.76	43.94	67.96

levels of the contaminant [Kabata-Pendias, Pendias 1999, Malik et al. 2010].

According to Wilk and Gworek [2009], cadmium has affinity for any components of sewage sludge and is relatively readily soluble, so that plants absorb it proportionally to the concentration of this element in environment regardless of the pH of soil. In our study, BCF for Cd (Figure 1) was the highest (on average of about 1.13) at the highest dose on applied sewage sludge (10 kg/m^2) from press with 5 l EM/m^2, while the lowest for control plots (0.37). The applied sludge from the press and the granules increased cadmium BCF in the aboveground parts of grass proportionally to the dose of biosolids. The addition of EM to sewage sludge increased the contend of Cd in plants in the plots with sludge from press, while in the plots with granules the BCF of cadmium in aboveground part of grasses was lower (0.42 and 0.47 respectively for plots with 1.5 and 3 kg/m^2 dose of biosolids with 5 l EM/m^2). Lead like cadmium is an element which is not required for the growth and development of plants. The BCF of Pb in this study was in the range of 0.17 – 0.74. One observed increased BCF in samples collected from plots where biosolids from the press with 5 l EM/m^2 were applied, while in the plots with granular sludge the BCF decreased.

Zinc and copper are microelements that are the most mobile in soil, accumulate in the upper layers of soil and they are readily soluble and available to plants [Węglarzy 2007]. In our study the BCF of Cu and Zn in the plots fertilized with sludge from press and granules sludge were in the range from 1.01 to 2.28 for Cu and from 1.35 to 2.42 for Zn. The addition of EM caused an increase of BCF in plots with sewage sludge from press in comparison with a pure sewage sludge without EM by 150% at dose of 5 kg/m^2, by 330% at dose of 10 kg/m^2 for copper, by 120% for single sewage sludge dose and by 60% at double dose of sludge for zinc. According to Węglarzy [2007], copper is hardly released from the soil, plants accumulate it mainly in roots but in contaminated environment its concentration increases in aboveground parts of plant. BCF of Cu and Zn decreased when sewage sludge with EM was used (Figure 1).

CONCLUSION

1. The fertilization of soil with sewage sludge from Municipal Wastewater Treatment Plant in Białystok at 5 and 10 kg/m^2 had a positive effect on the growth of lawn grasses biomass.
2. The use of sludge from the press and in a form of concentrated granules usually did not significantly affect the content of heavy metals both in the grass and in the soil.
3. The addition of EM to dewatered sludge from the press significantly decreased the content of heavy metals in the soil, and thus contributed to increase their absorption by plants. This resulted in the improvement of lawn grass remediation capacity for Cd, Cu, Pb and Zn.
4. The precipitate in the form of granules with addition of EM contributed to increased concentration of heavy metals in the soil and thus to the reduction of these metals in aboveground parts of plants.

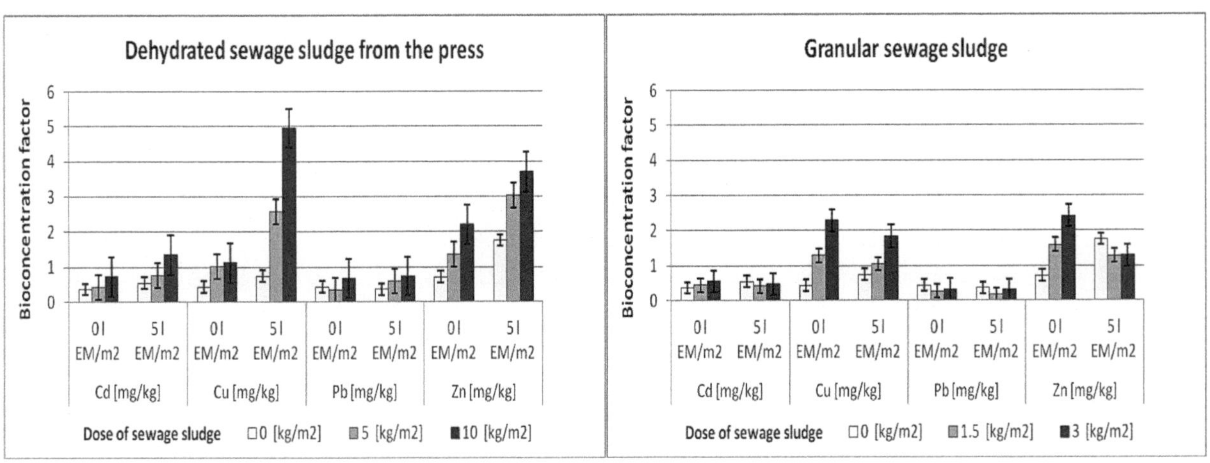

Figure 1. BCF of heavy metals (Cd, Cu, Pb and Zn) in lawn grasses depending on the dose and kind of sewage sludge with or without EM

5. BCF of heavy metals in plants was characterized by high variability, depending on the type of sediment and the addition of EM.

Acknowledgments

The research presented was supported financially by the project S/WBiIŚ/3/2011 and W/WBiIŚ/14/2014.

REFERENCES

1. Adriano D.C., Wenzel W.W., Vangronsveld J., Bolan N.S. 2004. Role of assisted natural remediation in environmental cleanup. Geoderma., 122, 121–142.
2. Alvarez E.A., Mochón M.C., Jiménze Sánchez J.C., Rodríguez M.T. 2002. Heavy metal extractable forms in sludge form wastewater treatment plants. Chemosphere, 47, 765–775
3. Bernacka J., Pawłowska L., 2000. Potentially toxic substances in sludge from municipal wastewater treatment. Wyd. IOŚ, Warszawa, 7–9.
4. Butarewicz A. 2013. Pathogenic organisms in sewage sludge – the detection and neutralization. The Publishing House of Bialystok University of Technology.
5. Dmochowski D., Gajkowska-Stefańska L., Dmochowska A., Presnarowicz R.K. 2011. Application of heavy metals speciation in assessing suitability of municipal post-fermentation sludge for producing roll-out grass. Prz. Nauk. Inż. Kszt. Środ., 53, 207–216.
6. Filipek T.A., Kaczor A., Badora A. 2003. Fundamentals and effects of chemicals in agroecosystems. Akademia Rolnicza, Lublin.
7. Fukas-Płonka Ł. 2011. Management of sewage sludge. Selected issues. Pol. Zrzeszenie Inżynierów i Techników Sanitarnych. Oddz. Wielkopolski. Poznań.
8. Gawroński S.W. 2009. Fitoremediacja a tereny zieleni. Zieleń Miejska, 10, 31.
9. Kabata-Pendias A., Pendias H. 1999. Biogeochemistry of trace elements, Państwowe Wydawnictwo Naukowe, Warszawa.
10. Karczewska A., Gersztyn L., Gałka B. 2012. The effects of various kinds of sewage sludge applied to soil on the amounts of soluble copper forms in polluted soils. Ochr. Środ. i Zasob. Natur., 51, 53–61.
11. Khaliq A, Abbasi MK, Hussain T. 2006. Effects of integrated use of organic and inorganic nutrient sources with effective microorganisms (EM) on seed cotton yield in Pakistan. Biores. Tech., 97, 967–972.
12. Lin Q., Chen Y.X., He Y.F., Tian G.M. 2004. Root-induced changes of lead availability in the rhizosphere of Oryza sativa L. Agr. Ecosyst. Environ., 104, 605–610.
13. Lis J., Pasieczna A. 1995. Geochemical atlas of Poland. Warszawa.
14. Pathak A., Dastidar M.G., Sreekrishnan T.R. 2009. Bioleaching of heavy metals from sewage sludge: A review. J. Environ Manage., 90(8), 2343–2353.
15. Paz-Ferreiro J., Gascó G., Gutiérrez B., Méndez A. 2012. Soil biochemical activities and the geometric mean of enzyme activities after application of sewage sludge and sewage sludge biochar to soil. Biol Fertil Soils., 48, 511–517.
16. PN-ISO 11047:2001. Soil quality. Determination of Cd, Cr, Co, Cu, Pb, Mn, Ni and Zn in extracts with aqua regia. Methods of flame and electrothermal atomic absorption spectroscopy.
17. Rosada J. 2007. Ekologiczne aspekty wykorzystania obszarów objętych oddziaływaniem emisji hut miedzi do upraw rolniczych. Postępy w Ochronie Roślin, 47(1), 119–127.
18. Singh A., Kuhad R.C., Ward O.P. 2009. Advances in applied bioremediation. Soil Biology 17, Springer-Verlag Berlin Heidelberg.
19. Skoczko I., Piekutin J. 2014. Photo-Fenton method usage to organic compounds degradation. Desalin. Water Treat. 52, 19–21.
20. Malik R.N., Husain S.Z., Nazir I. 2010. Heavy metal contamination and accumulation in soil and wild plant species from industrial area of Islamabad. Pakistan. Pak. J. Bot., 42(1), 291–301.
21. Węglarzy K. 2007. Heavy metals – a source of contamination and environmental impact. Wiad. Zootech., 45(3), 31–38.
22. Weiner R.F., Matthews R.A. 2003. Environmential Engineering, Elsevier Science, Burlington.
23. Wilk J., Wolańczyk F. 2008. Problems of the energy utilization of the waste from the sewage-treatment plant. Polityka Energetyczna, 11(2), 139–149.
24. Wilk M., Gworek B. 2009. Heavy metals in sewage sludge. Ochr. Środ. i Zasob. Natural., 39, 40–59.
25. Zhu S Q, Zang X P. 2001. On heavy metal pollution along Yangtze River stretches of urban areas in major cities. Yangtze River, 32(7), 23–25.

CHANGES IN THE MOISTURE CONTENT OF THE MIDDLE FEN SOILS IN THE ODRA RIVER VALLEY IN THE REGION OF BRZEG DOLNY IN THE VEGETATION PERIODS 2004–2009

Leszek Pływaczyk[1], Beata Olszewska[1]

[1] Institute of Environmental Development and Protection, Wrocław University of Environmental and Life Sciences, pl. Grunwaldzki 24, 50-363 Wrocław, Poland, e-mail: leszek.plywaczyk@up.wroc.pl; beata.olszewska@up.wroc.pl

ABSTRACT

In soils, where the water table is deeply located and has a minor impact on the moisture content of the surface layer, we are dealing with the precipitation-and-water type of water management. If underground water level is close to the surface, the top stratum of the soil, apart from precipitation, is additionally fed by water absorption from underground waters. Then we are dealing with ground-and-water type of management. We consider such types of water management of soil in the area of the left-bank valley of the Odra river, above and below the dam in Brzeg Dolny. The dominant soil types here are middle fen soils, based on middle clay and heavy clay as well as loam, which, in conditions of either excess or deficiency of moisture, are difficult to cultivate. The work compares water management of two soil profiles in vegetation periods between 2004 and 2009. The formation of underground waters, meteorological conditions and the course of the water reserves in the strata 0–50 cm and 0–100 cm were estimated with various supplying conditions of the active stratum of the soil. The volume of the supply with percolated water from underground water of the layer 50–100 cm on approximately 75–90 mm was also estimated. This value was mainly dependent on the depth of the retention of the water table of the soil profile above the level in Brzeg Dolny.

Keywords: moisture, soils, water, retention.

INTRODUCTION

River valley, natural between-field depressions and low peatbogs in Polish lowland account for about 11% of the total country area [25] and constitute specific habitats of characteristic water and soil proportions. They are characterized by a great production potential, their non-production role is growing including, inter alia, landscape and health qualities. Management and agricultural usage of river valleys normally require the management of water relations of the soils [4, 6, 14, 25]. The water factor, which is very changeable, determines the soil cover, plant cover and economic usage and is subject to changes according to human activities. Particularly important changes to the water factor are caused by river management, embankments, as well as exploitation of dams [18, 19, 22], retention reservoirs, fish ponds [8] and irrigation-and-drainage devices.

Rational water management is a key issue both for agriculture and other spheres of life. The dynamics of the soil moisture in the river valleys significantly depend on the process of meteorological and hydrological phenomena (river floods, infiltration from the riverbed, supply from underground waters) [17]. In many centres research is carried out regarding the evaluation of the impact of agro-meteorological factors on the formation of water resources of the soil, the impact of the depth of the underground water retention and the influence of water condition in watercourses on the volume of the water reserves in soil [5, 10, 12, 15, 16, 20, 23, 24]. The results of the works show that the fluctuations in the soil moisture mainly depend on the amount and distribution of precipitation. The meteorological conditions determine the dynamics of changes in the underground water depth as well as the volume of the water reserves in the upper soil strata [11, 13, 19]. Suitable types of cultivation or the methods of management of for example straw, do not have a significant impact on the level of moisture in soil [3].

The knowledge of management of soil and the dynamics of their moisture content mainly refers to the vegetation period, there are more and more works regarding variations in water resources in winter periods or identification of post-winter water resources in soil [1, 2, 9].

The main source of water resources in soil is precipitation. Plants absorb from the soil significant amounts of water, but only about 0.5% is used to build organic matter. Water in soil is not entirely available to, and utilized by, plants. Part of it evaporates directly from the soil, part soaks in below the root access or flows superficially. Also water contained in the roots' penetration zone is not always available for them. Plants can use only about 50% of the soil water [26]. In soils, where the water table is deeply situated and has only a slight impact on moisture content of the top stratum, we deal with a precipitation-and-water management of water. If the level of underground water is near to the ground level, the top stratum of soil, despite precipitation, is additionally supplied by wet soil from the underground waters. Then we deal with a ground-and-water type of management. We deal with such types of water management of soil in the area of the left-bank valley of the Odra river above and below the dam in Brzeg Dolny [22]. The dominant soil types here are middle fen soils, based on middle clay and heavy clay as well as loam, which, in conditions of either excess or deficiency of moisture, are difficult to cultivate. The aim of this work is the evaluation and comparison of water management in middle fen soils located in the Odra river valley below and above the dam in Brzeg Dolny against the meteorological conditions and the depth of the underground water retention as well as the course of water levels in the Odra below and above the dammed up level.

RESULTS OF THE RESEARCH

Changes in the content of moisture and the process of water reserves on selected soil profiles with various supply conditions of the active stratum of soil were subject to analysis. Two profiles were selected (marked as 2 – below water level and 5 – above the dam at Brzeg Dolny), distant from each other by about 5 km (Figure 1), wherein the content of moisture of the soil was measured on a monthly basis using a drier-and-weight method and the total of water reserves in strata

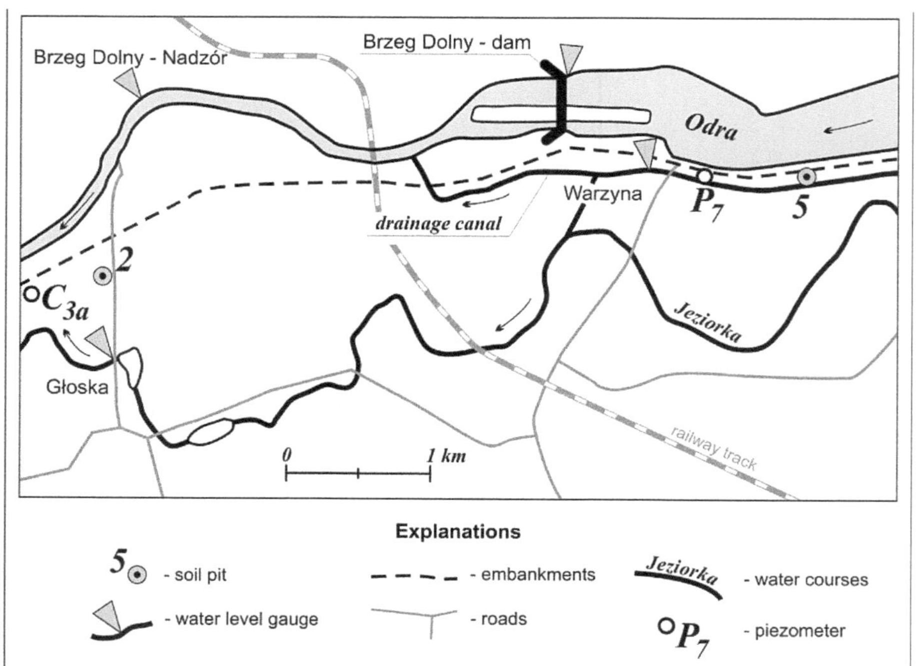

Figure 1. Localization of the selected soil profiles in the region of the Brzeg Dolny dam

0–50 and 0–100 cm were measured. Samples of soils were taken with a soil drill in three repetitions from strata 0–10, 20–30, 40–50, 65–75, 90–100 cm to the depth of the retention of the water table. Water hardly accessible for plants was determined from the difference between moisture with pF = 3.0 and pF = 4.2, whereas water that is easily accessible from the difference between pF = 3.0 and pF = 2.0 [22, 26]. The analysis was carried out for the vegetation periods between 2004 and 2009. On both research stations there were arable grounds. The selected profiles are built in the top stratum from compositions, hardly permeable to a depth of about 1.0 m (middle clay), underlain with permeable compositions (loose sand and loamy sand). The correct density of the analyzed soil adopts values between 2.51 and 2.66 $g \cdot cm^{-3}$, the porosity between 30.6 and 46.9%. In the top (one meter) stratum dominating are pores with diameter <3 µm, in deeper layers – pores with diameter of 300–30 µm [18].

Table 1 contains monthly sums of precipitation for the precipitation station in Brzeg Dolny and average monthly air temperatures for the station Wrocław-Strachowice in the periods IV–IX from 2004–2009 against corresponding average values for the period between 1954 and 2000. The precipitation station in Brzeg Dolny and the station in Wrocław are meteorological points situated nearest to the area on which the research of soil moisture were carried out. Table 1 also contains the evaluation of the periodical total of precipitation in the area, including summer half-years estimated as normal (2005 and 2006), medium dry (2004 and 2008), medium wet (2007) and wet (2009). With respect to thermal aspects, all analyzed periods, in relation to the period between 1954 and 2000, were estimated as warm.

River valleys are areas of close contact of underground and surface waters [4, 7]. Water interaction is intensified when we deal with surface water retention. In the valley above the barrage at Brzeg Dolny, the local conditions are of a pressure type, the underground water level is stable and mainly depends on the water condition in the Odra and in the drainage ditch and watercourse of Jeziorka (Figure 1). Below the dam, in the Odra bed, there are intensive processes of linear and local erosion. The erosion of the river bottom causes lowering of the water levels in the Odra which affects the water conditions in the adjacent valley [21].

Table 2 correlates monthly average depths of the retention of the water table in piezometers C_{3a} (below the dam) and P_7 (above the dam), located nearby soil pits and respectively approximately 150 and 90 m from the Odra bed (Figure 1). Whilst taking the measurements, in the periods between April and September in the years from 2004 to 2009, in piezometer P_7 the water table was in particular months on average between 0.42 and 0.92 m below the ground level. The water fluctuations were minor and depended mainly on the water levels in the Odra. However in piezometer C_{3a} the water table was significantly below 1.0 m from the ground surface, in particular months averaging between 2.21 m and 5.07 m – in the sands retaining under the top

Table 1. Monthly sums of precipitation P [mm] – including evaluation of the precipitation levels of the V–X half-year for the precipitation station in Brzeg Dolny and monthly average air temperatures t [°C] for the station Wrocław-Strachowice in the periods IV–IX from the years 2004–2009 against the values from the years 1954–2000

Specification		IV	V	VI	VII	VIII	IX	V-X	Evaluation
1954–2000	P [mm]	39	60	72	80	67	48	368	
	t [°C]	8.0	13.3	16.6	18.1	17.5	13.5	14.6	
2004	P [mm]	24.8	49.4	52.4	82.7	39.7	31.0	294.0	medium dry
	t [°C]	9.5	12.8	17	18.5	19.5	13.9	15.3	warm
2005	P [mm]	19.4	93.0	85.3	94.4	60.5	28.3	365.0	normal
	t [°C]	9.3	14.0	16.9	19.7	17.5	14.8	15.4	warm
2006	P [mm]	50.1	38.2	65.0	12.7	210.5	14.1	396.0	normal
	t [°C]	9.4	13.9	18.5	23.3	17.1	15.7	16.6	warm
2007	P [mm]	3.1	45.6	120.0	147.7	52.7	33.4	431.0	medium wet
	t [°C]	10.5	15.6	19.5	19.2	18.8	12.6	15.6	warm
2008	P [mm]	60.1	29.8	39.4	60.7	61.9	23.5	264.0	medium dry
	t [°C]	8.7	14.3	18.8	19.9	18.7	13.3	15.7	warm
2009	P [mm]	12.0	99.7	131.2	94.9	39.5	8.9	455.0	wet
	t [°C]	12.2	14.2	15.6	19.4	19.4	15.6	15.3	warm

Table 2. Monthly average depths of the retention of the water table [m] and the fluctuation amplitude [m] in piezometers C_{3a} (below the dam) and P_7 (above the dam)

Specification		IV	V	VI	VII	VIII	IX	Amplitudes
2004	C_{3a}	3.52	4.08	4.52	4.86	5.21	5.30	1.78
	P_7	0.70	0.60	0.51	0.49	0.53	0.51	0.21
2005	C_{3a}	3.76	4.14	4.43	4.69	4.86	4.78	1.10
	P_7	0.92	0.77	0.60	0.53	0.53	0.51	0.41
2006	C_{3a}	2.21	3.11	4.11	4.56	4.28	4.30	2.35
	P_7	0.77	0.74	0.67	0.71	0.68	0.58	0.19
2007	C_{3a}	4.07	4.64	4.88	4.95	5.07	4.54	1.00
	P_7	0.42	0.44	0.46	0.47	0.69	0.76	0.34
2008	C_{3a}	4.46	4.35	4.58	4.95	4.89	4.88	0.6
	P_7	0.62	0.61	0.61	0.57	0.61	0.54	0.07
2009	C_{3a}	3.64	4.42	4.43	3.48	4.15	4.71	1.07
	P_7	0.92	0.74	0.60	0.83	0.81	0.76	0.32

layer of the fen soils, having no effect on the water management of the active soil stratum.

The analysis of monthly average depths of the retention of the underground water from average monthly water levels in the Odra, made for the years 1976–2005 and for the areas distant by 120–550 m from the Odra bed, showed the following relations [21]:

- 120 m from the Odra bed
  $Hg = - 0.5042 \cdot H + 417.71$, $r = 0{,}87$;
- 220 m from the Odra bed
  $Hg = - 0.4083 \cdot H + 304.31$, $r = 0{,}79$;
- 550 m from the Odra bed
  $Hg = - 0.2565 \cdot H + 230.83$, $r = 0{,}56$;

where:

Hg– average monthly depth of retention of the water table [cm],

H – average monthly level of the water in the Odra according to the water-level gauge in Malczyce [cm],

r – correlation coefficient.

Tables 3 and 4 compile the values of the maximum and minimum water reserves measured in monthly average periods as well as their % participation in relation to the field water capacity in both analyzed pits (in strata 0–50 cm and 0–100 cm from the period IV–IX between 2004–2009). Figures 2 and 3 depict the water reserves in half a meter and a meter deep stratum of soil in the analyzed profiles and vegetation periods of the years 2004–2009 against the water reserves with the field water capacity and the capacity from a dry period.

In particular research periods of the analyzed years, the highest values of the water reserves in strata 0–50 cm and 0–100 cm in the soil profile below damming (2) were measured at the beginning of the vegetation period. Their volumes were affected by the post-winter retention values. In medium dry years the measured water reserves in the stratum 0–50 cm constituted 88–110% of the reserves equivalent to the values of the field water capacity, in normal years 83–104%, and in medium wet and wet years 108–114%. In the following months of the vegetation periods a stable reduction of the moisture content of the top stratum was observed. In a medium dry period of 2004, by the middle of April the reserves of water were below the value equivalent to the field water capacity, and in the period IV–IX 2008 in the middle of May. The lowest values of the moisture content and measured values of the water reserves were most often recorded in July and August. In medium dry periods the minimum values of the water reserves constituted only 48–50% of the value equivalent to the field water capacity, in normal periods 50–68%, in medium-wet and wet periods 63–77%. The dynamics of the moisture content and the process of water reserves in the stratum 0–50 cm were very diversified, and the changes in the reserves strictly depended on the amount and distribution of precipitation. The moisture content and the changes in the water reserves in the stratum 0–100 cm in the analyzed periods had a very similar process to the top half a meter stratum. In no measured period was a supply from the underground water recorded. The highest values of the water reserves were also recorded at the beginning of the considered periods, the lowest in July and August, particularly in the medium-dry periods of the years 2004 and 2008.

Table 3. Maximum and minimum water reserves and their percentage participation in relation to the field water capacity (PPW) in the 0–50 cm and 0–100 cm strata in the soil profile below the dam (2) in the periods IV–IX from the years 2004–2009

Period, year	Strata [cm]	Water reserves			
		Max. [mm]	%FWC	Min [mm]	%FWC
IV–IX 2004	0–50	128	88	72	50
	0–100	223	98	140	62
IV–IX 2005	0–50	151	104	98	68
	0–100	254	112	175	77
IV–IX 2006	0–50	120	83	73	50
	0–100	200	88	130	57
IV–IX 2007	0–50	157	108	112	77
	0–100	233	103	177	78
IV–IX 2008	0–50	160	110	69	48
	0–100	250	110	110	48
IV–IX 2009	0–50	166	114	91	63
	0–100	247	109	148	65

Table 4. Maximum and minimum water reserves and their percentage participation in relation to the field water capacity (PPW) in the 0–50 cm and 0–100 cm strata in the soil profile above the dam (5) in the periods IV–IX from the years 2004–2009

Period, year	Strata [cm]	Water reserves			
		Max. [mm]	%FWC	Min [mm]	%FWC
IV–IX 2004	0–50	186	101	156	85
	0–100	352	117	311	103
IV–IX 2005	0–50	179	97	158	86
	0–100	342	114	296	98
IV–IX 2006	0–50	178	97	149	81
	0–100	350	116	308	102
IV–IX 2007	0–50	182	99	160	87
	0–100	352	117	298	99
IV–IX 2008	0–50	183	99	158	86
	0–100	346	115	305	101
IV–IX 2009	0–50	186	101	128	70
	0–100	345	115	266	88

For the soil profile from the Odra valley above the dam at Brzeg Dolny (5), in the vegetation periods of the analyzed years, the calculated sums of the water reserves were characterized by a significantly lower changeability compared to the reserves in the soil profile from the areas below the dam. Regardless of the evaluation of the precipitation conditions in a given period, the water reserves in the stratum 0–50 cm were in the range of water easily accessible for plants. Maximum and minimum measured values of the water reserves were recorded in various periods in particular years and constituted respectively between 97 and 101% of the value of the field water capacity (maximum values) and between 70 and 87% of the value of the field water capacity (minimum values). Continuous supply from the underground waters had a significant effect on the volumes of the water reserves and the process of moisture content in a meter deep stratum of the soil profile. The water table was usually at the depth of 50–90 cm regardless of the precipitation level. Maximum and minimum measured values of the water reserves in a meter deep stratum of soil were respectively from 114 to 117% of the field water capacity and from 88 to 103% of the field water capacity. Low levels of precipitation and their unfavourable distribution did not have a remarkable

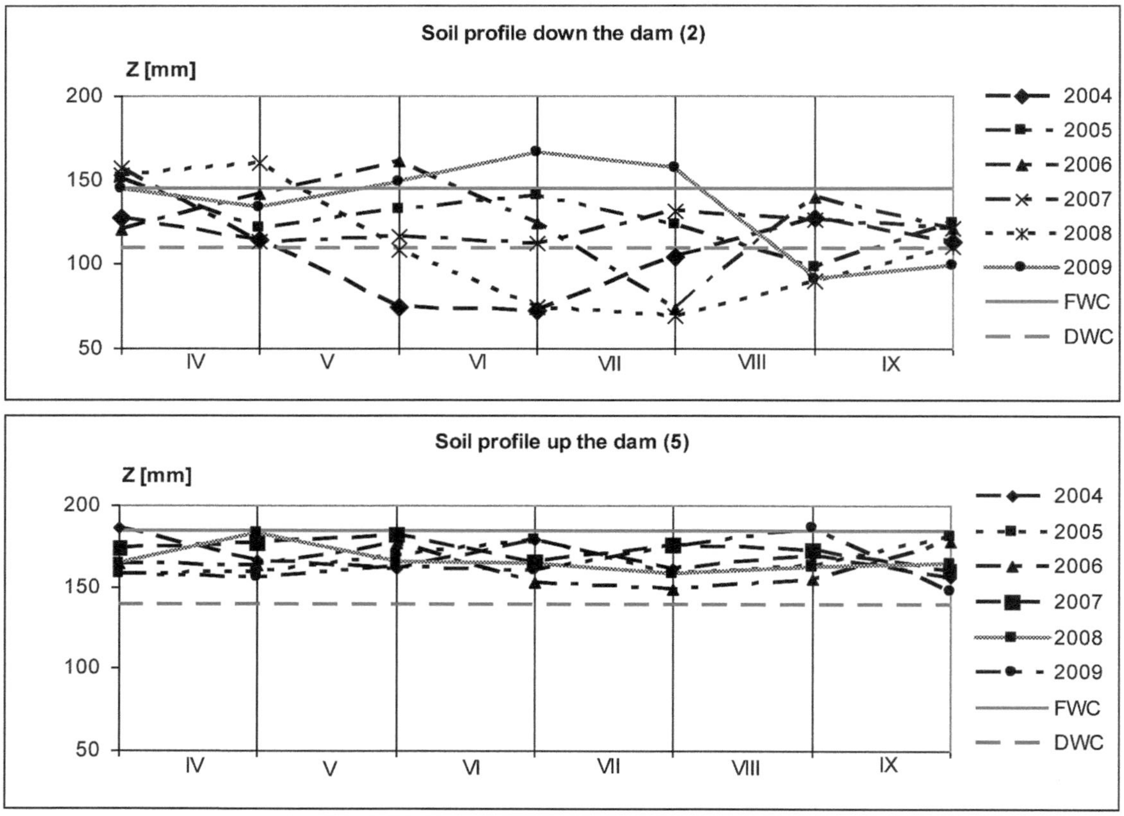

Figure 2. Water reserves in 0–50 cm soil stratum Z [mm] in soil profiles down and up the dam in vegetation periods in years 2004–2009 against the field water capacity (FWC) and drought water capacity (DWC).

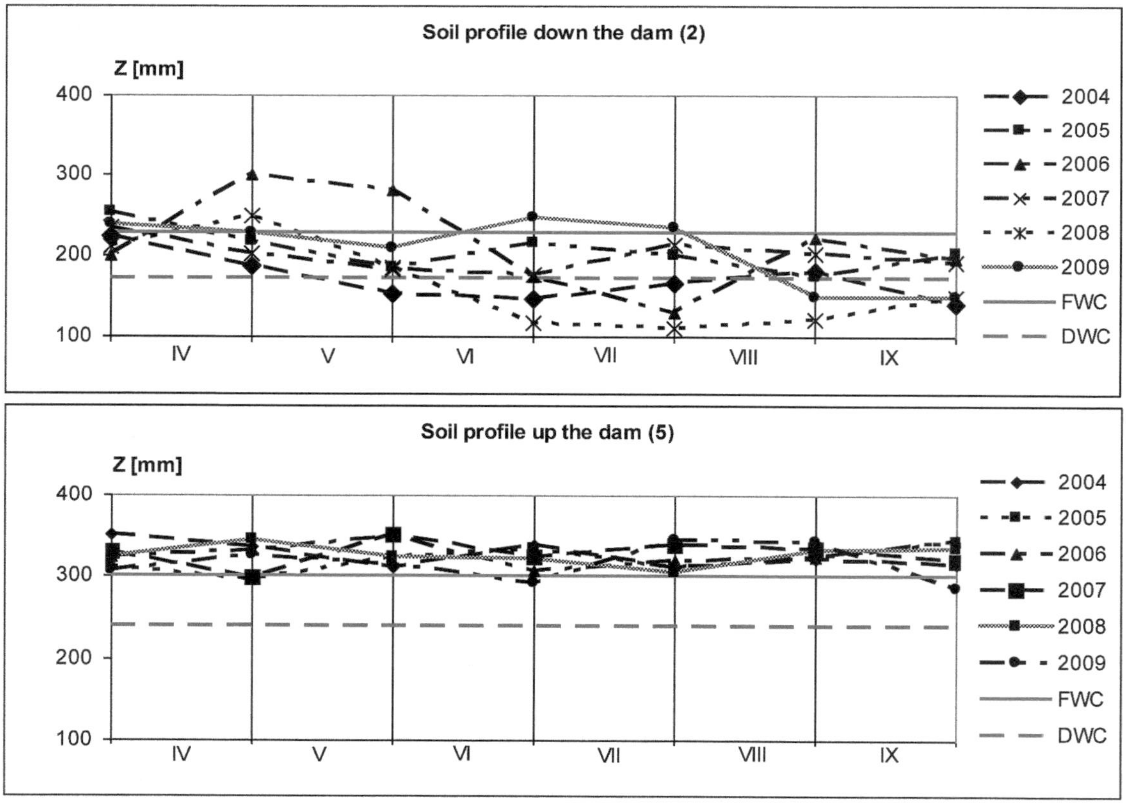

Figure 3. Water reserves in 0–100 cm soil stratum Z [mm] in soil profiles down and up the dam in vegetation period in years 2004–2009 against the field water capacity (FWC) and drought water capacity (DWC)

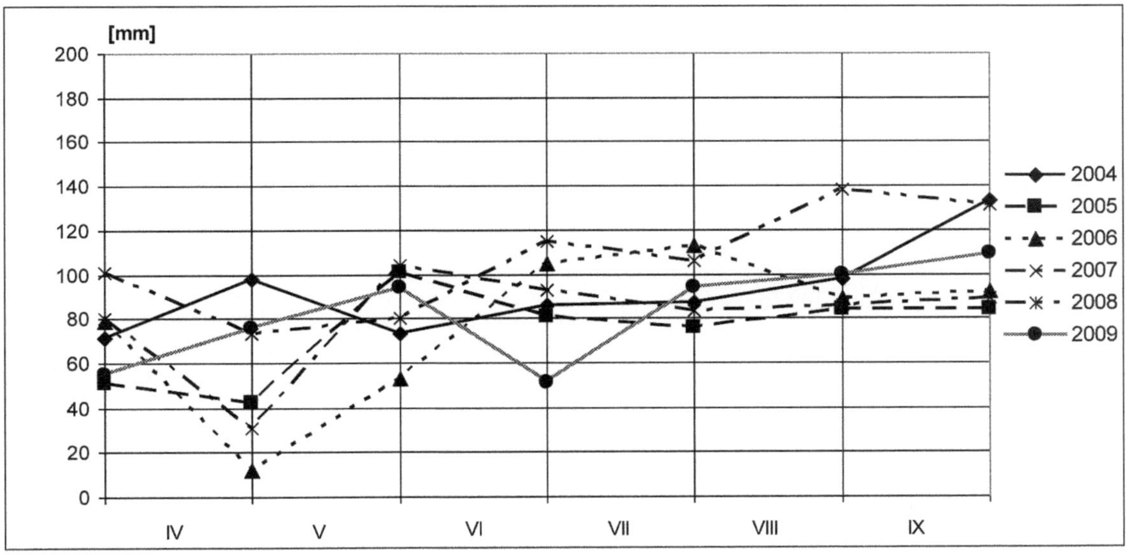

Figure 4. Differences in water reserves [mm] between soil profiles up and down the dam in 50–100 cm stratum

effect on the moisture content and changes in the water reserves in the top stratum of soil.

Differentiation in the volumes of water reserves in the analyzed middle fen soils of the Odra valley from Brzeg Dolny region, particularly in the 50-100 cm stratum, resulted from a different method of supply of the soil. Figure 4 represents the process of the differences in the soil reserves in the 50-100 cm stratum between the profiles under discussion for particular analyzed years. The smallest differences were usually recorded at the beginning of the vegetation period, in the following months the differences were getting greater, particularly in the medium-dry years. The analysis of the material gathered enabled the evaluation of the volume of the supply with wet soil from the underground waters of the 50-100 cm stratum on approximately 75-90 mm. This value mainly depended on the depth of the retention of the water table of the soil profile above the barrage at Brzeg Dolny.

CONCLUSIONS

1. The research carried out in the region of the dam at Brzeg Dolny showed that constant damming of a river has an effect on the depth of the underground water retention as well as on the water management of middle fen soils in the adjacent valley. In the analyzed vegetation periods of the years 2004–2009 the water table in the valley below the dam was retaining deeply, in particular months average from 2.21 to 5.30 m below the ground level. In the valley above the dam the present conditions were of a pressure type and the underground water level was stable, average from 0.42 to 0.92 m below the ground level.
2. The dynamics of the moisture content and the process of the water reserves in the soil of the valley below the dam in the strata 0-50 and 0–100 cm was dependent on the amount and the distribution of the rain water. In the analyzed periods and strata the water reserves were respectively from 69 to 120 mm and from 110 to 254 mm, often being below the value corresponding to the capacity of the dry period. However, an additional effect on the moisture content of the active stratum of the soil from the valley adjacent to the dammed up water of the Odra was the reaction of the water absorption from the underground water. The water reserves in the stratum 0–50 cm were from 128 to 178 mm, and in stratum 0–100 cm from 266 to 352 mm. The moisture content was usually equivalent to the values of the free water and water easily accessible for the plants.
3. The analysis and comparison of the volumes of the water reserves in both profiles enabled the evaluation of the volume of the supply with absorbed water from the underground water of stratum 50–100 cm on approximately 75–90 mm. This value mainly depended on the depth of the retention of the water table of the soil profile above the dam at Brzeg Dolny. The smallest differences were recorded at the beginning of the vegetation period (50–100 mm),

in the following months the differences were getting higher, in particular in the medium-dry years (80–135 mm).

REFERENCES

1. Bac S., Żyromski A. 2005. Spring soil water reserves as affected by hydrometeorological factors. Acta Agrophisica, 117, 5(3), 487–499 [In Polish].
2. Biniak-Pieróg M. 2008. Agrometeorological elements influence on soil water reserves variability during winter half-year. Monographs LVI, UP Wrocław, 129 [In Polish].
3. Biskupski A., Włodek S., Pabin J. 2002. Effects of the modes of tillage on changes in the soil moisture content. Rocz. AR Poznań CCCLXLII, no. 23, 13-20 [In Polish].
4. Brandyk T. 1990. Principles of moisture management for shallow water table soils. Treatises and Monographs SGGW-AR Warszawa, 120 [In Polish].
5. Bykowski J., Szafrański Cz., Fiedler M. 2005. Changes of soil water retention under condition of water storage in ameliorative ditch in vegetation periods with different sums of precipitation. Rocz. AR w Poznaniu CCCLXV, no. 26, 75–82.
6. Ciepielowski A. 1999. Principles of water management. Wyd. SGGW Warszawa, 326 [In Polish]
7. Głuchowska B., Pływaczyk L. 2008. Groundwater level in the Odra river valley downstream the Brzeg Dolny. Monographs V, UP Wrocław, 108 [In Polish].
8. Kamionka Sz. 1996. The role of fish ponds in the formation of retention resources of the adjacent areas.ZN AR Wrocław 283, Melioracja XLII, 195–224 [In Polish].
9. Koźmiński Cz., Michalska B. 2005. Variability of meteorological factors responsible for after-winter resources of water in light soil in Poland. Water-Environment-Rural Areas t. 5, z. 2(15), IMUZ Falenty, 67–92 [In Polish].
10. Liberacki D., Szafrański Cz., Stasik R., Korytowski M. 2006. Effect of meteorological conditions on water management of forest sites in Hutka river catchment. Acta Sci. Pol., Formatio Circumiectus 5 (2), 83–90 [In Polish].
11. Liberacki D., Plewiński D. 2001. Dynamics of water reserves in the soil of different habitat. Advances of Agricultural Science Problem Issues 476, Warszawa, 447–452 [In Polish].
12. Liberacki D. 2004. Groundwater levels and soil moisture content in the upper soil layers in a small forest catchment. Rocz. AR Poznań CCCLVII, N° 25, 305–311 [In Polish].
13. Liberacki D. Szafrański Cz. 2005. Moisture content and groundwater level changes in different lowland catchments. Wyd. Politechniki Koszalińskiej 61, 1–10 [In Polish].
14. Łoś M.J. 1998. The directions of changes in river valleys water reclamation systems. Wiad. Mel. i Łąk. 2, Vol. XLI, Warszawa, 61–64 [In Polish].
15. Łyczko W., Olszewska B., Pływaczyk L. Moisture dynamics of alluvial soils in the river valley in 1998-2000. Water-Environment-Rural Areas t. 2, z. 2, IMUZ Falenty, 47–58, 2002 [In Polish].
16. Nidzgorska-Lencewicz J. 2004. Dynamics of the moistening of uncultivated soil and soil in which rye and potato are grown. Acta Agrophysica, 104, Vol. 3(1), 125–132 [In Polish].
17. Nyc K., Pokładek R. 2009. Operation of land melioration systems as a foundation of water management in the natural-agricultural environment. Monographs LXX, UP Wrocław, 87 [In Polish].
18. Olszewska B. 1998. Influence of damming up structure on water conditions and selected element sof natura environment in valley – case study of Brzeg Dolny on the Odra. ZN AR Wrocław 349, 107–132 [In Polish].
19. Olszewska B. 2008. Water reserves in the medium alluvial soils in period IV–IX 2006 r. in the Odra valley in the Brzeg Dolny region. Advances of Agricultural Science Problem Issues 528, Warszawa: 97–104 [In Polish].
20. Orzepowski W., Pęczkowski G., Pływaczyk A., Kowalczyk T. 2008. Distribution of the reserves of moisture in profiles of black earth in the arable land area. Advances of Agricultural Science Problem Issues 528, Warszawa, 115–122 [In Polish].
21. Pływaczyk L., Olszewska B., Łyczko W., Klaus R. 2007. Impact of fall on Odra river on river bed and water conditions in the valley. Wiad. Mel. i Łąk. No. 3, Warszawa 106–114 [In Polish].
22. Pływaczyk L. 1997. The impact of the water damming up on the river valley – case study of Brzeg. Zesz. Nauk. AR Wrocław No. 311, Monographs XI, 47 [In Polish].
23. Przybyła Cz., Kozaczyk P., Sielska I., Bykowski J., Mrozik K. 2009. Changes of water content in soils of Nielegowo Polder during vegetation periods of years 2005–2007. Środkowo-Pomorskie Towarzystwo Naukowe Ochrony Środowiska, Tom 11, 64, 899–911 [In Polish].
24. Szafrański Cz. 1997. Dynamics of top soil layers moisture changes against course of meteorological conditions. Rocz. AR Poznań 294, 94–104 [In Polish].
25. Somorowski Cz. 1998. River valleys. Needs, demands and tendency to its development. Wiad. Mel. i Łąk. 2(377), Warszawa, 58–60 [In Polish].
26. Trybała M. 1996. Water management in agriculture. PWRiL Warszawa, 255 [In Polish].

23

EFFECT OF SUBSTRATES CONTAINING COMPOSTS WITH THE PARTICIPATION OF MUNICIPAL SEWAGE SLUDGE ON FLOWERING AND MACRONUTRIENT CONTENT IN THE LEAVES OF GARDEN PANSY (*VIOLA* × *WITTROCKIANA* GAMS.)

Agnieszka Zawadzińska[1], Piotr Salachna[1]

[1] Department of Horticulture, West Pomeranian University of Technology in Szczecin, Papieża Pawła VI 3, 71-459 Szczecin, Poland, e-mail: agnieszka.zawadzinska@zut.edu.pl; piotr.salachna@zut.edu.pl

ABSTRACT

The aim of the study was to assess the impact of substrates containing composts made from municipal sewage sludge and straw or sawdust on flowering, leaf greenness index and macronutrient content in the leaves of 'Fancy Carmine with Blotch' garden pansy. Two composts, with the following compositions: K I – 70% municipal sewage sludge, 30% rye straw; K II – 70% municipal sewage sludge, 30% coniferous tree sawdust, were used for the preparation of substrates. Six substrates of peat and compost at 25, 50 and 75% (by volume) were prepared. The control object was a peat substrate, pH 5.8–6.0, supplemented with Azofoska at 2.5 g·dm^{-3}. It has been found that leaf greenness index of pansy increased with an increasing share of compost. Composts involving municipal sewage sludge at the level of 50% had a more positive influence on the abundance of pansy flowering than at 25 and 75%. The content of total nitrogen, phosphorus, potassium and sulfur in the leaves increased with increasing the dose of compost. The content of total phosphorus in the leaves at the end of pansy cultivation in compost substrates ranged optimally for this species. In the experiment conducted in 2006, pansies contained more total nitrogen, potassium and sulfur than the flowering pansies planted in 2005.

Keywords: municipal sewage sludge, composts, SPAD, pansy, *Viola* × *wittrockiana* Gams.

INTRODUCTION

Garden pansy (*Viola* × *wittrockiana* Gams.) is a temperate climate plant and belongs to one of the most popular early spring and late summer ornamental plants in Europe, USA and Japan [Kessler 1998; Startek 2001; Wade and Thomas 2012]. For the cultivation of garden pansy, peat substrates are usually recommended, their physical properties are most suitable [Startek 2001; Di Benedetto et al. 2010]. Due to a large use of peat as a substrate for the cultivation of ornamental plants, including pansy, it is necessary to replace this material with alternative substrates, thanks to which one can reduce the use of peat [Chavez et al. 2008; Di Benedetto et al. 2010]. Previous studies indicate that pansy develops normally also in the substrate with the participation of clay and pine bark with nitrogen fertilizer [Anserimino et al. 1995; Sloan et al. 2004], in the substrate with rice husk in different proportions [Evans and Gauchukia 2004; Currey et al. 2010], and also in the substrate with the participation of coconut fiber [Janicka 2010]. In Poland, the possibility of using some of waste, such as dried shrimp [Placek 2009], municipal sewage sludge, potato pulp [Zawadzińska and Janicka 2007], as a substrate component for growing pansy was studied. Dried shrimp proved to be useful only in the case of putting a small part of it into the substrate, while the substrate containing composts with the participation of municipal sewage sludge in certain proportions turned out to be useful for the growing of cultivars of pansy of Thapsus group.

Municipal sewage sludge can be used naturally after having met the conditions contained in the Polish legislation [Regulation… 2008; Regulation… 2010]. Composts with the participation of municipal sewage sludge contain a lot of organic matter and are very rich in macro- and micronutrients [Kalembasa and Kuziemska 1999; Baran 2004]. However, the elevated pH and excessive salinity limit their use in the cultivation of ornamental plants [Abad et al. 2001; Moore 2004]. Mixing these composts with the ones containing a small amount of nutrient components, such as peat, reduces the salinity of the substrates. The use of composts as a component of substrates for the cultivation of ornamental plants has ecological and economic aspects, as it creates a nuisance to dispose of waste, while limiting the use of mineral fertilizers [Ribeiro et al. 2000; Zubilaga and Lavado 2001; Karoń and Pietr 2006].

The aim of this study was to evaluate the effect of substrates containing peat moss and two types of composts prepared from municipal sewage sludge and structure-forming components – straw and sawdust – on the leaf greenness index, flowering and macronutrient content in the leaves of the 'Fancy Carmine with Blotch' garden pansy.

MATERIAL AND METHODTS

The study was conducted in 2005–2006 in a plastic tunnel of West Pomeranian University of Technology in Szczecin. Plant material was 'Fancy Carmine with Blotch' garden pansy seedling from a German company Benary, it was obtained from sowing seeds on 1 June 2005 (for flowering during late summer) and 1 September 2005 (the spring blooming). Seeds were sown in peat substrate, following the instructions given in Startek's [2001] paper. After 4–5 weeks, the seedlings were quilted to substrates prepared 4 weeks earlier. For the preparation of substrates, two types of compost were used, after 7 and 12 months of decomposition. The process of composting is given in Krzywy et al. [2007] paper. The material composition of compost in dry weight was as follows: K I – 70% municipal sewage sludge, 30% rye straw; K II – 70% municipal sewage sludge, 30% coniferous tree sawdust. Six substrates based on peat moss and compost in three proportions by volume were prepared: 1 – 25% compost I + 75% sphagnum peat; 2 – 50% compost I + 50% sphagnum peat; 3 – 75% compost I + 25% sphagnum peat; 4 – 25% compost II + 75% sphagnum peat; 5 – 50% compost II + 50% sphagnum peat; 6 – 75% compost II + 25% sphagnum peat. The physical and chemical analysis of peat and substrates is shown in Table 1 and 2. Basing on the results, substrates too acidic for pansy were neutralized with chalk and dolomite to pH 5.8–6.0. Substrates with low potassium content were completed to the level of 300 mg $K \cdot dm^{-3}$, using potassium sulfate. The control object was sphagnum peat neutralized with chalk and dolomite to pH of 5.8–6.0, with the addition of Azofoska (13.6N+2.8 P+15.8K), at a dose of 2.5 $g \cdot dm^{-3}$.

Pansy seedlings were planted in prepared substrates into pots with a diameter of 10 cm and grown on tables in a plastic tunnel. In the cultivation, top

Table 1. Physical and chemical properties of substrates for garden pansy *(Viola × wittrockiana* Gams.*)* prepared for planting in 2005–2006

Substrate	2005				2006			
	PH_{H2O}	Dry matter (%)	Density ($g \cdot cm^{-3}$)	Salinity NaCl ($g \cdot dm^{-3}$)	PH_{H2O}	Dry matter (%)	Density ($g \cdot cm^{-3}$)	Salinity NaCl ($g \cdot dm^{-3}$)
Peat	3.60	59.8	0.18	0.35	3.60	59.8	0.18	0.35
1	4.50	36.9	0.26	2.85	4.45	36.8	0.26	2.98
2	5.35	36.3	0.39	3.50	5.15	36.3	0.38	3.66
3	5.70	45.2	0.40	4.34	5.40	45.1	0.38	4.37
4	5.25	38.8	0.33	1.55	5.10	37.5	0.31	1.65
5	5.80	38.1	0.39	1.69	5.80	36.1	0.39	1.69
6	6.30	70.8	0.42	1.78	6.20	59.0	0.40	1.66

Explanations: Composition of substrates: 1 – 25% compost I + 75% peat; 2–50% compost I + 50% peat; 3 – 75% compost I + 25% peat; 4 – 25% compost II + 75% peat; 5 – 50% compost II + 50% peat; 6 – 75% compost II + 25% peat.

Composition of composts: compost I – 70% sewage sludge, 30% rye straw; compost II – 70% sewage sludge, 30% saw dust from coniferous tree.

Table 2. Chemical characteristic of substrates for garden pansy *(Viola × wittrockiana* Gams.*)* prepared for planting in 2005–2006

Substrate	Total form content ($g \cdot kg^{-1}$ d.m.)						Available form content ($mg \cdot dm^{-3}$)				
	N	P	K	CA	MG	S	$N\text{-}NO_3$	P	K	CA	MG
2005											
Peat	8.90	1.60	1.60	2.10	0.30	0.50	17.0	20.0	6.00	42.0	27.0
1	15.9	13.3	2.20	5.70	0.55	4.20	252	340	398	1144	322
2	17.4	18.3	3.65	7.70	0.62	5.20	465	735	370	2110	430
3	22.4	19.3	3.80	7.90	0.65	6.80	590	850	360	2780	470
4	15.1	8.80	2.10	4.50	0.48	2.20	230	410	140	1225	270
5	15.3	12.0	2.20	6.10	0.60	4.50	332	540	256	2508	278
6	18.6	14.3	2.80	7.00	0.58	6.20	710	602	302	2996	366
2006											
Peat	8.90	1.60	1.60	2.10	0.30	0.50	17.0	20.0	6.00	42.0	27.0
1	16.3	15.3	2.40	5.70	0.56	4.00	364	460	289	1132	303
2	17.6	17.3	3.70	7.70	0.63	5.30	785	687	379	2265	500
3	25.2	19.1	3.70	8.10	0.64	6.70	800	847	385	3478	580
4	16.1	8.70	1.30	5.60	0,50	2.20	326	419	139	1430	236
5	16.6	13.0	2.20	6.70	0.56	4.70	346	671	224	2690	309
6	20.8	15.3	2.90	7.10	0.59	6.10	945	741	256	3080	327

Explanations: see table 1.

dressing was not used. Each experimental object occurred in four replications, 5 plants in each.

The air temperature while growing pansy in the summer months (VII–VIII) was controlled by intense airing and shading plants. While wintering, minimum air temperature in the plastic tunnel (I–III), as recommended, ranged from 2 to 5 °C.

At the beginning of the generative phase, the pansy leaf greenness index was measured with an optical device Chlorophyll Meter SPAD-502 (Minolta). According to the literature [Loh et al. 2002, Altland et al. 2003], leaf greenness index is highly correlated with chlorophyll content. This measurement method is frequently used to assess the degree of crop nutrition [Loh et al. 2002, Altland et al. 2003, Larcher et al. 2011]. This device performs a non-invasive measurement of the area of 6 mm^2 in SPAD (Soil Plant Analysis Development) units. Four measurements were performed on each mature pansy leaf, in the middle of the leaf blade. The results were averaged for each replication. In both years of the study, for 8 weeks of flowering, flowers were counted and then removed. The numbers of flowers are given per one plant.

After the cultivation, the total contents of nitrogen, phosphorus, potassium, calcium, magnesium and sulfur were determined in pansy leaves. Plant material for the analysis was obtained by taking samples from two replicates of each object, and then drying them at 60 °C. After grinding the material in a laboratory mill macronutrients were determined in them in two replications. Total nitrogen was determined with the Kjeldahl method after prior digestion of samples in concentrated sulfuric acid (VI) with the addition of a selenium mixture, total phosphorus with colorimetric method according to Barton, potassium and calcium with flame photometry and total sulfur with macro-sulfur elemental analyzer VarioMax. For the determination of total magnesium, the method of atomic absorption spectrometry (AAS) was used, after prior digestion of samples in a mixture of nitric acid (V) and chlorine acid (VII) in a 1:1 ratio.

The results concerning the number of flowers, SPAD readings and the total form of macronutrient content in the leaves were verified using the ANALWAR 4.0 analysis of variance for multifactor experiments. The significance of the average values was verified with Tukey's range test at a level of $\alpha = 0.05$.

RESULTS AND DISCUSSION

Garden pansy according to Hamrick [2003] has secondary cultivation requirements. The substrate for seeding as well as further cultivation of pansy should have good air-water properties.

As recommended by Kessler [1998] and Bailey [1998], the substrate pH for pansy is 5.4–5.8, while the pH recommended by German companies specializing in the cultivation and production of pansy is 5.5–6.2 [www.benary.com]. In our study, in the prepared substrates, the pH was adjusted according to the breeder of the tested cultivar – Benary (Table 1). Although pansy is produced in large quantities in Poland and in the world, in the available literature, there is no information about the limit numbers relating to the content of nutrients in substrates for its cultivation. In the fertilization of pansy, the form of nitrogen provided with a mineral fertilizer is important [www.benary.com]. Excess nitrogen in ammonium form can adversely affect the plant when it is grown at higher temperatures [Hamlin et al. 1999; Hamlin and Mills 2001]. The content of nitrate (V) and other components, especially phosphorus, potassium and calcium in the prepared substrates was quite high (Table 2), and in some, exceeded the recommended value for ornamental plants with high nutritional requirements [Strojny 1996]. Salinity of substrates (2.85–4.37 g KCl $\cdot dm^{-3}$) was significantly higher in objects containing compost I with the participation of municipal sewage sludge (70%) and straw (30%) than in the substrates containing compost II (municipal sewage sludge 70%, sawdust 30%) – Table 1. Despite the enhanced performance of substrates containing composts with municipal sludge, in a number of scientific studies in Poland and abroad [Ramahsamay 2004; Vabrit et al. 2007; Wraga and Zawadzińska 2007; Ahmad et al. 2012], it was demonstrated that ornamental plants grown in such substrates can grow and develop normally.

In the first year, the leaf greenness index of garden pansy in the early stages of flowering was higher in objects with substrates containing compost I, with 70% of the municipal sewage sludge and 30% of straw (Table 3). Most greenness was observed in pansies in objects containing 50%

Table 3. Leaf greenness index (SPAD) of 'Fancy Carmine with Blotch' garden pansy grown in substrates containing compost with municipal sewage sludge

Dose of compost in substrate [%]		2005			2006		
		Type of compost*		Mean	Type of compost		Mean
		K I	K II		K I	K II	
0		26.8	26.3	26.5	48.4	47.7	48.0
25		25.3	20.9	23.1	47.9	43.5	45.7
50		32.9	30.4	31.7	49.0	47.6	48.3
75		37.0	27.9	32.4	48.9	49.1	49.0
Mean		30.5	26.3	28.4	48.5	47.0	47.7
$LSD_{0,05}$ for:	compost	1.52			n.s.		
	dose	2.88			n.s.		
	compost × dose	3.05			n.s.		
	dose × compost	4.07			n.s.		

Explanations: see table 1; n.s. – no significant differences.

Table 4. Number of flowers of 'Fancy Carmine with Blotch' garden pansy grown in substrates containing compost with municipal sewage sludge

Dose of compost in substrate [%]		2005			2006		
		Kind of compost*		Mean	Kind of compost		Mean
		K I	K II		K I	K II	
0		31.7	31.3	31.5	32.4	32.0	32.2
25		23.6	21.6	22.6	28.8	25.9	27.3
50		30.6	29.1	29.8	30.5	30,8	30.6
75		25.9	22.0	24.0	30,0	26.7	28.3
Mean		27.9	26.0	27.0	30.4	28.9	29.6
$LSD_{0,05}$ for:	compost	n.s.			n.s.		
	dose	5.167			4.085		
	compost × dose	n.s.			n.s.		

Explanations: see table 1; n.s. – no significant differences.

and 75% of the compost in the substrate – respectively, 31.7 and 32.4 SPAD. In 2006, pansies in spring had a greater leaf greenness index than at the same stage of development in 2005. The type of compost and compost dose applied to the substrate did not affect leaf greenness index of pansy. Beneficial effects of the substrate with an addition of municipal sewage sludge or composts with the participation of municipal sludge on the leaf greenness index were demonstrated in experiments with geranium (*Pelargonium* × *hortorum*), chrysanthemum (*Chrysanthemum* × *grandiflorum*), impatiens (*Impatiens walleriana*) and small-sized pansies (*Viola* × *wittrockiana*) [Dudka et al. 1998; Wraga and Zawadzińska 2007; Zawadzińska and Janicka 2007; Zawadzińska and Dobrowolska 2009].

In both years of the study, the type of compost used for the preparation of substrates to grow pansy did not affect the number of flowers (Table 4). The influence of compost in the substrate on pansy flowering was shown. In the first year of the study, more flowers were obtained in the control object and objects containing compost at 50% - respectively, 31.5 and 29.8 of the flower.

In the second year of the study, most flowers were on the control plants (32.2 of the flower), but a lot of flowers were also present in the objects, where compost was at the level of 50% and 75% of the volume of the substrate (Table 4). Limitations on the number of flowers per plant by increasing the participation of municipal sludge in the substrate (from 25 to 100%) were observed in the Grigatti et al. [2007] study of constantly thriving begonia (*Begonia semperflorens*), *Mimulus hybridus*, sage (*Salvia splendens*) and marigold (*Tagetes* 'Zenith Lemon Yellow'). A positive effect on the flowering of the above-mentioned taxa was achieved by the lowest dose of municipal sludge. In our study, no interaction between the type of compost and its participation in the substrate was observed. In the spring term in 2006, pansies developed 9.6% more flowers than in the fall of 2005 (Table 4).

The type of compost and participation of compost in the substrate contributed to the total nitrogen content in the leaves of garden pansy in the first year of cultivation (Table 5). Pansies grown in the substrate containing compost II (70% municipal sewage sludge, 30% sawdust) had significantly more nitrogen in the leaves of the objects containing compost I (70% municipal sewage sludge, 30% straw). The greatest amount of total nitrogen was present in the leaves of pansy grown in the substrate containing 75% of compost. In 2006, pansies were not significantly different in nitrogen content in the objects containing composts. The least amount of nitrogen was present in the control object. Pansies flowering in 2006 contained more nitrogen in the leaves, on average by 46.3% compared to 2005. In both years, the interaction between the type of compost and its participation in the substrate occurred. Kessler [1998] reported that the nitrogen content in the leaves of a properly fertilized pansy should be 3.5–4.5% N d.m. On the other hand, according to Whipker et al. [2003], the total nitrogen content in the leaves of well-fertilized pansies ranges from 2.5 to 4.5% N d.m.. The authors probably studied nitrogen content in the indicator leaves, but this information was not included. In our study, the total nitrogen content was determined in all leaves at the end of the cultivation. In each test year, the total nitrogen content was lower than that recommended by the above-mentioned authors, but symptoms of nitrogen deficiency were seen only in plants flowering in 2005.

In both years of the study, the type of compost and compost content in the substrate affected the phosphorus content in the leaves of pansy (Table 5). More phosphorus was present in the plants grown in objects containing compost II (municipal sewage sludge 70%, sawdust 30%), 0.76–0.77% P d.m., than when the sawdust in the compost was replaced with straw - an average of 0.70% P d.m. The phosphorus content in a well-fertilized pansy according to Bailey [1998] and Whipker et al. [2003] is 0.25–1.00% P d.m., and according to Kessler (1998) 0.3–1.0% P d.m. In our study, the phosphorus content determined in all the leaves of the plant was within the range recommended by the above-mentioned authors.

The potassium content in the leaves of pansy in the first and second years of the study depended on both of the studied factors (Table 5). In the first year, pansies had more potassium in the leaves in the objects with compost II (municipal sewage sludge 70%, sawdust 30%) than those in the objects with compost I (municipal sewage sludge 70%, straw 30%). Most potassium was found in the leaves of pansy in the objects with the highest content of composts (1.32% K d.m.) and the lowest at the control object (0.76% K d.m.). In the second year of the study, most of potassium was present in the leaves of pansy grown in the substrate containing compost I. Pansy did not significantly differ in the potassium content between objects that contained composts, but in the control objects, plants contained the least potassium. Pan-

Table 5. The content of total nitrogen, phosphorus and potassium in the 'Fancy Carmine with Bloch' garden pansy leaves depending on the type and dose of compost in the substrate after cultivation

Dose of compost in substrate [%]		2005			2006		
		Total N [% d.m.]					
		Type of compost*		Mean	Type of compost		Mean
		K I	K II		K I	K II	
0		0.710	0.708	0.709	1.019	1.015	1.017
25		0.791	0.834	0.812	1.387	1.365	1.376
50		0.811	1.011	0.911	1.350	1.385	1.367
75		0.872	1.314	1.093	1.365	1.431	1.398
Mean		0.796	0.967	0.881	1.280	1.299	1.289
$LSD_{0,05}$ for:	compost	0.016			n.s.		
	dose	0.032			0.038		
	compost × dose	0.033			0.039		
	dose × compost	0.046			0.054		
Dose of compost in substrate [%]		P [% d.m.]					
		Type of compost		Mean	Type of compost		Mean
		K I	K II		K I	K II	
0		0.42	0.42	0.42	0.56	0.57	0.57
25		0.67	0.81	0.74	0.70	0.80	0.75
50		0,76	0.93	0.84	0.74	0.88	0.81
75		0.96	0.90	0.93	0.78	0.81	0.80
Mean		0.70	0.76	0.73	0.70	0.77	0.73
$LSD_{0,05}$ for:	compost	0.018			0.019		
	dose	0.035			0.037		
	compost × dose	0.035			0.038		
	dose × compost	0.049			0.052		
Dose of compost in substrate [%]		K [% d.m.]					
		Type of compost		Mean	Type of compost		Mean
		K I	K II		K I	K II	
0		0.76	0.76	0.76	1.44	1.44	1.44
25		0.90	1.21	1.05	1.71	1.62	1.67
50		1.18	1.24	1.21	1.75	1.63	1.69
75		1.29	1.35	1.32	1.71	1.70	1.70
Mean		1.03	1.14	1.09	1.65	1.60	1.62
$LSD_{0,05}$ for:	compost	0.023			0.018		
	dose	0.045			0.035		
	compost × dose	0.045			0.036		
	dose × compost	0.063			0.050		

Explanations: see table 1; n.s. – no significant differences.

sy blooming in spring 2006 contained more potassium in leaves – an average of 48.6% more than in 2005. In both years, the potassium content in the leaves of pansy was not within the range of 2.5–5.0% K d.m. as recommended by Bailey [1998] and Whipker et. al. [2003] and within the range of 3.0–4.5% K d.m. as given by Kessler [1998].

In 2005 and 2006, the examined factors did not affect the content of calcium and magnesium in the leaves of pansy (Table 6). Slightly more calcium in the leaves was present in the case of flowering pansies grown in 2005. The magnesium content in the leaves of pansy in the studied period was similar. In both years, the calcium content in all leaves ranged below the range of 0.6–3.0% Ca d.m. given by Bailey [1998] and Whipker et al. [2003], and 0.6–1.2% Ca d.m. given by Kessler [1998]. The magnesium content in the leaves of pansy was less than the optimal recommended by Bailey [1998] and Whipker et al. [2003].

The amount of total sulfur in the leaves of pansy did not depend on the type of compost (Table 6). The dose of compost affected the sulfur content in the leaves in both years of the study. In 2005, most sulfur (0.068% S d.m.) was present in the leaves of pansy grown in the substrate containing 75% of composts, while in 2006, there were no significant differences in the sulfur content between objects containing compost. In 2006, the plants contained an average of almost three times more sulfur than in 2005 in the leaves. According to Bailey [1998], in the properly nourished pansy, the sulfur content in the leaves should be within the range 0.2–0.7% S d.m. In the two years of study, the sulfur content was less than the optimum range recommended for pansy.

The applied substrates with the participation of composts were very rich in nutrients (Table 2). The small content of total nitrogen, potassium, calcium, magnesium and sulfur may be due to ion imbalance between the studied elements of the substrate and their antagonistic effects on one another, particularly between calcium and magnesium [Marschner 1993; Bailey et al. 1995;

Table 6. Calcium, magnesium and sulfur content in the leaves of 'Fancy Carmine with Blotch' garden pansy depending on the type and dose of compost in the substrate after cultivation

Dose of compost in substrate [%]		2005			2006		
		Ca [% d.m.]					
		Type of compost*		Mean	Type of compost		Mean
		K I	K II		K I	K II	
0		0.132	0.129	0.130	0.089	0.097	0.093
25		0.135	0.135	0.135	0.121	0.122	0.121
50		0.132	0.144	0.138	0.121	0.119	0.120
75		0.135	0.144	0.139	0.122	0.111	0.117
Mean		0.133	0.138	0.136	0.113	0.112	0.113
$LSD_{0,05}$ for:	compost	n.s.			n.s.		
	dose	n.s.			n.s.		
	compost × dose	n.s.			n.s.		
Dose of compost in substrate [%]		Mg [% d.m.]					
		Type of compost		Mean	Type of compost		Mean
		K I	K II		K I	K II	
0		0.087	0.086	0.086	0.079	0.075	0.077
25		0.085	0.084	0.084	0.087	0.087	0.087
50		0.086	0.082	0.084	0.087	0.085	0.086
75		0.087	0.087	0.087	0.083	0.086	0.084
Mean		0.086	0.084	0.085	0.084	0.083	0.083
$LSD_{0,05}$ for:	compost	n.s.			n.s.		
	dose	n.s.			n.s.		
	compost × dose	n.s.			n.s.		
Dose of compost in substrate [%]		S [% d.m.]					
		Type of compost		Mean	Type of compost		Mean
		K I	K II		K I	K II	
0		0.027	0.022	0.025	0.099	0.098	0.098
25		0.037	0.046	0.042	0.121	0.131	0.126
50		0.044	0.044	0.044	0.126	0.135	0.130
75		0.085	0.051	0.068	0.124	0.141	0.132
Mean		0.048	0.041	0.044	0.117	0.125	0.122
$LSD_{0,05}$ for:	compost	n.s.			n.s.		
	dose	0.010			0.026		
	compost × dose	0.010			n.s.		
	dose × compost	0.013			n.s.		

Explanations: see table 1; n.s. – no significant differences.

Lis-Krzyścin 2007; Francke 2010]. For pansy, the optimum ratio of Ca: Mg should be 5:2. The higher the ratio of these elements, the more their collection is hampered, and boron deficiency occurs [Laffe and Styer 1989; Bailey et al. 1995]. In compost substrates for growing pansy, the ratio Ca: Mg was much greater than 5:2, as recommended by Bailey et al. [1995]. The optimal macronutrient content provided by American researchers has probably been developed on the indicator leaves of pansy, for diagnostic purposes. Our study examined the content in all leaves, after completion of cultivation. This might be the reason for lower macronutrient content. Similar results with regard to the content of macronutrients in the leaves were obtained by Zawadzińska and Salachna [2013] in the experiment with *Impatiens walleriana* 'Fiesta Lavendr Orchid' grown in substrates with composts containing municipal sewage sludge.

In the control substrate, after pansy cultivation, the ingredients have been used for the construction of vegetative mass and the production of flowers. In the control object, pansies bloomed profusely. In addition, in 2006, pansies were better fed than in 2005. The research of Harris et al. [1998] and Hamlin and Mills [2001] shows that pansies grown at lower temperatures (12 °C) are better fed and their leaves contain more nutrients than pansies grown at higher temperatures (22 °C). In our own study, pansies blooming in spring in 2006 had more flowers, greater leaf greenness index, and a 46.3% increase in total nitrogen content in the leaves. In addition, in 2006, pansies were characterized by a 48.6% increase in the content of total potassium and contained almost three times more sulfur. Phosphorus and magnesium contents in the analyzed period were similar.

CONCLUSIONS

Composts of municipal sewage sludge (70%) and straw (30%) or sawdust (30%) mixed with peat substrate at a dose of 50–75% favourably influenced the intensity of leaf colour 'Fancy Carmine with Blotch' garden pansy. Leaf greenness index (SPAD) of plants while flowering increased with increasing the participation of composts in substrates. Composts added to the peat substrate at 50% had a more positive influence on the abundance of garden pansy flowering than composts at 25% and 75%.

Type of compost and percentage of compost in the substrate affected the content of total nitrogen, phosphorus, potassium and sulfur in the leaves of garden pansy. The content of total nitrogen, phosphorus and sulfur increased when increasingly higher doses of compost were used. Total phosphorus content in the leaves of garden pansy, after growing in the compost substrates, varied within the optimal range for this species.

In the experiment conducted in spring 2006 pansies better advantage of the nutritional components of substrates than in summer 2005. The contents of total nitrogen, potassium and sulhur in the leaves of pansies was greater in 2006.

REFERENCES

1. Abad M., Noguera P., Bures S. 2001. National inventory of organic wastes for use as growing media for ornamental potted plant production: case study in Spain. Bioresource Technol., 77, 197-200.
2. Ahmad I., Ahmad A., Ahmad S., Amjad A., Saleem M., Akram A. 2012. Effect of various agricultural substrates on biometric and qualitative characteristics of *Ruscus hypophyllum*. Int. J. Agric. Biol., 14(1), 116–120.
3. Altland J.E., Gilliam C.H., Keever G.J., Edwards J.H., Sibley J.L., Fare D.C. 2003. Rapid determination of nitrogen status in pansy. HortSci, 38(4), 537-541.
4. Ansermino S.D., Holcroft D.M., Levin J.B., Adams P., Hidding A.P., kipp J.A., Sonneveld C., Kreij C. 1995. A comparison of peat and pine bark as a medium for bedding plant pack production. Acta Hort., 401, 151-160.
5. Bailey D.A. 1998. Commercial pansy production. North Carolina Cooperative Extension Service, Horticulture Information, Leaflet 521, 6.
6. Bailey D.A., Baker J.R., Jones R.K. 1995. A penny for your penseés. North Carolina Flower Growers' Bulletin, 40(3), 1-18.
7. Baran S. 2004. Osady ściekowe w gospodarce rolno-środowiskowej. Zesz. Probl. Post. Nauk Rol., 499, 15-20.
8. Chavez W., di Benedetto A.G., Civeira G., Lavado R. 2008. Alternative soilless media for growing *Petunia × hybrida* and *Impatiens walleriana*: Physical behavior, effect of fertilization and nitrate losses. Bioresource Technol., 99, 8082-8087.
9. Currey C.J., Camberato D.M., Torres A.P., Lopez R.G. 2010. Plant growth retardant drench efficacy is not affected by substrate containing parboiled rice hulls. HortTechnology, 20(5), 863-866.
10. Di Benedetto A., Petracchi J.C., Marcella G., Montaron P., Chavez W. 2010. Evaluation of alternative substrates for bedding plants. Int. J. Agric. Res. 5(11), 1020-1029.

11. Dudka S., Das K.C., Miller W.P. 1998. Blends of composted biosolids and bottom ash as potting media to grow ornamentals. Composting in the Southeast, The University of Georgia, Proceedings of the Conference, September 9-11 Athens, Georgia.
12. Evans M.R., Gachukia M. 2004. Fresh parboiled rice hulls serve as an alternative to perlite in greenhouse crop substrates. HortSci, 39(2), 232-235.
13. Francke A. 2010. The effect of magnesium fertilization on the macronutrient content of pepino dulce (*Solanum muricatum* Ait.) fruit. J. Elem., 15(3), 467-475.
14. Grigatti M., Giorgioni M.E., Ciavatta C. 2007. Compost-based growing media: Influence on growth and nutrient use of bedding plants. Bioresource Technol., 98, 3526-3534.
15. Hamlin R.L., Mills H.A. 2001. Pansy florel development and nutrient absorbtion as influenced by temperature, nitrogen form and stage of plant development. J. Plant Nutr., 24(12), 1975-1985.
16. Hamlin R.L., Mills H.A., Randle W.M. 1999. Growth and nutrition of pansy as influenced by N-form ratio and temperature. J. Plant Nutr., 22(10), 1637-1650.
17. Hamrick D. 2003. Ball Redbook. 2. Crop production. Ball Publishing, Batavia, Ill., USA, 678-684.
18. Harris C.E., Simonne E.H., Eakes D.J. 1999. Effect of nitrogen form ratio on pansy growth and nutrition and the palatability to white-tailed deer. J. Plant Nutr., 22(12), 1807-1814.
19. Janicka D. 2010. Wpływ podłoży organicznych oraz chitozanu na cechy morfologiczne i wartość dekoracyjną bratka ogrodowego (*Viola x wittrockiana* Gams.) z grupy Patiola. Praca doktorska, Zachodniopomorski Uniwersytet Technologiczny w Szczecinie.
20. Kalembasa S., Kuziemska B. 1999. Wpływ dawki i terminu stosowania osadów na plon siana oraz wartości współczynników wykorzystania azotu i fosforu z osadów. Fol. Univ. Agric. Stein., 200, Agriculture, 77, 121-124.
21. Karoń B., Pietr S.J. 2006. Wartość nawozowa kompostu z osadu ściekowego. Zesz. Probl. Post. Nauk Rol., 512, 305-313.
22. Kessler J.R. 1998. Pansy production and marketing. Alabama A&M and Auburn Universities. ANR-596.
23. Krzywy E., Zawadzińska A., Klessa M. 2007. Badania przydatności podłoży z udziałem kompostów z komunalnego osadu ściekowego do uprawy roślin ozdobnych. Zesz. Probl. Post. Nauk Rol., 518, 101-110.
24. Laffe S.R., Styer R.C. 1989. Too B or not to B: learn to recognize boron deficiency. Grower Talks, 53(8), 66.
25. Larcher F., Berruti A., Gullino P., Scariot V. 2011. Reducing peat and growth regulator input in camellia pot cultivation. Hort. Sci. (Prague) 38(1), 35-42.
26. Lis-Krzyścin A. 2007. Zastosowanie nawozów Osmocote i Vitrofosmak w uprawie aksamitki rozpierzchłej. Rocz. AR Pozn. CCCLXXXIII, Ogrodn. 41, 109-114.
27. Loh F.C., Grabosky J.C., Bassuk N.L. 2002. Using the SPAD 502 Meter to assess chlorophyll and nitrogen content of benjamin fig and cottonwood leaves. HortTechnology, 12(4), 682-686.
28. Marschner H. 1993. Mineral nutrition of higher plants. Academic Press, London-Toronto.
29. Moore K.K. 2004. Growth of bedding plants in substrates amended with compost and fertilized with three different release rates of a controlled-release fertilizer product. Hort Technology, 14 (4), 474-478.
30. Placek M. 2009. Możliwości wykorzystania biostymulatorów w uprawie nowych grup i odmian rodzaju *Viola*. Praca doktorska, Zachodniopomorski Uniwersytet Technologiczny w Szczecinie.
31. Ramahsamay K.D. 2004. Oil palm wastes and sewage sludge composts as potting media for chrysanthemum. Univ. Putra Malaysia. Degree of Master of Science.
32. Regulation of the Minister for Agriculture and Rural Deevlopment of 18 June 2008 on execution of some regulations contained in the act on fertilisers and fertilisation, Journal of Law, 2008, No. 119, item 765. Dz.U. 08.119.765 (in Polish).
33. Regulation of the Minister of Environment of 13 July 2010 on communal sewage sludge. Journal of Law, 2010, No. 137, item 924. Dz.U. 10.137.924 (in Polish).
34. Ribeiro H.M., Vasconcelos E., dos Santos J.Q. 2000. Fertilization of potted geranium with a manicipal solid waste compost. Bioresource Technol., 73, 247-249.
35. Sloan R.C., Harkess R.L., Kingery W.L. 2004. Nitrogen and cotton gin waste enhance effectiveness of pine bark soil amendment. HortTechnology, 14(2), 212-217.
36. Startek L. 2001. Wpływ retardantów wzrostu oraz miejsca uprawy i zimowania na cechy morfologiczne bratka ogrodowego (*Viola × wittrockiana* Gams.). Akademia Rolnicza, Szczecin 201.
37. Strojny Z. 1993. Nawożenie roślin ozdobnych pod osłonami. Centrum Ogrodnicze, Skierniewice.
38. Vabrit S., Urman K., Põldma P., Jõepera K. 2007. Ornamental quality of *Bidens ferulifolia* 'Golden Star' pot plants dependend of composted sewage sludge as a component of growing media. Acta Hortic., 755, 367-372.
39. Wade G.L., Thomas P.A. 2012. Success with pansies in the winter landscape: A guide for landscape

professionals. University of Georgia. Cooperative Extension Bulletin 1359, 1-7.

40. Whipker B., Cavins T., Gibson J., Dole J., Nelson P., Fonteno W., Bailey D. 2003. Water, media and nutrition testing. In: Hamrick D. (ed), Ball Red Book, 17th ed. Ball Publ. Batavia, 47-62.

41. Wraga K., Zawadzińska A. 2007. Ocena wpływu podłoża z komunalnego osadu ściekowego na kwitnienie i walory dekoracyjne chryzantemy wielkokwiatowej (*Chrysanthemum × grandiflorum* (Ramat. Kitam.). Ochrona Środowiska i Zasobów Naturalnych, 31, 249-254.

42. www.benary.com (last accessed 08.04.2013).

43. Zawadzińska A., Dobrowolska A. 2009. Wpływ podłoży z dodatkiem kompostów z komunalnego osadu ściekowego na wzrost i kwitnienie niecierpka Walleriana (*Impatiens walleriana*). Ochrona Środowiska i Zasobów Naturalnych, 40, 608-616.

44. Zawadzińska A., Janicka D. 2007. Effects of compost media on growth and flowering of parviflorous garden pansy. Part I. Plant growth and conformation. Acta Agrobot., 60(2), 161-166.

45. Zawadzińska A., Salachna P. 2013. The effect of substrates containing municipal sewage sludge compost on the accumulation of macrocomponents in *Impatiens walleriana* Hook. J. Elem., DOI 10.5601/jelem.2013.18.4.545.

46. Zubillaga M.S., Lavado R.S. 2001. Biosolid compost component of potting media for bedding plants. Gartenbauwissenschaft, 66, 304-309.

SEASONAL CHANGES OF NITROGEN AND PHOSPHORUS CONCENTRATION IN SUPRAŚL RIVER

Mirosław Skorbiłowicz[1], **Piotr Ofman**[1]

[1] Technical University in Bialystok, Wiejska 45A, 15-351 Białystok, Poland, e-mail: m.skorbilowicz@pb.edu.pl

ABSTRACT

The purpose of the study was to determine seasonal changes of ammonia nitrogen, nitrate nitrogen, total nitrogen, phosphate and total phosphorus in the years 2003–2009 in the river Supraśl. For this purpose a monitoring network was created with selected three control points on the basis of which the assessment was made. Control points were located near Gródek, Nowodworce and Dzikie. The research results were averaged for each month. On this basis the assessment of seasonal changes of nitrogen and phosphorus forms in the river Supraśl has been carried out. Research and analysis showed that the concentration of nitrogen and phosphorus in waters of Supraśl were characterized by a clear seasonal variation, the highest concentration of nitrogen was observed in January and February and the lowest in June in each of the analyzed points, the course of changes in the concentration of nitric and total nitrogen in March were different from the general course of reigning changes in their concentrations in surface waters.

Keywords: nitrogen, phosphorus, seasonality, river.

INTRODUCTION

Chemical composition of surface waters is very diverse and depends on the type of surrounding areas usage. In case of small agricultural catchment, main components that run across rivers are nutrients, mainly in forms of nitrogen and phosphorus [Koc 2006, Czyżyk et al. 2008, Sapek 2008]. A number of changes associated with a decrease or increase in the concentration of each nutrient is associated not only with human activity, but also with the processes of natural mineral components migration into surface waters as a result of the prevailing climatic conditions.

The succession of the seasons is an important factor influencing on the chemistry of surface waters. Individual periods are characterized by different intensity of environmental factors, which in conjunction with the development cycles of organisms affect the intensity of their life processes, and therefore, the uptake, release and migration of nutrient elements [Vymazal 1995, Gergel et al. 1999, Bucka and Wilk-Woźniak et al. 2002].

Air temperature is not without significance, it directly affects the temperature of water in a given period. Water temperature, in turn, determines the rate of the natural changes occurring in the aquatic environment, including the ability of rivers self-purification [Jarosiewicz 2007, Kozłowski 2009].

Supraśl River is located in north-eastern Poland in the poviat of Białystok. Its catchment area is about 1800 km^2. Supraśl catchment includes such tributaries as: Sokołda, Płoska, Dzierniakówka, Pilinica, Biała, Czarna, Grzybówka i Słoja. The river (Supraśl) is located among the areas used primarily for agricultural purposes. In the catchment area small settlement units are located and there is no industry that might affect significantly the qualitative changes of the aquatic environment [Skorbiłowicz 2003].

The purpose of this paper was the determination of seasonal changes in the concentration of nitrogen and phosphorus in waters of Supraśl over six years research cycle.

RESEARCH METHODS

To assess the concentration changes of nitrogen and phosphorus forms the research results conducted by the Regional Inspectorate for Environmental Protection in Bialystok from 2003 to 2009 were used. Examination of water samples was performed in three control points located in the vicinity of: Gródek (CP 1), Nowodworce (CP 2) and Dzikie (CP 3) (Figure 1). Arithmetic mean and standard deviation (SD) of test results for each month in the period 2003–2009 were calculated using the statistical packet Statistica 10 for each control point.

The water samples were studied for the following components: ammonium nitrogen, nitrate nitrogen, total nitrogen, phosphates and total phosphorus, in addition air temperature was measured. Different parameters in the water samples were indicated by an UV-VIS spectrometry in accordance to the following standards and research procedures:

- ammonium nitrogen – PN-ISO 7150-1:2002,
- nitrate nitrogen – PN-82/C-04576.08,
- total nitrogen – PB–52 ed. 4 from day 04.06. 2007, in accordance to PN-73/C-04576/14,
- phosphates – PN-EN ISO 6878:2006,
- total phosphorus – PN-EN ISO 6878:2006.

RESULTS AND DISCUSSION

From the average temperature distribution (Figure 2), during the period of the study, results that the lowest temperature was observed in each of the control points in January, and the highest in July. The temperature distribution achieved in the measurements, is characteristic for Polish lowland areas (Web portal: poland.gov.pl).

The highest concentrations of ammonia nitrogen were observed in January, while the lowest in June. Comparing the average monthly temperature (Figure 2), with changes of ammonium nitrogen (Figure 3), a pattern can be noted. Namely, the concentration of this component in water increases with simultaneous temperature of ambient air decrease, which also was pointed by Blanket (2006), Sojka et al. (2007) and Czyżyk et al. (2008). The concentration of nitrate nitrogen (Figure 4) underwent changes, as the concen-

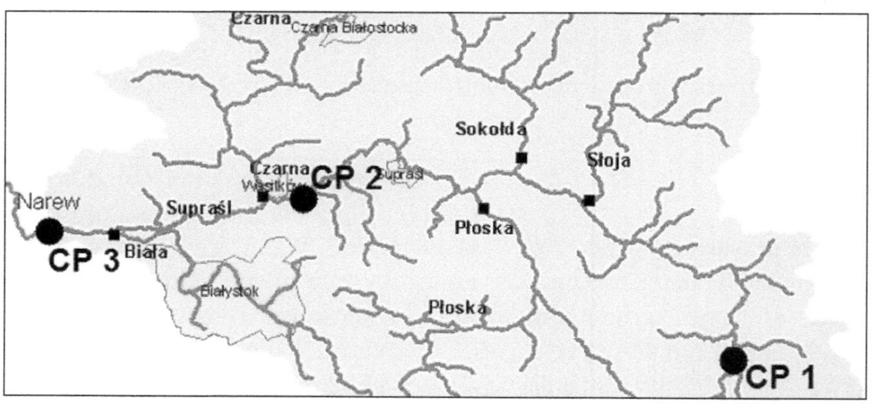

Figure 1. Location of control points on the river Supraśl [Ocena... WIOŚ 2010]

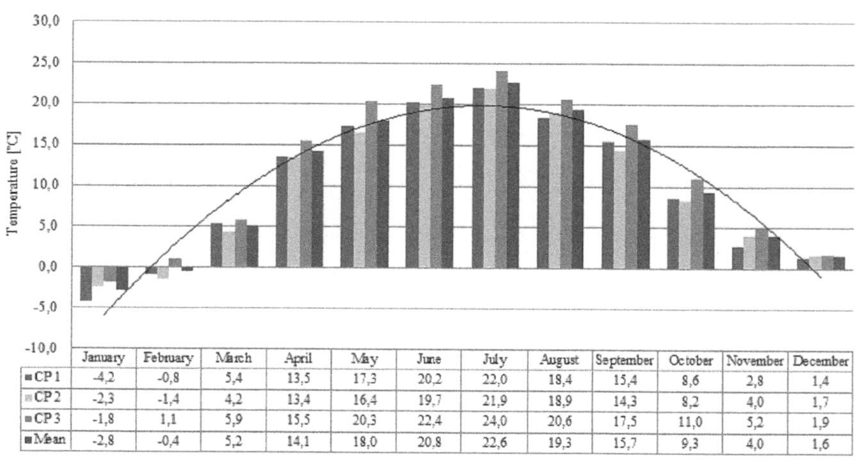

	January	February	March	April	May	June	July	August	September	October	November	December
CP 1	-4,2	-0,8	5,4	13,5	17,3	20,2	22,0	18,4	15,4	8,6	2,8	1,4
CP 2	-2,3	-1,4	4,2	13,4	16,4	19,7	21,9	18,9	14,3	8,2	4,0	1,7
CP 3	-1,8	1,1	5,9	15,5	20,3	22,4	24,0	20,6	17,5	11,0	5,2	1,9
Mean	-2,8	-0,4	5,2	14,1	18,0	20,8	22,6	19,3	15,7	9,3	4,0	1,6

Figure 2. Average temperature distribution of atmospheric air at control points on the river Supraśl [Dane... WIOŚ]

tration of ammonia nitrogen in the same period. Maximum concentration of this component was observed in March and a minimum in July. The exception from the general trend of increased concentrations is the results achieved in all control points in March. Possibly in this case, the overall variability of nitrate nitrogen concentration was disturbed by external (anthropogenic) source of this component in the studied waters. The spring period, which begins in March, is characterized by increased runoff from agricultural fields. Those waters are enriched in increased amounts of nitrate nitrogen unused by plants, as indicated in studies carried out by Banaszuk et al. [2009].

The concentration of total nitrogen (Figure 5) was characterized by a similar trend of changes as the nitrate nitrogen. Maximum concentration of total nitrogen was observed in March and a minimum in August. The concentration changes rhythm of total nitrogen is related to concentration changes of nitrate nitrogen. Phosphate concentration (Figure 6) was the highest in July and lowest in November. A characteristic changes trend was observed in all control points- the gradual increase in the concentration of phosphates with its beginning in May and ending in July.

The concentration of total phosphorus (Figure 7) showed an analogical changes trend as the concentration of dissolved phosphorus in the water. The maximum concentration of total phosphorus was observed in all control points in July, while the minimum in April. The data collected in CP 1 (Figure 3–7) indicate, the greatest concentration of nitrogen present in the winter and in the first decade of the spring – January, February, March. In the rest of the analyzed period, the nitrogen concentration has not changed significantly. It is strictly related to changes of atmospheric air temperature and oxygen conditions improvement in watercourse occurring during the period of temperatures above 0 °C [Koc 2006, Sojka et al. 2007, Czyżyk et al. 2008]. In the phosphorus case, high-

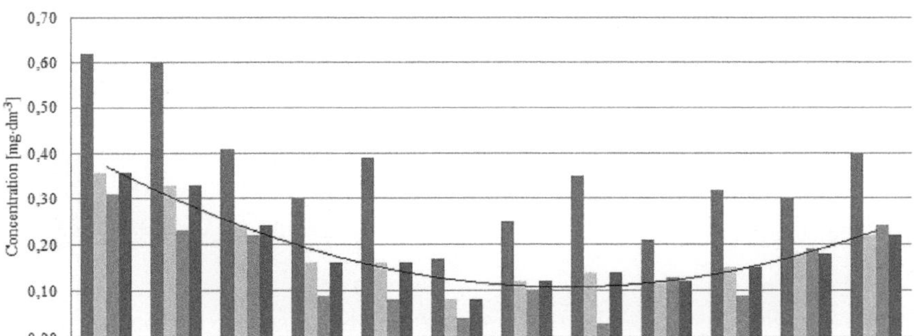

	January	February	March	April	May	June	July	August	September	October	November	December
CP 1	0,62	0,60	0,41	0,30	0,39	0,17	0,25	0,35	0,21	0,32	0,30	0,40
CP 2	0,36	0,33	0,24	0,16	0,16	0,08	0,12	0,14	0,12	0,15	0,18	0,22
CP 3	0,31	0,23	0,22	0,09	0,08	0,04	0,10	0,03	0,13	0,09	0,19	0,24
Mean	0,36	0,33	0,24	0,16	0,16	0,08	0,12	0,14	0,12	0,15	0,18	0,22

Figure 3. The concentration of ammonia nitrogen in the river Supraśl [Dane... WIOŚ]

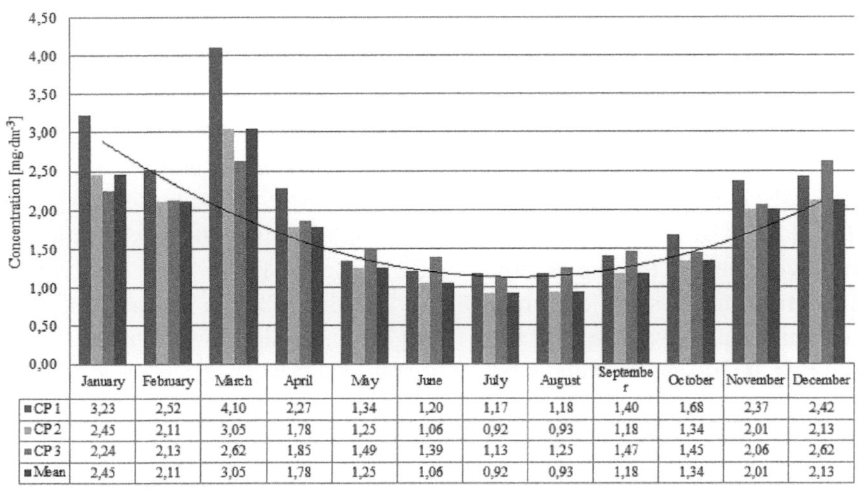

	January	February	March	April	May	June	July	August	September	October	November	December
CP 1	3,23	2,52	4,10	2,27	1,34	1,20	1,17	1,18	1,40	1,68	2,37	2,42
CP 2	2,45	2,11	3,05	1,78	1,25	1,06	0,92	0,93	1,18	1,34	2,01	2,13
CP 3	2,24	2,13	2,62	1,85	1,49	1,39	1,13	1,25	1,47	1,45	2,06	2,62
Mean	2,45	2,11	3,05	1,78	1,25	1,06	0,92	0,93	1,18	1,34	2,01	2,13

Figure 4. The concentration of nitrate nitrogen in the river Supraśl [Dane... WIOŚ]

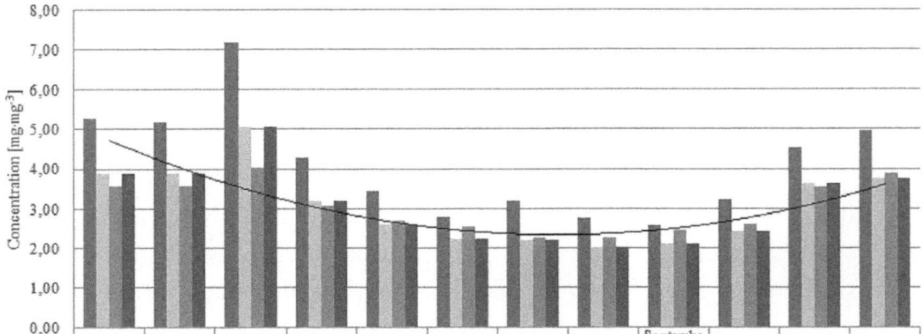

	January	February	March	April	May	June	July	August	September	October	November	December
CP 1	5,26	5,19	7,16	4,27	3,45	2,79	3,20	2,77	2,57	3,22	4,54	4,96
CP 2	3,88	3,88	5,07	3,18	2,61	2,22	2,21	2,02	2,10	2,41	3,64	3,76
CP 3	3,57	3,57	4,04	3,07	2,70	2,53	2,26	2,26	2,46	2,59	3,53	3,87
Mean	3,88	3,88	5,07	3,18	2,61	2,22	2,21	2,02	2,10	2,41	3,64	3,76

Figure 5. The concentration of total nitrogen in the river Supraśl [Dane... WIOŚ]

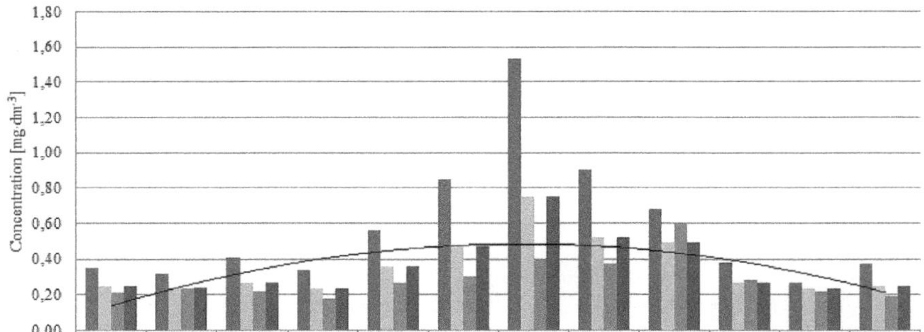

	January	February	March	April	May	June	July	August	September	October	November	December
CP 1	0,35	0,32	0,41	0,34	0,56	0,85	1,53	0,90	0,68	0,38	0,27	0,37
CP 2	0,25	0,24	0,27	0,23	0,36	0,47	0,75	0,52	0,49	0,27	0,23	0,25
CP 3	0,21	0,23	0,22	0,18	0,27	0,30	0,40	0,37	0,60	0,28	0,22	0,19
Mean	0,25	0,24	0,27	0,23	0,36	0,47	0,75	0,52	0,49	0,27	0,23	0,25

Figure 6. The concentration of phosphorus in the river Supraśl [Dane... WIOŚ]

est concentration was recorded in July. Then, a gradual increase in the concentration of phosphorus was observed, which begins in May and ends in October. As pointed by Banaszuk [2009] and Sapek [2011] the highest concentration of studied forms of phosphorus occurs in July.

The data collected from CP 2 (Figures 3–7), shows that the greatest concentration of nitrogen is observed in winter and early spring, and phosphorus in July – as in CP 1. Average monthly concentrations of the individual indicators are lower than in CP 1. It is worth mentioning that the area between points CP 1 and CP 2 is largely covered with forests. Therefore, the nutrient supply from anthropogenic origin is relatively small. In comparison to the results obtained from CP 2 samples it can be concluded that a significant part of the ingredients introduced to Supraśl in CP 1 is bioremediated in this stretch. Research carried out by Kozłowski [2009] confirms the above described mechanism.

Total phosphorus concentration in the control point CP 3 is not significantly different in any of the analyzed months. Phosphate levels in the studied waters, as in other points reaches their maximum in July. The concentration of phosphate in the waters of CP 3 was increased during the growing season. The data presented in Figures 3–7 that the concentration of various forms of nitrogen are characterized by a similar trend of changes-maximum occurrence was observed in winter and early spring, while the lowest concentration was observed in July, which is also indicated by Dąbrowska [2008]. Concentrations of phosphates and phosphorus are the highest during the growing season, which is typical for a small agricultural catchment according to Kuźniar et al. [2008].

From the results summarized in Table 1, the largest dispersion in standard deviations occurs in increased concentrations periods of each indicator.

Similar results were achieved by Skorbiłowicz [2005], who showed that seasonal changes in selected components of the waters of Narew were primarily dependent upon climatic factors such

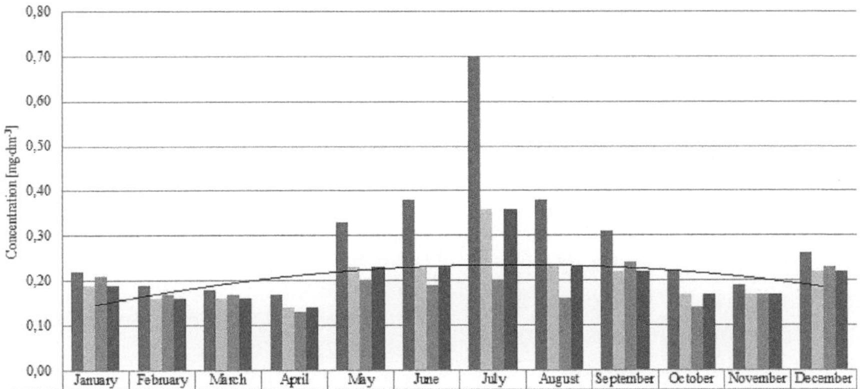

	January	February	March	April	May	June	July	August	September	October	November	December
CP 1	0,22	0,19	0,18	0,17	0,33	0,38	0,70	0,38	0,31	0,22	0,19	0,26
CP 2	0,19	0,16	0,16	0,14	0,23	0,23	0,36	0,23	0,22	0,17	0,17	0,22
CP 3	0,21	0,17	0,17	0,13	0,20	0,19	0,20	0,16	0,24	0,14	0,17	0,23
Mean	0,19	0,16	0,16	0,14	0,23	0,23	0,36	0,23	0,22	0,17	0,17	0,22

Figure 7. The concentration of total phosphorus in the river Supraśl [Dane... WIOŚ]

Table 1. The standard deviations (SD) of test results from the years 2003–2009

Indicator	Point	January	February	March	April	May	June	July	August	September	October	November	December
Temperature °C	CP 1	10.11	4.52	2.86	3.72	5.70	5.58	5.07	2.38	5.28	4.13	5.00	4.05
	CP 2	9.56	6.56	2.93	3.49	9.85	3.56	3.16	2.84	3.32	4.97	3.79	4.35
	CP 3	9.53	5.08	2.91	3.56	7.22	4.45	5.05	4.42	4.05	3.60	2.93	4.33
Ammonium nitrogen mg N×dm^{-3}	CP 1	0.47	0.38	0.35	0.12	0.25	0.11	0.34	0.42	0.10	0.08	0.15	0.18
	CP 2	0.08	0.06	0.10	0.10	0.02	0.04	0.04	0.03	0.03	0.11	0.02	0.01
	CP 3	0.20	0.11	0.32	0.09	0.06	0.05	0.12	0.03	0.17	0.00	0.72	0.86
Nitrate nitrogen mg N×dm^{-3}	CP 1	0.75	0.23	2.13	0.46	0.20	0.66	0.59	0.43	0.29	0.33	1.14	1.20
	CP 2	0.30	0.11	1.15	0.41	0.16	0.17	0.19	0.26	0.35	0.11	0.52	0.06
	CP 3	0.32	0.38	0.78	0.39	0.26	0.58	0.46	0.81	0.44	0.43	0.20	1.36
Total nitrogen mg N×dm^{-3}	CP 1	0.48	1.55	2.48	0.69	0.58	0.50	1.06	0.45	0.63	0.17	1.13	1.36
	CP 2	0.28	0.61	1.56	0.57	0.10	0.33	0.24	0.50	0.41	0.58	0.78	0.11
	CP 3	0.51	0.60	0.89	0.65	0.42	0.78	0.76	0.90	0.71	0.84	0.80	2.26
Phosphates mg PO_4×dm^{-3}	CP 1	0.20	0.16	0.36	0.06	0.12	0.28	1.10	0.59	0.28	0.09	0.04	0.21
	CP 2	0.04	0.02	0.05	0.05	0.01	0.08	0.09	0.13	0.05	0.03	0.04	0.03
	CP 3	0.02	0.07	0.08	0.05	0.03	0.07	0.08	0.14	1.08	0.17	0.06	0.02
Total phosphorus mg P×dm^{-3}	CP 1	0.15	0.07	0.10	0.05	0.07	0.13	0.52	0.24	0.12	0.05	0.03	0.13
	CP 2	0.10	0.02	0.05	0.05	0.03	0.03	0.05	0.06	0.03	0.08	0.06	0.06
	CP 3	0.09	0.04	0.07	0.06	0.10	0.10	0.05	0.05	0.34	0.05	0.10	0.21

as air temperature and precipitation, and that the concentration of nitrate nitrogen was characterized by the largest seasonal variability.

Research carried out by Kulbika and Kowalik [2002] also showed that nitrate concentrations in rivers are reaching the summer minimum (July and August) and two winter maxima (January and March).

Dąbkowski and Pawlata-Zakrzywaj [2003] studied the river Raszynka and found that the purity status changed during each month of the research year. The growing rate of quality changes in the hydrological year pointed to that. Their research showed maximum concentration of nitrogen in the period January-April (snow melt), and the minimum- in the period July-October (full vegetation and low outputs).

According to the results obtained by the Trender [2010] and Pulikowski et al. [2011] nitrogen concentrations show a similar trend of changes taking place in other streams of surface water. In small agricultural catchments nitrogen concentration reaches its maximum mostly in February or March. In case of phosphorus maximum is observed most commonly during the growing season [Kiryluk 2011].

CONCLUSIONS

1. The concentration of nitrogen and phosphorus in waters of Supraśl were characterized by a notable seasonal variation.
2. The largest concentration of nitrogen was observed in January and February and the lowest in June in each of the analyzed points.
3. The changes course of nitrate and total nitrogen concentrations in March was different from the generally prevailing changes course of their concentrations in surface waters.
4. Probably in some areas the concentration of phosphorus in surface waters increases with the beginning of the growing season and decreases thereafter.

Acknowledgement

The project was funded by S/WBiIŚ/4/2011.

REFERENCES

1. Banaszuk P., Krasowska M., Kamocki A. 2009. Źródła azotu i fosforu oraz drogi ich migracji podczas wezbrania roztopowego w małej zlewni rolniczej. Woda-Środowisko-Obszary wiejskie, 9(4): 5-26.
2. Bucka H., Wilk-Woźniak E. 2002. Gatunki kosmopolityczne ubikwistyczne wśród glonów pro- i eukariotycznych występujących w zbiornikach wodnych Polski południowej. Monografia. ZBW PAN, Kraków.
3. Czyżyk F., Strzelczyk M. 2008. Zanieczyszczenie środowiska wodnego pod wpływem stosowania gnojowicy na glebie lekkiej i ciężkiej. Woda-Środowisko-Obszary wiejskie, 8(1): 61-68.
4. Dane środowiskowe WIOŚ od 2003 do 2009 dla rzeki Supraśl.
5. Dąbkowski Sz., Pawłat-Zakrzywaj A. 2003. Jakość wód Raszynki i jej dopływów. Wiad IMUZ, 3(6): 111-123.
6. Dąbrowska J. 2008. Ocena zawartości związków azotu i fosforu w wodach rzeki Trzemy. Infrastruktura i ekologia terenów wiejskich, 7: 57- 68.
7. Gergel S.E., Turner M.G., Kratz T.K. 1999. Dissolved organic carbon as an indicator of the scale of watershed influence on lakes and rivers. Ecological Application, 9(4): 1377-1390.
8. Kiryluk A., Rauba M. 2011. Wpływ rolnictwa na stężenie fosforu ogólnego w wodach powierzchniowych zlewni rzeki Śliny. Inżynieria Ekologiczna, 26: 122-132.
9. Koc J., Solarski K. 2006. Wpływ systemów melioracyjnych na wymywanie związków azotu i fosforu ze zlewni użytkowanych rolniczo. Woda-Środowisko-Obszary wiejskie, 6(1): 195-205.
10. Kowalik P., Kulbik M. 2002. Wpływ pokrywy glebowej w zlewni na kształtowanie się obszarowego spływu niektórych zanieczyszczeń do wód powierzchniowych. Wiad. IMUZ, t. 2, 1(4): 211-223.
11. Kozłowski S., Zielewicz W. 2009. Obecność azotu azotanowego w roślinach i w wodzie – przeszłość i przyszłość. Woda- Środowisko-Obszary wiejskie, 9(2): 125- 137.
12. Kuźniar A., Twardy S., Kowalczyk A. 2008. Przyczyny zmian stężenia związków azotu i fosforu w wodach powierzchniowych górnej zlewni Sanu w latach 1990-2005. Woda-Środowisko-Obszary wiejskie, 8(1): 185-196.
13. Web portal: www.poland.gov.pl
14. Pulikowski K. et al. 2011. Sezonowe zmiany wielkości ładunku azotu odpływającego z mikrozlewni użytkowanych rolniczo, Infrastruktura i Ekologia Terenów Wiejskich, 10: 161-171.
15. Sapek A. 2008. Nawożenie fosforem a jego skutki w środowisku, Woda-Środowisko-Obszary wiejskie, tom 8, z. 2b(24): 127-137.
16. Sapek A., Sapek B. 2011. Fosfor w opadzie atmosferycznym, Ochrona Środowiska i Zasobów Naturalnych, 50: 122-133.
17. Skorbiłowicz E. 2003. Ocena stanu zanieczyszczenia małych cieków Puszczy Knyszyńskiej. Acta Agrophysica, 1(20): 311-320.
18. Skorbiłowicz M. 2005. Zmiany sezonowe stężenia związków azotu, fosforu i makroelementów w górnym biegu Narwi. Monografie, PAN Lublin, Komitet Inżynierii Środowiska, 30: 997-1004.
19. Sojka M., Murat-Błażejewska S., Kanclerz J. 2007. Ocena czasowej zmienności ładunków związków azotu i fosforu wymywanych ze zlewni rolniczej. Inżynieria Ekologiczna, 18: 153-154.
20. Trender W., Kowalczyk W. 2010. Sezonowe zmiany zawartości makroelementów w wodzie rzeki Rawki w latach 2003-2009. Zeszyty Naukowe Instytutu Sadownictwa i Kwiaciarstwa im. Szczepana Pieniążka, 18: 155- 165.
21. Vymazal J. 1995. Algae and Element Cycling in Wetlands. Lewis Publishers, Boca Raton, Ann. Arbor, London, Tokyo.
22. Ocena stanu czystości wód w zlewni rzeki Supraśl w 2010 roku. Wojewódzki Inspektorat Ochrony Środowiska w Białymstoku. Białystok 2011.
23. Jarosiewicz A. 2007. Proces samooczyszczania w ekosystemach rzecznych, Słupskie Prace Biologiczne, 4: 27- 41.

ESTIMATION OF WHEAT STRAW AND RAPESEED USEFULNESS FOR WHEAT FERTILISATION IN ARABLE FARMING

Sławomir Stankowski[1], Bożena Michalska[2], Janusz Smagacz[3], Marzena Gibczyńska[4]

[1] Department of Agronomy, West Pomeranian Technological University in Szczecin, Papieża Pawła VI 3, 71-459 Szczecin, Poland, e-mail: sstankowski@hoga.pl

[2] Department of Meteorology and Green Areas Management, West Pomeranian Technological University in Szczecin, Papieża Pawła VI 3, 71-459 Szczecin, Poland, e-mail: bozena.michalska@zut.edu.pl

[3] Institute of Soil Science and Plant Cultivation, State Research Institute, Czartoryskich 8, 24-100 Puławy, Poland, e-mail: smagacz@iung.pulawy.pl

[4] Department of General and Ecological Chemistry, West Pomeranian Technological University in Szczecin, 71-434 Szczecin, Słowackiego 17, Poland, e-mail: marzena.gibczynska@zut.edu.pl

ABSTRACT

The aim of the research was the estimation of wheat and rapeseed straw usefulness for fertilising plants in arable farming. The subject matter of the study was the analysis of the influence of soil fertilisation with wheat and rapeseed straw on the yield and changes in the content of phosphorus, potassium, calcium and magnesium in winter wheat grain cultivars: Finezja and Ludwig. The static field experiment was conducted at Agricultural Experiment Station, Grabów, Institute of Soil Science and Plant Cultivation – State Research Institute (51°21′ N, 21°40′ E) in the years 2008–2010 on Haplic Luvisol. The effect of fertilisation with wheat and rapeseed straw did not have a significant effect on grain yield of winter wheat cultivars 'Finezja' and 'Ludwig'. The introduction of complementary nitrogen fertilisation caused an increase in grain yield of wheat cultivar Ludwig. Organic fertilisation with wheat and rapeseed straw as well as complementary nitrogen fertilisation did not significantly differentiate the content of phosphorus, potassium, calcium and magnesium in winter wheat grain cultivars 'Finezja' and 'Ludwig'.

Keywords: 'Finezja' and 'Ludwig' wheat, magnesium, phosphorus, potassium, calcium, soil fertilization.

INTRODUCTION

Straw is ripe or dry stalk of cereal plants as well as of legumes, flax or rapeseed. The regulation of the Minister of Environment on R10 recovery operation [Dz.U. 2011, 86, 476] includes categorises of straw, as a kind of waste product, as plant-tissue waste with a code: 02 01 03. The appendix to the Regulation specifies the conditions of R10 recovery operation and classifies waste for distribution on the surface of soil in order to fertilise or ameliorate soil. The regulation provides for the condition that waste is used in soil in which the admissible concentration values of substances, specified in the Regulation of the Minister of Environment, dated 9 September 2002 on soil quality standards and land quality standards, are not exceeded [Dz.U. 2002, 165, 1359]. However, taking into consideration the decreasing content of organic matter in soils in Poland, the most favourable solution is recirculation of organic matter and nutrients, contained in crop residue, in arable fields [Ministry of agriculture and rural development Warsaw 2010]. Straw is a basic organic fertiliser in stock-free farming. Straw of cereals can be a good organic fertiliser under the condition that it is milled into particles sized most preferably less than 10 cm and ploughed immediately after grain gathering. After ploughing of straw the nutrients contained therein circulate in soil and additionally constitute a good humus creating material. The nutrients contribute to greater utilisation and efficacy of mineral fertilisers. Wheat straw is characterised by the content of phosphorus, potassium calcium and magnesium: 1.3, 6.8, 5.6 and 1,3 $g \cdot kg^{-1}$ d.m. respectively [Buraczyńska and Ceglarek 2011a]. Rapeseed straw which contains particularly high amount of calcium and potassium as well as other components is valued more than that of basic cereals. One tonne of rapeseed straw contains approximately 1.3 kg of phosphorus, 16.8 kg of potassium, 11.1 kg of calcium and 0.8 kg of magnesium [Szewczuk 2010].

The aim of the research was the estimation of wheat and rapeseed straw usefulness for fertilising plants in arable farming. The subject matter of the study was the analysis of the influence of soil fertilisation with wheat and rapeseed straw on yield and changes in the content of phosphorus, potassium, calcium and magnesium in winter wheat grain cultivars 'Finezja' and 'Ludwig'.

MATERIALS AND METHODS

The conditions of the experiment

The static field experiment was conducted at Agricultural Experiment Station, Grabów, Institute of Soil Science and Plant Cultivation – National Research Institute (51° 21′ N, 21° 40′ E) in the years 2008–2010 on Haplic Luvisol [FAO 2006]. It was based on an ongoing, since 1987, static crop rotation experiment with different variants of straw use, set up according to the randomized block method in four replications. The size of crop plots was 45 m^2. In its present setup, the experiment has been conducted since the autumn of 1997 and the presented results regard 11th –13th years of the experiment. The crop rotation held on all the plots were: winter rapeseed – winter wheat – winter wheat, with varying frequency of ploughing the straw. In this two factor experiment, the following factors were taken into consideration: I – winter wheat cultivars 'Finezja' and 'Ludwig', II – straw fertilisation. The experiment was conducted over the period of three years. Table 1 presents the variant number and fertilisation mode.

Rapeseed straw was applied in the amount of 3 $t \cdot ha^{-1}$, and wheat 5 $t \cdot ha^{-1}$ before conventional ploughing for winter cereals. Additionally, in order to balance C:N proportion in ploughed straw, nitrogen in a form of ammonium nitrate was used in plots 3 and 4 in the amount of 30 kg $N \cdot ha^{-1}$. Winter wheat was sown in an optimal term for this region of the country in the third decade of September in the amount of 4.5 million of germinating seeds from a hectare. Wheat cultivars cultivated in the experiment were Finezja and Ludwig, both classified as quality wheat. Annual average phosphorus and potassium fertilisation in the amount of 70 kg $P \cdot ha^{-1}$ and 120 kg $K \cdot ha^{-1}$ was applied in the experiment before conventional ploughing for winter cereals. Top dressing fertilisation of winter wheat with nitrogen was applied in the amount of 90 kg $N \cdot ha^{-1}$, yet ⅔ of the dose, that is 60 kg $N \cdot ha^{-1}$, was applied in the early spring during the beginning of vegetation. The remaining amount (30 kg $N \cdot ha^{-1}$) was applied in the stem elongation phase (BBCH 31–32).

Plant protection against weeds and diseases was congruent with the guidelines of the Plant Protection Institute – State Research Institute in Poznań. Maxim 025 FS seed treatment in the amount of 200 and 400 cm^3 of water for 100 kg of seed were used for seed treatment of both cultivars of winter wheat in 2008. In the following years Kinto Duo 080 FS in the amount of 200 and 400 cm^3 of water were used for 100 kg of seed. Each autumn, winter wheat plants in the growth stage of 2–3 unfolded leaves (BBCH 12–13) were sprayed against monocot and dicot weeds. Expert Met 56 WG in the amount of 0.35 $dm^3 \cdot ha^{-1}$ was used in 2008, and Alister Grande 190 OD was used in the autumn of 2009 in the amount of 0.9 $dm^3 \cdot ha^{-1}$. In the growth stage of the first node (BBCH 31) the plants were sprayed each year against stem base diseases and powdery mildew of cereals and grasses using Alert 375 SC in the amount of 1 $dm^3 \cdot ha^{-1}$. The spraying was combined with the use of growth regulator (Cycocel 460 SL in the amount of 1.7 $dm^3 \cdot ha^{-1}$). At the beginning of flowering (BBCH 61) the plants were protected against leaf and ear diseases of winter wheat using Charizma 207 EC in the amount of 1.5 $dm^3 \cdot ha^{-1}$ in 2008, and Artea 330 EC in the amount of 0.5 $dm^3 \cdot ha^{-1}$ in 2009 and 2010. Finezja winter wheat cultivar was registered in 2002. This cultivar, originated at "Danko" Plant Breeders Ltd., and is suitable for growing all over Poland and tolerant to slightly acid soils. Ludwig winter wheat cultivar from Probstdofer Saatzucht is quality wheat (A class) with high yield potential, very high protein and gluten content and is suitable for growing all over Poland.

Table 1. Fertilisation variant number and components used

Variant	Applied fertilisation
1	Control – without ploughing the straw
2	Straw ploughed one time in rotation, rapeseed only
3	Straw ploughed two times in rotation – rapeseed and winter wheat straw sown after rapeseed + 30 kg N·ha^{-1}
4	Straw ploughed three times in rotation – rapeseed straw, wheat straw sown after rapeseed and wheat straw sown after wheat + 30 kg N·ha^{-1}
5	Straw ploughed three times in rotation – rapeseed straw, wheat straw sown after rapeseed and wheat straw sown after wheat

Methodology of chemical analyses

In order to determine the total macro component content, the grain samples were mineralised in the mixture of nitric(V) acid and chloric(VII) acid in 1:1 proportion. After mineralisation of the samples the phosphorus content was determined using colimetric determination, and the contents of potassium, calcium and magnesium were determined with the use of atomic absorption spectrometer SOLAAR AA SERIES.

The analysis of the obtained results was made with the use of analysis variance. The values of LSD were calculated using Tukey's test at the statistical significance of $\alpha = 0.05$. The experimental error for testing effects in experiment was means square of interaction C × F × Y. Calculations were performed using Statistica version 10.

Climatic conditions

During the period of three years thermal conditions were highly varied: positive deviations which resulted in an increased mean annual temperature in the subsequent years – 0.9°, 0.8° and 0.4 °C respectively were predominant, in comparison with multi-year values (7.9 °C) – Table 2. The greatest positive deviation in air temperature was recorded in the winter and at the beginning of spring of 2007/2008 – in February the temperature exceeded the average by 4.4 °C. In the period 2008/2009 nearly all months were warmer, with the greatest deviation (2.6 °C) recorded in April. The period 2009/2010 was characterised by low temperatures in winter, it was particularly chilly in January as the mean air temperature was below the average by 5.9 °C.

Table 2. Deviation of the multiannual values (1976–2010) of air temperature and total precipitation according to months and seasons in the period of 2007/2008 and 2009/2010

Deviation of the norm	Years	09	10	11	12	01	02	09–08
Air temperature, °C	2007/2008	-0.2	-0.6	-1.8	0.3	3.5	4.4	0.9
	2008/2009	-0.5	1.5	1.9	2.2	-0.5	0.6	0.8
	2009/2010	1.9	-1.3	2.3	-0.3	-5.9	-0.4	0.4
Norm, °C	**1976–2010**	**13.0**	**8.2**	**3.0**	**-1.0**	**-2.7**	**-1.8**	**7.9**
Precipitation totals, %	2007/2008	119	30	117	28	164	64	99
	2008/2009	109	126	92	121	75	113	111
	2009/2010	50	21	173	146	70	100	113
Norm, mm	**1976–2010**	**64.9**	**42.9**	**41.7**	**37.6**	**31.6**	**26.4**	**637.4**

Deviation of the norm	Years	03	04	05	06	07	08	09–08
Air temperature, °C	2007/2008	1.4	0.9	-0.6	1.0	0.4	1.1	0.9
	2008/2009	0.1	2.6	-0.2	-0.2	1.2	0.3	0.8
	2009/2010	0.9	0.9	0.2	1.0	3.0	2.1	0.4
Norm, °C	**1976–2010**	**2.1**	**8.1**	**13.7**	**16.6**	**18.5**	**17.8**	**7.9**
Precipitation totals, %	2007/2008	161	160	140	65	99	67	99
	2008/2009	197	1	92	150	140	92	111
	2009/2010	65	46	182	64	62	191	113
Norm, mm	**1976–2010**	**39.0**	**44.8**	**62.6**	**78.8**	**86.0**	**81.1**	**637.4**

The other months were warmer with the greatest positive deviation (3.0 °C) recorded in July.

Annual precipitation was within average during the period 2007/2008 and in the remaining two periods it was slightly above the average value (Table 3). Humidity conditions were more varied in individual months, especially in the year 2009 – in March precipitation amounted to 200% of the average values, and in April only 1%. The characteristics of the humidity conditions [Michalska and Kalbarczyk 2005] in the two-months-long periods from March to August is given in Table 3.

The spring of 2008 is regarded as moist, even extremely moist at its beginning, while the remaining periods were within the average. The vegetation season of 2009 was characterised by interchangeable dry and moist periods, and the beginning of spring of 2010 (03–04) as well as the beginning of summer (06–07) were dry – the remaining two-month-long periods were within average.

RESULTS AND DISCUSSION

Grain yield

The yield was the greatest in the first year of the experiment during which the most favourable weather-conditions occurred. The dependence of the yield on the meteorological conditions is confirmed by the numerous publications [Gaj 2012; Järvan et al. 2011; Joniec 2011; Masilionytė & Maikštėnienė 2011]. In the two subsequent years, the grain yield was lower than in the first year of the experiment. Therefore, it is most valid to analyse the mean results from the three-year experiment. Taking into consideration the mean yield values from the period of three years, the grain yield of wheat cultivar Ludwig was higher by 0.7 $t \cdot ha^{-1}$. Variations in the fertilisation variants had little effect on wheat grain yield and according to the calculations the effect was not significant. There were differences in the reaction of the cultivars. The yield of

Table 3. Estimation humidity conditions in the two-month periods, spring and summer

Years–months	03–04	04–05	05–06	06–07	07–08
2008	very moist	moist	normal	normal	normal
2009	normal	dry	normal	wet	normal
2010	dry	normal	normal	dry	normal

Table 4. Winter wheat grain yield ($t \cdot ha^{-1}$) depending on cultivar (C), fertilization (F)variant and years (Y)

Cultivar	Fertilization variant	Years			Average
		2008	2009	2010	
Finezja	1	6.36	5.92	5.84	6.04
	2	6.52	5.98	5.84	6.11
	3	6.42	5.90	5.88	6.07
	4	6.43	6.16	5.74	6.11
	5	6.74	5.96	5.87	6.19
Average		6.48	5.98	5.83	6.10
Ludwig	1	7.30	7.10	6.08	6.83
	2	7.34	6.57	5.98	6.63
	3	7.33	6.86	6.28	6.82
	4	7.04	7.51	6.24	6.93
	5	7.12	6.62	6.12	6.62
Average		7.22	6.93	6.14	6.77
Average	1	6.83	6.51	5.96	6.43
	2	6.93	6.27	5.91	6.37
	3	6.87	6.38	6.08	6.44
	4	6.74	6.83	5.99	6.52
	5	6.93	6.29	5.99	6.40
Total average		6.85	6.46	5.99	6.43
$LSD_{0.05}$ for:		C – 0.117, F – n.s., Y – 0.178, C × F – n.s., C × Y – 0.252, F × Y – n.s.			

winter wheat cultivar 'Finezja' was similar regardless of the fertilisation variant. The analysis of mean values of wheat cultivar Ludwig shows that straw fertilisation caused a slight decrease in yield. Higher yield achieved using additional nitrogen fertilisation can be attributed to the fact that, according to other authors [Buraczyńska and Ceglarek 2011b; Wysokiński and Kalembasa, 2011; Czyżyk and Rajmund. 2013; Janušauskaitė et al. 2013], straw fertilisation causes an unfavourable phenomenon of phosphorus regression in soil which results in the decrease in yield, which can be compensated by introducing complementary doses of nitrogen (Table 4).

Phosphorus (P)

Phosphorus is a macrocomponent which as a non-metal makes up organic compounds (phytins, phospholipids and nucleic acids) found in wheat grain. The average phosphorus content in wheat grain is approximately 4 g $P \cdot kg^{-1}$ d.m. The amount of phosphorus in wheat grain cultivar Finezja ranged from 3.36 to 4.27 $g \cdot kg^{-1}$ d.m. The phosphorus content in wheat grain cultivar Ludwig ranged from 4.01 to 4.80 $g \cdot kg^{-1}$ d.m. (Table 5). The phosphorus content in wheat grain cultivars 'Finezja' and 'Ludwig' was not significantly influenced by organic fertilization with straw or by complementary fertilization with nitrogen, which has been proved in numerous studies on the topic [Wacławowicz et.al. 2005; Wiśniowska-Kielian and Klima 2006; Wilczewski and Skinder 2011].

Potassium (K)

Potassium is found in plants as potassium ions (I). Contrary to phosphorus it does not form organic compounds, yet it plays a fundamental part in many biochemical and physiological processes such as, among others, the transport of organic and inorganic compounds in vascular plant tissue. The average potassium content in winter wheat grain is 3.5 $g \cdot kg^{-1}$ d.m. [Bobrecka-Jamro et al. 2013]. The amount of potassium in wheat grain cultivar 'Finezja' ranged from 2.76 to 3.88 $g \cdot kg^{-1}$ d.m., and in cultivar Ludwig the amount of potassium was similar and ranged from 2.96 to 3.89 $g \cdot kg^{-1}$ d.m. (Table 6). The comparison of grain richness in individual years shows that the content of potassium was the highest in the second year of the experiment. Precipitation in June and July of that year exceeded manifold the mean

Table 5. Phosphorus content in wheat grain (g $P \cdot kg^{-1}$ d.m.) depending on cultivar (C), fertilization (F) variant and years (Y)

Cultivar	Fertilization variant	Years			Average
		2008	2009	2010	
Finezja	1	4.01	4.27	3.75	4.01
	2	4.10	4.01	4.27	4.10
	3	3.36	4.01	4.01	3.80
	4	4.01	4.27	4.10	4.13
	5	3.75	4.10	4.01	3.95
Average		3.89	4.13	4.03	4.00
Ludwig	1	4.01	4.27	4.10	4.13
	2	4.10	4.10	4.53	4.24
	3	4.27	4.10	4.53	4.30
	4	4.10	4.27	4.80	4.39
	5	4.27	4.27	4.53	4.36
Average		4.15	4.20	4.50	4.28
Average	1	4.01	4.27	3.92	4.07
	2	4.05	4.05	4.40	4.17
	3	3.81	4.05	4.27	4.06
	4	4.05	4.27	4.45	4.26
	5	4.01	4.18	4.27	4.15
Total average		3,99	4.17	4.26	4.14
$LSD_{0.05}$ for:		C – 0.133, F – n.s., Y – 0.202, C × F, C × Y and F × Y – n.s.			

Table 6. Potassium content in wheat grain (g K·kg^{-1} d.m.) depending on cultivar (C), fertilisation (F) variant and years (Y)

Cultivar	Fertilization variant	Years			Average
		2008	2009	2010	
Finezja	1	3.13	3.87	3.27	3.42
	2	2.96	3.75	3.20	3.30
	3	2.81	3.63	3.23	3.22
	4	3.15	3.88	3.47	3.50
	5	2.76	3.58	3.03	3.12
Average		2.96	3.74	3.24	3.31
Ludwig	1	2.96	3.81	3.27	3.35
	2	3.32	3.86	3.20	3.46
	3	3.15	3.59	3.23	3.32
	4	3.18	3.89	3.47	3.51
	5	3.05	3.89	3.03	3.32
Average		3.13	3.81	3.24	3.39
Average	1	3.04	3.84	3.27	3.38
	2	3.14	3.80	3.20	3.38
	3	2.98	3.61	3.23	3.27
	4	3.16	3.88	3.47	3.50
	5	2.90	3.73	3.03	3.22
Total average		3.05	3.77	3.24	3.35
$LSD_{0.05}$ for:		C – n.s.; F – 0.198; Y – 0.127, C × F, C × Y and F × Y – n.s.			

precipitation values from long-term period. This could foster an increased dissolution of potassium compounds and therefore its greater accessibility for the cultivated wheat. Although wheat and rapeseed straw contain several times more potassium, the definite influence of straw fertilisation on potassium content in wheat cultivar Finezja and Ludwig was not reported in the experiment. The stability of potassium content in wheat grain depending on organic and mineral fertilisation is ascertained in many studies on the topic [Wacławowicz et al. 2005; Wiśniowska-Kielian and Klima, 2006; Rachoń and Szumiło, 2009; Wilczewski and Skinder, 2011].

Calcium (Ca)

Calcium, as a basic macro component for plants. It regulates the activity of many enzymes; influences plant cell division, growth and proper functioning of chloroplasts, mitochondria and chromosomes. It also plays a part in water management of plants and maintenance of mono- and divalent cation balance. The content of calcium in wheat grain is relatively diversified. On the basis of the comparison of the content of macro components in wheat grain from organic and conventional faming in Małopolska Voivodeship, [Wiśniowska-Kielan and Klima 2006] state that wheat grain contains 0.40 g·kg^{-1}. In the study by Brzozowska [2008] macroelement content in winter wheat grain is affected by cultivation and nitrogen application that calcium content in dry matter of wheat grain ranges from 0.29 to 0.70 g·kg^{-1} d.m. The content of calcium in wheat grain cultivars ‘Finezja’ and ‘Ludwig’ was similar and ranged from 0.45 to 0.99 g·kg^{-1} d.m. (Table 7). The experiment did not show a significant influence of straw fertilization on calcium content in wheat grain cultivars ‘Finezja’ and ‘Ludwig’. The explanation of this fact could be that wheat and rapeseed straw are characterised by a similar content of calcium.

Magnesium (Mg)

Magnesium is an indispensable element of chlorophyll and plays a role in various physiological processes such as photosynthesis, metabolism of carbohydrates, proteins and lipids [Bose et al. 2010; Hao and Papadopoulos 2004]. Magnesium deficiency causes, above all, a decrease in yield. As compared to other plants, cereals accumulate moderate amounts of magnesium. The average magnesium content in wheat grain is approximately 0.8 g Mg·kg^{-1} d.m. [Wiśniowska-

Table 7. Calcium content in wheat grain (g Ca·kg^{-1} d.m.) depending on cultivar (C), fertilisation (F) variant and years (Y)

Cultivar	Fertilization variant	Years			Average
		2008	2009	2010	
Finezja	1	0.45	0.51	0.53	0.50
	2	0.59	0.46	0.71	0.59
	3	0.41	0.55	0.55	0.50
	4	0.56	0.99	0.52	0.69
	5	0.60	0.71	0.57	0.63
Average		0.52	0.64	0.58	0.58
Ludwig	1	0.77	0.77	0.57	0.70
	2	0.56	0.66	0.48	0.57
	3	0.47	0.66	0.66	0.60
	4	0.75	0.50	0.49	0.50
	5	0.53	0.55	0.53	0.54
Average		0.62	0.63	0.55	0.60
Average	1	0.61	0.64	0.55	0.60
	2	0.57	0.56	0.59	0.58
	3	0.44	0.60	0.60	0.55
	4	0.65	0.74	0.50	0.63
	5	0.56	0.63	0.55	0.58
Total average		0.70	0.64	0.56	0.59
$LSD_{0.05}$ for:		C × F, C × Y and F × Y – n.s. C × F, C × Y and F × Y – n.s.			

Table 8. Magnesium content in wheat grain (g Mg·kg^{-1} d.m.) depending on cultivar (C), fertilisation (F) variant and years (Y)

Cultivar	Fertilization variant	Years			Average
		2008	2009	2010	
Finezja	1	0.94	0.86	0.85	0.88
	2	0.91	0.82	0.84	0.87
	3	0.91	0.83	0.88	0.87
	4	0.95	0.98	0.77	0.90
	5	0.90	0.82	0.84	0.85
Average		0.92	0.86	0.84	0.87
Ludwig	1	1.06	1.02	0.92	1.00
	2	1.03	0.97	0.99	1.00
	3	1.06	0.87	0.97	0.97
	4	1.09	0.91	1.01	1.00
	5	1.01	0.94	0.88	0.94
Average		1.05	0.94	0.95	0.98
Average	1	1.00	0.94	0.88	0.94
	2	0.97	0.89	0.91	0.95
	3	0.98	0.85	0.92	0.92
	4	1.02	0.94	0.89	0.95
	5	0.95	0.88	0.86	0.90
Total average		0.99	0.90	0.89	0.93
$LSD_{0.05}$ for:		C – 0.051, F – n.s., Y – 0.077, C × F, C × Y and F × Y – n.s.			

Kielan and Klima 2006]. Magnesium content in wheat grain cultivar Finezja ranged from 0.77 to 0.98 g·kg^{-1} d.m. Wheat grain cultivar Ludwig contained greater amounts of magnesium which ranged from 0.87 to 1.09 g·kg^{-1} d.m. (Table 8). The comparison of wheat grain content in individual years of the experiment shows that the highest content of magnesium was recorded in the first year of the experiment. In the subsequent years the magnesium content gradually decreased in wheat grain of both cultivars. Wheat and rapeseed fertilisation used in the experiment did not influence any changes in magnesium content of winter wheat grain (Table 8). The lack of diversity in magnesium content in wheat grain could result from similar magnesium content in straw. In their analyses many authors support the findings which indicate the lack of significant differences in magnesium content in wheat grain as a result of diversified fertilisation [Wiśniowska-Kielian and Klima, 2006; Kraska and Pałys, 2009; Gondek and Gondek 2010; Wilczewski and Skinder 2011].

The mean content of phosphorus, potassium, calcium and magnesium in wheat grain cultivar Finezja and Ludwig cultivated in the experiment amounted to: 4.14, 3.35, 0.93 and 0.59 kg^{-1} of dry matter respectively, and did not differ significantly from the results of other research studies. Comparing the content of the macro components in question in wheat grain, the following disparity can be deduced: P > K > Mg > Ca. Wiśniowska-Kielian and Klima [2006] report the same relationship in their comparison of the content of macro components in wheat grain from organic and conventional farming.

CONCLUSIONS

1. The effect of fertilisation with wheat and rapeseed straw did not have a significant effect on grain yield of winter wheat cultivars 'Finezja' and 'Ludwig'.
2. The introduction of complementary nitrogen fertilisation caused an increase in grain yield of wheat cultivar 'Ludwig'.
3. Organic fertilisation with wheat and rapeseed straw as well as complementary nitrogen fertilisation did not significantly differentiate the content of phosphorus, potassium, calcium and magnesium in winter wheat grain cultivars 'Finezja' and 'Ludwig'.

REFERENCES

1. Bobrecka-Jamro D., Kruczek G., Romaniak M., Jarecki W., Buczek J. 2013. Effect of the dose and method of top-dressing with nitrogen on the yield and quality of winter wheat grain. Acta Scientiarum Polonorum, Agricultura, 12 (4), 19–30.
2. Bose J., Babourina O., Rengel Z. 2010. Role of magnesium in alleviation of aluminium toxicity in plants. Journal of Experimental Botany, 62 (7), 2251–2264.
3. Brzozowska I. 2008. Macroelement content in winter wheat grain as affected by cultivation and nitrogen application methods. Acta Agrophysica, 11 (1), 23–32.
4. Buraczyńska D., Ceglarek F. 2011a. previous crop value of post-harvest residues and straw of spring wheat, field pea and their mixtures for winter triticale part i. weight and chemical composition of post-harvest residues and straw. Acta Scientiarum Polonorum, Agricultura, 10 (2), 3–18.
5. Buraczyńska D., Ceglarek F. 2011b. Previous crop value of post-harvest residues and straw of spring wheat, field pea and their mixtures for winter triticale. Part II. Winter triticale yield. Acta Scientiarum Polonorum, Agricultura, 10 (2), 19–32.
6. Czyżyk F., Rajmund A. 2013. Przenikanie azotu do środowiska wodnego wskutek nawożenia gleby lekkiej. Inżynieria Ekologiczna, 29–34.
7. Dz.U.02.165.1359. 2002. Rozporządzenie Ministra Środowiska z dnia 9 września 2002 r. w sprawie standardów jakości gleby oraz standardów jakości ziemi [Regulation of the Minister of Environment dated 9 September 2002 on soil quality standards and land quality standards]. (Dz. U. z dnia 4 października 2002 r.) (in Polish).
8. Dz.U.2011.86.476. 2011. Rozporządzenie Ministra Środowiska z dnia 5 kwietnia 2011 r. w sprawie procesu odzysku R10. [The regulation of the Minister of Environment on R10 recovery operation] (Dz. U. z dnia 22 kwietnia 2011 r.) (in Polish).
9. FAO. 2006. Guidelines for soil description. Food and Agriculture Organization of the United Nations, 1–97 (4th ed.).
10. Gaj R. 2012. The effect of different phosphorus and potassium fertilization on plant nutrition in critical stage and yield of winter triticale. Journal of Central European Agriculture, 13 (4), 704–716.
11. Gondek K., Gondek A. 2010. Wpływ nawożenia mineralnego na plonowanie i zawartość wybranych makro i mikroelementów w pszenicy jarej. [The influence of mineral fertilization on the yield and content of selected macro and microelements in spring wheat]. Journal of Research and Applications in Agricultural Engineering, 55 (1), 30–36 (in Polish).

12. Janušauskaitė D., Arlauskienė A., Maikštėnienė S. 2013. Soil mineral nitrogen and microbial parameters as influenced by catch crops and straw management. Zemdirbyste-Agriculture, 100 (1), 9–18.

13. Järvan M., Edesi L., Adamson A. 2012. Effect of sulphur fertilization on grain yield and yield components of winter wheat. Acta Agriculturae Scandinavica, Section B - Soil & Plant Science, 62 (5), 401–409.

14. Joniec A. 2011. Plonowanie pszenicy ozimej w zależności od systemu uprawy roli. [Effect of the tillage and reduced system on the yield of winter wheat]. EPISTEME, 12 (I), 161–167 (in Polish).

15. Hao X., Papadopoulos A.P. 2004. Effects of Calcium and Magnesium on Plant Growth, Biomass Partitioning, and Fruit Yield of Winter Greenhouse Tomato. HORTSCIENCE, 39 (3), 512–515.

16. Masilionytė L, Maikštėnienė S. 2011. The effect of agronomic and meteorological factors on the yield of main and catch crops. Zemdirbyste-Agriculture, 98 (3), 235–244.

17. Michalska B, Kalbarczyk E. 2005. Longterm changes in air temperature and precipitation on Szczecińska Lowland. EJPAU Ser. Environmental Development, vol. 8, iss. 1 (www.media.ejpau.pl).

18. Ministry of agriculture and rural development. 2010. Wytyczne w zakresie wykorzystania produktów ubocznych oraz zalecanego postępowania z odpadami w rolnictwie i przemyśle rolno-spożywczym. [Guidelines for the use of by-products and recommended behavior of waste in agriculture and agri-food industry] (in Polish).

19. Rachoń L. Szumiło G. 2009. Comparison of chemical composition of selected winter wheat species. Journal of Elementology, 14 (1), 135–146.

20. Szewczuk Cz. 2010. Zanim sprzedasz słomę – bilans substancji odżywczych. [Before the sale of straw - the balance of nutrients]. Wiadomości Rolnicze Polska, 1, 1–3 (in Polish).

21. Wacławowicz R., Parylak D. Śniady R. 2005. Następczy wpływ nawożenia organicznego oraz mineralnego azotowego na plonowanie oraz wybrane cechy jakościowe ziarna pszenicy jarej. [Residual effect of organic and nitrogen fertilization on the yields and selected quality features of spring wheat grain]. Pamiętnik. Puławski, 139, 277–288 (in Polish).

22. Wilczewski E., Skinder Z. 2011. Wartość przedplonowa roślin niemotylkowatych uprawianych w międzyplonie ścierniskowym dla pszenicy jarej cz. II. Zawartość ważniejszych makroskładników w ziarnie i słomie. [Previous crop value of non-legumes grown in stubble catch crop for spring wheat part II. content of major macronutrients in grain and straw]. Fragmenta Agronomica, 28 (1), 107–114 (in Polish).

23. Wiśniowska-Kielian B., Klima K. 2006. Comparison of macroelement contents in the winter wheat grain from organic and conventional farms, Journal of Research and Applications in Agricultural Engineering, 51 (2), 204–208.

24. Wysokiński A., Kalembasa S. 2011. The Yield and Utilization Coefficient of Nitrogen by Plants after Applying of Fresh and Composted Sewage Sludge with Mineral and Organic Additions. Polish Journal of Environmental Studies. 20 (6), 1617–1625.

EFFECT OF CHITOSAN ON PLANT GROWTH, FLOWERING AND CORMS YIELD OF POTTED FREESIA

Piotr Salachna[1], **Agnieszka Zawadzińska**[1]

[1] Department of Horticulture, Faculty of Environmental Management and Agriculture, West Pomeranian University of Technology in Szczecin, Papieża Pawła VI 3, 71-459 Szczecin, Poland, e-mail: piotr.salachna@zut.edu.pl; agnieszka.zawadzinska@zut.edu.pl

ABSTRACT

The research was aimed at determining the influence of chitosan with different molecular weights on the growth and yield of flowers and corms of 'Gompey' freesia cultivated in pots. Freesia corms were soaked for 20 minutes in 0.5% chitosan solutions with low molecular weight (2 kDa), medium molecular weight (50 kDa) and high molecular weight (970 kDa). The average deacetylation level of the chitosans used was 85%. The plants were cultivated under controlled conditions (18/16 °C day/night, relative air humidity of 60%, quantum irradiance of 90 $\mu mol \cdot m^{-2} \cdot s^{-1}$ with a light/dark cycle of 14/8 hours). The research results obtained have shown that chitosan is used as a biostimulator in the cultivation of potted freesia. Regardless of the molecular weight of the compound, the chitosan-treated plants had more leaves and shoots, flowered earlier, formed more flowers and corms. The application of medium- and high-molecular-weight chitosan resulted in higher plants with a higher relative chlorophyll content (SPAD). The highest increase in the corm weight was observed as a result of treating plants with high-molecular-weight chitosan. No chitosan influence on the length of the main inflorescence shoot or the inflorescence length was observed.

Keywords: Freesia, ecological polymer, plant-growth enhancer, molecular weight.

INTRODUCTION

Chitosan is a biopolymer, a chitin derivative, a compound which is completely safe for the environment. This compound is characterized by unique properties, such as bioactivity and biocompatibility [Dias et al. 2013]. The results from the literature show that, when used in plants, chitosan can increase the yield [Mondal et al. 2012], reduce transpiration [Dzung et al. 2011] and induce a range of metabolic changes as a result of which, plants become more resistant to viral, bacterial and fungal infections [Al-Hetar et al. 2011]. Moreover, plants treated with chitosan may be less prone to stress evoked by unfavourable conditions, such as drought, salinity, low or high temperature [Lizarraga-Pauli et al. 2011, Jabeen and Ahmad 2013, Pongprayoon et al. 2013]. Chitosan stimulates vital processes of plants on every level of biological organization, from single cells and tissues, through physiological and biochemical processes, to changes on the molecular level related to expression of genes [Limpanavech et al. 2008, Hadwiger 2013, Nguyen Van et al. 2013].

On an industrial scale, chitosan is obtained as a result of chemical or enzymatic chitin deacetylation. Chitin occurs mostly in crustacean armours. Chitosan parameters are mostly determined by three parameters: morphological structure, the degree of deacetylation and the molecular weight [Aranaz et al. 2009]. Some of the aforementioned properties of chitosan may specifically stimulate plant reactions and the impact on microorganisms and the molecular weight seems to be one of the most important factors affecting the biological activity of this biopolymer [Kulikov et al. 2006, Li et al. 2011].

Freesia (*Freesia* Eckl. ex Klatt) is one of the ornamental plants, which are of the greatest economic importance grown for cut flowers. On

Dutch markets, in 2012 almost 300 million cut freesia flowers of the value of 50 million euro were sold and a sales growth of 6.5% in comparison to 2011 was observed [FloraHolland 2012]. As a result of growth, the assortment of freesia cultivars is extended to include new genetically low cultivars which are becoming more and more popular as pot plants [Ehrich et al. 2009]. Basic and still insufficiently solved problems in freesia production include improvement of cultivation methods, corm reproduction and health of plants [Vaira et al. 2009, Ruamrungsri et al. 2011].

There is no information on the use of chitosan in the cultivation of pot freesia, which is why the research was aimed at establishing how this polymer with various molecular weight affects the growth, flowering and health of freesia plants.

MATERIALS AND METHODS

The material included freesia corms (5-6 cm in circumference) of the 'Gompey' cultivar, which belongs to the Easy Pot group of cultivars characterised by genetically low height. Before planting, the corms were stored for 14 weeks at 28–30 °C and the relative humidity of 85%.

Freesia corms were soaked for 20 minutes in 0.5% chitosan solutions with low molecular weight (2 kDa), medium molecular weight (50 kDa) and high molecular weight (970 kDa). Corms soaked in water were the control. Biopolymers were purchased from Center of Bioimmobilisation and Innovative Packaging Materials at West Pomeranian University of Technology in Szczecin, Poland. Native chitosan supplied by Yuhuan Ocean Biochemical Co. Ltd. (China). The products have been obtained using the free-radical degradation process [Bartkowiak 2001]. The chitosan have been purified (filtration), and characterized using HPLC method (SmartLine Knauer, Germany; Tessek Separon HEMA-BIO 40 column – Tessek, The Tschech Republik). The average deacetylation degree of the products was 85%.

After drying, the single corms were planted in pots with a diameter of 12-cm in peat substrate with a pH value of 6.5–6.8. Directly before planting, the slow-release fertiliser Osmocote Plus 5–6 (15N-10P-12K plus micronutrients) was added to the substrate, at a dose of 5.0 g·dm^{-3} of the substrate. The plants were placed under controlled conditions in the phytotron at 18 °C during the day and 16 °C at night, relative air humidity of 60%, quantum irradiance of 90 μmol·m^{-2}·s^{-1} with a light/dark cycle of 14/8 hours. The plants were cultivated in the phytotron for 16 weeks in two cycles.

At full blossom of the plants, when half of the main inflorescences were developed, the following parameters were defined: the height of plants, the number of shoots and leaves, the number of inflorescence shoots, the length of main inflorescence stems, the length of main inflorescence. The relative chlorophyll content in the leaves was measured using the Chlorophyll Meter SPAD 505 (Minolta, Japan). This chlorophyll meter measures the leaf transmittance at approximately 650 nm and 940 nm. Three fully developed leaves of each plant were used for the measurement. After the cultivation period was over, the corms were taken out of the pots and after being slightly dried, they were counted and weighed to calculate the coefficients of increase in weight and number.

Tests for the occurrence of freesia mosaic virus (FMV) were performed in three samples for each corm variant by means of the DAS-ELISA serological method [Clark and Adams 1977], with a set of γ-globulins and γ-globulin conjugates containing alkaline phosphatase [Kamińska 1991]. Corm samples were homogenised with the extract buffer at a 1:10 ratio. The results of the reaction were defined using a Multiscan II spectrophotometer (Labysystems) with a 405 nm filter. The result was considered to be positive if the absorbance value (A_{405}) was at least twice as high as the average value A_{405} for the negative control.

The experiment was conducted in a complete randomisation system as a one-factor experiment. Each experimental variant consisted of 40 corms, 10 in each replication. The obtained feature values as the average of two cycles of cultivation were statistically verified by means of a variance analysis method (ANOVA) and the differences between the means were assessed using Tukey's test at the significance level of $\alpha = 0.05$.

RESULTS AND DISCUSSION

The research showed that soaking 'Gompey' freesia corms in a chitosan solution before planting had a stimulating effect on the plant height, the number of shoots and leaves and the relative chlorophyll content (Table 1). On the basis of statistical analysis, it was found that as a result of chitosan use with medium- and high-molecular-weights, plants were by 16.8% higher as compared with the control. The chitosan-treated plants had more shoots and leaves, by respectively 45.9% and 35.1%. It was not confirmed that the number of freesia shoots or leaves depend on the molecular weight of the biopolymer used. A similar positive influence of chitosan on plant vegetative features was observed in multiple genera from family *Orchidaceae*, such as *Cymbidium* [Nahar et al. 2012] or *Dendrobium* [Tantasawat et al. 2010]. The research by Lee et al. [2005] showed a positive chitosan influence on soy seedling growth and the stimulating action of chitosan was directly proportional to the molecular weight of the compound used in the experiment. The authors quoted above suggest that soaking seeds in a chitosan solution with a high molecular weight (>1,000 kDa) has the most beneficial effect on the growth of soy seedlings. On the other hand, in the research by Luan et al. [2005], the stimulating effect of low-molecular-weight chitosan (16 kDa) was shown. It was added *in vitro* to the medium culture, which increased the height of shoots and fresh biomass *Chrysanthemum morifolium*, *Limonium latifolium*, *Eustoma grandiflora* and *Fragaria ananasa*. The observed discrepancies could result from various application methods and chitosan concentrations and conditions for plant cultivation.

As a result of chitosan treatment with medium and high molecular weight, the relative chlorophyll content in the leaves increased significantly by 13.4%, as compared with the control plants (Table 1). An increase in the chlorophyll content in plants, as a result of chitosan used, has been confirmed by numerous authors. Dzung et al. [2011] report that as a result of spraying of coffee seedlings with chitosan solutions of the molecular weight of 2 kDa, the content of chlorophylls and carotenoids in leaves increased by 15.36% in plants grown in the field and by 46.38–73.5% in plants grown in the greenhouse in comparison to the control. The increase of the chlorophyll content as a result of application of chitosan may be caused by plants' enhanced uptake of nutrients, which occurred in the studies by Nguyen Van et al. (2013) on coffee seedlings. The authors demonstrated that after spraying the seedlings with chitosan of the molecular weight of 600 kDa three times, an increase of the content of nitrate, phosphorus and potassium in leaves by respectively 9.8–27.4%, 17.3–30.4% and 30–45% was observed. The experiment also proved that seedlings sprayed with a 10–50 ppm chitosan solution were characterized by increased intensity of net photosynthesis.

Soaking freesia corms in chitosan solutions before planting accelerated the plant flowering (by 11 days on average) and increased the number of inflorescences per plant (1.75 on average) (Table 2). It was not found that the number of days until flowering and the number of inflorescences per plant depend on the chitosan molecu-

Table 1. Effect of molecular weight of chitosan on the growth of freesia 'Gompey'

Molecular weight of chitosan	Plant high (cm)	No. of shoots	No. of leaves	Relative chlorophyll content (SPAD)
Control	51.2 b	1.12 b	5.23 b	65.7 b
Low (2 kDa)	54.2 ab	1.55 a	7.50 a	71.7 ab
Medium (50 kDa)	58.5 a	1.58 a	7.25 a	74.3 a
High (970 kDa)	61.1 a	1.41 a	8.14 a	74.5 a

Comments: Values followed by different letters are significantly different according to Tukey's multiple range test at $P < 0.05$.

Table 2. Effect of molecular weight of chitosan on the flowering of freesia 'Gompey'

Molecular weight of chitosan	No. of days to flowering	No. of inflorescence shoot	Length of inflorescence shoot (cm)	Length of inflorescence (cm)
Control	75.7 a	3.75 b	15.1a	6.15a
Low (2 kDa)	64.0 b	5.75 a	15.2a	6.47a
Medium (50 kDa)	65.5 b	5.50 a	14.7a	6.57a
High (970 kDa)	64.7 b	5.25 a	16.3a	6.25a

Comments: Values followed by different letters are significantly different according to Tukey's multiple range test at $P < 0.05$.

lar weight. The results obtained are confirmed by the results of the research by Ohta et al. [1999], where it was shown that lisianthus (*Eustoma grandiflora*) cultivated in a substrate with addition of chitosan flowered 15 days earlier than the control plants. Earlier flowering resulting from chitosan application was also found in the following species: *Exacum affine*, *Lobelia erinus*, *Mimulus×hybridus*, *Sinningia speciosa* and *Torenia fournieri* [Ohta et al. 2004]. A stimulating effect of chitosan on the number of flowers was observed in plants such as lisianthus [Ohta et al. 1999], gerbera [Wanichpongpan et al. 2001] and gladioli [Ramos-Garcia et al. 2009]. Also Limpanavech et al. [2008], applying chitosan of various physicochemical properties in growing of *Dendrobium* 'Eiskul', found that plants treated with chitosan of the molecular weight of 45 kDa and deacetylation degree of >90% flowered earlier and produced more inflorescences. The research results showed that chitosan did not affect the length of the inflorescence shoot and the length of the 'Gompey' freesia inflorescence (Table 2). According to Win et al. [2005], spraying *Dendrobium* 'Missteen' plants with chitosan significantly increased the length of the inflorescence but did not affect the size of flowers.

The stimulating effect of chitosan on the increase in the weight and number of 'Gompey' freesia corms was confirmed statistically (Table 3). The use of high-molecular-weight chitosan made it possible to obtain the highest coefficient of corm weigh increase by 31.6% higher than in control plants. Regardless of the chitosan molecular weight, an increase in the number of corms in chitosan-treated plants was on average 40.4% higher, as compared to the plants which were not treated with chitosan. Also, Ramos-Garcia et al. [2009] found that the number of gladiolus corms increased as a result of using the chitosan-containing Biorend® preparation. A similar positive effect of chitosan on the yield of potato minituber was observed by Asghari-Zakaria et al. [2009] in *in vitro* cultures. Plants grown on a medium with the addition of 500 mg·dm^{-1} chitosan produced 3.33 minituber on average, while plants that were not treated with chitosan formed 2.44 minituber. Hasegawa et al. [2005] reported that corms with an increased diameter and height are obtained as a result of *Arisaema ternatipartitum* cultivation in a substrate with an addition of chitosan.

In this study, serological tests on 'Gompey' freesia corms rendered positive results for the FMV in all the experimental variants (Table 3). No influence of the chitosan molecular weight on the degree of freesia corm infection was found. Research on the bean mosaic virus [Kulikov et al. 2006] showed that the highest antiviral activity of chitosan was exhibited by the fractions with molecular weight 2.2 and 1.2 kDa but chitosan monomers did not exhibit any antiviral activity. Pospieszny and Atabekov [1989] reported that chitosan may exhibit antiviral activity in plants depending on the species and environmental conditions. As freesia's infections with viruses and phytoplasmas are still a serious problem [Vaira et al., 2009], it is necessary to conduct further, detailed studies involving different concentrations and methods of chitosan's applications, for example spraying or drenching plants.

CONCLUSIONS

The presented research results have shown chitosan usefulness as a biostimulator in the cultivation freesia 'Gompey' as a pot plant. As a result of soaking corms in a chitosan solution before planting the plants were higher, had more leaves and shoots, flowered earlier, had more flowers and corms. It was observed that among the chitosan types tested in the experiments, the applica-

Table 3. Effect of molecular weight of chitosan on corms yield of freesia 'Gompey'

Molecular weight of chitosan	Coefficient of corm weight increase	Coefficient of corm number increase	Value at 405 nm (detection of FMV)*
Control	2.28 ^{c}	4.87 ^{c}	0.16 ^{a} (positive)
Low (2 kDa)	2.94 ^{ab}	6.62 ^{ab}	0.15 ^{a} (positive)
Medium (50 kDa)	2.70 ^{b}	6.87 ^{ab}	0.17 ^{a} (positive)
High (970 kDa)	3.00 ^{a}	7.02 ^{a}	0.15 ^{a} (positive)

Comments: Values followed by different letters are significantly different according to Tukey's multiple range test at $P < 0.05$.

Explanations: *buffer (negative control) = 0.07; absorbance values above 0.14 were considered positive outcomes (samples containing FMV-infected).

tion of medium- and high-molecular-weight chitosan resulted in higher plants with leaves contained more chlorophyll. Soaking freesia corms with chitosan, especially high-molecular-weight, seems to be an effective way to enhance of coefficient of corm weigh increase.

Acknowledgements

The investigation is being conducted in the frame of own research Nr 2PO6R07227 financed by Polish Ministry of Science and Higher Education.

REFERENCES

1. Al-Hetar M.Y., Zainal Abidin M.A., Sariah M., Wong M.Y. 2011. Antifungal activity of chitosan against *Fusarium oxysporum* f. sp. *cubense*. J. Appl. Polym. Sci., 120: 2434-2439.
2. Aranaz I., Mengibar M., Harris R., Panos I., Miralles B., Acosta N., Galed G., Heras A. 2009. Functional characterization of chitin and chitosan. Curr. Chem. Biol., 3(2): 203-230.
3. Asghari-Zakaria R., Maleki-Zanjani B., Sedghi E. 2009. Effect of *in vitro* chitosan application on growth and minituber yield of *Solanum tuberosum* L. Plant Soil Environ., 55(6): 252-256.
4. Bartkowiak A. 2001. Binary polyelectrolyte microcapsules based on natural polysaccharides. Wyd. PS Szczecin.
5. Clark M.F., Adams A.N. 1977. Characteristics of the microplate method of enzyme linked immunosorbent assay for the detection of plant viruses. Gen. Virol., 34: 475-483.
6. Dias A.M.A., Cortez A.R., Barsan M.M., Santos J.B., Brett C.M.A., De Sousa H.C. 2013. Development of greener multi-responsive chitosan biomaterials doped with biocompatible ammonium ionic liquids. ACS Sustainable Chem. Eng., 1(11): 1480-1492.
7. Dzung N.A., Khanh V.T.P., Dzung T.T. 2011. Research on impact of chitosan oligomers on biophysical characteristics, growth, development and drought resistance of coffee. Carbohydr. Polym., 84: 751-755.
8. Ehrich L., Gruneberg H., Ulrichs C. 2009. Growth rhythms of South African *Iridaceae* forced as pot plants. Acta Hort., 813: 177-182.
9. FloraHolland 2012. Fact and Figures 2012. http://www.floraholland.com/media/1525240/KengetallenEN2012.pdf
10. Hadwiger L.A. 2013. Plant science review: Multiple effects of chitosan on plant systems: Solid science or hype. Plant Sci., 208: 42-49.
11. Hasegawa A., Kanechika R., Oguni S. 2005. Effect of low temperature and chitosan on dormancy breaking and growth of young corms of three *Arisaema* species. Acta Hortic., 673: 603-609.
12. Jabeen N., Ahmad R. 2013. The activity of antioxidant enzymes in response to salt stress in safflower (*Carthamus tinctorius* L.) and sunflower (*Helianthus annuus* L.) seedlings raised from seed treated with chitosan. J. Sci. Food Agr., 93(7): 1699-1705.
13. Kamińska M (1991). Preparation and testing of antiserum to freesia mosaic virus (FMV). Phytopath. Polonica, XII: 5-7.
14. Kulikov S.N., Chirkov S.N., Ilina A.V., Lopatin S.A., Varlamov V.P. 2006. Effect of the molecular weight of chitosan on its antiviral activity in plants. Appl. Biochem. Micro., 42: 200-203.
15. Lee Y.S., Kim Y.H., Kim S.B. 2005. Changes in the respiration, growth, and vitamin C content of soybean sprouts in response to chitosan of different molecular weights. HortScience, 40: 1333-1335.
16. Li M, Chen X, Liu J, Zhang W, Tang X (2011). Molecular weight-dependent antifungal activity and action mode of chitosan against *Fulvia fulva* (cooke) ciffrri. J Appl. Polym. Sci., 119: 3127-3135.
17. Limpanavech P., Chaiyasuta S., Vongpromek R., Pichyangkura R., Khunwasi C., Chadchawan S., Lotrakul P., Bunjongrat R., Chaidee A., Bangyeekhun T. 2008. Chitosan effects on floral production, gene expression, and anatomical changes in the *Dendrobium* orchid. Sci. Hortic., 116(1): 65-72.
18. Lizarraga-Pauli E.G., Torres-Pacheco I., Moreno-Martinez E., Miranda-Castro S.P. 2011. Chitosan application in maize (*Zea mays*) to counteract the effects of abiotic stress at seedling level. Afr. J. Biotechnol., 10(34): 6439-6446.
19. Luan L.Q., Ha V.T.T., Nagasawa N., Kume T., Yoshii F., Nakanishi T.M. 2005. Biological effect of irradiated chitosan on plants *in vitro*. Appl. Biochem., 41: 49-57.
20. Mondal M.M.A., Malek M.A., Puteh A.B., Ismail M.R., Ashrafuzzaman M., Naher L. 2012. Effect of foliar application of chitosan on growth and yield in okra. Aust. J. Crop, 6: 918-921.
21. Nahar S.J., Shimasaki K., Haque S.M. 2012. Effect of different light and two polysaccharides on the proliferation of protocorm-like bodies of *Cymbidium* cultured *in vitro*. Acta Hort., 956: 307-313.
22. Nguyen Van S., Dinh Minh H., Nguyen Anh D. 2013. Study on chitosan nanoparticles on biophysical characteristics and growth of Robusta coffee in green house. Biocatal. Agric. Biotechnol., 2(4): 289-294.
23. Ohta K., Morishita S., Suda K., Kobayashi N., Hosoki T. 2004. Effects of chitosan soil mixture treatment in the seedling stage on the growth and flowering of several ornamental plants. J. Jpn. Soc. Hortic. Sci., 73: 66-68.

24. Ohta K., Taniguchi A., Konishi N., Hosoki T. 1999. Chitosan treatment affects plant growth and flower quality in *Eustoma grandiflorum*. HortScience, 34: 233-234.

25. Pongprayoon W., Roytrakul S., Pichayangkura R., Chadchawan S. 2013. The role of hydrogen peroxide in chitosan-induced resistance to osmotic stress in rice (*Oryza sativa* L.). Plant Growth Regul., 70(2):159-173.

26. Pospieszny H., Atabekov J.G. 1989. Effect of chitosan on the hypersensitive reaction of bean to alfalfa mosaic virus. Plant Sci., 62: 29–31.

27. Ramos-Garcia M., Ortega-Centeno S., Hernandez--Lauzardo A.N., Alia-Tejacal I., Bosquez-Molina E., Bautista- Baños S., 2009. Response of gladiolus (*Gladiolus* spp) plants after exposure corms to chitosan and hot water treatments. Sci. Hortic., 121: 480-484.

28. Ruamrungsri S., Bundithya W., Potapohn N., Ohtake N., Sueyoshi K., Ohyama T. 2011. Effect of NPK levels on growth and bulb quality of some geophytes in substrate culture. Acta Hortic., 886: 213-218.

29. Tantasawat P., Wannajindaporn A., Chantawaree C., Wangpunga C., Poomsom K., Sorntip A. 2010. Chitosan stimulates growth of micropropagated *Dendrobium* plantlets. Acta Hortic., 878: 205-212.

30. Vaira AM, Hansen MA, Murphy C, Reinsel MD, Hammond J (2009). First report of freesia sneak virus in *Freesia* sp. in Virginia. Plant Dis., 93: 965-965.

31. Wanichpongpan P, Suriyachan K, Chandrkrachang S (2001). Effect of chitosan on the growth of gerbera flower plant (*Gerbera jamesonii*). In T. Uragami, K. Kurita, T. Fukamizo (Eds.), Chitin and Chitosan in Life Science, Yamaguchi: 198-201.

32. Win N.K.K., Jitareerat P., Kanlayanarat S. 2005. Preharvest chitosan spraying on leaf spot disease and growth of orchid (*Dendrobium* 'Missteen'). Proceedings of APEC. Symposium on assuring quality and safety of fresh produce, Bangkok: 457-461.

CHANGES IN THE MIDDLE SECTION OF THE NAREW RIVER (BETWEEN OSTROŁĘKA AND ŁOMŻA) IN VIEW OF NATURAL FACTORS AND ANTHROPOGENIC PRESSURES

Bożena Grabińska[1], Józef Koc[1], Ireneusz Cymes[1], Sławomir Szymczyk[1]

[1] Department of Land Improvement and Environmental Management, University of Warmia and Mazury in Olsztyn, Pl. Łódzki 2, 10-719 Olsztyn, Poland, e-mail: katemel@uwm.edu.pl

ABSTRACT

Contemporary and archive cartographic materials illustrating horizontal profile of the middle channel of the Narew River (between Ostrołęka and Łomża) were compared in the GeoMedia application. The results of the analysis revealed that natural and anthropogenic factors had not significantly contributed to changes in the channel profile over a period of two centuries. Narew is a meandering river, and its activity is associated with intensified erosion of right-bank bends and lateral movement along its flat-bed channel. The most noticeable changes include reduced area of oxbow lakes, mainly due to succession, and suburban settlement in floodplains that had been regarded as unfit for development in the 20th century.

Keywords: Middle Narew River, horizontal profile, development of the riverbed, cartographic analysis.

INTRODUCTION

Narew is one of the best preserved and most beautiful European rivers. Its catchment area is a manifestation of cultural landscape where natural processes and human activities are balanced or nearly balanced [Grabińska 2012]. The catchment area of the Narew River is characterized by a mosaic landscape, extensive farming practices, high forest cover and low levels of habitat conversion, in particular in the segment of the valley that is bound by the outwash plain of the Kurpie Plain in the north [Koncepcja szlaków… 2006]. The discussed landscape incorporates surviving fragments of the Kurpiowska Forest, extensive meadows, peatlands and agricultural biocenoses with traditional farming systems.

The Middle Narew Valley has been included in the Natura 2000 network, on account of its unique ecological value. It has also been designated as a Special Protection Area under the Birds Directive (area code: PLB 140014). The areas stretching along rivers Omulew and Pisa, right tributaries of the Middle Narew, play a similar role. The Narew River with a strip of meadows and wetlands forms an international ecological corridor as a part of the ECONET Ecological Network. The corridor borders core areas: the above mentioned tributaries of the Narew River in the north and the Lower Bug River Valley in the south.

The natural resources in the catchment area of the Narew River stimulate local development, create opportunities for economic growth and constitute a source of economic and ecological indicators for continued development [Grabińska 2012].

The aim of this study was to evaluate changes in the horizontal profile of the Middle Narew channel (along a section of 56.8 km bounded by hydrometric profiles in Ostrołęka and Łomża) induced by natural and anthropogenic factors.

MATERIALS AND METHODS

The usefulness of archive source materials and contemporary maps for evaluating changes in the natural environment, in particular in aquatic habitats, has been emphasized by numerous authors [Myga-Piętek 2003, Kaniecki 2007].

A comparative analysis was performed based on archive and contemporary maps of the middle section of the Narew River between hydrometric profiles in Ostrołęka and Łomża. The hydrometric profiles are described in Table 1.

Source materials were acquired in accordance with standard procedures and were processed based on standard criteria for analyses of the type [Kistowski and Iwańska 1997, Bornette et al. 1998]. The following materials were used in the analysis:

- topographic maps in the 1:100 000 scale, developed by the former Military Institute of Geography (WIG) in the 1930s,
- topographic maps in the 1:100 000 scale, developed in the 1980's,
- topographic maps in the 1:100 000 scale, developed in the 1990's,
- digital map of Poland in the 1:200 000 scale, developed by the Institute of Geodesy and Cartography – used for control and comparison,
- cartographic materials (including archive resources) and technical data concerning hydraulic structures on the Narew River – supplied by the Regional Water Management Authority in Warsaw.

Source maps were processed in the GeoMedia Professional environment. Maps of the Narew river channel were vectorized based on georeferenced topographical raster maps in the 1:100 000 scale (WIG maps developed in the 1930s and maps dating back to the 1980s). If sections of the river overlapped in two different maps, raster tracing (tracing the existing sections) and snapping (creating new sections in the database) functions were used during the vectorization process. This approach was used to preserve the topology of the introduced objects and to ensure the cohesion of data relating to river sections.

Due to the large volume of the analyzed cartographic materials, only the section of the river characterized by the most distinctive changes in profile was presented and discussed in this study (Figures 1 and 2).

Technical data concerning hydraulic structures on the Narew River and the results of field observations were also used. The results of the analysis were compared with the published findings of other research institutions.

RESULTS AND DISCUSSION

Natural factors

The development the middle channel of the Narew River has been influenced by numerous natural factors, including the geological structure of the stream bed, successive glacial periods, meltwater and the effect of changes in air temperature on river flow and plant cover.

The geological structure of the stream bed has a major impact on the river network in the Polish Lowlands [Cesnulevicius and Morkunaite 2003, Falkowski 2010]. The development of the Middle Narew valley was influenced by the presence of an extensive basin in the sub-quaternary surface that had been deformed during successive glacial periods and under the influence of meltwater [Kondracki 2000, Grabińska and Kubeł 2011].

The left bank of the analyzed section of the Middle Narew valley is characterized by postglacial landform that was shaped during the Middle Pleistocene and transformed by periglacial and postglacial processes. Today, it is a nearly flat plain with slopes not exceeding 2%. The monotonous landform of the Łomża Interfluve is broken by the morainic hills of Czerwony Bór, flat and extensive dead-ice depressions, fluvial and denudation valleys. The left bank of the Middle Narew features distinctive edge zones with slopes greater than 20%. Outside developed areas, the landscape on the left bank of the river is formed by small farms and forest enclaves, mostly pine [Kondracki 2000].

Table 1. Hydrometric profiles limiting the middle section of the Narew River

Parameter	Hydrometric profile in Ostrołęka	Hydrometric profile in Łomża
Location	below the city	above the city
Kilometers along the river's course [km]	146.8	203.6
Catchment area [km²]	21 862.0	15 296.5
Share of total catchment area (excluding the Bug River) [%]	77.3	54.1

Source: Own elaboration based on *Atlas hydrologiczny Polski* [1987].

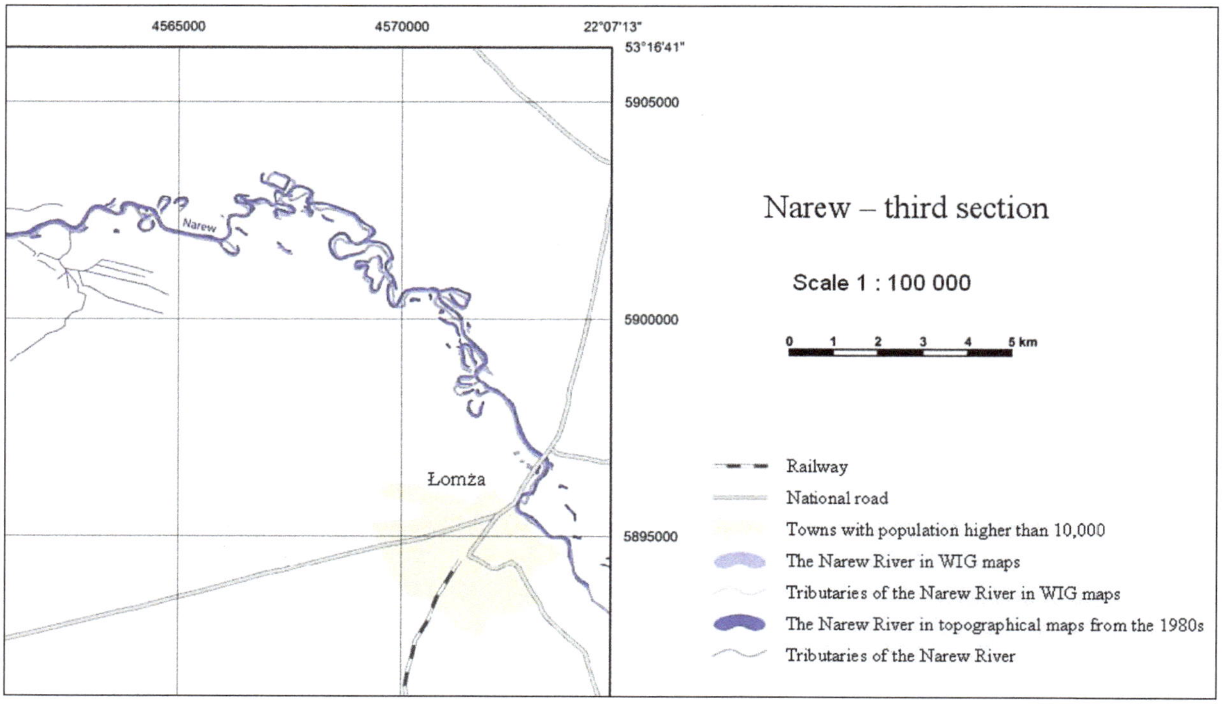

Figure 1. Horizontal profile of the Narew river channel below Łomża in topographic maps from the 1930s and 1980s
Source: own elaboration based on cartographic materials supplied by the Institute of Geodesy and Cartography in Warsaw

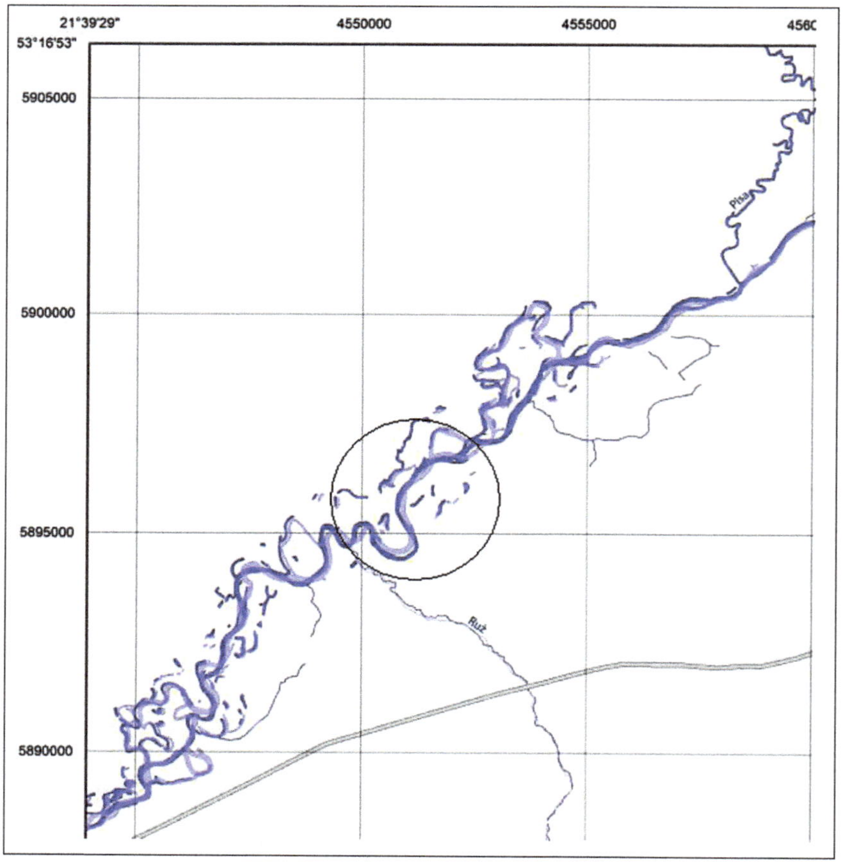

Figure 2. Horizontal profile of the Narew River channel above Łomża (right bank – municipality of Lelis, left bank – municipality of Rzekuń) in topographic maps from the 1930s and 1980s
Source: own elaboration based on cartographic materials supplied by the Institute of Geodesy and Cartography in Warsaw. Refer to Figure 1 for a list of symbols

On the right bank, the Narew River valley is enclosed by outwash plains and the Kolno morainic plateau. Narew's tributaries have flat and often water-logged stream beds. The areas between tributaries feature numerous aeolian formations, mostly dune ridges overgrown by pine forests. The local landscape is complemented by vast grasslands, patches of extensively farmed land on low-quality soils as well as fallow land [Kondracki 2000, Grabińska 2010].

The landscape of the river corridor is replete with clusters of osiers. The increase in the area occupied by this broad-spreading shrub indicates that evaluations of the local landscape should account for the influence of natural and anthropogenic processes on the growth of the above species [Kowalska 2009]. The floodplains of the Narew River are characterized by a predominance of osier and alder streamside forests, osier shrubs, rush plants and birch coppices on drier land. Floodbanks are largely covered by meadows and pastures, which are also found on the edge of forests and in wetland basins. Forests and meadows are overgrown by shrubs and herbaceous plants (burdock, mugwort, nettle and currant communities) [Ochrona Środowiska 2011].

The results of our field survey were largely consistent with the findings of Adynkiewicz-Piragas and Lejcuś [2009] who analyzed the continuity Narew's corridor and demonstrated that the river section between Ostrołęka and Łomża is characterized by undisturbed flow.

Anthropological factors

The landform of the analyzed fragment of the Middle Narew valley has undergone considerable transformations resulting from human activity, including hydraulic structures, land improvement systems, road and bridge embankments, moats surrounding post-war fortifications, sand and gravel pits, and settlements [Stan czystości wód... 1998].

In the past two centuries, extensive drainage and regulation works had been carried out in the catchment area of the Middle Narew to convert wetlands (riverside swamps) into farmland. Those efforts involved the construction of drainage canals as well as seasonal irrigation channels to improve meadow productivity. Substantial parts of the catchment area were completely deforested. The most recent transformations in the landscape of Narew's corridor involve valley drying, which is caused by an increase in mean winter air temperatures [Grabińska 2010] and changes in river flow resulting from the construction of the Siemianówka dam reservoir [Gospodarowanie wodą... 2002].

In the analyzed section of the Narew River and in the section between Różan and Ostrołęka [Grabińska and Szymczyk 2012], hydraulic structures have been developed to stabilize and direct the current and prevent erosion. Flood walls with total length of 3.5 km and spur dikes were built in the discussed section of the Narew River (Table 2).

Twenty-one side weirs have been built along the Narew River, of which the shortest has the length of 100 m, and the longest has the length of 750 m and is additionally reinforced with a diversion embankment of similar length. Side weirs have the combined length of 9.8 km. Passenger and cargo shipping operations do not take place on the Narew River. The middle and upper sections of the river cannot be used as transport corridors due to shallow waters and an absence of hydraulic structures [Mikulski 2000].

The picturesque landscape of Narew's corridor, in particular its edge zone, contributes to the development of human settlements and recreational facilities, which leads to the gradual elimination of agricultural land. The areas extending above the escarpment of the Narew valley in municipalities of Rzekuń, Miastkowo and Nowogród enjoy the greatest popularity on account of their

Table 2. Hydraulic structures in the section of the Narew River between Ostrołęka and Łomża

Type of structure	Number of structures	Length [m]		
		minimum	maximum	total
Spur dikes	93	10	110	5059
Flood walls	21	100	750	3520
Earth dams	2	35	90	125
Side weirs	21	70	710	9853
Diversion embankments	1	–	–	690

Source: Own elaboration based on 2011 data supplied by the Regional Water Management Authority in Warsaw.

attractive landform, climate and scenery. Residential and commercial construction projects are initiated on territories that had been regarded as unfit for development in the 20th century, including in the Lower Pisa Valley and the Narew Valley where Pisa feeds into Narew, their floodplains and extensive water-logged meadows on the right bank of the Narew River. Residential estates require the construction of embankments containing rubble and soil excavated from building sites. Catch basins are filled, shrubs and trees are cleared, and the area of oxbow lakes is reduced. Those measures lead to considerable changes in the structure of the Narew Valley.

Water quality analyses revealed that settlements have a minor impact on environmental conditions in the Narew Valley [Grabińska 2010], and much higher pollution levels were reported in the Warta River [Ilnicki et al. 2001]. The annual pollutant loading from soluble substances in Narew was comparable to that noted in catchment areas covered by forests and farmland [Glińska-Lewczuk and Koc 2003]. The catchment area of the Narew River is free of heavy-metal contamination [Grabińska 2010]. Water quality in the catchment differed across years and seasons subject to weather conditions, flow volume and land-use structure. Agricultural production on drained land with mostly light soils and intensive animal farming in the valleys of Narew's tributaries led to a deterioration in water quality [Mioduszewski 2006, Grabińska 2012].

According to 16th century maps presented by Niedziałkowska [2002], Middle Narew was a braided river in the region of Ostrołęka, and its channel was intersected by numerous mid-channel bars (islets). Similarly to other rivers in the Polish Lowlands, the transformation of the Middle Narew from a branched to a meandering river was prompted not only by natural factors (changes in flow rate), but also by anthropogenic influences because rivers played an essential role in economic development throughout history [Czaja 1997, Kalicki 2006]. Despite the above, a comparison of digital images illustrating changes in the course of the Narew River resulting from the development of hydraulic structures, infrastructure and human settlements did not reveal significant differences in the horizontal profile of its middle section. Irregular loops were formed, meander necks were cut off and the area of oxbow lakes was considerably reduced, mainly due to ecological succession. Distinctive changes in the stream channel were observed above Ostrołęka (Figure 2) where the river intensively meanders and contributes to erosion of right-bank bends. The course of the river shifted laterally across the channel bed in the area of a well-developed floodplain. Erosion controls have been implemented on the edge of the left bank of the Narew River.

CONCLUSIONS

1. Contemporary and archive cartographic materials were compared to evaluate changes in the horizontal profile of the Middle Narew channel along the section bounded by hydrometric profiles in Ostrołęka and Łomża. The analysis revealed that the transformation of the Middle Narew from a branched to a meandering river resulted mainly from human activities, including settlement, river regulation, hydraulic structures and agriculture.
2. The natural factors that have exerted the greatest influence on the development the middle channel of the Narew River include the geological structure of the stream bed, successive glacial periods, meltwater and valley drying due to changes in mean winter and spring air temperatures.
3. At present, Narew freely meanders in the river valley, and its flow is not restricted. River activity is associated with intensified erosion of right-bank bends and lateral movement along its flat-bed channel. The most noticeable changes include reduced area of oxbow lakes, mainly due to plant succession, suburban settlement and commercial development.
4. Patches of lush vegetation along river banks (in particular the right bank) and hydraulic structures, which have been developed over a period of two centuries, stabilize the Narew channel. Narew is not used as a water route. Agricultural practices in its catchment area are characterized by low productivity and do not deteriorate water quality.
5. Geomorphological changes (edge erosion) are increasingly frequently observed in the Middle Narew. The analyzed section of the river is renowned for its scenic beauty, which contributes to the development of settlements and commercial facilities in areas that had been regarded as unfit for development in the 20th century.

REFERENCES

1. Adynkiewicz-Piragas M., Lejcuś I. 2009. Ocena ciągłości ekologicznej koryta rzecznego na przykładzie Nysy Łużyckiej na odcinku Porajów - Sobolice. Nauka Przyr. Technol., t. 3, z. 3: 78-87.
2. Bornette G., Amoros C., Piegay H., Tacht J., Hein T. 1998. Ecological complexity of wetlands within river landscape. Biological conservation, 83: 35-45.
3. Cesnulevicius A., Morkunaite R. 2003. Correlation of genetic and morphometric types of Sventoji River catchment relief (north-east Lithuania). Geol. Quart. Vol. 47(2): 149-154.
4. Czaja S. 1997. Antropogeniczne przeobrażenia powierzchniowej sieci hydrograficznej w zlewni Rawy w latach 1801-1994. Kształt. Środ. Geogr. i Ochr. Przyr., 24: 12-18.
5. Falkowski T. 2010. Basement of the alluvia influence on the channel pattern in example of selected reach of the Pilica River. Annals of Warsaw University of Life Sciences – SGGW. Land Reclamation, 42(1): 105-114.
6. Glińska-Lewczuk K., Koc J. 2003. Przyrodnicze uwarunkowania odpływu substancji ze zlewni potoku leśnego (Pojezierze Mazurskie). [W:] Kształtowanie i ochrona środowiska leśnego. Wyd. AR w Poznaniu: 100-112.
7. Gospodarowanie wodą w łęgowej dolinie Górnej Narwi 2002. (red. W. Mioduszewski). Wydawnictwo IMUZ, Falenty, ss. 77.
8. Grabińska B., 2010. Zmiany jakości wód rzecznych na tle zróżnicowanych warunków przyrodniczych oraz sposobu użytkowania zlewni. [W:] Ochrona zasobów i jakości wody w krajobrazie wiejskim (red. J. Koc). Olsztyn: 223-252.
9. Grabińska B., 2012. Potencjał przyrodniczy podstawą rozwoju lokalnego na przykładzie zagospodarowania zlewni Narwi. [W:] Problemy gospodarki wodnej na obszarach wiejskich (red. S. Szymczyk). Olsztyn: 147-161.
10. Grabińska B., Kubeł S. 2011. Geneza doliny Narwi i terenów bezpośrednio przyległych w badaniach geologiczno-geomorfologicznych Polski NE. Zesz. Nauk., Ostrołęckie Towarzystwo Naukowe im. A. Chętnika, Ostrołęka, 25: 53-61.
11. Grabińska B., Szymczyk S. 2012. Przyrodnicze i antropogeniczne uwarunkowania rozwoju koryta Narwi (wielkie zakole poniżej Różana). Inż. Ekolog., 31: 27-37.
12. Ilnicki P., Kaleta-Więckowska M., Marciniak M., Mikołajewska E., 2001. Zanieczyszczenia rzeki Warty przez źródła punktowe w latach 1993-1998. Zesz. Probl. Post. Nauk Roln., 476: 125-131.
13. Kalicki T., 2006. Zapis zmian klimaty oraz działalności człowieka i ich rola w holoceńskiej ewolucji dolin środkowoeuropejskich. Prace Geograficzne IGiPZ PAN, z. 204, ss. 348.
14. Kaniecki A., 2007. Atrakcyjność starych przekazów kartograficznych dla współczesnych badań środowiskowych. Mat. V Konf., 'Wiosna w geodezji i kartografii", 19-21 kwietnia, Poznań-Jeziory: 15-18.
15. Kistowski M., Iwańska M., 1997. Systemy informacji geograficznej. Podstawy techniczne i metodyczne. Przegląd pakietów oprogramowania i zastosowań w badaniach środowiska przyrodniczego. Bogucki Wyd. Nauk., Poznań, ss. 190.
16. Koncepcja szlaków turystyczno-kulturowych Pasmo Narwi. 2006. Analizy i studia, z. 4, Warszawa: ss. 66.
17. Kondracki J., 2000. Geografia regionalna Polski. Wyd. Nauk. PWN, Warszawa, ss. 440.
18. Kowalska A., 2009. Zmiany sposobu użytkowania terenów rolniczych a zanikanie przyrodniczo cennych zbiorowisk roślinnych na przykładzie doliny środkowej Wisły. Polskie Krajobrazy Wiejskie Dawne i Współczesne, Prace Komisji Krajobrazu Kulturowego Nr 12, PTG, Sosnowiec: 166-177.
19. Mikulski Z., 2000. Polskie drogi wodne wobec wymogów europejskich. Gospodarka Wodna, 6: 213-217.
20. Mioduszewski W. 2006. Woda na obszarach wiejskich. Woda Środ. Obsz. Wiej., 6, 1(16): 277-295.
21. Myga-Piętek U. 2003. Przykłady zastosowań map dawnych w analizach geośrodowiskowych z użyciem narzędzi GIS. [In:] Quick reference Guide (eds. Gajos M., Michalski A., Myga-Piętek U., Stylińska M.). International Conference and Exhibition GIS SILESIA 2003, 22-26 Sept., Sosnowiec: 107-108.
22. Niedziałkowska Z. 2002. Ostrołęka. Dzieje miasta. OTN im. A. Chętnika Ostrołęka: ss. 487.
23. Ochrona Środowiska. GUS, Warszawa 2011, ss. 514.
24. Stan czystości wód powierzchniowych obszaru Zielonych Płuc Polski. 1998. PIOŚ, WIOŚ, Biblioteka Monitoringu Środowiska, Warszawa: 15-185.

EFFECT OF MUSHROOM SUBSTRATE ON THE FEED QUALITY FROM THE PERMANENT MEADOW

Beata Wiśniewska-Kadżajan[1]

[1] Department of Grassland and Landscape Architecture, Siedlce University of Natural Sciences and Humanities, B. Prusa 14. 08-110 Siedlce, Poland, e-mail: laki@uph.edu.pl

ABSTRACT

Fodder nutritive value of permanent grassland is dependent on the concentration of individual nutrients, the degree of digestibility and energy concentration. In a grassland study on the base crude fiber content and relevant conversion formulas, digestibility of dry matter and net energy concentration of the feed can be calculated, without the need for specialized equipment. The aim of the study was to determine the effect of mushroom substrate and farmyard manure supplemented NPK, on the digestibility and energy concentration in the sward of permanent meadows in three-year field experiment. In the study, overall digestibility and net energy concentration of feed derived from permanent grassland remained at a high level. Overall digestibility of yield dry matter of grass varies significantly, depending on the collected cuts and year of study, while the net energy concentration did not show differences under the influence of experimental factors studied. Mushroom substrate used in the experiment (alone and completed mineral NPK) caused no significant effect on the digestibility and net energy concentration of the tested feed with regard to both the farmyard manure used as standard and with NPK.

Keywords: mushroom substrate, crude fiber, net energy, feed digestibility, permanent meadow.

INTRODUCTION

In view of the natural fertilization deficient one should be looking for new ways to increase organic matter in cultivated soils . One of them is the use of waste organic materials as substrate after mushrooms cultivation, whose annual capacity in Poland amounted 1 500 000 tones [Kalembasa, Wiśniewska 2004, 2008 a and b].

The spent mushroom substrate is a potentially a good material for fertilizer and a particularly valuable source of organic matter and available forms of nutrients. Also has a generally neutral reaction, the preferred C: N ratio and low heavy metal content [Kalembasa, Wiśniewska 2001, Jankowski et al. 2005]. Mushroom substrate is often characterized by a large diversity of chemical composition, which is mainly related to the composition of the starting materials, production technology and the size of mushrooms yield. Therefore, for fertilization it can be require a continuous analysis of the chemical composition.

The study shows that mushroom substrate can be used to fertilize at arable land and permanent grasslands in horticulture, or in orchards, and to the establishment and maintenance of green areas [Rak et al. 2001, Kalembasa, Wiśniewska 2004, 2006 and 2008, Jankowski et al. 2004, 2005, Ciepiela et al. 2007, Maszkiewicz 2010, Jankowski et al. 2012].

Meadows and pastures fertilization of determinates a key factor in the value food - feed from pasture. According to many authors the type of fertilizer as well as its level has a huge impact on the botanical composition of the sward, the level of nutrients utilization implemented in these fertilizers, which is reflected in the food value of received feed [Wesołowski, Jankowska-Huflejt 1999, Niczyporuk, Jankowska-Huflejt 2000].

The productivity measure of fodder from permanent grasslands is the size of obtained yields of dry matter, protein and oat units, while about the value of food determines the concentration of each nutrient and digestibility degree [Sosnowski 2012b].

For the purposes of grassland experiments, or when comparing values of different feed from grasslands is proposed to adopt a simple, easy-to- calculate patterns, based on the current analysis of the crude fiber content [Pawlak 1988]. Based on these models, and on the content of total fiber and crude protein in feed can easily determine the value of crops.

The aim of this paper was to determine the effect of the mushroom substrate and also with manure and also with NPK supplemented on digestibility and energy concentration in the sward of permanent meadow in three-year field experiment.

MATERIALS AND METHODS

The experiment was established in the spring of 1999 with randomized block design in four replications on permanent meadow. Plots with an area of $9m^2$ (1.5 m × 6 m) were separated by 1 meter wide paths. Paths and edge of experiment were mowed several times in order to maintain a low sward. The experiment was located on the glial ground soil.

The tested soil in humus level was slightly alkaline solution both in KCl and in H_2O, and showed a high content of nitrogen, manganese and iron, the average content of magnesium and very low content of phosphorus and potassium.

The object of the study experiment was to organic fertilization sward applied fertilization. In the study the following fertilizer objects (in the tables marked with abbreviations in parentheses) were highlighted:

- control object, without fertilization (O),
- manuring (Ob),
- fertilization with mushroom substrate (Pp),
- manuring with NPK fertilization (Ob + NPK),
- mushroom substrate fertilization with NPK (Pp + NPK) .

Fertilization with manure and mushroom substrate were applied once early in the spring of 1999 in the amount of 10 t · ha^{-1} of fresh weight. Mineral fertilization was used in each study year, nitrogen at a dose of 180 kg · ha^{-1} (ammonium nitrate), phosphorus – 48 kg · ha^{-1} (TSP), potassium – 125 kg · ha^{-1} (potassium salt). The annual amount of nitrogen and potassium were divided into three equal parts for each regrowth, phosphorus was applied once in the spring.

In each growing season three regrowth were collected. Immediately after cutting, sward from each plot were weighed to the determined yield and were collected 0.5 kg sample of green matter to determine the drying coefficient of and to chemical analysis.

On the basis of a crude fiber content estimated by the Henneberg and Stohman method modified by Lepper, stated in an earlier publication of Ciepiela et al. [2007] and proposed by Pawlak [1988] the conversion formulas dry matter digestibility and concentration of net energy were calculated:

$$K = 1.50 - 0.02\,x$$

where: K – net energy concentration in 1 kg D.M.
x – crude fiber content in % D.M.
%. D.M. $S = 103 - 1.2\,x$ D.M.
S – dry matter digestibility in %.

The results were statistically evaluated using the analysis of variance. Calculations were performed using STATISTICA, and for means comparisons were used Tukey's test at a significance level $\alpha \leq 0.05$.

According to the meteorological station ZSD Zawady in the studied growing seasons (IV - IX) average air temperatures were 16.5 ^{0}C (average of several years 12.5 ^{0}C) and precipitation totals were 16.5 mm and compared to multi-year were higher by 66.7 mm.

RESULTS AND DISCUSSION

Statistical analysis showed significant differences of total digestibility of obtained dry matter yield of grass depending on the regrowth yield and the study year, whereas the type of fertilization did not affect significantly the value of the study characteristic (Table 1).

Total digestibility (average from years, objects, and cuttings) of dry matter sward obtained from permanent meadow was at a relatively high level of 69.9%.

Analyzing the hay yield (average from the years and cuts) from each fertilizing objects by higher digestibility characterized the forage harvested from objects fertilized with mushroom

Table 1. Influence the kind of fertilization, cut and years of research on the digestibility of feed for permanent meadow (%)

Years	Cut	Fertilized object					Mean
		O	Ob.	Pp.	Ob. + NPK	Pp + NPK	
1999	1	71.0	68.2	70.9	70.3	69.4	69.9
	2	71.9	69.5	71.1	68.7	69.9	70.2
	3	70.3	67.9	70.2	69.1	68.1	69.1
	x	70.9	68.5	70.2	69.4	69.1	69.6
2000	1	70.9	73.2	69.5	67.9	70.5	70.4
	2	69.4	69.2	71.8	65.1	68.2	68.7
	3	73.8	73.3	73.0	73.5	75.7	73.8
	x	71.3	71.8	70.4	68.9	71.5	70.8
2001	1	68.8	67.0	71.2	68.4	69.4	68.9
	2	69.1	69.5	67.3	69.1	69.3	68.8
	3	68.8	69.9	69.7	70.3	70.4	69.8
	x	68.9	68.8	69.4	69.3	69.7	69.2
Mean	1	70.2	69.4	70.5	68.9	69.8	69.8
	2	70.1	69.4	69.1	67.6	69.1	69.1
	3	70.7	70.4	70.5	71.0	71.4	70.8
	x	70.4	69.7	70.0	69.2	70.1	69.9

$LSD_{0,05}$ A – fertilization A = n.s. B = 1.09 C = 1.09
B – cut A/B = n.s. B/A = n.s. A/C = n.s.
C – years C/A = n.s. B/C = 1.88 C/B = 1.88

substrate and NPK (70.1%) and also with mushroom substrate used alone (70.0%), than from the objects of manure and NPK (69.2%) or the manure (69.7%).

The highest digestibility of tested feed (average for three years) was recorded for the third cut in all fertilizers combinations. Higher digestibility of hay were recorded on the objects fertilized with manure and mushroom substrate supplemented with mineral fertilization NPK (respectively 71.0 and 71.4%) than in the objects fertilized with the manure (70.4%)or with the substrate after mushrooms cultivation (70.5%).

Average from fertilizer in all objects in the first study year, the highest digestibility of feed were obtained in the second cut (70.2%), in the second year for the third cut (73.8%) and in the third year for the third cut (69.8%).

The results obtained in the present study, were similar to those obtained by Pawlak [1988]. Jankowski, et. al. [2000] also noted a high digestibility in compound grass legume mixtures harvested at different terms.

The statistical analysis showed no significant differences for concentration of the net obtained in the energy yield of grass dry matter, depending on the cut study year, as well as the type of fertilization (Table 2).

The concentration of energy is dependent on the crude fiber content, what is decisive about the amount in the feed of digestible organic matter processed for livestock products. Nutritioners argue that the energy concentration should not be less than 0.7 of oats units per 1 kg of dry matter.

The concentration of net energy (average for years objects and cuts) in the dry matter of the sward obtained from permanent meadow was at a relatively high level (0.94 o.u./kg D.M.). The concentration of net energy (average from year and cuts) in the feed obtained from objects fertilized with the mushrooms substrate used alone and supplemented with mineral fertilization NPK was identical (0.95 o.u./kg DM) but energy concentration obtained from the object fertilized with the manure was slightly higher (0.94 u.o./kg D.M.) than in the feed from the object where the manure with NPK (0.93 u.o./kg D.M.) was applied.

Like for the digestibility, the highest energy concentration of the tested feed (average from three years) was recorded in the third cut in all fertilizing combinations except the object fertilized with manure. Greater energy concentration in the hay obtained in the cut has reported on objects fertilized with manure and with mushroom substrate supplemented with mineral fertilizers NPK (respectively 0.96 and 0.97 o.u./kg D.M.)

Table 2. The influence of the kind of fertilization, cut and years of research on the net energy concentration in the feed from permanent meadow (o.u./kg DM)

Years	Cut	Fertilized object					Mean
		O	Ob.	Pp.	Ob. + NPK	Pp. + NPK	
1999	1	0.96	0.92	0.96	0.95	0.94	0.94
	2	0.98	0.94	0.92	0.92	0.94	0.94
	3	0.95	0.91	0.92	0.93	0.91	0.92
	x	0.96	0.92	0.95	0.94	0.93	0.94
2000	1	0.96	1.30	0.94	0.91	0.95	1.01
	2	0.94	0.93	0.93	0.86	0.91	0.91
	3	1.01	1.00	1.00	1.00	1.04	1.01
	x	0.93	0.98	0.95	0.93	0.97	0.95
2001	1	0.93	0.90	0.97	0.92	0.94	0.93
	2	0.93	0.94	0.90	0.93	0.93	0.93
	3	0.93	0.94	0.94	0.95	0.95	0.94
	x	0.93	0.93	0.94	0.93	0.94	0.93
Mean	1	0.95	0.93	0.95	0.93	0.94	0.94
	2	0.95	0.94	0.93	0.91	0.93	0.93
	3	0.96	0.86	0.95	0.96	0.97	0.94
	x	0.95	0.94	0.95	0.93	0.95	0.94
$LSD_{0,05}$ A – fertilization B – cut C – years	A = n.s. A/B = n.s. C/A = n.s.	B = n.s. B/A = n.s. B/C = n.s.	C = n.s. A/C = n.s. C/B = n.s.				

than on the objects fertilized with manure (0.86 o.u./kg D.M.) and with the mushrooms substrate after cultivation (0.95 o.u./kg D.M.).

Average from all fertilizer objects largest net energy concentration (1.01 o.u./kg D.M.) was recorded in the first and third cut in the second study year; the smallest digestibility of feed was recorded in the second cut (0.91 o.u./kg D.M.) in the same study year.

Digestibility and energy concentration are very important indications of the biological value of forage. These ratios depend largely on crude fiber content, which largely determines the amount in feed and the digestible organic matter mass processed into animal products [Borowiecki in 2000, Sosnowski 2012 b].

Our results clearly show that the digestibility and energy level of tested feed derived from permanent meadow was very high. They are reflected in similar studies conducted by Jankowski et al. [2000] and Ziliotto et al. [2003].

CONCLUSIONS

1. Total digestibility of feed derived from permanent meadow was at a high level and showed significant differences depending on the crop cut and the study year.
2. The concentration of net energy in the yield of dry matter of feed received was very high and statistical analysis showed no significant influence of study factors.
3. Gowing substrate mushrooms used in the experiment and supplemented with NPK did not cause a significant impact on the digestibility and net energy concentration of the tested feed for manure used separately and with NPK applied as standard.

REFERENCES

1. Borowiecki J. 2002. Wpływ nawożenia azotem na plon i wartość pokarmową Festulolium Braunii odm. Felopa. Pamiętnik Puławski, 131, 39-48.
2. Ciepiela G.A., Jankowska J., Jankowski K, Kolczarek R. 2007. Wpływ niekonwencjonalnych nawozów organicznych na jakość paszy z łąki trwałej. Fragmenta Agronomica, 1(93,)14 – 24.
3. Jankowski K., Ciepiela G.A., Kolczarek R. 2000. Jakość plonu mieszanek motylkowo-trawiastych w zależności od terminu zbioru pierwszego pokosu. Zesz. Nauk. AP w Siedlcach, 57, 75-82.

4. Jankowski K., Ciepiela G.A., Jodełka J., Kolczarek R. 2004. Możliwość wykorzystania kompostu popieczarkowego do nawożenia użytków zielonych. Annales UMCS. 59, 4, 1763–1770.
5. Jankowski K., Czeluściński W., Jankowska J., Sosnowski J., Wiśniewska-Kadżajan B. 2012. Wpływ zróżnicowanej dawki odpadu popieczarkowego na cechy jakościowe muraw trawnikowych. Folia Pomeranie Universitatis Technologie Stetinensis, 295(22), 13–20.
6. Jankowski K., Jodełka J., Ciepiela G.A. 2005. Wpływ nawożenia łąki trwałej kompostem popieczarkowym na zawartość wybranych mikroelementów w runi łąkowej. Łąkarstwo w Polsce, 8, 81–85.
7. Kalembasa D., Wiśniewska B. 2006. Zmiany składu chemicznego gleby i życicy wielokwiatowej pod wpływem stosowania podłoża popieczarkowego. Zesz. Probl. Post. Nauk Roln., 512: 265-276.
8. Kalembasa D., Wiśniewska B. 2004. Wykorzystanie podłoża popieczarkowego do rekultywacji gleb. Roczniki Gleboznawcze, 55, 2, 209-217.
9. Kalembasa D., Wiśniewska B. 2004. Wykorzystanie podłoża popieczarkowego do rekultywacji gleb. Roczniki Gleboznawcze, 55, 2, 209-217.
10. Kalembasa D., Wiśniewska B. 2008. Wpływ nawożenia podłożem popieczarkowym na plon i zawartość wybranych makroelementów w życicy wielokwiatowej. Zesz. Probl. Post. Nauk Rol., 526, 191-198.
11. Kalembasa S., Wiśniewska B. 2001. Skład chemiczny podłoży po produkcji pieczarek. Zesz. Probl. Post. Nauk Roln., 475, 295-300.
12. Maszkiewicz J. 2010. Zużyte podłoże popieczarkowe jako nawóz i paliwo. In: Pieczarki. Biuletyn Producenta Pieczarek. Wyd. Hortpress, 1: 59-60.
13. Niczyporuk A., Jankowska-Huflejt H. 2000. Kształtowanie się szaty roślinnej w okresie 50-letniego różnego użytkowania i nawożenia trwałego użytku zielonego. Zesz. Nauk. AR w Krakowie, 368, 235-246.
14. Pawlak T. 1988. Propozycja nowych metod oznaczania produkcyjności użytków zielonych. Wiad. Mel. i Łąk., 1, 29–31.
15. Rak J., Koc G., Jankowski K. 2001. Zastosowanie kompostu popieczarkowego w regeneracji runi łąkowej zniszczonej pożarem. Pam. Puł., 125: 401-408.
16. Sosnowski J. 2012a. Wartość produkcyjna, energetyczna i pokarmowa Festulolium braunii (K. Richt) A. Camus zasilanej mikrobiologicznie i mineralnie. Fragm. Agron., 29(2), 115-122.
17. Sosnowski J. 2012b. Wpływ użyźniacza glebowego stosowanego w uprawie Lolium perene L., Dactylis glomerata L., Festuca pratensis Huds. na wartość RFV uzyskanej paszy.
18. Wesołowski P., Jankowska-Huflejt H. 1999. Ocena wpływu nawożenia organicznego i mineralno-organicznego na produkcyjność łąk. Mat. Sem. IMUZ, 44, 52-62.
19. Ziliotto U.D., Ottavio P., Scotton M. 2003. Productivity and persistence of different cultivars of Italian ryegrass (Lolium multiflorum Lam.). Optimal Forage System for Animal Production and Environment. EGF, vol. 8, 429-432.

FLOOD RISK FACTORS IN SUBURBAN AREA IN THE CONTEXT OF CLIMATE CHANGE ADAPTATION POLICIES – CASE STUDY OF WROCLAW, POLAND

Szymon Szewrański[1], Jan Kazak[1], Marta Szkaradkiewicz[1], Józef Sasik[1]

[1] Department of Spatial Economy, Wroclaw University of Environmental and Life Sciences, Grunwaldzka 53, 50-357 Wrocław, Poland, e-mail: szymon.szewranski@up.wroc.pl; jan.kazak@up.wroc.pl; marta.szkaradkiewicz@up.wroc.pl

ABSTRACT

The uncontrolled sprawl of urban development exerts environmental impact in rural areas. The aim of this study is to identify areas vulnerable to climate change in the context of implementation of policies adapting to climate change at the local level. Such areas can be defined as those where the negative implication of flesh flood overlapping with soil sealing is observed. The study areas composed of municipalities which are influenced by the urban sprawl process of the city of Wroclaw, Poland. The analyses were performed using publicly available spatial data from aerial orthophotomaps from 2004–2012, the satellite images; archival and current land use maps. The database CORINE 1990, 2000, 2006; Urban Atlas and geodatabase of the European Environment Agency were also of an important usage for this study.

Keywords: climate change adaptation, water management, flood risk, soil sealing.

INTRODUCTION

The main climatic factors of floods, among other, include precipitation (volume of rainfall, its intensity and duration) or air temperature (especially important for the formation of winter floods) [Kundzewicz 2014]. Whereas the non-climatic factors include catchment topography, slopes, soil type, land use, density of the river network, etc. [De Roo et al. 2003, 2001]. Analysis of these conditions is necessary in order to correctly identify areas vulnerable to climate change, which in the case of Polish include mainly floods and droughts.

Historical records and ongoing hydro-meteorological observations prove that there have been approx. 90 recorded floods in Oder river basin (an average of 1 over 10 years). Most of them were small floods. Big floods occurred less frequently [Dubicki et al. 2005]. Floods were not significantly limited by hydro-technical treatments (river training, levees, flood-control reservoirs etc.). The size of economic and social losses was often the result of improper land management in the catchment area.

Flood events are expected to increase in the future [Falloon and Betts 2010]. Integrated management of water resources requires the development strategy of both mitigation and adaptation to upcoming changes in the environment. The choice of the mitigation and adaptation solution depends on the availability of economic resources and social acceptance [Kundzewicz and Matczak 2012].

MATERIALS AND METHODS

The research area consists of the city of Wroclaw and rural municipalities near Wroclaw: Kostomłoty, Miękinia, Oborniki Slaskie, Wisznia Mała, Czernica, Długołęka, Katy Wroclaw, Kobierzyce, Siechnice and Żórawina. Within these administrative bodies, 266 cadastral units were carefully investigated. The analyses were performed using publicly available spatial data from aerial orthophotomaps from 2004–2012, the satellite images; archival and current land use maps. The database CORINE 1990, 2000, 2006; Urban Atlas and geodatabase of the European Environment

Agency were also of an important usage for this study. Having the above sources analysed, changes in land use in the suburban area of Wrocław from 1990 to 2006 were identified. The analyses were performed using open GIS systems. The area covered by the study is subject to strong structural and spatial transformations. Spreading residential zone, increasing the number of investments in infrastructure and new industrial facilities have caused changes in the quality of the natural environment [Szewranski et al. 2013].

The holistic analysis of the precipitation, microclimate and hydrological conditions can be useful in choosing the location of investments in the context of adaptation to climate change. These conditions were compared with the current trends in the spatial development of real estates. The conclusions of these studies enable answering the question whether the current planning system requires more intense inclusion of an environmental impact assessment of the policy. Such analyses seem to be necessary in the context of the implementation of The National Strategy for Adaptation to Climate Change 2020 [SAP2020].

RESULTS

In order to identify the cadastral units where the risk of flooding is greatest, hydrological maps based on floods' greatest spatial distribution in the case study area have been investigated (Figure 1).

The greatest risk of flooding occurs in the Odra valley, particularly in the area of the city of Wroclaw, due to large amount of water (cadastral unit: Wroclaw, Paniowice, Kotowice, Blizanowice–Trestno) and numerous tributaries of the river. The smallest number of areas being at risk of flooding are located in the valley of the river Ślęza (cadastral units: Wroclaw, Żerniki Great Galowice), a bit more in the valley of the Bystrzyca (cadastral units: Wroclaw, Samotwór, Skalka, Romnów, Sadowice, Krobielowice, Gniechowice, Zachowice, Kamionna and Czereńczyce) and the largest in the catchment of the Olawa (cadastral units: Wroclaw, Mokry Dwór, Groblice, Zębice, Grodziszów, Siechnice–Prawocin, Święta Katarzyna) and the Widawa (cadastral units: Szymanów, Psary, Krzyżanowice, Wilczyce, Wroclaw).

This can be explained by downslopes of each sub-catchment and soil permeability. If the downslopes are larger and soils are more compact peak flow occurs faster. That was a cause which triggered the fact that such a wave usually occurs much earlier in the area of Wroclaw, than in the Oder River. There were two main reasons for the catastrophic floods in the Odra River in July 1997: an intense flow of water and impeded drainage of water.

The average annual precipitation in the study area is about 550–660 mm. A flood type that prevails in the Oder River is widespread rainfall floods caused by prolonged rains of large scale and considerable intensity. A flood type in the

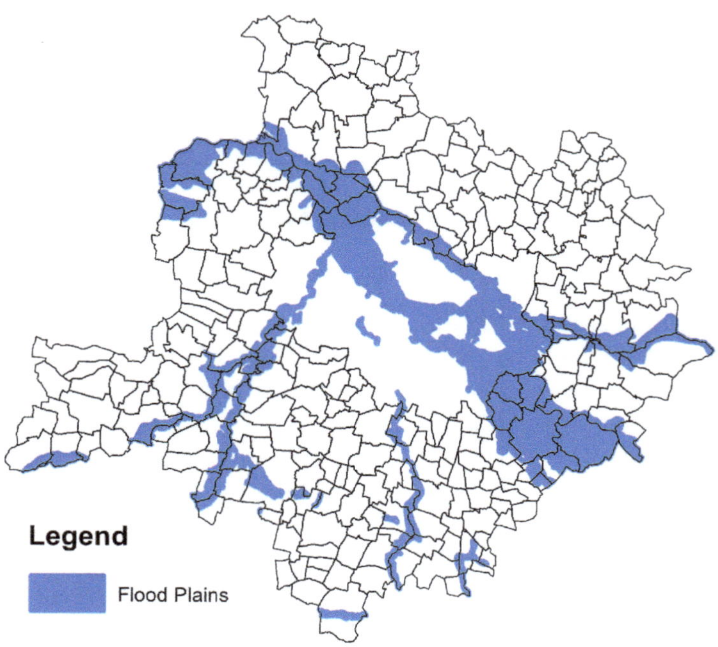

Figure 1. Flood plains in the case study area

tributaries of the Oder River (river: Bystrica, Olawa, Ślęza, Widawa, etc.) is of torrential rainfall caused among many factors by convective storms in river basins. When analysing the effect of precipitation on flood creation one should take into consideration not only annual, monthly or ten-day sums but also their distribution in time and space. In the study area there are rarely embolic type of flooding caused by local rising due to ice floes.

It is assumed that by 2030, the biggest threat to the economy and society will be extreme weather and climate (torrential rains, flooding, landslides, heat waves, droughts, hurricanes, etc.). These phenomena will occur with increasing frequency and intensity, and will cover more and more areas of the country. In the case study area heavy precipitation will increase in the first half of the summer, with intensities exceeding 5 mm/min, with a probability of ≥10%. The change in precipitation patterns meaning a rapid increase in the number of days with high intensity precipitation (daily precipitation ≥50 mm for 2 days/decade (SAP2020, 2013) will occur on the majority of the land in Poland (including the study area).

Climatic conditions of Wroclaw exhibits characteristics typical of large urban and industrial agglomerations, where the impact of the economic activity is reflected by a change in the physical characteristics of the surface, air pollution and artificial heat emission produced in municipal and industrial processes. Whereas non-urbanized areas are characterized by significant deviations of particular climate components - in particular precipitation and temperature. The increased rainfalls in Wroclaw were identified for the city centre, the Old Town and Karłowice and less intensive rainfall in the eastern districts (Biskupin, Swojec), as well as western – the average annual total precipitation is about 30–50 mm larger [Schmuck 1967].

The study area can be characterized by the average annual temperature of approx. 8.5 °C, and the duration of the growing season is 220 days, predominate winds from the west. Fairly high temperatures during the winter half-year time result in rare spring embolic flooding causing the water daming on ice or snow jam. The greatest effect of air temperature can be observed during the formation of the summer floods. For the last thirty years there has been an increase in air temperature (with the exception that it is much stronger in winter and weaker in summer). Among the adverse harmful thermal effects to the environment, society and economy, severe heat waves must be mentioned (continuous days with daily maximum temperature ≥30 °C persisting for at least 3 days) and hot days (the temperature ≥30 °C), the most common in the region of south-western part of the Poland.

In comparison to suburban are, solar radiation in Wroclaw is weakened, the number of hours of sunshine reduced, night cooling is less intense. The result is an increase of air temperature, known as the urban heat island, which is easily recognised during clear nights in summer when there is more than 5 °C. The average annual relative humidity in the city centre is approx. 6% lower than outside the city during clear nights in the summer by up to 40%.

The geological structure and terrain affect the development of the river network, and indirectly the development of hydrographical networks and the formation of river floods. Areas which have a large river network density are characterized by high rainfall, low evaporation losses, varied terrain and impermeable surface.

From the hydrological point of view, the most important geological and soil factor that has got impact on the formation of the outflow is the permeability of the ground. This feature determines the seepage of rainwater and depends on the constant physical properties of the soil. The phenomenon which is very important is the growth of built-up areas, especially in urban areas. This is done at the expense of green spaces, reducing the area of land covered with natural cover. Wroclaw in hydrological terms resembles a mountain basin, where because of the intensive real-estate development, increased runoff occurs (Figure 2).

Cadastral units with the highest soil sealing (above 5%) are Wilkszyn, Jeszkowice, Sośnica-Różaniec, Ratowice, Żórawina, Łany, Pietrzykowice-Rybnica, Suchy Dwór, Nowa Wieś Kącka, Szymanów, Tyniec Mały, Biestrzyków-Radomierzyce, Mokronos Dolny, Wilczyce, Czernica, Iwiny, Nowa Wieś Wrocławska, Psary, Gajków, Święta Katarzyna, Kobierzyce, Ślęza, Kamieniec, Pustków Żurawski, Wysoka, Rolantowice, Oborniki Śląskie, Smolec, Kiełczów, Siechnice-Prawocin, Żerniki Wrocławskie, Mirków, Radwanice, Długołęka, Kąty Wrocławskie, Bielany Wrocławskie, Biskupice Podgórne. The vast majority of these areas are located in the Wroclaw county, south and east of the centre of Wroclaw.

In hydrological practice, a parameter characterizing the permeability of the basin is mostly

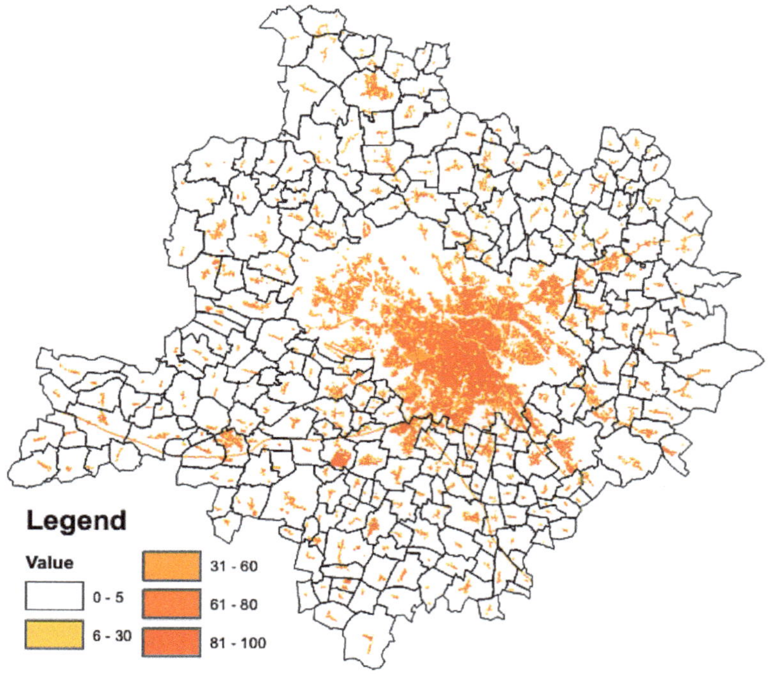

Figure 2. Soil sealing in % [European Environment Agency]

expressed by density of the river network. On permeable soils river network is poorly developed because there are good conditions for infiltration. There are not small, periodically drying up watercourses, but there are large streams leading water during low periods. Infiltration is low in catchments with impervious surface, which increases surface runoff. In research process the length of the watercourses and the density of the river network in the selected cadastral units were analysed. The results showed that the longest watercourses occurs in the area of Wroclaw and much less in the catchments inflowing to the Oder River. If that was calculated per unit area, it gives the standard density of a river network of 0.8 m/m^2. The highest density occurs in the Odra valley, a bit smaller in the catchment of Widawa and definitely the lowest in the basins of Bystrica and Ślęza.

The factor that had a significant influence on the runoff formation and its distribution over time is land cover. The biggest impact on runoff catch play forests, that by increasing the roughness of the surface, cause a reduction in the rate of runoff and facilitate infiltration. Moreover, forests retard the process of snow melting which increases the retention capacity of the catchment. The forest cover in the case study area is lower than the national average, which affects the formation of flood. There is, however, a relatively large proportion of orchards in the areas. Low degree of forest cover in the study area is caused by the high quality soils turned into arable land, which have smaller impact on the runoff catch. Runoff volume depends on the type of crop, development stages of plants and applied agrotechnical treatments.

When suburbian districts spring up rapidly and new infrastructure is being built the share of impervious surface is growing, which increases the risk of floods. In terms of surface area the largest new Urban Morpholocial Zone (UMZ) zones were created in 1990–2006 at the borders of districts: Kamieniec, Kotowice, Zacharzyce, Bielany Wrocławskie, Łany, Radwanice, Iwiny, Blizanowice-Trestno, Biestrzyków-Radomierzyce, Wilczyce, Krzeptów, Mirków, Żerniki Wrocławskie, Dobrzykowice, Wysoka, Krępice, Domaszczyn, Paniowice, Psary, Zabrodzie, Brzezina, Wróblowice, Szymanów, Biskupice, Lutynia, Wilkszyn, Mokronos Dolny, Krzyżanowice, Mokronos Górny, Galów, Ramiszów, Smolec. Newly built-up areas are also more vulnerable to the phenomenon of flash floods. It is assumed that areas vulnerable to climate change could be characterised by both high risk of flooding and excessive soil sealing (Figure 3).

Identified cadastral units are located where the impact of suburbanization is very high and at the same time flood risk. Beside Wrocław city there are: Szymanów, Psary, Wilczyce, Siechnice-Prawocin as well as Święta Katarzyna.

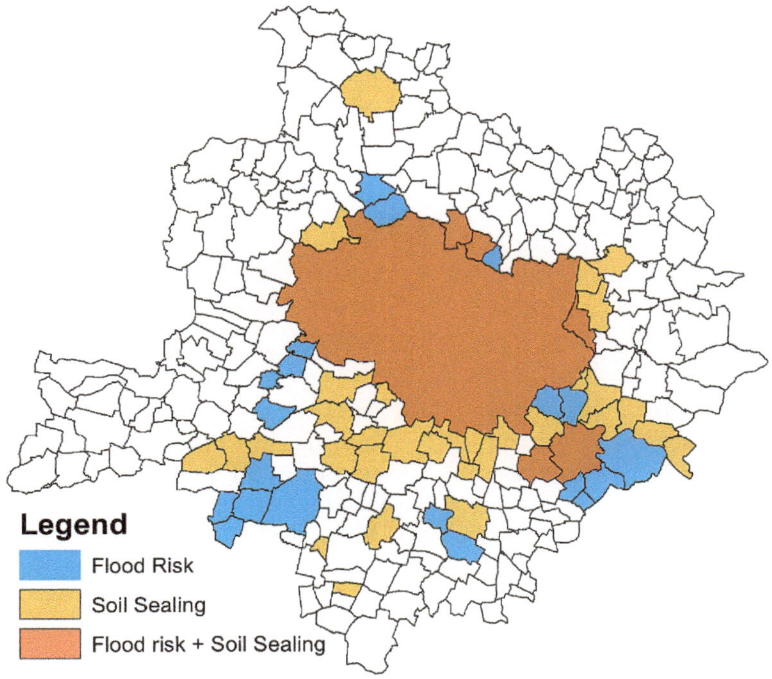

Figure 3. Cadastral units vulnerable to climate change because of high flood risk and excessive soil sealing

CONCLUSIONS

Changing climate conditions will have an impact not only on the size and availability of water resources, but also on the frequency and extent of flooding. The SAP2020 includes the analysis of the loss and cost of removing the damage caused by climate change. It proves that the phenomena that might cause the greatest damage in Poland are mainly related to floods. This problem applies to all sectors of the economy, especially to infrastructure located in the floodplain. Floods in small scale happen every year and cause losses of 0,08-0,1% of GDP. Once a dozen or so years, however, flooding appears causing above-average losses valued respectively of 0.5 and 0.9% of GDP.

It is, though, extremely important to adopt such an approach to the spatial management that assess the impact and minimize the risk and socio-economic losses of investments in early stage of their planning. Proper location of public or private investment should correspond to contemporary environmental challenges too. New development projects should be planned in a way to minimize negative impacts on the environment (e.g. the impact of soil sealing on water infiltration).

It is assumed that in order to be fully efficient spatial planning and water management should be integrated. Investments in water management infrastructure (including flood control activities) may not be only an addition to the spatial development plans. On the contrary, future solutions should be integrated, multicriterial and strategic. Development of flood protection programs should begin by setting priorities: protection of big cities, protection of agricultural land, etc. After determining the hierarchy of objectives, protective and adaptation measures could be addressed. The aim should be to integrate the activities of water management, spatial planning and environmental protection. It is necessary to limit the losses and compensate for lost benefits for those most affected by the implementation of the plan. In order to achieve that, it is vital to establish a strategy for plan implementation, indicate responsible actors and set indicators to monitor and evaluate the objectives' fulfilment.

Acknowledgments

The study was carried out in the framework of the research project NCN No. N N305 384838 "Indicator-based assessment of environmental changes caused by unsustainable sprawl of large cities".

REFERENCES

1. De Roo, A., Odijk, M., Schmuck, G., Koster, E., Lucieer, A., 2001. Assessing the effects of land use changes on floods in the meuse and oder catch-

ment. Phys. Chem. Earth Part B Hydrol. Oceans Atmosphere 26, 593–599.

2. De Roo, A., Schmuck, G., Perdigao, V., Thielen, J., 2003. The influence of historic land use changes and future planned land use scenarios on floods in the Oder catchment. Recent Dev. River Basin Res. Manag. 28, 1291–1300.
3. Dubicki, A., Malinowska-Małek, J., Strońska, K., 2005. Flood hazards in the upper and middle Odra River basin – A short review over the last century. 11th Magdebg. Semin. Oct. 2004 Waters Cent. East. Eur. Assess. Prot. Manag. 35, 123–131.
4. Falloon, P., Betts, R., 2010. Climate impacts on European agriculture and water management in the context of adaptation and mitigation—The importance of an integrated approach. Spec. Sect. Integrating Water Agric. Manag. Clim. Change 408, 5667–5687.
5. Kundzewicz, Z.W., 2014. Adapting flood preparedness tools to changing flood risk conditions: the situation in Poland. Oceanologia 56, 385–407.
6. Kundzewicz, Z.W., Matczak, P., 2012. Natural risks: mitigation and adaptation. Ecohydrol. Hydrobiol. 12, 3–8.
7. Schmuck, A., 1967. Wpływ miasta na opady atmosferyczne (na przykładzie Wrocławia). Przegląd Geofiz. 12 (20), 293–310.
8. SAP2020, 2013. The National Strategy for Adaptation to Climate Change 2020. Poland.
9. Szewranski, S., Kazak, J., Sasik, J., 2013. Suburbanisation processes and their environmental effects in a zone of uncontrolled spread of a large city. Prace Naukowe Uniw. Ekon. we Wrocławiu, 170–179.

OCCURRENCE OF ENTOMOPATHOGENIC FUNGI IN SOILS FROM *FESTUCA PRATENSIS* HUDS. CROP

Roman Kolczarek[1], Kazimierz Jankowski[1]

[1] Institute of Agronomy, Faculty of Natural Sciences, Siedlce University of Natural Sciences and Humanities, B. Prusa 14, 08-110 Siedlce, Poland, e-mail: laki@uph.edu.pl

ABSTRACT
Entomopathogenic fungi are the largest group of microorganisms existing in the soil environment. Occurrence and pathogenicity of entomopathogenic fungi in soil is dependent on many factors affecting the soil environment. The aim of this study was to compare the species composition and the intensity of the occurrence of entomopathogenic fungi in soils from monoculture crops of *Festuca pratensis* Huds. The study material consisted of soil samples taken from the experiment conducted in two experimental stations of the Research Centre for Cultivars Study. The insecticides fungi were isolated from soil using a method of the selective substrate. Three species of entomopathogenic fungi *Beauveria bassiana, Isaria fumosorosea* and *Metarhizium anisopliae* were isolated from the study soils using the selective medium.

Keywords: entomopathogenic fungi, soil, meadow fescue.

INTRODUCTION

Entomopathogenic fungi, also known as insecticides, are the largest group of micro-organisms present in the soil environment. Their main advantage is the ability to reduce pests population of agricultural crops, horticultural crops and under cover [Miętkiewski et al., 1991; Quesada-Moraga et al. 2007].

Mostly insect pathogen infection occurs in soil, which is a natural habitat for their existence [Bidochka et al. 1998]. However, existing and pathogenicity of entomopathogenic fungi in soil is dependent on a number of factors affecting the soil environment include temperature, humidity, soil type, structure, and the manner of its use [Tkaczuk, Renella, 2003]. It is essential that the humidity of the soil which determines the possibility of infection, and the temperature which often results in the spreading of some fungi species.

The species composition of fungi as well as their intensity depend on the environment, from which is the soil, and also from the season of year (spring and autumn) and crop species [Gottwald, Tedders 1984; Bałazy 2006].

Insecticides fungi as one of the first pathogens were used in biological pest control in plants. The aim of this study was to compare the species composition and the intensity of the occurrence of entomopathogenic fungi in soils of monoculture crops of *Festuca pratensis L.*

MATERIAL AND METHODS

The study material consisted of soil samples taken from scientific experiments conducted in two experimental stations of the Research Centre for Cultivar Study (COBORU) in Krzyżewo (Podlaskie province) and Uhnin (Lublin province). The samples were taken in two periods: spring and autumn.

The experiment was carried out in two locations and concerned the assessment of yielding of different varieties of tall fescue. Detailed information on the experiment design and soil conditions has shown in Table 1. Samples were taken at random from a 10-th points on a plot. The soil was collected using soil canes to a depth of 15 cm. From the material collected from the area mixed

Table 1. Conditions of experimentation – *Festuca pratensis* Huds.

Lp.	Name of place	Height above sea level (m)	Soil					Forecrop
			Agriculture Value in 100 point scale IUNG	Complex of agricultural suitability	Type	Species	pH 1 n KCl	
	1	2	3	4	5	6	7	8
1	Krzyżewo	130	70	4	A	gl	6,4	Spring barley
2	Uhnin	155	80	1z	M	N	6,6	meadow

Col. 4; 4 – rye very good, 1z – grassland very good and good.
Col. 5: A – podzolic; M – muck.
Col. 6; gl – loam, n – deep peat with thickness of organic matter >30 cm.

sample was performed and stored in plastic bags at 0–4 °C.

Insecticides fungi were isolated from soil using a method of selective substrate developed by Strasser et al. [1996]. The substrate is commonly used for the isolation of entomopathogenic fungi in the soil [Keller et al. 2003, Tkaczuk 2008] .

From each sample originating from a given plot, 2 g was weighed and then 18 ml of distilled water with the addition of 0.05 ml of a solution of Trithon X-100 preparation was added, and vigorously shaken about 35 seconds. Then 0.1 ml of the soil solution was poured in and spread using a glass spatula on a surface of the selective substrate in three Petri dishes, which were replicates. The dishes were placed in incubators at 22 °C and after 8–10 days colonies were counted in each of fungi species. Results were expressed as the number of infectious units (CFU) of entomopathogenic fungi in 1 g of soil.

RESULTS

Using a selective substrate from meadow soils three species of entomopathogenic fungi *Beauveria bassiana, Isaria fumosorosea* and *Metarhizium anisopliae* were isolated. Species composition and units density of the colony-forming of these fungi were varied, depending on the variety and the place from which soil samples were derived. In the soil samples collected in the spring time in Krzyżewo from varieties of tall fescue there were 3 species of fungi. The dominant species was *M. anisopliae* (Figure 1), which formed an average of 1.1×10^3 g^{-1} CFU in 1 gram of soil. *B. bassiana* formed an average of 0.3×10^3 g^{-1}, and *I. fumosorosea* occurred only in soil under Anturka variety. The greatest concentration of CFU of entomopathogenic fungi was observed in the soil from Pasja variety (1.7×10^3 g^{-1}) and Limosa (1.5×10^3 g^{-1}) (Table 2).

In the soil samples derived from the cultivation of tall fescue in Krzyżewo in the second period (autumn), also the dominance of *M. anisopliae* was noted, which created an average of 1.8×10^3 g^{-1} CFU in 1 g of soil (Figure 1). No presence of the fungus *B. bassiana*, and *I. fumosorosea* was noted, it was appeared in the soil under two of the four varieties, forming an average of 0.3×10^3 g^{-1} infectious units per 1 gram of the soil (Table 3, Figure 1).

In the soil samples collected in the spring at Uhnin from tall fescue cultivation a higher density of CFU of the fungus *M. anisopliae* was reported than in autumn. This fungus formed an average of 1.8×10^3 g^{-1} of colony forming units in 1 g of soil, and *B. bassiana* average of 1.3×10^3 g^{-1} (Figure 2). *I. fumosorosea* fungus appeared only in the soil under Anturka vari-

Table 2. The density of infectious units (CFU × 10^3 g^{-1}) of entomopathogenic fungi in soil under cultivation of different varieties of tall fescue (Krzyżewo spring)

Variety	Species of fungi			
	Beauveria bassiana	*Isaria fumosorosea*	*Metarhizium anisopliae*	Total
Limosa	–	–	1.5f	1.5
Pasja	0.7b	–	1.0d	1.7
Anturka	0.2ab	0.2ab	0.7c	1.1
And 1009	0.2ab	–	1.2d	1.4

The value marked with the same letters do not differ significantly.

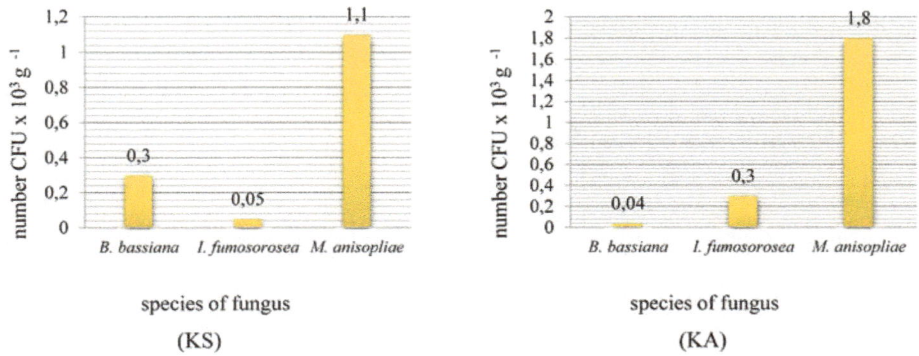

Figure 1. Average concentration of infectious units (CFU × 10^3 g^{-1}) of individual entomopathogenic fungi in soil from cultivation of tall fescue (Krzyżewo – K, spring – S, autumn – A)

ety. Insecticides fungi formed together the most CFU in the soil from plots where varieties Limosa (6.5×10^3 g^{-1}) and Pasja (3.8×10^3 g^{-1}) were cultivated (Table 4).

In the soil samples collected in autumn in Uhnin under cultivation of different varieties of tall fescue the most CFU fungus *B. bassiana* was formed. This species occurred in all soil samples from four cultivated varieties and formed an average of 2.7×10^3 g^{-1} CFU in 1 g of soil (Figure 2).

The fungus *M. anisopliae* formed their colonies only in two from four tested soil samples and *I. fumosorosea* was found only in one sample from the cultivation of Limosa. Insecticides fungi form together the most infectious units in soils under crop varieties Pasja and and 1009 (Table 5).

DISCUSSION

Isolated from meadow soil, the species of insecticides fungi were the most common in the soil environment. This is confirmed by numerous studies carried out by other authors [Keller et al 2003, Bidochka et al 1998, Klingen 2000, Ali-Shtayeh et al 2002, Tkaczuk and Renella 2003, Quesada-Moraga et al 2007, Sun et al 2008].

Using a selective substrate from the soils, on which various species and varieties of grasses were cultivated, three species of entomopathogenic fungi: *Beauveria bassiana, Isaria fumosorosea* and *Metarhizium anisopliae* were isolated. Species composition and units density of the colony-forming of these fungi was varied and depending on the species, variety of grasses, seasons of the

Table 3. The density of infectious units (CFU × 10^3 g^{-1} soil) of entomopathogenic fungi in soil under cultivation of different varieties of tall fescue (Krzyżewo, autumn)

Variety	Species of fungi			
	Beauveria bassiana	*Isaria fumosorosea*	*Metarhizium anisopliae*	Total
And 1009	–	–	3.7h	3.7
Limosa	–	0.4b	0.4b	0.8
Pasja	–	0.2ab	1.4ef	1.6
Anturka	–	–	1.4ef	1.4

The value marked with the same letters do not differ significantly.

Table 4. The density of infectious units (CFU × 10^3 g^{-1}) of entomopathogenic fungi in the soil under cultivation of tall fescue (Uhnin, spring)

Variety	Fungi species			
	Beauveria bassiana	*Isaria fumosorosea*	*Metarhizium anisopliae*	Total
Pasja	3.5h	–	0.3b	3.8
And 1009	1.3ef	–	–	1.3
Limosa	–	–	6.5i	6.5
Anturka	0.3b	0.2ab	0.3b	0.8

The value marked with the same letters do not differ significantly.

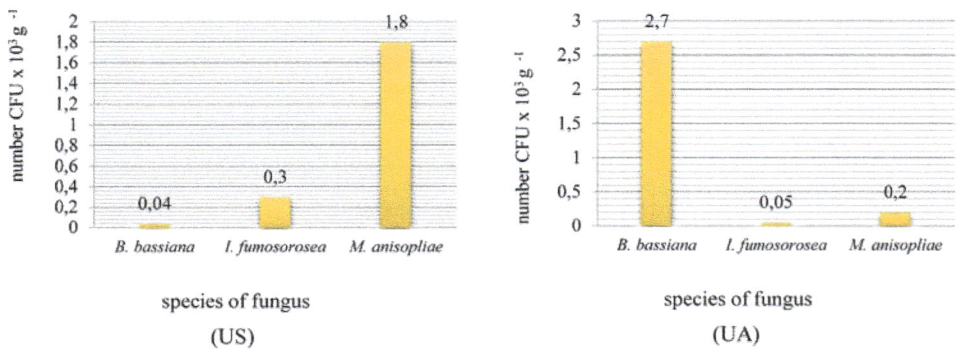

Figure 2. Average concentration of infectious units (CFU × 10^3 g^{-1}) of individual entomopathogenic fungi in soil from cultivation of tall fescue (Uhnin – U, spring – S, autumn – A)

Table 5. The density of infectious units (CFU × 10^3 g^{-1}) of entomopathogenic fungi in the soil under cultivation of tall fescue (Uhnin, autumn)

Variety	Fungi species			
	Beauveria bassiana	*Isaria fumosorosea*	*Metarhizium anisopliae*	Total
Pasja	3.7h	–	–	3.7
And 1009	3.7h	–	–	3.7
Limosa	1.4ef	0.2ab	0.4b	2.0
Anturka	1.9g	–	0.4b	2.3

The value marked with the same letters do not differ significantly.

year and the place from which soil samples were derived. In the soil samples taken on two dates in Krzyżewo, from all grass species and varieties, the fungus *M. anisopliae* was dominated. In Tkaczuk's research [2008], in the national meadow and pasture soils the most common is fungus *M. anisopliae* (present in 96.3% of samples), and then *B. bassiana* and *I. fumosorosea* (50% of the samples). *M. anisopliae* also struck the most *G. mellonella* larvae in these soils. This species seems to be characteristic for meadow and pasture environments and was previously listed as dominant in the soils for meadows and pastures in the UK [Chandler et al 1997], Switzerland [Keller et al 2003] and Australia [Rath et al 1992]. In the soil samples collected in spring and autumn from different species and varieties of grasses in Station of Variety Assessment in Uhnin the most CFU fungus *Beauveria bassiana* was formed. It should be noted that this soil was characterized by a high content of organic matter, in contrast to light and mineral soil from Station Variety Assessment COBORU in Krzyżewo. Research of some authors point to a higher abundance of grassland soils in insecticides fungi than adjacent soils of farmland. Miętkiewski et al. [1992] conducted a study to compare the existing of entomopatogenic fungi in the soil arable and meadow.

In the soil derived from meadow showed significantly greater intensity of *M. anisopliae* than in the soil of cultivated fields. Also, other species with the exception of *P. fumosoroseus*, were more numerous in the meadow soil. The differences were not statistically significant, but the tendency to higher occurrence of entomopatogenic fungi in the meadow soil compared with topsoil was significant. Analyzing the density of CFU in the meadow soils in two seasons – spring and fall, it was found that some species of fungi were present in greater numbers in the spring, while others in fall. Tkaczuk [2008] performing a similar analysis of soils, but using a trap method showed that, depending on the season of year (spring and autumn), there are some changes in the frequency of isolation and in the percentage share of each species of entomopathogenic fungi in the mortality of larvae of *G. mellonella* lined to soils from studded environments in Poland.

Treating all the environments and species of fungi, no significant differences in the frequency of their isolation from the soil in spring and autumn was stated. Fungus *B. bassiana* was significantly more frequently isolated from environments soils in spring (93.2% of samples) than in autumn (73.5%), while in the case of the fungus *M. anisopliae* an opposite tendency was found.

Looking at the separate individual environments only in relation to soils from meadows and pastures was found that insecticides fungi including were significantly more isolated in the spring. The fungus *B. bassiana* existed more numerous in soils of cultivated fields in the spring but the other species were isolated with similar frequency in both seasons. Fungi *B. bassiana* and *I. fumosorosea* were recovered significantly more often in spring than in autumn from soil samples of meadows and pastures

CONCLUSIONS

1. Species composition and density of the colony-forming units varied depending on the variety, season of the year and the place from which originate the tested soil samples.
2. From the study soils using the selective medium three species of entomopathogenic fungi *Beauveria bassiana, Isaria fumosorosea* and *Metarhizium anisopliae* were isolated.
3. Regardless of the soil type the most infectious units were created by fungi *Beauveria bassiana* and *Metarhizium anisopliae.*

REFERENCES

1. Ali-Shtayeh M.S., Mara'I A.B.B.M., Jamous R.M. 2002. Distribution, occurrence and characterization of entomopathogenic fungi in agricultural soil in the Palestinian area. Mycopathologia, 156: 235-244.
2. Bałazy S. 2006. Diagnosis and attempt to estimate the role of entomopathogenic fungi in the stands. Studies and Research Centre for Education Natural Sciences and Forest, 4(14): 154-165.
3. Bidochka M.J., Kasperski J.E., Wild G.A.M. 1998. Occurrence of the entomopathogenic fungi *Metarhizium anisopliae* and *Beauveria bassiana* in soils from temperature and near-northern habitats. Can. J. Bot., 76: 1198-1204.
4. Chandler D., Hay D., Reid A.P. 1997. Sampling and occurrence of entomopathogenic fungi and nematodes in UK soils. Appl. Soil. Ecol. 5: 133-141.
5. Gottwald T.R., Tedders W.L. 1984. Colonization, transmission and longevity of *Beauveria bassiana* and *Metarhizium anisopliae* (*Deuteromycotina, Hyphomycetes*) on pecan weevil larvae (*Coleoptera, Curculionidae*) in the soils. Environ. Entomol., 13: 557-560.
6. Keller S., Kessler P., Schweizer C. 2003. Distribution of insect pathogenic soil fungi in Switzerland with special reference is *Beauveria brongniartii* and *Metarhizium anisopliae*. Biocontrol, 48: 307-319.
7. Keller S., Zimmermann G. 1989. Macopathogen of Siol insects. [In:] Insect – Fungus Interactions. Wilding N., Collins N.M., Hammon P.H., Webber J.F/ (eds), Acadamic Press, London: 204-270.
8. Kessler P., Enkerli J., Schweizer C., Keller S. 2004. Survival of *Beauveria brongniartii* in the soil after application as a BioControl agent against the European cockchafer *Melolontha melolontha*. Biocontrol 49: 563-581.
9. Klingen I. 2000. Natural occurrence of insect pathogenic fungi and their pathogenicity on different host species with emphasis on *Delia Floralis*. Ph.D. Thaesis, Agricultural University of Norway.
10. Miętkiewski R. Tkachuk C., Badowska-Czubik T. 1992. Entomogenous fungi isolated from strawberry plantation soil infested by *Otiorhynchus ovatus L.* Annals of Agricultural Sciences Ser. E, 22(1/2) 39-46.
11. Miętkiewski R., Tkachuk, C., Zurek M., Bałazy S. 1991. Occurrence entomopatogennych fungi in soil arable and forest litter. Rocz. Agricultural Sciences. Ser. E, 21(1/2): 61-68.
12. Quesada-Moraga E., Navas-Cortes J.A., Maranhao A.A., Ortiz-Urquiza A.S., Santiago-Alvarez C. 2007. Factors affecting the occurrence and distribution of entomopathogenic fungi in natural and cultivated soils. Mycol. Res., 111: 947-966.
13. Rath C., Koen T.B., Yip H.Y. 1992. The influence of Abiotic factors on the distribution and abundance of *Metarhizium anisopliae* in Tasmanian pasture soils. Mycol. Res., 96: 378-384.
14. Strasser H., Forrer A., Schinner F. 1996. Development of media for the selective isolation and maintenance of virulence of *Beauveria brongniartii.* [In:] Microbial control of soil dwelling pests. (eds T.A /Jackson and T.R/ Glare), AgResearch, Lincoln, New Zealand: 125-130.
15. Sun B.D., Yu H.Y., Chen A.J., Liu X.Z. 2008. Insect-associated fungi in soils of field crops and orchards. Crop Prot., 27: 1421-1426.
16. Tkaczuk C. 2008. Occurrence and potential infectious fungal insecticides in soils and environments seminatural agrocenos in the agricultural landscape. Thaesis No. 94, Edited University of Podlasie in Siedlce.
17. Tkaczuk C., Renella G. 2003. Occurrence of entomopathogenic fungi in soils from Central Italy under different management. IOBC/WPRS Bulletin, 26(1): 85-89.

MOLLUSC SPECIES PROTECTED IN POLAND AND THREATENED IN EUROPE RECORDED IN STEPNICA RIVER (NW POLAND)

Małgorzata Raczyńska[1], Juliusz C. Chojnacki[1], Monika Hałupka[1]

[1] Department of Marine Ecology and Environmental Protection, Faculty of Food Sciences and Fisheries, West Pomeranian Technological University in Szczecin, Kazimierza Królewicza 4, 71-550 Szczecin, Poland, e-mail: malgorzata.raczynska@zut.edu.pl; juliusz.chojnacki@zut.edu.pl

ABSTRACT
In the course of the study two bivalve species protected in Poland were found in the river Stepnica: *Sphaerium solidum* and *Sphaerium rivicola*. Moreover, the study material collected from the river contained gastropod and bivalve specimens representing the following species from the IUCN Red List of Threatened Species: *Theodoxus fluviatilis*, *Pisidium henslowanum*, *Pisidium casertanum* and *Pisidium pseudosphaerium*.

Keywords: Bivalvia, Gastropoda, threatened species, protected species.

INTRODUCTION

Molluscs are the only group among the numerous groups of animals comprising 130 000 species distributed all over the world which inhabit both aquatic and terrestrial environments. Representatives of two taxonomic classes of the Mollusca live in freshwater environments, namely the Gastropoda (slugs and snails) and the Bivalvia [Jura 2007]. As they are not particularly mobile and often even sedentary animals (mainly in the case of bivalves), they are reliable long-term indicators of local pollution, understood as a sum of biotic and abiotic influences in environment [Protasowicki & Niedźwiecki 1995]. For instance, it is thanks to molluscs that periodical discharge of toxic wastewater is possible to detect [Chojnacki 1998]. The ability of molluscs to serve as pollution indicators allows for evaluating a relative level of water contamination with toxic substances, assuming that concentrations of these substances in mollusc tissues are proportional to their concentrations in water [Jurkiewicz-Karnkowska 1998, 2003, 2004, 2011, Oertel 1998]. Therefore, molluscs have proved particularly useful as bioindicators of water contamination with toxic substances and acidization of environment. They are also important for so-called emergency population warnings, being used for detection of unfavorable changes in water quality, for instance in waterworks systems [Jurkiewicz-Karnkowska 1998, 2004].

Unfortunately, progressively increasing pollution of aquatic ecosystems, caused mainly by human activity, contributes to a drop in the abundance of this group of invertebrates. Presently, several mollusc species are threatened with extinction and many species are legally protected. Therefore, conducting faunistic studies on this group of invertebrates in Poland is recommended, especially in regions where little is known about native mollusc fauna (e.g. in small water courses of Pomerania). Such studies will not only allow to make a list of areas where protected species of freshwater molluscs occur, but will also facilitate undertaking appropriate measures to protect these species.

MATERIAL AND METHODS

Mollusc material discussed in the present study was collected at monthly intervals during the period of research on macrobenthic fauna conducted from 21 December 2004 to 12 December 2006. The material was collected in the upper reaches of the river, using a bottom scraper, at four research stations along a river stretch situated 10 km away from the headstreams in the neighborhood of the settlement of Wisławie, as far as the town of Maszewo (Figure 1). The collected material was transferred to a laboratory, sifted through 0.45 mm mesh sieves and preserved with 70% buffered alcohol. The samples prepared in this way were arranged according to systematic groups and then qualitative identification was conducted. Detailed taxonomic identification of the molluscs was based on conchological features used in the studies by Piechocki [1979] and Piechocki & Dyduch-Falniowska [1993].

RESULTS

Malacofauna of the river Stepnica was represented by 3 genera and 14 species of the Mollusca (Table 1), and its abundance in the years 2004–2006 amounted to 332 individuals. The most diverse taxonomically and at the same time the most abundant fauna of the Mollusca was found at the research station no. 2, situated in the town of Maszewo. At the remaining stations only representatives of the Gastropoda were recorded; they were encountered in the highest numbers at the research station no. 3, in the town of Radzanek, whereas only two gastropod species were recorded at the station no. 1 (Wisławie), and only one at the station no. 4 (Maszewo) (Table 1).

Among the recorded molluscs there were both freshwater species common in Poland and protected species, as well as species threatened with extinction in Poland and in Europe. The species threatened with extinction in Europe found in the river Stepnica included one gastropod species: *Theodoxus fluviatilis* (12 individuals), and three bivalve species: *Pisidium henslowanum* (37 individuals), *Pisidium casertanum* (42 individuals) and *Pisidium pseudosphaerium* (22 individuals). Furthermore, taxonomic composition of the Bivalvia in the river Stepnica included also two species protected in Poland: *Sphaerium solidum* and *Sphaerium rivicola*, represented by 16 and 2 individuals respectively (Table 1).

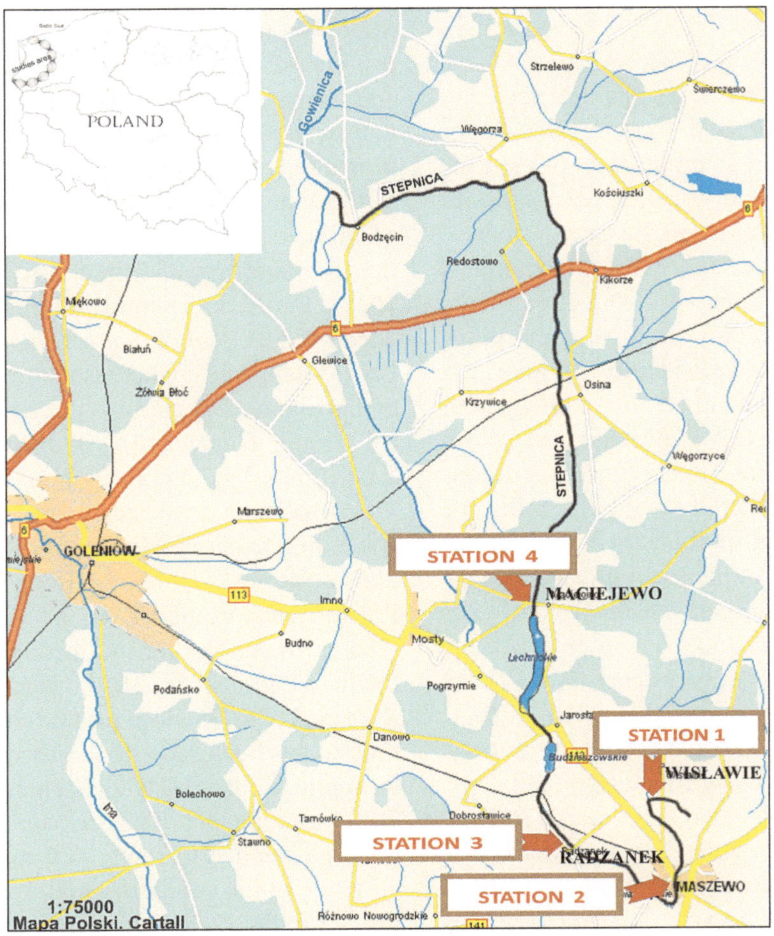

Figure 1. Location of research stations

Table 1. Average abundance of gastropods and bivalves in the river Stepnica in the years 2004–2006

	Station 1	Station 2	Station 3	Station 4	Total
		Gastropoda			
Theodoxus fluviatilis		8	4		12
Valvata piscinalis			42		42
Bithynia tentaculata	6	15			21
Lymnaea stagnalis			2		2
Radix peregra		4	1		5
Planorbis carinatus		15	21	3	39
Planorbarius corneus	14	1	1		16
Anisus sp.			1		1
		Bivalvia			
Sphaerium sp.		2			2
Sphaerium rivicola		2			2
Sphaerium solidum		16			16
Pisidium sp.		8			8
Pisidium amnicum		13			13
Pisidium casertanum		42			42
Pisidium henslowanum		37			37
Pisidium milium		52			52
Pisidium pseudosphaerium		22			22

DISCUSSION

The upper reaches of the river Stepnica, where the study was conducted, were characterized by specific environmental conditions. According to Raczyńska et al. [2013], from the point of view of hydromorphological quality, ecological condition of the four research stations on the river Stepnica ranged from poor to bad and did not meet standards of the Water Framework Directive. This was due to a low diversity of natural morphological elements and a considerable amount of alterations caused by human activity taking place on the river in the past. A hydrochemical study of the river revealed that its ecological condition was less than good, especially at the research station no. 3, in the town of Radzanek, and at the research station no. 4, in the town of Maciejewo. During the studies, hydrogen sulfide was detected at the Radzanek station (no. 3); the presence of the compound had a negative impact on the river biocenosis [Raczyńska et al. 2013]. The impact of these facts on the molluscs was noted during the studies, since the most abundant population of these animals was found at the research station no. 2 in the town of Maszewo, where water quality was the highest. It was also the only station where the presence of bivalves was recorded in the course of the studies (Table 1).

The two bivalve species recorded solely at the Maszewo research station (no. 2), i.e. the solid orb mussel *Sphaerium solidum* [Normand 1844] and the river orb mussel *Sphaerium rivicola* [Lamarck 1818] are on the list of animals under strict protection in Poland, pursuant to the Ordinance of the Minister of Environment on the protection of animal species [Journal of Laws of the Republic of Poland No. 237, item 1419, of 12 October 2011]. Furthermore, on the Polish list of threatened species, *Sphaerium solidum* is classified as EN (Endangered), whereas on the IUCN Red List of Threatened Species it is put under the category NT (Near Threatened) [Dyduch-Falniowska & Zając 2004, Van Damme et al. 2012].

S. solidum is a species encountered in some rivers in northern Poland, which in the 19th and 20th centuries was relatively abundant in larger Polish rivers, especially in the middle and lower reaches of the Vistula and the Oder [Piechocki & Dyduch-Falniowska 1993, Dyduch-Falniowska & Zając 2004]. In more recent times, the solid orb mussel has been recorded in merely 10 localities throughout Poland [Dyduch-Falniowska &i Zając 2004], and the fact of its occurrence in the Stepnica has expanded its range in the country. The occurrence of the solid orb mussel and the river orb mussel in the river Stepnica at the Maszewo station may be connected with the presence of substrates preferred by the species. *S. solidum* prefers sandy bottoms with a slight admixture of mud, or even with coarse-grained sand, while *Sphaerium rivicola* prefers rivers with muddy sediments and low water current [Wolff 1970]. Exactly this type of sandy and muddy substrate was present at the station in question. *S. solidum* is an exceptionally reophilous and oxygen-loving species [Glöer et al. 1992], and its preferred habitats are deep sections of rivers and channels [Van Damme et al. 2011], rather unlike its habitats in the river Stepnica. Although the amount of dissolved oxygen at the Maszewo station was characteristic for the first water quality class in Poland [Raczyńska et al. 2013], such characteristics of the river as the flow (imperceptible) and average depth of about 20 cm during the study period did not provide optimal conditions for the development of the discussed species. Even though *Sphaerium rivicola* has a very low tolerance to pollution, it has sometimes been encountered in a slightly polluted river current [Wolff 1970, Van Damme et al. 2011]. Water quality at the station no. 2 ranged between the first and second class, but the concentrations of total suspended solids, as well as nitrogen and phosphorus compounds indicated that the actual water quality was less than good [Raczyńska et al. 2013], which might limit population abundance of the discussed species.

The remaining molluscs included in the IUCN Red List of Threatened Species, i.e. *Theodoxus fluviatilis, Pisidium henslowanum, Pisidium casertanum* and *Pisidium pseudosphaerium* belonged to the category LC (Least Concern). Although disturbances caused by human activity and local water pollution are potential threats to these species, presently there is no need to undertake any protective measures [Categories & Criteria, Killeen 2011, Cordeiro & Killeen 2011, Budha 2012, Kebapçı & Van Damme 2012]. All of the above mentioned species are common in their preferred habitats in Polish waters. So far, no declining tendency has been observed with respect to their populations in Poland and they have not been classified under any category of threatened Polish species.

CONCLUSIONS

Unexpectedly, in the river Stepnica, a minor water course in Western Pomerania, strongly altered by human activity and vulnerable to pollution, mollusc species have been found. They are not only legally protected in Poland but also featured on the IUCN Red List of Threatened Species. The results of the present study lead to a conclusion than in spite of locally occurring ecological disasters, populations of natural inhabitants are still present in the studied river and they may easily and dynamically expand their ecological structures, thus lead to increased biodiversity.

Acknowledgements

The authors would like to thank Professor Lucjan Agapow for his assistance with mollusc identification.

REFERENCES

1. Budha P.B., 2012. *Pisidium casertanum.* In: IUCN Red List of Threatened Species. Version 2013.2, www.iucnredlist.org
2. Categories & Criteria: (version 3.1) http://www.iucnredlist.org/static/categories_criteria_3_1#categories
3. Chojnacki J.C., 1998. Podstawy ekologii wód. Wyd. Akad. Roln. Szczecin, 1–177.
4. Cordeiro J., Killeen I., 2011. *Pisidium henslowanum.* In: IUCN Red List of Threatened Species. Version 2013.2, www.iucnredlist.org
5. Dyduch-Falniowska A., Zając K., 2004. W: Polska czerwona księga zwierząt – bezkręgowce. Wersja 2004–2009, http://www.iop.krakow.pl/pckz/opis.asp?id=126&je=pl
6. Journal of Laws of the Republic of Poland No. 237, item 1419, of 12 October 2011 (Dz.U. Nr 237, poz. 1419 z dnia z dnia 12 października 2011 r.): Rozporządzenie Ministra Środowiska z dnia 12 października 2011 r. w sprawie ochrony gatunkowej zwierząt.
7. Glöer P., Meier-Brook C., Ostermann O., 1992. Süßwassermollusken. Aufl. Hamburg, DJN, 5, 81.
8. Jura Cz., 2007. Bezkręgowce. Podstawy morfologii funkcjonalnej, systematyki i filogenezy PWN.
9. Jurkiewicz-Karnkowska E., 1998. Reakcje mięczaków na zanieczyszczenie środowisk wodnych metalami ciężkimi i możliwości ich wykorzystania w bioindykacji. Wiad. Ekol. 44(3), 217–234.
10. Jurkiewicz-Karnkowska E., 2003. Application possibilities of snails from the ge-nus Lymnaea as bioindicators of heavy metals. Ekologija 2, 28–32.
11. Jurkiewicz-Karnkowska E., 2004. Malacocoenoses of large lowland dam reservoirs of the Vistula River basin and selected aspects of their function. Folia Malacologica 12, 1–56.

12. Jurkiewicz-Karnkowska E., 2011. Accumulation of heavy metals in selected Gastropod species - lower Bug river and its floodplain water bodies. Ochr. Środ. Zas. Nat. Nr 48, 316–323.

13. Kebapçı U., Van Damme D., 2012. *Theodoxus fluviatilis*. In: IUCN Red List of Threatened Species. Version 2013.2, www.iucnredlist.org.

14. Killeen I., 2011. *Pisidium pseudosphaerium*. In: IUCN Red List of Threatened Species. Version 2013.2, www.iucnredlist.org

15. Oertel N., 1998. Molluscs as bioindicators of heavy metals in a side-arm system of the River Danube disturbed by engineering activity. Verh. Internat. Verein. Limnol. 26, 2120–2124.

16. Piechocki A., 1979. Mięczaki (Mollusca). Ślimaki (Gastropoda). Fauna Słodkowodna Polski, 7. PWN, Warszawa-Poznań.

17. Piechocki A., Dyduch-Falniowska A., 1993. Mięczaki (Mollusca). Małże (Bivalvia). Fauna Słodkowodna Polski, zesz. 7 A. PWN, Warszawa.

18. Protasowicki M., Niedźwiecki E., 1995. Zanieczyszczenie osadów dennych ujścia Odry metalami ciężkimi w świetle wieloletnich badań. W: Europejski ład ekologiczny (red: Chojnacki J., Pałyga J.E). Bibl. Fund. A.F. Modrzewskiego 14, Folia humanitaties 6, 122–127.

19. Raczyńska M., Machula S., Grzeszczyk-Kowalska A., 2013. Stan ekologiczny rzeki Stepnicy (Pomorze Zachodnie). Inżynieria Ekologiczna (Ecological Engineering), 35, 46–59.

20. Van Damme D., Killeen I., von Proschwitz T., 2012. *Sphaerium solidum*. In: IUCN Red List of Threatened Species. Version 2013.2, www.iucnredlist.org

21. Wolff W.J., 1970. The Mollusca of the estuarine region of the rivers Rhine, Meuse and Scheldt in relation to the hydrography of the area. IV. The genus Sphaerium. Basteria 34(3–4), 75–90.

CHARACTERISTICS OF SELECTED PHYSICOCHEMICAL INDICES OF UPLAND CARBONATE STREAM WATER WITH COARSE-GRAINED SUBSTRATE

Agnieszka Policht-Latawiec[1], Włodzimierz Kanownik[1], Ksymena Grubka[2]

[1] Department of Land Reclamation and Environmental Development, University of Agriculture in Krakow, Mickiewicza 24-28, 30-059 Krakow, Poland, e-mail: a.policht@ur.krakow.pl; rmkanown@cyf-kr.edu.pl

[2] Graduate student of the University of Agriculture in Krakow, Mickiewicza 24-28, 30-059 Krakow, Poland

ABSTRACT

Hydrochemical research was conducted in 2011 at three measurement-control points on the Sztoła river, a left-bank tributary to the Biała Przemsza river. Water temperature and pH, dissolved oxygen concentration, oxygen saturation and electrolytic conductivity were measured directly in the field. 11 selected physicochemical indices of the Sztoła surface water were assessed in the laboratory. The analysis of the research results revealed that the water of the Sztoła along the analyzed river reach was first class quality. The analysis of water usable qualities proved that at point 1 it can be used for water supply after high-performance physical and chemical treatment, whereas at other points after a typical physical and chemical treatment, due to high manganese concentrations. Water does not meet the requirements of natural habitat for the Salmonoid fish only at point 1 because of exceeded standard values for nitrites. The Cyprinid fish have the proper habitat conditions along the examined river reach. Among 15 analysed physiochemical indices of water, 7 were statistically higher in the lower course, but they did not affect worsening of the Sztoła river water quality. Greater water pollution in the central and lower river reach is caused by the anthropogenic pressure connected mainly with the operations of extraction industry in the catchment.

Keywords: Sztoła river, water quality, physicochemical indices.

INTRODUCTION

Water quality is determined by many natural factors, such as: topography, soil properties, intensity of lack of atmospheric [Xiao et al. 2014] and anthropogenic precipitations [Kanownik et al. 2013], which contribute to a considerable water loading with nutrients [Wang et al. 2014], pesticides, fertilizers [Monaghan et al. 2005] and other inorganic and organic substances [Wilcock et al. 2006]. Agricultural and forest land management influence water quality [Pierzgalski et al. 2005, Grajewski et al. 2009, Grzywna 2014], as nitrogen and phosphorus supplied with surface runoffs contribute to eutrophication and degradation of waters [Cooper and Thomsen 1988, McDowell and Wilcock 2004, Rodgers et al. 2010].

Pollution from point sources [Policht-Latawiec et al. 2013] is far easier to document in comparison with the area pollution [Kanownik and Pijanowski 2007], which greatly contribute to water contamination in the catchment [Alexander et al. 2002, Piirainen et al. 2013]. Therefore, an assessment of the efficiency of catchment management policy is so important [Ferraz et al. 2013], especially based on increasingly better applications for modelling and forecasting ecological hazard to waters, including both climatic and anthropogenic factors [Lai et al. 2013]. Understanding ecological processes is very important for sustainable management of surface and ground waters in a catchment [Di and Cameron 2005].

MATERIAL AND METHODS

The hydrochemical research of the Sztoła river, situated in the north-western part of the Malopolska voivodeship was conducted in 2011. The Sztoła river, 13.4 km long and with catchment area of 154km^2 is a left bank tributary of the Biała Przemsza. According to the abiotic typology it belongs to an upland carbonate stream with coarse-grained substratum. The riverhead is located at the foot of Kozłowa Góra hill in the area of the Olkusz Upland. Initially the river flows in a deep valley in the area of Dolinki Krakowskie Landscape Park towards north-west and several kilometers further flows into "Leśny Dwór" reservoir in Bukowno. The headwater section is almost totally devoid of water because of depression pit formed as a result of the operation of the zinc and lead mines in Olkusz, but also due to drainage of sand deposits exploited by the Szczakowa opencast mine. The Baba watercourse, originating near Olkusz, inflows to the Sztoła river in Bukowno. The watercourse is not fed by the primary springs, which are totally dry, but almost in 100% by the mining wastewater from the Pomorzany Mine in Olkusz. From Bukowno the Sztoła river flows to the west through pine forests, strongly meandering, in some places forming inaccessible wetlands and finally flows into the Biała Przemsza in Sławków.

Water for the analyses was sampled in 3 measurement-control points: **point 1** at km 13+250, **point 2** – at km 10+040, **point 3** – at km 7+370 of the watercourse (Figure 1). The upper part of the catchment is almost wholly managed as a forest land, whereas residential areas and mining industry centres are located only below point 2 and these may pose a potential pollution hazard to the analyzed river.

Assessments of water quality indices of the Sztoła river were conducted using referential methods [Rozporządzenie MŚ 2011b]. In the field pH was determined by CP-104 pH meter, electrolytic conductivity (EC) by CC-102 conductometer and the water temperature, dissolved oxygen contents and degree of oxygen saturation using CO-411 oxygen meter. The following assessments were made in the laboratory: soluble solids by evaporation, sulphates (SO_4^{2-}) by precipitation method; Ca^{2+}, Mg^{+}, Mn^{2+} ions and Fe^{2+}, Fe^{3+} (total iron) using atomic absorption spectrometry (ASA) on UNICAM SOLAR 969 spectrometer. Ammonium ($N\text{-}NH_4^{+}$) and nitrate ($N\text{-}NO_3^{-}$) nitrogen concentrations, as well as phosphates (PO_4^{3-}) and chlorides (Cl^{-}) were assessed by means of colorimetric flow analysis on FIAstar 5000 apparatus. Concentrations of NO_2^{-} and NO_3^{-} ions were computed from the nitrogen forms assessed in the laboratory.

Elaboration of the results comprised determining the minimum and maximum values of individual indices and computing the arithmetic means. Statistical inference about the significance of differences of indices values between the measurement–control points was conducted by means of non-parametric Mann-Whitney U test on the significance level $\alpha = 0.05$. The test was chosen because of the lack of distribution normality for a majority of analyzed indices, according to the results of Shapiro-Wilk test and the lack of the equality of variance determined by the Fisher-Snedecor test. Water quality was assessed according to the Decree of the Minister of the Natural Environment dated 9 November 2011 [Rozporządzenie MŚ 2011a], whereas the usable values – by comparison of the assessment results with the values admissible for water intended for water supply to people [Rozporządzenie MŚ 2002b] and for fish habitat under natural conditions [Rozporządzenie MŚ 2002a].

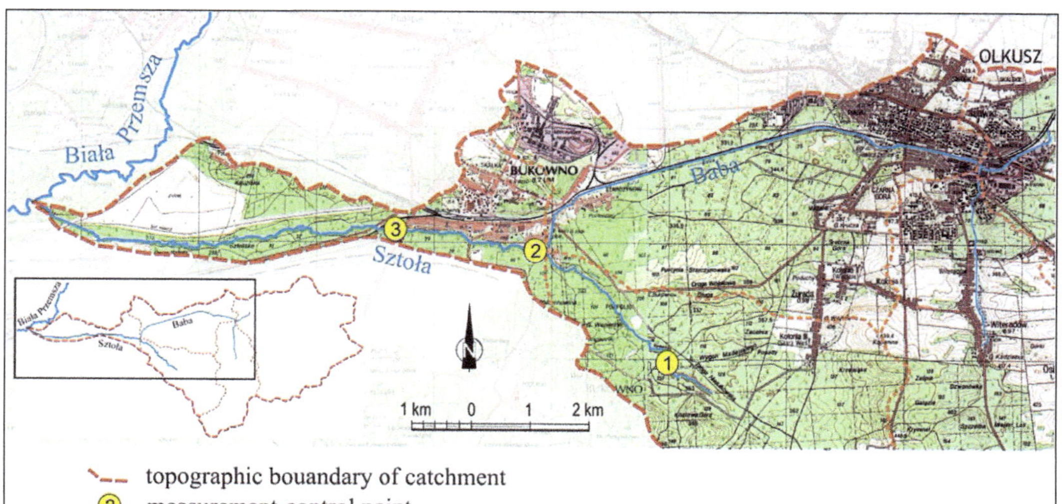

Figure 1. Location of measurement-control points in the Sztoła river catchment

RESULTS

During the period of investigations water temperature in the Sztoła river ranged from 7.3 to 14.9 °C and its pH from acid (pH 6.6) to alkaline (pH 8.3). Electrolytic conductivity did not exceed 582 μS·cm^{-1} and concentration of soluble solids was a maximum of 488 mg·dm^{-3} (Table 1). These values were lower than the limit values for water quality class I, which evidences a low mineral pollution of water. During the whole period of the investigations water at the analyzed river reach was characterized by very good oxygen conditions – the saturation level was over 100% and did not exceed 133%, whereas dissolved oxygen concentration was higher than 7 mg·dm^{-3}. Low concentrations of nitrites and nitrates were registered along the whole length of the analyzed watercourse, whereas ammonium nitrogen and phosphates occurred in trace amounts. The values of biogenic indices fell within the water quality class I [Rozporządzenie MŚ 2011a]. Salinization of the analyzed Sztoła reach was on a low level; the highest sulphate concentration (SO_4^{2-}) was 113 mg·dm^{-3}, chloride (Cl^-) – 25 mg·dm^{-3}, calcium (Ca^{2+}) – 155 mg·dm^{-3} and manganese (Mg^{2+}) – 30 mg·dm^{-3}. The highest total iron concentration (0.36 mg·dm^{-3}) was registered in point 1, but its mean value was the lowest, reaching 0.10 mg·dm^{-3}. In other points total iron concentrations did not exceed 0.30 mg·dm^{-3}, with the mean value of 0.13 mg·dm^{-3}. Concentration of manganese (Mn^{2+}) fluctuated from 0.01 to 0.12 mg·dm^{-3} and an average value of the concentration along the whole watercourse length did not exceed 0.04 mg·dm^{-3} (Table 1).

Assessment of the water usability for drinking water supply revealed that among the 18 analyzed indices, only 12 have been included in

Table 1. Range and mean values of analyzed physicochemical indices and quality class, and usable value of the Sztoła river water

Index	Value minimum – maximum			Mean value			Water usability: For water supply [Rozporządzenie 2002b]			Water usability: As fish habitat [Rozporządzenie MŚ 2002a] salmonid			cyprinid		
	Measurement-control point														
	1	2	3	1	2	3	1	2	3	1	2	3	1	2	3
Temperature [°C]	7.3–12.5	7.7–14.9	7.6–13.9	9.7*	11.4	11.2	A1	A1	A1	yes			yes		
Reaction (pH)	6.6–8.3	7.2–8.3	7.5–8.3	7.3	7.9	7.2	A1	A1	A1	yes			yes		
Total suspended solids [mg·dm^{-3}]	0.1–4.2	0.2–3.6	0.4–4.1	1.1	2.3	2.8	A1	A1	A1	yes			yes		
Dissolved oxygen [mgO_2·dm^{-3}]	7.4–12.7	7.5–13.2	7.1–11.8	9.6	10.1	10.0	–	–	–	yes			yes		
Oxygen saturation degree [%]	91–122	89–133	90–128	98	101	100	A1	A1	A1	–			–		
Nitrites [mgNO_2^-·dm^{-3}]	0–0.02	0–0.01	0–0.01	0.003	0.001	0.001	–	–	–	no	yes		yes		
Nitrates [mgNO_3^-·dm^{-3}]	2.3–9.7	4.3–13.3	3.6–13.3	4.4	9.7	9.6	A1	A1	A1	–					
Nitrate nitrogen [mgN-NO_3^-·dm^{-3}]	0.5–2.2	1.0–3.0	0.8–3.0	1.0	2.2	2.2	–	–	–	–					
Ammonium nitrogen [mgN-NH_4^+·dm^{-3}]	Trace amounts			Trace amounts			A1	A1	A1	yes					
Phosphates [mgPO_4^{3-}·dm^{-3}]	Trace amounts			Trace amounts			A1	A1	A1	yes					
Electrolytic conductivity [μS·cm^{-1}]	380–582	512–560	528–550	487	542	541	A1	A1	A1	–					
Soluble solids [mg·dm^{-3}]	312–428	404–488	380–480	345	430	429	-	-	-	–					
Chlorides [mgCl^-·dm^{-3}]	4–22	17–25	6–25	9	19	18	A1	A1	A1	–					
Sulphates [mgSO_4^{2-}·dm^{-3}]	32–98	90–113	72–113	64	105	104	A1	A1	A1	–					
Magnesium [mgMg^{2+}·dm^{-3}]	28–30	22–28	21–26	29	24	24	–	–	–	–					
Calcium [mgCa^{2+}·dm^{-3}]	45–100	70–155	74–129	62	89	87	–	–	–	–					
Manganese [mgMn^{2+}·dm^{-3}]	0.01–0.12	0.02–0.07	0.02–0.09	0.03	0.04	0.04	A3	A2	A2	–					
Total iron [mg$Fe^{3+/2+}$·dm^{-3}]	0.02–0.36	0.04–0.26	0.02–0.30	0.10	0.13	0.13	A2	A1	A1	–					

* – I quality class [Rozporządzenie 2011a].

A1 – water requiring simple treatment.

A2 – water requiring typical physical and chemical treatment.

A3 – water requiring high performance physical and chemical treatment.

the Decree of the Minister of the Natural Environment [2002b]. A majority of the researched indices (the temperature, pH, electrolytic conductivity, degree of oxygen saturation, nitrates, ammonium nitrogen, phosphates, chlorides and sulphates) in all measurement-control points classified the Sztoła river water to category A1, i.e. water requiring simple physical treatment (Table 1). High concentration of total iron and manganese makes difficult the Sztoła river water treatment for consumption. The highest manganese concentration – 0.12 mg·dm^{-3} was registered at point 1. It caused that the Sztoła river water could be drawn for drinking water supply to people after a high performance physical and chemical treatment (category A3). At the other analyzed points, the river water would require a typical treatment (category A2).

Maximum total iron concentration (0.36 mg·dm^{-3}), like in case of manganese, occurred at point 1. Water of the Sztoła river at this point fulfilled the requirements for category A2 and would need only a typical physical and chemical treatment. At the other two points the water was in category A1. High concentrations of these metals in watercourses is typical for the surface waters of the Malopolska voivodeship. Their origin is natural and connected with the geological structure of the catchment, chemical composition of the rocks and soils in the catchment, and the river beds [Kanownik et al. 2013].

At measurement–control point 1, the Sztoła river waters did not meet the requirements for the salmonoid fish habitat due to exceeded admissible value of nitrites (0.01 mg·dm^{-3}) [Rozporządzenie MŚ 2002b]. On the other hand for cyprinids all analyzed indices meet the requirements for the freshwaters which are the habitats for these fish (Table 1).

Statistical analysis conducted using Mann-Whitney U test revealed that values of a majority of the tested indices differed significantly between points (Table 2).

Statistically lower values of pH, nitrates, electrolytic conductivity, soluble solids, chlorides, sulphates and calcium were registered, as well higher concentrations of magnesium in the upper course of the stream (point 1) in comparison with its lower course (points 2 and 3).

Table 2. Significance of differences of water indices among the individual measurement-control points – difference between medians and probability of Mann-Whitney U test

Point	Temperature			pH			Total suspended solids			Dissolved oxygen		
	1	2	3	1	2	3	1	2	3	1	2	3
1		0.207	0.208		**0.046**	**0.012**		0.267	0.245		0.753	0.875
2	1.6		0.793	**0.52***		0.752	0.8		0.875	1.13		0.916
3	1.4	-0.3		**0.61**	0.09		0.9	0.1		1.45	0.32	
	Saturation degree			Nitrites			Nitrates			EC		
	1	2	3	1	2	3	1	2	3	1	2	3
1		0.561	0.631		0.644	0.927		**0.009**	**0.009**		**0.036**	**0.049**
2	3		0.916	0		0.487	**7.0**		0.793	**71**		0.400
3	4	1		0	0		**6.8**	-0.2		**66**	-5	
	SS			Chlorides			Sulphates			Magnesium		
	1	2	3	1	2	3	1	2	3	1	2	3
1		**0.005**	**0.004**		**0.006**	**0.016**		**0.001**	**0.002**		**0.001**	**0.001**
2	**90**		0.875	**12.0**		1	**46**		0.916	**-5.6**		0.875
3	**98**	8		**12.5**	-0.5		**47**	1		**-5.8**	-0.2	
	Calcium			Manganese			Iron			EC – electrolytic conductivity SS – soluble solids		
	1	2	3	1	2	3	1	2	3			
1		**0.009**	**0.009**		0.338	0.203		0.167	0.269			
2	**23**		0.713	0		0.507	0.09		0.958			
3	**23**	0		0.02	0.02		0.08	-0.01				

Test probability (p)

Difference between medians

* Statistical values in bold mean statistically significant differences at $p < 0.05$.

CONCLUSIONS

1. On the basis of the analyzed physicochemical indices supporting biological elements the ecological state of water along the examined length of the Sztoła river was determined as very good.
2. The Sztoła river water may be used for drinking water supply for people, however, due to high concentrations of manganese and total iron, it must be subjected to an appropriate physical and chemical treatment.
3. In the headwater part of the Sztoła river, its water does not fulfill the requirements for the salmonoid fish habitat because of exceeded nitrite concentrations. Along the whole length of the analyzed watercourse, the cyprinid have proper habitat conditions.
4. Among the 15 analyzed physicochemical water indices, values of 7 were statistically higher in the upper course of the river due to a higher anthropopressure, mainly due to the mining industry. However, it did not affect worsening of the Sztoła river water.

REFERENCES

1. Alexander R.B., Elliott A.H., Shankar U., McBride G.B. 2002. Estimating the sources and transport of nutrients in the Waikato River Basin, New Zealand. Water Resour. Res., 38, 1268–1290.
2. Cooper A.B., Thomsen C.E. 1988. Nitrogen and phosphorus in streamwaters from adjacent pasture, pine and native forest catchments. N.Z. J. Mar. Freshwater Res., 22, 279–291.
3. Di H.J., Cameron K.C. 2005. Reducing environmental impacts of agriculture by using a fine particle suspension nitrification inhibitor to decrease nitrate leaching from grazed pastures. Agric. Eco. Environ., 109, 202–212.
4. Ferraz S.F.B., Lima W.P., Rodrigues C.B. 2013. Managing forest plantation landscapes for water conservation. Forest Ecology and Management, 301, 58–66.
5. Grajewski S., Krysztofiak A., Miler A.T. 2009. Potential water retention capacity of forest areas – an example of the Experimental Forest Range Unieszów. Journal of Water and Land Development, 13a, 71–84.
6. Grzywna A. 2014. Chemiczne wskaźniki jakości wody w zlewni Lasów Parczewskich. Inżynieria Ekologiczna, 36, 120–127.
7. Kanownik W., Kowalik T., Bogdał A., Ostrowski K. 2013. Quality categories of stream water included in a Small Retention Program. Pol. J. Environ. Stud., 22 (1), 159–165.
8. Kanownik W., Pijanowski Z. 2007. A load of selected chemical components carried away by the outflow from mountain catchments. Polish J. of Environ. Stud., 16 (2A), 297–303.
9. Lai Y.C., Tu Y.T., Yang C.P., Surampalli R.Y., Kao C.M. 2013. Development of a water quality modeling system for river pollution index and suspended solid loading evaluation. Journal of Hydrology, 478, 89–101.
10. McDowell R.W., Wilcock R.J. 2004. Particulate phosphorus transport within stream flow of an agricultural catchment. J. Environ. Qual., 33, 2111–2121.
11. Monaghan R.M., Paton R.J., Smith L.C., Drewry J.J., Littlejohn R.P. 2005. The impacts of nitrogen fertilisation and increased stocking rate on pasture yield, soil physical condition and nutrient losses in drainage from a cattle-grazed pasture. N. Z. J. Agric. Res., 48, 227–240.
12. Pierzgalski E., Tyszka J., Stolarek A. 2005. Water outflow during drought years in watersheds with various forest cover. Ann. of Warsaw Agricultural University, 36, 21–28.
13. Piirainen S., Domisch T., Moilanen M., Nieminen M. 2013. Long-term effects of ash fertilization on runoff water quality from drained peatland forests. Forest Ecology and Management, 287, 53–66.
14. Policht-Latawiec A., Kanownik W., Łukasik D. 2013. Wpływ zanieczyszczeń punktowych na jakość wody rzeki San. Infrastruktura i Ekologia Terenów Wiejskich, 4 (1), 253–269.
15. Rodgers M., O'Connor M., Healy M.G., O'Driscoll C., Asam Z.-ul-Z., Nieminen M., Poole R., Müller M., Xiao L. 2010. Phosphorus release from forest harvesting on an upland blanket peat catchment. Forest Ecology and Management, 260, 2241–2248.
16. Rozporządzenie Ministra Środowiska z dnia 4 października 2002 roku w sprawie wymagań, jakim powinny odpowiadać wody śródlądowe będące środowiskiem życia ryb w warunkach naturalnych (Dz. U. 2002 Nr 176, poz. 1455). [2002a]
17. Rozporządzenie Ministra Środowiska z dnia 27 listopada 2002 r. w sprawie wymagań, jakim powinny odpowiadać wody powierzchniowe wykorzystywane do zaopatrzenia ludności w wodę przeznaczoną do spożycia (Dz. U. 2002 Nr 204, poz. 1728). [2002b]
18. Rozporządzenie Ministra Środowiska z dnia 9 listopada 2011 roku w sprawie sposobu klasyfikacji stanu jednolitych części wód powierzchniowych oraz środowiskowych norm jakości dla substancji priorytetowych (Dz. U. Nr 257, poz. 1545). [2011a]

19. Rozporządzenie Ministra Środowiska z dnia 15 listopada 2011 roku w sprawie form i sposobu prowadzenia monitoringu jednolitych części wód powierzchniowych i podziemnych (Dz. U. Nr 258, poz. 1550 z późn. zm.). [2011b]
20. Wang Y., Li Y., Liu F., Li Y., Song L., Li H., Meng C., Wu J. 2014. Linking rice agriculture to nutrient chemical composition, concentration and mass flux in catchment streams in subtropical central China. Agriculture, Ecosystems & Environment, 184, 9–20.
21. Wilcock R.J., Monaghan R.M., Quinn J.M., Campbell A.M., Duncan M.J., Thorrold B.S., McGowan A.W., Betteridge K. 2006. Land use impacts and water quality targets in the intensive dairying catchment of the Toenepi Stream, New Zealand. N. Z. J. Mar. Freshwater Res., 40, 123–140.
22. Xiao J., Jin Z., Wang J. 2014. Geochemistry of trace elements and water quality assessment of natural water within the Tarim River Basin in the extreme arid region, NW China. Journal of Geochemical Exploration, 136, 118–126.

33

EFFECT OF THE APPLICATION OF SEWAGE SLUDGE COMPOST ON THE CONTENT AND LEACHING OF ZINC AND COPPER FROM SOILS UNDER AGRICULTURAL USE

Teresa Bowszys[1], Jadwiga Wierzbowska[1], Paweł Sternik[1], Marta Karolina Busse[1]

[1] Department of Agricultural Chemistry and Environmental Protection, University of Warmia and Mazury in Olsztyn, Oczapowskiego 8, 10-744 Olsztyn, Poland, e-mail: teresa.bowszys@uwm.edu.pl

ABSTRACT

Municipal sewage sludge can be used in agriculture provided that the permissible levels of heavy metals are not exceeded in either the sewage sludge or in the top layer of soil to be amended by this substance, and that its application does not deteriorate the soil quality. The purpose of this study was to determine the effects of different forms of sewage sludge on the content of Cu and Zn in soil and in soil leachate. The study comprised 2 rotations (potato, spring barley, winter oilseed rape, winter wheat). Each series was composed of the following treatments: NPK, FYM, municipal sewage sludge composted with straw, dried and granulated municipal sewage. FYM and composted sewage were applied once in the rotation (under potato) in a dose of 10 Mg d.m. ha^{-1} and twice (under potato and under winter oilseed rape) in a dose of 5 Mg d.m. ha^{-1}. In the other years (under spring barley and winter wheat), soil received only mineral fertilization. In order to evaluate the effect of the composts on the leaching of Cu and Zn from soil, a lysimetric experiment was conducted under controlled conditions. Before the experiment, the soil content of Cu was low (1.47 mg kg^{-1}) and Zn was medium (10.11 mg kg^{-1}). The content of copper in the composts ranged from 4.5 to 340.1 mg kg^{-1} d.m. and that of zinc was from 109.5 to 1310.1 mg kg^{-1} d.m. The composted sewage sludge significantly raised the soil content of available forms of Cu and Zn, but did not change the soil nutrient abundance class. Fertilization modified the content of the microelements in the soil leachate.

Keywords: sewage sludge, composts, zinc, copper, soil leachate.

INTRODUCTION

With the continually growing amounts of municipal sewage sludge, the problem of its recycling becomes more and more important. Sewage sludge is a noxious but unavoidable by-product of sewage and wastewater treatment. According to the Main Statistial Office (GUS), the total amount of sewage sludge generated in Poland in 2012 was 533.3 thousand tons of d.m., of which 115 thousand tons were then used in agriculture [GUS 2013]. Apart from farming, sewage sludge can also be used to rehabilitate degraded land (50.3 thousand tons of dry matter) and to cultivate plants grown for compost (33.3 thousand tons of d.m.). One of the most rational ways to utilize this waste is their use in composted forms for agricultural purposes. Raw and composted sewage sludge is a valuable source of organic matter and nutrients, hence its use as an unconventional soil amending substance [Sądej et al. 2007, Bowszys et al. 2009a]. However, such products must satisfy a number of safety standards to avoid environmental pollution. As the civilization advances, the amount of waste increases and its chemical composition changes. Depending on its place of origin, waste may contain large amounts of harmful substances, mainly heavy metals, PAHs, PCBs, microbial and parasitic contaminants [Iżewska 2007, Milinovic et al. 2014]. To gain deep and thorough knowledge of the ecological consequences of the incoproration of composted waste to soil, it is necessary to complete a broad range of basic and experimental studies, which will enable us to work out guidelines for rational and safe application of composts.

The purpose of this study was to determine the influence of different forms of sewage sludge on the content of Cu and Zn in soil and in soil leachate.

MATERIAL AND METHODS

The experiments comprised 2 rotation cycles: potato, spring barley, winter oilseed rape, winter wheat, run in the years 2004–2007 and 2008–2011. The following treatments were tested: NPK; FYM10, 2×5 Mg d.m.·ha^{-1}; municipal sewage sludge composted with straw 10, 2×5 Mg d.m.·ha^{-1}; sewage sludge composted alone 10, 2×5 Mg d.m.·ha^{-1}; dried and granulated sewage sludge 10, 2×5 Mg d.m.·ha^{-1}. FYM and sewage sludge composts used in a dose of 10 Mg d.m.·ha^{-1} were applied once in a rotation cycle, under potato (in 2004 and 2008), while the doses of 5 Mg d.m.·ha^{-1} were given twice, under potato (2004, 2008) and under winter oilseed rape (2006, 2010). In other years, the crops were nourished only with mineral fertilizers: spring barley in 2005 and 2009, and winter wheat in 2007 and 2011. On the plots with FYM and organic substances, nitrogen was balanced depending on the N-total content in the soil. The effect of composts on the potential volume of leached Cu and Zn was assessed in a model lysimetric experiment, carried out under controlled conditions. Soil samples from each rotation were collected after wheat harvest, from three layers of the soil horizon: 0–30 cm, 31–60 cm and 61–90 cm. The total amount of water used for leaching corresponded to the average annual amount of atmospheric precipitation in the Province of Warmia and Mazury (605 mm).

The experiment was set up on proper brown podzolic soil developed from light boulder clay, which belonged to good wheat complex (2A gp. gc:gl) [*Systematyka* ... 2011]. Before the experiment, the soil was low in available (in 1 mol HCl·dm^{-3}) forms of copper (1.47 mg·kg^{-1}) and moderately high in zinc (10.11 mg·kg^{-1}). The content of Cu in composts ranged from 4.5 to 340.1 mg·kg^{-1} d.m., Zn from 109.5 to 1310.1 mg·kg^{-1} d.m., and its pH was 5.04 to 1 mol KCl·dm^{-3}.

Methods of chemical analyses: Zn and Cu in soil after extraction in 1 mol HCl·dm^{-3} and in leachate were determined with the atomic absorbance spectrophotometric (AAS) method.

The results of chemical analyses were processed statistically with a software package Statistica 10°. The significance of differences between the data was verified by the Tukey's HSD (Honestly Signifficant Differences) test at the level of significance α=0.05.

RESULTS AND DISCUSSION

The content of zinc in soil in the subsequent years of the research was similar and ranged on average from 6.41 to 15.14 mg·kg^{-1} (Tables 1, 2).

In both rotations, the highest concentration of zinc in soil was found after the harvest of winter oilseed rape (15.14 mg·kg^{-1} and 13.59 mg·kg^{-1}), which had been treated with 5 Mg d.m.·ha^{-1} of FYM and composts. In the first rotation, soil fertilized with dried and granulated sewage sludge was significantly the most abundant of Zn among all the treatments where organic materials had been applied. It was also found out that after each four years of the research, compared to the NPK and FYM fertilization regimes, significatly more Zn was found in soil fertilized with composted sewage sludge (with and without straw) and with dried and graulated sewage sludge. In turn, the frequency of soil application of the examined materials, i.e. once or twice in a rotation cycle, did not have any significant effect on the zinc content.

In the first rotation, the content of copper in soil was differentiated by the year and type of applied fertilizers and composts, ranging from 1.30 to 2.94 mg·kg^{-1} (Table 3).

In the second rotation cycle, however, there were much fewer significant dependences between the experimental factors, and the cotent of copper ranged from 1.38 to 2.76 mg·kg^{-1} (Table 4). In the first year of the experiment, the average content of copper in soil was 1.51 mg Cu·kg^{-1}, increasing significantly up to 1.98 mg·kg^{-1} in the fourth year of the cycle. In 2008 (1st year of the second cycle), the soil already contained an average of 1.65 mg Cu·kg^{-1}, and a significant increase relative to this value was not achieved until four years afterwards (2.27 mg·kg^{-1}). After the first four years of the research, significantly more Cu – relative to its content in soil fertilized with FYM – was found in soil enriched with sewage sludge composts (with straw +4% and without straw +16%) and with dried and granulated

Table 1. Content of available forms of zinc in soil in the first rotation cycle (mg Zn·kg^{-1})

Year	Frequency of application	NPK	FYM	Compost from sewage sludge and straw	Compost from sewage sludge	Dried and granulated sewage sludge	Mean
2004	a	11.21	8.53	10.07	12.52	15.84	11.63
	b		9.40	10.48	10.50	11.60	10.64
	mean		8.97	10.28	11.51	13.72	11.14
2005	a	4.76	5.56	7.03	8.13	7.88	6.67
	b		5.77	6.45	6.81	6.98	6.16
	mean		5.67	6.74	7.47	7.43	6.41
2006	a	9.01	11.06	19.41	11.16	20.49	14.23
	b		11.50	14.18	18.56	27.04	16.06
	mean		11.28	16.80	14.86	23.77	15.14
2007	a	5.55	8.90	7.70	9.53	15.06	9.35
	b		6.51	8.38	9.66	24.66	10.95
	mean		7.71	8.04	9.59	19.86	10.15
Mean	a	7.63	8.51	11.05	10.33	14.82	10.47
	b		8.30	9.88	11.38	17.57	10.95
	mean		8.41	10.46	10.86	16.19	–

HSD P<0.05
year of experiment 1.21
fertilizer 1.49
frequency of application n.s.
year*fertilizer 3.84
year* frequency of application 1.72
fertilizer* frequency of application 2.11
year*fertilizer* frequency of application 5.43

a – once in a rotation; b – twice in a rotation

Table 2. Content of available Zn forms in soil in the second rotation cycle (mg Zn·kg^{-1})

Year	Frequency of application	NPK	FYM	Compost from sewage sludge and straw	Compost from sewage sludge	Dried and granulated sewage sludge	Mean
2008	a	9.76	12.10	12.83	11.96	17.30	12.79
	b		10.57	15.91	17.33	12.45	13.21
	mean		11.33	14.37	14.65	14.88	13.00
2009	a	7.02	8.79	11.62	9.73	11.52	9.74
	b		10.80	9.48	11.08	10.29	9.73
	mean		9.80	10.55	10.40	10.91	9.74
2010	a	12.60	10.36	13.03	14.69	18.72	13.88
	b		10.06	13.22	14.59	16.03	13.30
	mean		10.21	13.13	14.64	17.38	13.59
2011	a	12.62	8.40	13.10	13.09	14.27	12.30
	b		10.89	10.56	10.26	12.41	11.35
	mean		9.65	11.83	11.68	13.34	11.82
Mean	a	10.50	9.91	12.65	12.37	15.46	12.18
	b		10.58	12.29	13.32	12.80	11.90
	mean		10.25	12.47	12.84	14.13	–

HSD P<0.05
year of experiment 1.52
fertilizer 1.87
frequency of application n.s.
year*fertilizer 4.80
year* frequency of application n.s.
fertilizer* frequency of application n.s.
year*fertilizer* frequency of application n.s.

a – once in a rotation; b – twice in a rotation

Table 3. Content of available Cu forms in soil in the first rotation cycle (mg Cu·kg^{-1})

Year	Frequency of application	NPK	FYM	Compost from sewage sludge and straw	Compost from sewage sludge	Dried and granulated sewage sludge	Mean
2004	a	1.49	1.43	1.67	1.49	1.61	1.54
	b		1.17	1.15	1.67	1.97	1.49
	mean		1.30	1.41	1.58	1.79	1.51
2005	a	1.48	1.53	1.72	1.79	1.83	1.67
	b		1.52	1.46	1.70	1.75	1.58
	mean		1.53	1.59	1.74	1.79	1.63
2006	a	1.33	1.56	1.63	1.74	1.95	1.64
	b		1.55	1.58	1.83	1.92	1.64
	mean		1.55	1.60	1.79	1.93	1.64
2007	a	1.53	1.87	1.77	1.87	2.21	1.85
	b		1.61	1.72	2.06	3.67	2.12
	mean		1.74	1.74	1.96	2.94	1.98
Mean	a	1.46	1.60	1.70	1.72	1.90	1.67
	b		1.46	1.48	1.82	2.33	1.71
	mean		1.53	1.59	1.77	2.11	–

HSD P<0.05
year of experiment 0.11
fertilizer 0.13
frequency of application n.s.
year*fertilizer 0.34
year* frequency of application 0.15
feartilizer* frequency of application 0.18
year*fertilizer* frequency of application 0.47

a – once in a rotation; b – twice in a rotation

Table 4. Content of available Cu forms in soil in the second rotation cycle (mg Cu·kg^{-1})

Year	Frequency of application	NPK	FYM	Compost from sewage sludge and straw	Compost from sewage sludge	Dried and granulated sewage sludge	Mean
2008	a	1.46	1.39	1.73	1.76	1.96	1.66
	b		1.37	1.57	1.91	1.93	1.65
	mean		1.38	1.65	1.83	1.94	1.65
2009	a	1.55	1.55	1.93	2.03	2.30	1.87
	b		1.45	1.73	2.21	2.16	1.82
	mean		1.50	1.83	2.12	2.23	1.85
2010	a	1.60	1.47	1.89	1.70	2.11	1.75
	b		1.28	1.50	1.36	3.42	1.83
	mean		1.37	1.70	1.53	2.76	1.79
2011	a	2.33	1.82	2.50	2.25	2.53	2.29
	b		2.33	2.21	2.05	2.38	2.26
	mean		2.08	2.36	2.15	2.45	2.27
Mean	a	1.74	1.56	2.01	1.94	2.22	1.89
	b		1.61	1.75	1.88	2.47	1.89
	mean		1.58	1.88	1.91	2.35	–

HSD P<0.05
year of experiment 0.38
fertilizer 0.47
frequency of application n.s.
year*fertilier n.s.
year* frequency of application n.s.
fertilizer* frequency of application n.s.
year*fertilizer* frequency of application n.s.

a – once in a rotation; b – twice in a rotation

sewage sludge (+38%). In the second rotation cycle, a significat increase in the Cu content in soil (+49%) was found only in response to dried compost. The frequency of application did not have any significant effect on the modification of the soil abundance of copper. In both rotations, a positive residual effect on this characteristic was demonstrated for the applied NPK fertilizers, FYM and composts.

During the treatment of wastewater and sewage, heavy metals are removed and accumulated in the sewage sludge. Their total content does not mirror the potential threat which they pose to the environment. Metals bound with aluminum silicates, metal sulphates and strong metal-organic bonds are immobile forms and do not create toxicological risk. Ion-exchangable forms, oxides and cabonates can permeate into soil and water environments, where they can be a source of nutrients for plants. However, in high concentrations, they can threaten plants as well as the soil and water environment [Gawdzik 2012]. Results of numerous investigations suggest some increase in the content of heavy metals observed in soils fertilized with sewage sludge, but it is small enough to leave the soil pollution class assignment as zero pollution soils [Bowszys et al. 2009b, Greinert et al. 2009, Niedźwiecki et al. 2009]. According to Sienkiewicz and Czarnecka [2012], in alkaline soils, a rise in the content of available forms of Cu, Zn and Mn in a soil fertilized with large doses (up to 280 t·ha^{-1}) did not threaten the environment, but just improved the nutrition of plants with these microelements. In another study, Sienkiewicz et al. [2009] showed that the content of available forms of copper, zinc and manganese in soil regularly fertilized with FYM was much higher than in soil with sole mineral fertilization. Gondek [2010] determined that fertilization with FYM and sewage sludge in the first year did not cause significant mobilization of mobile forms of zinc in soil. However, as a result of the mineralization of organic materials and progressing soil alkalization, in the second and third year of the experiment, the content of mobile zinc forms increased, although to a lesser extent than in response to mineral fertilization.

Trace minerals, if present in sewage sludge in high concentrations, can be a threat to the soil and water environment [Witczak, Adamczyk 1995, Rozporządzenie... 2008]. The volume of microelements lost by leaching is highly varied. The application of fertilizer components in doses highly exceeding the crops' nutritional demands may lead to changes in the ionic balance of the soil solution and cause the transfer of nutrients to groundwater [Gondek 2009]. A single application of even a large dose of sewage sludge will not trigger a distinct increase in leaching heavy metals from soil relative to FYM or mineral NPK fertilization. On the other hand, due to the positive balance of these elements in soil, long-term application of sewage sludge could be problematic, especially when soil acidity is raised and, consequently, the mobility of elements as well as their leachability are higher [Sevel et al. 2014]. Milinovic et al. [2014] claim that the drying of sewage sludge prior to its application to soil in general reduces the leaching of zinc but increases copper loss by leaching from the sludge. According to Page et al. [2014], the soil reaction cannot be treated as an exclusive indicator in the assessment of the mobility of heavy metals in soil, their phytoavailability and the risk of their migration

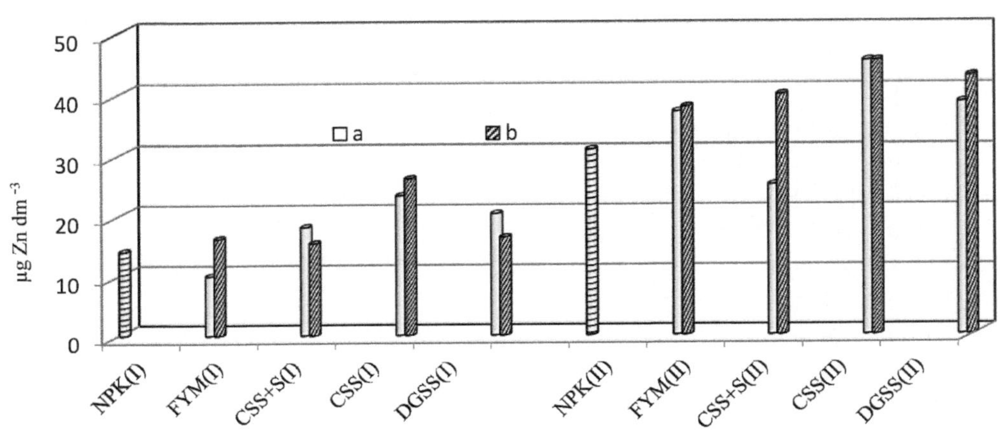

Figure 1. Content of zinc in leachate (css+s – compost from sewage sludge and straw; css – compost from sewage sludge; dgss – dried and granulated sewage sludge; (I) – first rotation; (II) – second rotation; a – once in a rotation b – twice in a rotation)

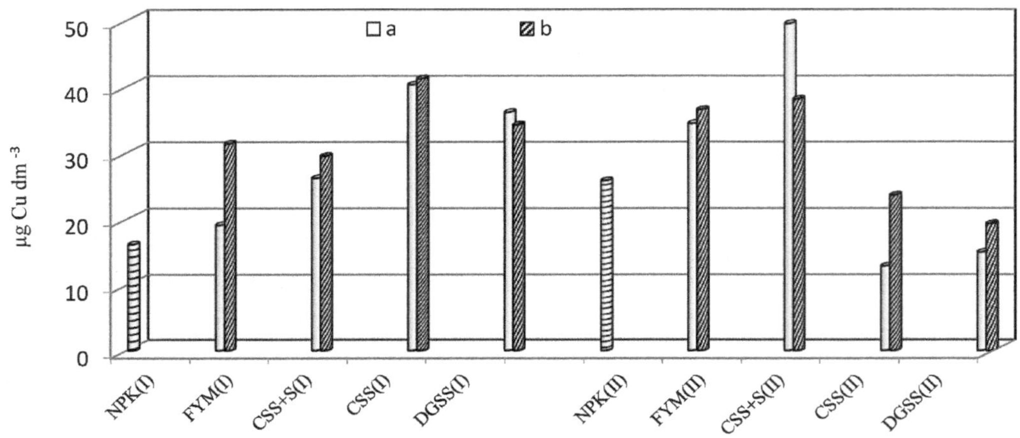

Figure 2. Content of copper in leachate (css+s – compost from sewage sludge and straw; css – compost from sewage sludge; dgss – dried and granulated sewage sludge; (I) – first rotation; (II) – second rotation; a – once in a rotation b – twice in a rotation)

to water environment. Kopeć et al. [1991] demonstrated that the use of organic fertilization reduced the transfer of microelements deep into the soil profile.

The content of zinc in leachate from soils fertilized with NPK, FYM, composts and dried sewage sludge was significantly varied depending on the rotation (Figure 1). After the first rotation, the amount of zinc in leachate ranged from 10.0 to 26.1μg·dm^{-3}, and after the second one was twice as high (from 25.0 to 45.4 μg·dm^{-3}). The highest loss of zinc by leaching in the first and second rotation (on average 24.7 and 45.4 μg·dm^{-3}) was found in the soil leachate from the treatment fertilized by compost from sewage sludge (containing 1310 mg Zn·kg d.m.). Fertilization of soil with compost from sewage sludge and straw or with dried and granulated sewage sludge modified the content of zinc in leachate differently depending on the frequency of application. In both rotations, the leaching of Zn from soil was higher following the application of composts, which were 3- to 7-fold richer in this element than FYM.

The hydrogeochemical background characteristic for the groudwaters in Poland varies from 0.01 to 0.5 mg Cu·dm^{-3} [Rozporządzenie… 2008]. Among the organic materials used for soil fertilization, sewage sludge composted without straw contained most copper (340 mg·kg^{-1}s.m.). This explains why the leachate collected after the first rotation from this treatment also had the highest copper content (on average 40.7 μg·dm^{-3}) (Figure 2). Less copper was leached after the appliction of dried sewage sludge (on average 35.1 μg·dm^{-3}) and compost with straw (on average 27.8 μg·dm^{-3}). The least copper was lost by leaching from soil with mineral fertilization alone (16.2 μg dm^{-3}). In the second rotation, the content of copper in leachate from soil fertilized by sewage sludge composted without any additives was the lowest (above 17 μg·dm^{-3}), and the highest one was determined after the application of compost with straw (on average 43.8 μg·dm^{-3}).

CONCLUSIONS

1. Application of sewage sludge composted without straw one or twice during a rotation cycle to fertilize soil significantly increases the soil content of available forms of zinc and copper, but retains the same soil abundance class (medium for Zn and low for Cu)
2. The content of zinc and copper in soil leachate varies depending on the rotation and applied fertilization. Most copper in both rotations (24.7 and 45.4 μg·dm^{-3}) is leached from soil fertilized with composted sewage sludge. In the first rotation, most copper (40.7 μg·dm^{-3}) was also leached from the soil fertilized with this compost, but in the second rotation the highest loss of copper by leaching occurred from soil fertilized with sewage sludge composted with straw (43.8 μg·dm^{-3}).

REFERENCES

1. Bowszys T., Wierzbowska J., Bowszys J. 2009a. Content and removal of Cu and Zn with harvested crops grown on soil fertilized with composted municipal sewage sludge. J. Elem. 14(1), 23–32.

2. Bowszys T., Wierzbowska J., Bowszys J., Bieniek A. 2009b. Modifications in the content of available zinc and copper in soil fertilized with bio-waste compost. J. Elementol., 14(1), 33–42
3. Gawdzik J., 2012. Mobilność metali ciężkich w osadach ściekowych na przykładzie wybranej oczyszczalni ścieków. Inżynieria i Ochrona Środowiska, 15, 1, 5–15.
4. Gondek K. 2009. Wpływ nawożenia na zawartość mobilnych form wybranych mikroelementów w glebie oraz ich wymycie w doświadczeniu wazonowym. Acta Agroph. 13(1), 89–101.
5. Gondek K. 2010. Zinc and cadmium accumulation in maize (*Zea mays*L.) and the concentration mobile forms of these metals in soil after application of farmyard FYM and sewage sludge. J. Elementol., 15(4), 639–652.
6. Greinert A., Drab M., Węclewski S. 2009. Ryzyko chemicznego zanieczyszczenia gleb nawożonych osadami ściekowymi. Zesz. Prob. Post. Nauk Rol. 537, 135–144.
7. Iżewska A. 2007. Wpływ nawożenia obornikiem, osadem ściekowym i kompostem i osadów ściekowych na właściwości gleby. Zesz. Probl. Post. Nauk Rol. 518, 85–92.
8. Kopeć S., Nowak K., Smoroń S. 1991. Straty składników nawozowych przez wymywanie w zależności od nawożenia i uprawianej rośliny. Roczn. Glebozn. XLII, 3/4, 109–114.
9. Milinovic J., Vidal M., Lacorte S., Rigol A. 2014. Leaching of heavy metals and alkylphenolic compounds from fresh and dried sewage sludge. Environ Sci Pollut Res, 21, 2009–2017.
10. Niedźwiecki E., Wojcieszczuk T., Meller E., Fedorowicz J. 2009. Oddziaływanie kompostów wytworzonych w zakładzie utylizacji odpadów w Gorzowie Wielkopolskim na skład chemiczny gleb. Zesz. Probl. Post. Nauk Rol. 537, 249–254.
11. Page K., Harbottle M.J., Cleall P.J., Hutchings T.R. 2014. Heavy metal leaching and environmental risk from the use of compost-like output as an energy crop growth substrate. Science of the Total Environment 487, 260–271.
12. Rocznik Statystyczny 2013. GUS, Ochrona środowiska, http://stat.gov.pl/.
13. Rozporządzenie Ministra Środowiska z dnia 23 lipca 2008 r. w sprawie kryteriów i sposobów oceny wód podziemnych. Dz.U z 2008 r. Nr 143, poz. 896.
14. Sądej W., Bowszys T., Wierzbowska J. 2007. Kształtowanie właściwości fizykochemicznych gleby płowej nawożonej osadami ściekowymi. Zesz. Probl. Post. Nauk Rol., 520, 363-369.
15. Sevel L., Ingerslev M., Nord-Larsen T., Jørgensen U., Holm P. E., Schelde K. 2014. Fertilization of SRC Willow, II: Leaching and Element Balances, Bioenerg. Res., 7, 338–352
16. Sienkiewicz S., Czarnecka M. 2012. Content of available C, Zn and Mn in soil amended with municipal sewage sludge. J. Elem. 17(4), 649–657.
17. Sienkiewicz S., Wojnowska T., Krzebietke S., Wierzbowska J., Żarczyński P. 2009. Content of available forms of some micronutrients in soil after long-term differentiated fertilization. J. Elem. 14(4), 787–794.
18. Systematyka gleb Polski. 2011. Rocz. Glebozn. LXII (3), 1–193.
19. Wieczorek J., Frączek K. 2013. Assessment of possibility of agricultural use of municipal sewage sludge generated in selected small sewage treatment plants from Kraków district. J. Ecol. Eng. 14, 2, 36–42.
20. Witczak S., Adamczyk A. 1995. Katalog wybranych fizycznych i chemicznych wskaźników zanieczyszczenia wód podziemnych i metod ich oznaczania. T. 2, Biblioteka Monitoringu Środowiska PIOŚ, Warszawa.

CHEMICAL COMPOSITION OF EDIBLE POTATO TUBERS IN RETAIL OUTLETS IN EAST-CENTRAL POLAND

Krystyna Zarzecka[1], Marek Gugała[1], Iwona Mystkowska[1], Magdalena Zarzecka[2]

[1] Chair of Plant Cultivation, University of Natural Sciences and Humanities in Siedlce, B. Prusa 14, 08-110 Siedlce, Poland, e-mail: kzarzecka@uph.edu.pl; gugala@uph.edu.pl

[2] Medical University of Białystok, J. Kilinskiego 1, 15-089 Białystok, Poland

ABSTRACT
The work was an attempt to determine and describe certain quality characteristics of edible potato on sale in Siedlce and Międzyrzec Podlaski. Potato tubers for examination were purchased in 3 kinds of retail outlets (ten shops of each kind) in east-central Poland. A total of 90 samples were collected to determine tuber size, protein and vitamin C content, reducing sugars and sucrose content. Potatoes purchased in all kinds of outlets met the standards set for tuber size. Potatoes bought in supermarkets had a better chemical composition than tubers purchased in groceries or fruit and vegetable shops. All the potato tubers were a good staple.

Keywords: potato, vitamin C, reducing sugars, sucrose, total protein.

INTRODUCTION

Potato is the major crop plant which, alongside wheat, maize and rice, is the staple food worldwide [Zgórska 2013]. Although the area of potato grown has been declining, Poland remains a significant potato producer in Europe. The country is a leader of potato production in the European Union surpassed by Germany only [Rynek ziemniaka 2013].

Appropriate nutrition, that is supplying sufficient levels of energy, nutrients and minerals, is a precondition determining the right functioning of the organism. Potato is one of major food crops and recently there has been an increasing interest in potato-based diets due to the crop's nutritional and dietetic aspects as well as medicinal use [Jansky 2010; Jansen et al. 2001; Stypa, Zgórska 2010].

Potato is a central and distinctive element of everyday diet in Poland due to its high nutritional and culinary value. Moreover, it is available all year round and consumed in large quantities. Per capita potato consumption in recent years has ranged from 107 to 130 kg, being 107 kg per year in the 2012/2013 season [Rynek ziemniaka 2013]. It should also be emphasised that a wide range of cultivars has been available with as many as 129 cultivars being registered with the National List in 2011, including 101 edible varieties [Lista opisowa odmian 2013].

Potato nutritional value is determined by tuber content of chemical components which are important for human nutrition (starch, protein, vitamins, total sugars, reducing sugars and minerals) as well as low concentration or lack of toxic compounds (glycoalkaloids, nitrates, heavy metals, pesticides) [Kolasa 1993; Lisińska 2006; Love, Pavek 2008]. Protein is one of major potato tuber components which, although not abundant, is a key element determining tubers' nutritional value due to the so-called 'pure protein' (making up 35–65% nitrates), which contains all the exogenous amino acids [Leszczyński 2012; Tajner-Czopek 2006]. The biological value of potato is higher than that of the majority of plant proteins, for example pea, wheat and rice proteins [Bártová et al. 2009; Öztürk et al. 2010; Pęksa 2003]. Ascorbic acid and dehydroascorbic acid, which are vitamin C constituents, are other potato tuber components. Vitamin C content is very much different and is mainly affected by potato cultivar and environmental conditions [Leszczyński 2012; Love, Pavek 2008; Tajner-Czopek 2006]. What is more, vitamin C is the strongest water-soluble antioxidant protecting the organism against cancer, particularly stomach and oesophagus cancer [Czerwiecki 2009; Grajek 2004; Kris-Etherton et al. 2002]. Sugars are yet another potato components affecting tuber suitability for cooking and processing. They include total sugars, reducing sugars and sucrose. Increased concentrations of these compounds result in edible tubers having a palpable, sweeter taste and their flesh blackening more rapidly, which is a very undesirable process particularly in tubers destined for chips and crisps [Sawicka, Pszczółkowski 2005; Tajner-Czopek 2006; Zgórska, Grudzińska 2012].

As healthy and "convenient" food and consumption of unprocessed potato are increasingly popular with consumers, more and more studies of edible tuber quality are conducted. Furthermore, a change in the way and place of purchasing edible tubers has been recently observed, whether for immediate consumption or for storage throughout autumn and winter months. Instead of purchases made on markets or directly from farmers, potatoes are more and more frequently bought in groceries, fruit and vegetable shops or supermarkets [Chotkowski, Rembeza 2005]. This, coupled with the fact that more and more supermarkets offer a wide range of cultivars, necessitates the studies to examine whether appealing appearance of the product is accompanied with internal quality traits meeting the specifications set for edible potato.

MATERIALS AND METHODS

Samples of edible potato tubers were purchased in three kinds of outlets, all of them operating either in Siedlce or Międzyrzec Podlaski, east-central Poland: supermarkets, grocery shops and fruit and vegetable shops. Three packets of potato tubers (three replicates during laboratory analyses), each weighing 2.0–2.5 kg, were purchased in 10 shops representing each kind of outlet (Table 1). A total of 90 samples were analysed. Tuber size and share (by weight) in the overall sample weight was determined following the Regulation of the Minister of Agriculture and Rural Development (MAiRD) [Rozporządzenie 2003]. Percentage of tubers with diameter above 50 mm was determined, too. Chemical analyses were conducted on dry (total nitrogen) and fresh material (vitamin C, sugars) in three replicates. Total protein content was calculated based on total nitrogen content using the conversion value of 6.25. Total nitrogen was determined using the Kjeldahl method on a 2300 Kjeltec Analizer Unit, and vitamin C content was analysed following the Pijanowski method. Reducing sugars (glucose + fructose) and total sugars (glucose + fructose + sucrose) were determined by means of the Luff-Schoorl method, sucrose content being calculated as total sugars after hydrolysis minus reducing sugars ×0.95 [Ostrowska et al. 1991]. The results were statistically analysed by one-way variance analysis at the significance level of 0.05.

RESULTS AND DISCUSSION

The size of all the tubers collected from supermarkets, groceries and fruit and vegetable shops met the specifications set in the Regulation of MAiRD [Rozporządzenie 2003]. Their diameter was over 35 mm (Table 2). The percentage of tubers with the diameter of over 50 mm was the highest in samples taken in supermarkets (75.6% on average), which probably resulted from the applied technologies in potato cultivation and the presence of foreign cultivars.

According to Lisińska [2006], protein content in fresh matter ranged from 1.7 to 2.3% whereas Jansen et al. [2001], who examined 466 cultivars, reported the range of 4.5 to 13.6% d.m., the average concentration amounting to 8.12%. In the present study, protein content fell within the range of 11.57–14.08% d.m. The highest av-

Table 1. Methodological data

Kind of outlet	Place of purchase	Potato cultivars
Supermarket	Biedronka (3 shops), Stokrotka (2), Carefour (1), Tesco (1), Topaz (2), Lidl (1)	Melody (1), Augusta (1), Milva (1), Satina (2), Asterix (1), Sante (2)*
Grocery shop	Delikatesy (4 shops), PSS Społem (3), Carlos (2), Mini Market (1)	Irga (8), Lord (1), Tajfun (1)
Fruit and vegetable shop	Small local shops (10)	Irga (6), Lord (2), Vineta (2)

* Two cultivars were not labelled.

Table 2. Size of potato tubers and their total protein and vitamin C contents

Kind of outlet	Percentage share of tubers, diameter		Total protein content (% dry matter)		Vitamin C content (mg·kg^{-1} fresh matter)	
	>35 mm	>50 mm	mean	range min-max	mean	range min-max
Supermarket	100.0	75.6	13.82a	13.28–14.07	214.4a	201–221
Grocery shop	100.0	37.7	13.76a	13.44–14.08	205.4b	198–209
Fruit and vegetable shop	100.0	51.6	12.07b	11.57–12.59	206.2b	200–211

a, b – mean values denoted by the same letter do not differ significantly at $p = 0.05$.

erage concentration was determined in tubers purchased in supermarkets and significantly the lowest in potatoes collected from fruit and vegetable shops, 13.82 and 12.07%, respectively (Table 2). Similar values were reported by Zarzecka and Gugała [2006] but their results were affected by agronomical practices and weather conditions during the study years. In turn, Czech and Turkish authors [Bártová et al. 2009; Öztürk et al. 2010] reported lower protein contents ranging from 6.84 to 10.20%.

Vitamin C content of potato tubers averaged 100–250 mg/kg of fresh matter [Kita, Lisińska 2007] and was the highest in samples of potato purchased in supermarkets, 214.4 mg/kg, on average. The average vitamin C contents were significantly lower for the two remaining kinds of shops. Other authors [Love, Pavek 2008] reported similar values. There are many reports mentioning vitamin C loses ranging from 20 to 80% during heat treatment of tubers, the highest following boiling or frying in oil [Kita, Lisińska 2007; Navarre et al. 2010]. Daily consumption of 250–300 g potatoes provides the human body with 40 to 70% vitamin C so consumers should pay close attention to the cultivar and cooking method they choose [Pęksa 2003].

Regulations set the following standards pertaining to the quality characteristics of potato for direct cooking or for chips and crisps: tubers should contain, respectively, less than 0.5, 0.25 to less than 0.5 and less than 25% of tuber fresh matter [Zgórska, Frydecka-Mazurczyk 2002]. Grudzińska and Zgórska [2008] reported low reducing sugars (less than 25%) in tubers immediately after harvest and after a period of storage. In the present work, tubers for direct consumption purchased in all kinds of outlets met the specifications set for reducing sugars (Table 3). Least reducing sugars were accumulated by potatoes sold in supermarkets (0.370%, on average). They were significantly higher in groceries as well as fruit and vegetable shops. Also, sucrose content of tubers was low in the three kinds of outlets but the between-mean differences were not statistically significant. Sawicka and Pszczółkowski [2005] reported similar, though slightly lower, reducing sugars and sucrose content, which was probably due to the fact that the determinations were made just after harvest. During storage starch breaks down into sucrose and then to reducing sugars (glucose + fructose), after-cooking darkening appears and tubers taste sweeter [Leszczyński 2012]. Based on correlation coefficients, Grudzińska and Zgórska 2008] concluded that starch content and reducing sugars were significantly and negatively correlated, which means that an increase in reducing sugars was followed by reduction in starch content.

CONCLUSIONS

1. Potatoes purchased in all three types of outlets in Siedlce and Międzyrzec Podlaski met standards set for edible potato tuber size. Samples taken in supermarkets had the highest percentage of large tubers.

Table 3. Reducing sugars and sucrose content in potato tubers

Kind of outlet	Reducing sugars (%)		Sucrose (%)	
	mean	range min-max	mean	range min-max
Supermarket	0.370a	0.361–0.379	0.245a	0.238–0.256
Grocery shop	0.376b	0.345–0.391	0.245a	0.208–0.276
Fruit and vegetable shop	0.375b	0.360–0.390	0.242a	0.209–0.266

a, b – mean values denoted by the same letter do not differ significantly at $p = 0.05$.

2. Total protein content, vitamin C content and reducing sugars in the tubers were affected by the place of purchase. Again, tubers bought in supermarkets had the best chemical composition.
3. The chemical composition of edible potatoes was good, regardless of the kind of outlet they were collected from.

REFERENCES

1. Bártová V., Bárta J., Diviš J., Švajner J., Peterka J. 2009. Crude protein content in tubers of starch processing potato cultivars in dependence on different agro-ecological conditions. Journal Central European Agriculture, 10(1), 57–66.
2. Chotkowski J., Rembeza J. 2005. Preferencje konsumentów i zmiany w spożyciu ziemniaków. Roczniki Naukowe Stowarzyszenia Ekonomistów Rolnictwa i Agrobiznesu, VII, 8, 42–49.
3. Czerwiecki L. 2009. Współczesne poglądy na rolę przeciwutleniaczy roślinnych w profilaktyce chorób cywilizacyjnych. Roczniki Państwowego Zakładu Higieny, 60(3), 201–206.
4. Grajek W. 2004. Rola przeciwutleniaczy w zmniejszaniu ryzyka wystąpienia nowotworów i chorób układu krążenia. Żywność Nauka Technologia Jakość, 1(38), 3–11.
5. Grudzińska M., Zgórska K. 2008. Wpływ zawartości cukrów w bulwach ziemniaka na barwę chipsów. Żywność Nauka Technologia Jakość, 5(60), 107–115.
6. Grudzińska M., Zgórska K. 2010. Wpływ efektywności zabiegu rekondycjonowania wybranych odmian bulw ziemniaka na barwę frytek. Nauka Przyroda Technologie, 4(2), 1–8.
7. Jansky S.H. 2010. Potato flavor. American Journal of Potato Research, 87, 209–217.
8. Jansen G., Flamme W., Shüler K., Vandrey M. 2001. Tuber and starch quality of wild and cultivated potato species and cultivars. Potato Research, 44, 137–146.
9. Jemison J.M. Jr., Sexton P., Ellen M., Camire M.E. 2008. Factors influencing consumer preference of fresh potato varieties in Maine. American Journal of. Potato Research, 85, 140–149.
10. Kita A., Lisińska G. 2007. Ocena składu chemicznego i jakości organoleptycznej mrożonych produktów ziemniaczanych pochodzących z sieci handlowej. Żywność Nauka Technologia Jakość, 3(52), 15–27.
11. Kolasa K.M. 1993. The potato and human nutrition. American Potato Journal, 70, 375–384.
12. Kris-Etherton P.M., Hecker K.D., Bonanome A., Coval S.M., Binkoski A.E., Hilpert K.F., Griel A.E. 2002. Bioactive compounds in foods: their role in the prevention of cardiovascular disease and cancer. American Journal Medicine, 113, 71–88.
13. Leszczyński W. 2012. Żywieniowa wartość ziemniaka i przetworów ziemniaczanych (Przegląd literatury). Biul. IHAR, 266, 5–20.
14. Lisińska G. 2006. Wartość technologiczna i jakość konsumpcyjna polskich odmian ziemniaka. Zeszyty Problemowe Postępów Nauk Rolniczych, 511, 81–94.
15. Lista opisowa odmian. Rośliny rolnicze. 2013. Cz. 2. Wyd. COBORU, Słupia Wielka, 1–202.
16. Love S.L., Pavek J.J. 2008. Positioning the potato as a primary food source of vitamin C. American Journal of Potato Research, 85 (4), 277–285.
17. Navarre D.A., Shakya R., Holden J., Kumar S. 2010. The effect of different cooking methods on phenolics and vitamin C in developmentally young potato tubers. American Journalof Potato Research, 87, 350–359.
18. Ostrowska A., Gawliński S., Szczubiałkowa Z. 1991. Metody analizy i oceny właściwości gleb i roślin. Wyd. Inst. Ochr. Środ. Warszawa.
19. Öztürk E., Kavurmaci Z., Kara K., Polat T. 2010. The effects of different nitrogen and phosphorus rates on some quality traits of potato. Potato Research., 53, 309–312.
20. Pęksa A. 2003. Białko ziemniaczane – charakterystyka i właściwości. Postępy Nauk Rolniczych, 5, 79–94.
21. Rozporządzenie Ministra Rolnictwa i Rozwoju Wsi w sprawie szczegółowych wymagań w zakresie jakości handlowej ziemniaków. Dz. U. nr 194 poz. 1900 z 2003 roku.
22. Rynek ziemniaka. Stan i perspektywy. Analizy rynkowe. 2013. Wyd. IERiGŻ-PIB, ARR, MRiRW, 40, 1–24.
23. Sawicka B., Pszczółkowski P. 2005. Dry matter and carbohydrates content in the tubers of very early potato varieties cultivated under coverage. Acta Scientarium Polonorum Hortorum Cultus, 4(2), 111–122.
24. Stypa I., Zgórska K. 2010. Ziemniak nasz powszedni. Wyd. IHAR-PIB, Bonin, 1–24.
25. Tajner-Czopek A. 2006. Metodyka określania wartości technologicznej i jakości konsumpcyjnej ziemniaka. Zeszyty Problemowe Postępów Nauk Rolniczych, 511, 95–103.
26. Zarzecka K. 2009. Potato as a global plant nutritional, dietary and medicinal values. Rozprawy Naukowe PWSZ im. Jana Pawła II w Białej Podlaskiej, T. III, 163–175.
27. Zarzecka K., Gugała M. 2006. Zawartość białka ogólnego i właściwego w bulwach ziemniaka w zależności od sposobów uprawy roli i odchwasz-

czania. Acta Scientarium Polonorum Agricultura, 5(2), 107–115.

28. Zarzecka K., Gugała M., Mystkowska I. 2009. The effect of agricultural treatments on the content of total and protein nitrogen in potato tubers. Ecological Chemistry and Engineering, A 16(8), 1065–1070.

29. Zgórska K. 2013. Wykorzystanie ziemniaka do celów spożywczych i przemysłowych. Inżynieria Przetwórstwa Spożywczego, 3/4, 7, 5–9.

30. Zgórska K., Frydecka-Mazurczyk A. 2002. Normy i wymagania jakościowe ziemniaków jadalnych i do przetwórstwa spożywczego. W: Produkcja i rynek ziemniaków jadalnych. Red. J. Chotkowski. Wyd. Wieś Jutra, Warszawa, 183–192.

31. Zgórska K., Grudzińska M. 2012. Zmiany wybranych cech jakości bulw ziemniaka w czasie przechowywania. Acta Agrophysica, 19(1), 203–214.

BIOLOGICAL CHARACTERIZATION OF *NEOMYSIS INTEGER* (LEACH, 1815) FROM THE POMERANIAN BAY IN 2006–2007

Anna Grzeszczyk-Kowalska[1], Juliusz C. Chojnacki[1], Małgorzata Raczyńska[1], Mariusz Raczyński[2]

[1] Department of Marine Ecology and Environmental Protection, Faculty of Food Sciences and Fisheries West Pomeranian University of Technology Szczecin, Kazimierza Królewicza str. 4, 71-550 Szczecin, Poland, e-mail: agrzeszczykowal@zut.edu.pl; juliusz.chojnacki@zut.edu.pl; mraczynska@zut.edu.pl

[2] Department of Fisheries Management, Faculty of Food Sciences and Fisheries West Pomeranian University of Technology Szczecin, Kazimierza Królewicza str. 4, 71-550 Szczecin, Poland

ABSTRACT

The aim of the study was to determine the biological characters typical of *Neomysis integer*, to investigate and update issues of its biology, ecology, and distribution in the southwestern Baltic Sea, and to compare biological properties of *N. integer* collected during eight study seasons in 2006–2007 in the region from Świnoujście to Darłowo. No link was established between *N. integer* size and sample collection site. The population attained sizes within the range of 2.64 to 18.84 mm. The size at which females achieved sexual maturity varied seasonally. A mean of 34 eggs was noted in the marsupium. The mean wet weight of *N. integer* was 0.011 g. Three generations were confirmed at most of the study sites.

Key words: Mysidacea, *Neomysis integer*, Pomeranian Bay, Baltic Sea.

INTRODUCTION

The genus Neomysis (Czerniawski) occurs in waters adjacent to all of the continents [Mees et al. 1994]. One of the most commonly occurring representatives of Mysidacea of the genus Neomysis in European coastal waters is *Neomysis integer* (Leach, 1815). Many aspects of this species have been investigated by Chojnacki [1991], Jensen et al. [1985], Rudstam et al. [1986], and Wiktor [1961]. Möbius was the first to observe and describe this species in the Pomerania Bay in 1873 [Jażdżewski et al. 2005]. *N. integer* is a crustacean species with an elongated, almost transparent body. Its head and most of its thorax segments are covered by a chitinized carapace, fused to the head and the proximal thorax segments. The eyes of these mysids are on stalks. The first pair of antennae is divided into two, and on the second pair there is characteristic sharp antennal scale, which is very important for the identification of this species. The abdomen comprises eight segments with biramous appendages, the first set of which function as maxillipeds. The abdomen is highly developed and ends with a narrowing, wedged tail plate with the telson and biramous appendages forming the tail fan. Statocysts, which are the primary balance organs, are located on the inner sides of the tail appendages [Żmudziński 1990]. The mean length of this species in the Pomeranian Bay is 12.5 mm [Chojnacki 1991].

N. integer is a typical euryhaline and eurythermal species [Mees et al. 1994]. It occurs at high abundance in estuarine water zones [Tattersall and Tattersall 1951, Mees et al. 1994]. It undertakes daily feeding migrations: detritus dominates its stomach contents at night indicating it feeds at the bottom, while in the daytime the dominant component is phytoplankton, which indicates this species feeds in the water column [Węsławski 1981]. *N. integer* is an omnivorous crustacean that consumes phytoplankton, zooplankton, and dead organic material [Fockedey and Mees 1999, Jerling and Wooldridge 1995, Mauchline1980, Tattersall and Tattersall 1951].

This mysid filters particles of potential food as it swims, but it can also hunt actively. It is an important link in the food chain as a consumer of secondary production, but it is also a component of food base of benthic and pelagic fish, as well as of epibenthic crustaceans [Mauchline 1980]. In the coastal waters of the Baltic sea, this species is a food component of juvenile Platichthyidae, Gobiidae, and sea trout, herring, sprats, perch, and pikeperch [Lindén et al. 2003, Mauchline 1971, Rudstam and Hannson 1990, Shvetsova et al. 1992, Szypuła et al. 1997, Węsławski 1981, Wiktor 1961].

As other mysid species do, *N. integer* females deposit eggs into the marsupium, which is formed by oostegites located on the thoracic legs on sections 2–8 of the trunk [Mauchline 1980, Mees et al. 1994]. When the young exit the marsupium they are fully formed but immature [Wittmann, 1984]. This provides them with a measure of protection from predators during development. The optimal conditions for larval development are a temperature of 15°C and a salinity range of 14–17 PSU [Fockedey et al. 2006]. The number of eggs in the marsupium and their size depends largely on the size of a female, which changes seasonally [Kinne 1955, Mees et al. 1994, Parker and West 1979].

Available literature indicates that Mysidacea are more susceptible to toxic substances than are many other studied species [Hunt et al. 2002, Morton et al. 1997, Roast et al. 1998, Verslycke et al. 2003], and as such they have been used for more than twenty years as bioindicators of environmental toxicity [Brandt et al. 1993, Emson and Crane 1994, Gaudy et al. 1991, Hunt et al. 1997, Jacobs and Grant 1974, Khan et al. 1992, Langdon et al. 1996, Lussier et al. 1985; 1999, Martin et al. 1989, McKenney 1998, Roast et al. 1999; 2001, Wildgust and Jones 1998].

The aim of the current study was to determine the biological characters typical of *Neomysis integer*, to investigate and update issues of its biology, ecology, and distribution in the southwestern Baltic Sea, and to compare the biological properties (weight, length, sex ratio, number of eggs, etc.) of *N. integer* collected during eight study seasons in 2006–2007 in the region from Świnoujście to Darłowo.

This paper includes the results of investigations of *N. integer* length and weight, the number of generations occurring annually, sex ratio, fecundity, and proximal body composition.

MATERIALS AND METHODS

Twelve sampling stations distributed along the coastal zone of the Pomeranian Bay in the southern Baltic Sea were designated for qualitative studies of *N. integer* populations. These sites were located near the mouths of the more important rivers of West Pomerania, or, in the case of the locality of Międzyzdroje, in the vicinity of marine constructions. The geographic coordinates of the locations of the sampling sites are presented in the table below (Table 1) and in Figure 1. Depending on whether or not the Lake Liwia Łuża Canal was linked with the sea, there was either just one sampling site at Niechorze or two (western and eastern).

Table 1. Geographic coordinates of the 12 sampling stations as determined with a GPS device by Garmin

Station	Geographic coordinates	
	N	E
Świnoujście west. (Ś.z.)	53°55,246'	14°16,593'
Świnoujście east. (Ś.w.)	53°55,126'	14°17,906'
Międzyzdroje (M.)	53°55,912'	14°26,662'
Dziwnów east. (D.w.)	54°01,416'	14°43,828'
Niechorze (N.)	54°06,089	15°05,973'
Mrzeżyno west. (M.z.)	54°08,792'	15°17,043'
Mrzeżyno east. (M.w.)	54°08,834'	15°17,164'
Dźwirzyno west. (Dź.z.)	54°09,546'	15°23,269'
Dźwirzyno east. (Dź.w.)	54°09,550'	15°23,291'
Kołobrzeg east. (K.w.)	54°11,238'	15°33,209'
Darłowo west. (Da.z.)	54°26,308'	15°33,213'
Darłowo east. (Da.w.)	54°26,488'	16°22,972'

The study material was collected at a depth range of 0.5 to 1 m with a drag that was constructed for this purpose (Photograph 1). The mesh bar length in the netting was 1 mm. The material collected was rinsed on a 0.5 mm mesh sieve and then preserved in 30% ethanol denatured with pyridine.

The material collected was sorted, classified into different systematic groups, and then identified to the genus, and, when possible, to the species. These tasks were performed with a Nikon C-DSS230 stereo microscope. The next stage of the study was performed on organisms from the group Mysidacea. Depending on density, the Mysidacea samples were reduced two- to fourfold using a Folsom plankton splitter [McEvan et al. 1954] to obtain a representative sub-sample of at

Figure 1. Sampling sites in the Pomeranian Bay

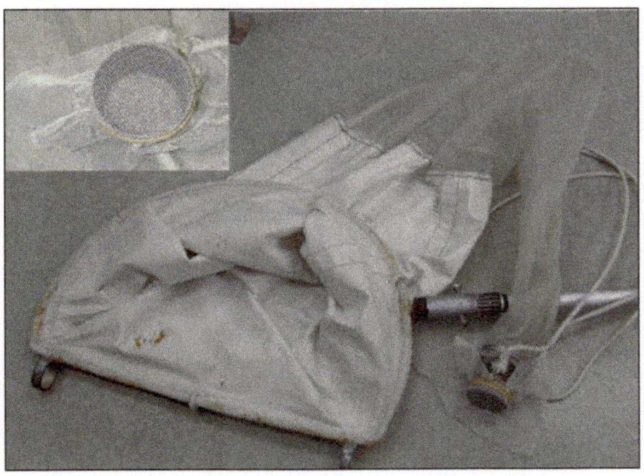

Photo 1. Sample collecting drag – constructed by the staff of the Department of Marine Ecology and Environmental Protection of the West Pomeranian University of Technology Szczecin

least 100 individuals. All of the caught *N. integer* specimens were identified using keys by Köhn et al. [1992] and Mańkowski [1955], in both of which the principle criterion for qualification to a particular species is the structure of the telson and scales.

The total number of *N. integer* individuals used for further study and analysis was 2894. As stated previously, the study material came from 13 sites which was collected in eight seasons in the 2006–2007 period. The precise collection dates of the samples is presented in Table 2.

The subsequent stage of the laboratory study was to take linear measurements of each specimen and to determine weight and sex. The description of the biological character of the *N. integer* specimens in the current work was done based on the following formula that was used to calculate the total length-weight dependency:

$$W = a \cdot L^b$$

where: W – weight

L – total length

a and b – coefficients characteristic of given populations presented graphically.

The measurement methodology used in the current study was that proposed by Mauchline [1971] with which *N. integer* total length is determined as the distance between the eye stalk and the posterior end of the uropods.

Individual weight (wet weight) was measured for each *N. integer* specimen in the sample to the nearest 0.00001 g on an electronic Radwag XA 110 jewelery scale after the specimens had been blotted on filter paper until all traces of wetness had been removed. Weights were measured in a closed laboratory without drafts and at a constant temperature.

In each of the samples, the eggs or embryos in the female marsupium were counted after weighing.

In order to obtain more complete information regarding the biology and properties of *N. integer* from the coastal zone of the southern Baltic Sea, the proximate composition of a pilot sample (100 grams of *N. integer* – about 10000 individuals) of the studied specimens was analyzed. This mate-

Table 2. Precise details of sampling events (for legend key see Table 1)

Station	Sampling dates in different study seasons.							
	Spring 2006 (W2006)	Summer 2006 (L2006)	Autumn 2006 (J2006)	Late autumn 2006 (PJ2006)	Winter – spring 2007 (ZW2007)	Spring 2007 (W2007)	Summer 2007 (L2007)	Autumn 2007 (J2007)
Ś.z.	12.05.	29.06.	17.09.	20.10.	07.03.	25.04.	21.06.	15.09.
Ś.w.	12.05.	29.06.	17.09.	20.10.	07.03.	25.04.	21.06.	15.09.r
M.	12.05.	29.06.	17.09.	20.10.	07.03.	25.04.	21.06.	15.09.
D.w.	12.05.	29.06.	17.09.r	20.10.	07.03.	25.04.	21.06.	15.09.
N.w.	12.05.	29.06.	17.09.	20.10.	07.03.	25.04.	21.06.	15.09.r
N.	12.05.	29.06.	17.09.	20.10.	07.03.	25.04.	21.06.	15.09.
M.z.	13.05.	28.06.	16.09.	19.10.	06.03.	24.04	20.06.	14.09.
M.w.	13.05.	28.06.	16.09.	19.10.	06.03.	24.04.	20.06.	14.09.
Dź.z.	13.05.	28.06.	16.09.	19.10.	06.03.	24.04.	20.06.	14.09.
Dź.w.	13.05.	28.06.	16.09.	19.10.	06.03.	24.04.	20.06.	14.09.
K.w.	13.05.	28.06.	16.09.	19.10.	06.03.	24.04.	20.06.	14.09.
Da.z.	13.05.	28.06.	16.09.	19.10.	06.03.	24.04.	20.06.	14.09.
Da.w.	13.05.	28.06.	16.09.	19.10.	06.03.	24.04.	20.06.	14.09.

rial was from the eastern Dźwirzyno site that was caught in the fall season of 2007. The proximate composition of *N. integer* was analyzed according to AOAC 1990 (Official Methods of Analysis, 15th Ed. Association of Official Analytical Chemists, Washington DC): water content was determined with the desiccation method; nitrogen content was determined with the Kjeldahl method; total fat content was determined with the ether extraction method. The fatty acid methyl esters were separated with gas chromatography. The percentage share of fatty acids in the tissues was determined according to AOAC 991.39 (Fatty Acids in Encapsulated Fish Oils and Fish Oil Methyl and Ethyl Esters).

RESULTS AND DISCUSSION

Origin and quantity of material

The study material was collected from areas that are typical habitats of *N. integer*: near hard substrates, such as marine constructions, boulders and smaller rocks, and sheltered areas near river mouths [Mauchline 1971, Żmudziński 1974]. The study sites were at least a few kilometers apart; however, within the areas of most sites a secondary division was made because of water flow into the sea (to the west or east, or to the left or right banks of the rivers). This choice of study sites stemmed from the desire to verify whether or not this species, which is known to be very widely distributed [Kinne 1955, Mees et al. 1994, Wiktor 1961], forms a morphological population with, for example, a distinct body build within a relatively small region (157 km of Baltic coastline). Unfortunately, the origin of the material collected on the day of catch was unknown, since the life cycle of this species is known to encompass varied migrations, including those of significant distance. It is known that environmental factors have a substantial impact on individual characters [Bremer and Vijverber 1982, Toda et al. 1984, Wiktor 1961]. It is plausible that the specimens of *N. integer* caught could have been transported by the rivers or they could have originated from the open sea, just as they could also have originated from a population that had inhabited the catch site for many generations. Indirect confirmation of this fact was the abundance of the *N. integer* specimens caught at the different sites in each of the seasons. It was easy to differentiate different stations since this species occurred only periodically, for example at Świnoujście and Dźwirzyno (Table 3). It was also significant that the highest densities of this species were observed in late fall and in early spring. In a certain sense, Mauchline [1971] confirmed these results since he observed three annual generations of *N. integer* (spring, summer, fall). He also emphasized that the continuity of one generation depends on the presence of all the generations. Consequently, in the current study, if at a given site no *N. integer* was noted in spring, or even worse, in summer, then the appearance of this species in fall was attributed exclusively to migration or inflow. In their studies of the southern Baltic Sea, Chojnacki [1991] and Pawlicka [1978] observed the occurrence of just two generations, but the general principle was the same: the spring generation should provide the individuals at a given site in the summer. Additionally, Wiktor [1961] noted that at low water temperatures, such as in late fall, these crustaceans migrate out to sea to depths of about 20 m, and then in the spring, when the water temperature increases, they return to the coastal zone. In late fall, when high densities of this species were noted, the mean water temperature was 12°C. Thus, it can be assumed that the specimens occurring in the littoral zone in early spring originate from the population occurring at a given site in late fall. Additionally, it can also be assumed that the *N. integer* specimens that were obtained for the study originate from the population of shallow waters in the littoral zone of the sea [Mauchline 1971, Węsławski 1981].

Obviously, the impact of many other factors, such as environmental conditions, feeding, lethal factors, that can affect the frequency of occurrence of *N. integer* at the study sites cannot be excluded [Chojnacki 1991, Fockedey et al. 2006, Roast et al. 1998, Roast et al. 1999, Roast et al. 2000, Roast et al. 2001a, Roast et al. 2001b, Verslycke and Jansen 2002, Verslycke et al. 2003, Verslycke et al. 2004, Wildgust and Jones 1998]. In laboratory studies, Winkler and Greve [2002] confirmed that juvenile *N. integer* specimens grow about threefold faster at temperatures of 15°C than they do at 10°C. This suggests that the nektobenthic *N. integer*, similarly to fish, can seek out better habitation conditions. During sampling for the current study, water temperature was always measured; however, no immediate dependence between water thermal conditions and the quantity of specimens caught was identified at any given sampling site. However, in late fall one of the distinct limiting factors was the strong and frequent water wavy motion in the littoral zone of the sea.

One of the factors that had an undoubtedly great impact on the occurrence of *N. integer* in the Baltic coastal zone, and about which, although it was witnessed during sampling, nothing is written, is the physical interference of humans. People were seen conducting commercial catches of this species for sale to aquarium shops, and in the summer season the presence of people and animals on the beach can disturb and effectively scare off this crustacean.

Total length and weight

The data in the literature indicates that *N. integer* specimens caught during studies performed in the Pomeranian Bay were of the following total length ranges: 7–23 mm [Wiktor 1961]; 4–19 mm [Jansen 1985]; 5–16 mm [Chojnacki 1991]. However, these authors used a different *N. integer* body segment as the measure of total length. In the current study, the smallest juvenile specimen caught was 2.64 mm; thus, it may be presumed that at this length it was possible for it to have

Table 3. Details of the abundance of *N. integer* specimens in study seasons at different sites (for legend key see Tables 1 and 2)

Station	Study seasons							
	W2006	L2006	J2006	PJ2006	ZW2007	W2007	L2007	J2007
Ś.z.	28	0	0	0	0	28	32	128
Ś.w.	11	0	0	25	51	0	66	78
M.	4	58	0	37	58	143	0	5
D.w.	69	17	1	32	77	116	0	18
N.w.	0	0	6	0	0	0	0	0
N.	0	1	0	3	13	0	0	18
M.z.	92	0	0	153	13	0	0	0
M.w.	133	0	68	5	78	38	0	0
Dź.z.	101	0	0	86	15	0	0	0
Dź.w.	71	9	5	234	261	0	0	0
K.w.	0	0	8	47	0	56	0	0
Da.z.	0	0	0	0	21	0	0	0
Da.w.	0	0	0	10	188	5	0	0

exited the marsupium. In a study in the brackish Lake Ferring, which is 100 m from the North Sea, Aaser et al. [1995] reported that juvenile stages exit the marsupium of female *N. integer* at lengths of 2-3 mm. In turn, Mees [1994] reported cohorts of individuals with a mean length of 3 mm in the months of April and May in the brackish mouth of the Westerschelde River. In the present results, the mean length of all the specimens analyzed ranges from 6.3 to 13.8 mm, while the largest specimen noted has a length of 18.84 mm. The female mean length was 9.9 mm, while that of the males was 9.2 mm. The length range of *N. integer* was, thus, similar to that reported by Wiktorowa [1961] and Jansen [1985]. Remerie et al. [2005] reported mean female lengths that are similar to the data that is the foundation of this work, i.e. 10.29 mm (Atlantic estuarine zone). Mees et al. [1994] and other authors have reported female lengths of 12 mm and male lengths of 10.5 mm, which are confirmed by the current observations of this species in the southern Baltic Sea that the mean size of females is larger than that of the males, and this is confirmed by data from other authors.

As was mentioned in the previous section, measurements were also taken of gravid females. It was confirmed that gravid females reached a mean body length range of 9.9–10.8 mm in the summer and of 10.8–12.3 mm in the fall. Along with the beginning of the winter months, it was noted that females of body lengths of about 15 mm had full marsupia. In spring, the length of the gravid females oscillated around 10.4–11.8 mm. It is plausible, thus, that the size at which females achieve sexual maturity fluctuates seasonally and is largely dependent on temperature. Summer generations achieve sexual maturity more quickly than do the others [Astthorson and Ralph 1984; Kuhlmann 1984; Mauchline 1985, Verslycke et al. 2004, Winkler and Greve 2002]. Muchline [1973] reported the mean body length of gravid females in the waters of the northeast Atlantic as 13.9 mm, which is higher than that determined in the current study (11.8 mm); this is likely linked to the higher salinity of open waters.

The mean wet weight of *N. integer* specimens in the current study is 0.011 g, which is lower than the figure reported by Orav-Kotta et al. [2009] in their studies conducted in the northern Baltic (0.016 g). Irvine et al. [1995] reported a wet weight range of 0.013 to 0.014 g. The studies by Verslycke et al. [2004] indicate that this species can attain much higher weights than those reported (mean from 0.014 g in winter to 0.027 g in spring), and this is confirmed by the results published by Mees et al. [1994]; however, significant factors in these instances might include water salinity and thermal regimes. The lightest specimen noted in the current study weighed 0.00013 g, and within this range of weight it is possible for juvenile forms of *N. integer* to exit the marsupium. The heaviest specimens in the current study weighed 0.124 g. The weight ranges of specimens from the current study were much wider than those reported in the literature. The comparison of female to male weights indicates that the latter are heavier in most instances, which is absolutely understandable since their weight is largely dependent on the weight of the marsupium.

Issue of the number of genreations

According to Winkler and Greve [2002], every *N. integer* population comprises overlapping cohorts from subsequent hatchings. It is reported in the literature that two to three mysid generations occur annually: three have been reported by Vorstman [1951], Mauchline [1971], Borghouts [1978], Parker and West [1979], Bremer and Vijverberg [1982], Mees et al. [1994], and Verslycke et al. [2004]. Two annual generations have been described in the mouths of the River Ythan by Astthorsson and Ralph [1984] and the Eider-Ring by Kinne [1955], while Kinne [1955], Wiktor [1961], Jansen [1985], Rudstam et al. [1986], Chojnacki [1991], and Aaser et al. [1995] all write about two annual generations in the coastal waters of the Baltic Sea. However, Mees et al. [1994] reports that in lower geographic zones populations of *N. integer* reproduce nearly throughout the year. Mauchline [1971] observed year-round mysid reproduction in Loch Etive, but that during winter it was not abundant. The analysis of the weight-length dependence of *Neomysis integer* in the current study leads to the observation that in many instances the graphs indicate there is more than one generation present. Two populations occurring at the same time in a single study season are visible in the population from summer 2007 at the east Świnoujście site and in winter-spring 2007 at the east Dźwirzyno site (Figures 2 and 3).

Considering the information that from the spring season *N. integer* lives for three months only [Verslycke et al. 2004], it is plausible that in one of the other annual seasons a subsequent generation, the third, appears. The weight-length

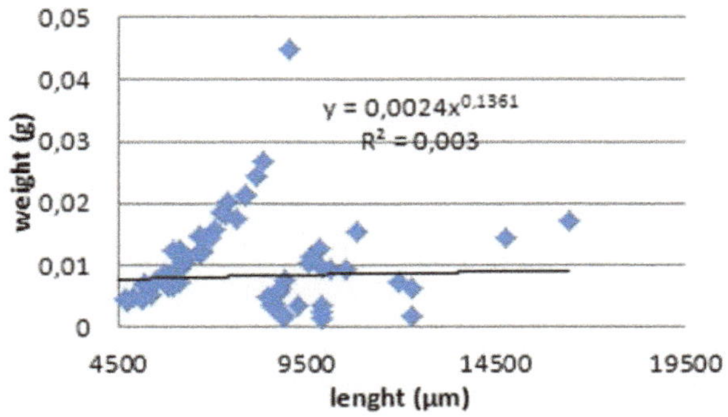

Figure 2. Weight-length dependence for the whole sample of *Neomysis integer* from the eastern Świnoujście site in the summer season 2007

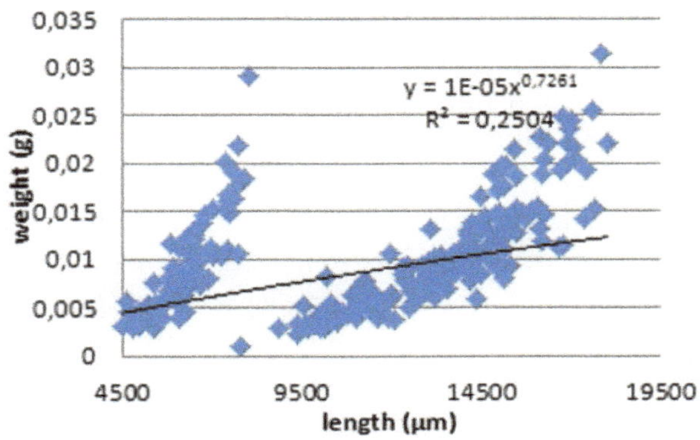

Figure 3. Weight-length dependence for the whole sample of *Neomysis integer* from the eastern Dźwirzyno site in the winter-spring season 2007

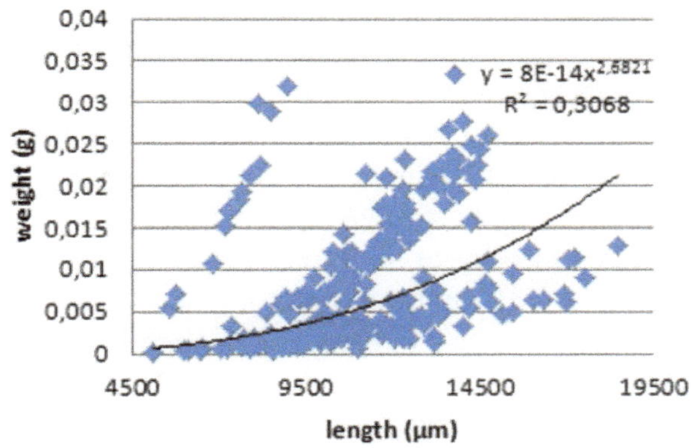

Figure 4. Weight-length dependence for the whole sample of *Neomysis integer* from the western Mrzeżyno site

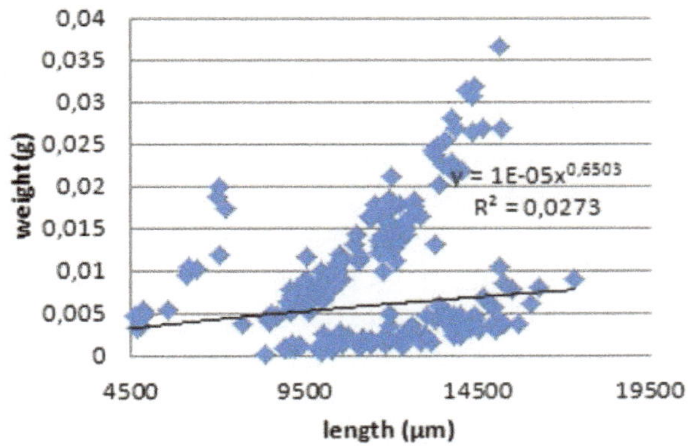

Figure 5. Weight-length dependence for the whole sample of *Neomysis integer* from the western Dźwirzyno site

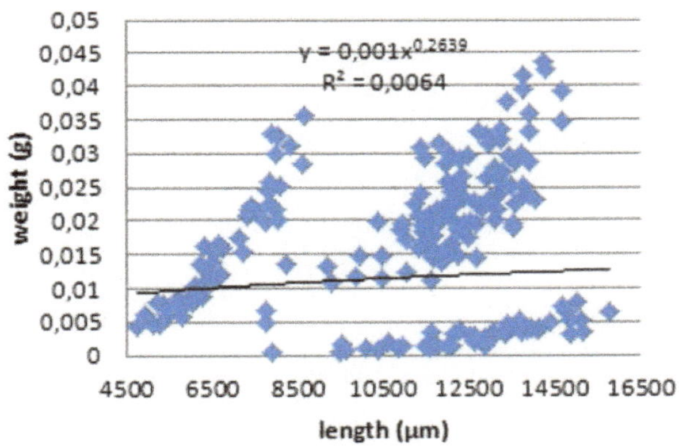

Figure 6. Weight-length dependence for the whole sample of *Neomysis integer* from the eastern Kołobrzeg site

dependence graphs of specimens from the sites at western Mrzeżyno, western Dźwirzyno, and eastern Kołobrzeg unequivocally illustrate three generations of *N. integer* (Figures 4, 5, 6). Thus, the study results presented in this paper indicate that three, not two, annual generations of *N. integer* occur in the coastal waters of the Pomeranian Bay. The increased number of annual generations as compared to that in previous studies [Kinne 1955, Wiktor 1961, Jansen 1985, Rudstam et al. 1986, Chojnacki 1991, Aaser et al. 1995] might be evidence that the environmental conditions for *N. integer* population habitation in the Pomeranian Bay have improved.

Numerical sex ratio

The sex ratio in the samples studied (Table 4) was varied: in spring the number of females was higher than the number of males (from 1.03:1 at the eastern Dźwirzyno site to 3.5:1 at the western Świnoujście site); in summer the dominance of females was even higher (sporadically 45.5:1 at the eastern Kołobrzeg site); but in fall the difference between the number of females and males decreased (1.1:1 at the eastern Mrzeżyno site). This difference began to increase again in late fall with figures of 3.9:1 at the eastern Dźwirzyno site and 23.6:1 at the western Mrzeżyno site. In the winter at the eastern Świnoujście, Międzyzdroje, and Dziwnów sites the number of males was higher than the number of females, while at the other sites the number of females was higher.

In subsequent study seasons, the predominance of females was observed again. Periods of an equalized sex ratio of 1:1 mature adult females to males is thought to be optimal for reproduction [Chojnacki 1991, Vortsman 1951]. The spring 2006 and 2007 study seasons were close to this optimum, when females were only slightly predominant over males, thus, it can be hypothesized that spring is the principal reproduction period for *N. integer* in the littoral of the southern Baltic. Węsławski [1981], however, believes that during the most intense reproduction period females significantly outnumber males, which was also reflected at several sampling sites in the present study. Grieze [1972] also reports a similar phenomenon with regard to several species of Amphipoda. The lesser number of males in the samples might also lend confirmation to the opinion that after copulation some of the males expire [Węsławski, 1981], or even that following copulation the females require more nutrition and consume the males (cannibalism among *N. integer* is described in the work of Verslycke and Janssen 2002). All of these hypotheses indicate indirectly that the periods of summer and fall in the waters of the Pomeranian Bay in the southern Baltic Sea are ones of intense reproduction. The predominance of males in the season of ZW2007 may confirm Verslycke's [2004] belief that the fall generation does not reproduce until the spring. Similar conclusions were drawn by Jansen (1985), who observed the predominance of males in the month of December. Mees et al. [1994] underscore that the at temperatures of under 10°C *N. integer* reproduction ceases.

Fecundity

A significant challenge when comparing populations of the *N. integer* is the number of embryos carried by one female [Węsławski 1981]. This number is undoubtedly dependent on the size of the female and varies seasonally [Fockedey et al. 2006]. Mauchline [1971], Pawlicka [1978], and Wiktor [1961] report that the maximum number of embryos in the marsupium is from 70 to 74, while Kinne [1955] reports the maximum to be 55 and Węsławski [1981] just 46. The results of the present study differed from the data in the literature and ranged from 14 embryos to a maximum of 44, at a mean of 34 in one marsupium. The lower number of embryos in the marsupia of female *N. integer* in the current study in comparison with the results of other authors does not necessarily indicate the lesser abundance of these organisms in the Pomeranian Bay. It might just indicate that abundance was higher. The results of the current total body length measurements of *N. integer* do not differ to a great degree from the measurement results reported by other crustacean researchers studying this species in the Pomeranian Bay and the waters of the southern Baltic coastal zone. Considering that egg size depends on female body size [Kinne, 1955, Parker and West 1979, Mees et al., 1994] and that there were fewer eggs in the crustaceans examined in the present study, one might conclude that the eggs were larger; thus there is the possibility that both embryo survival and that of the future progeny is significantly higher [Kolding and Fenchel 1981, Wehrtmann and Lopez 2003]. Mees et al. [1994] note that among females of the same size, those

Table 4. Sex ratio of *N. integer* females and males at different study sites (for legend key see Table 1 and 2)

Station	W2006	L2006	J2006	PJ2006	ZW2007	W2007	L2007	J2007
Ś.z.	1.6:1	5.2:1	–	–	–	–	–	1.2:1
Ś.w.	3.5:1	–	–	20:1	1.1:1	–	1.6:1	1.4:1
M.	1:3	2:1	–	4.8:1	1:1.11	1.1:1	–	5;0
D.w.	1.8:1	2.4:1	0:1	25;0	1:1.8	1.9:1	–	1:2.6
N.w.	–	–	2:1	–	–	–	–	–
N.	–	1:0	–	3:0	1.7:1		–	1.6:1
M.z.	3.2:1	–	–	23.6:1	1.2:1	–	–	–
M.w.	1.8:1	–	1.1:1	5:0	1.3:1	1:1.05	–	–
Dź.z.	1.15:1	–	–	16:1	2;1	–	–	–
Dź.w.	1.03:1	7:1	1.5:1	3.9:1	1.2:1	–	–	–
K.w.	–	45.5:1	8:0	11:1	–	5.9:1	–	–
Da.z.	1;2	–	–	–	2.5:1	–	–	–
Da.w.	–	–	–	9;1	2.3:1	1.5:1	–	–

which hibernate have, on average, more eggs in spring. Verslycke et al. [2004] reports, however, that the spring generation lives for only three months, and the large number of eggs in the marsupia might result in the birth of weaker specimens. In the present study the mean number of eggs in the marsupia in spring was higher than that in the summer with a mean of 36 in spring 2006 and a mean of 31 eggs in summer 2006.

The current results also confirm conclusions of other researchers regarding the growth rate of individuals in the marsupium. Fockedey et al. [2006] and Mauchline [1980] note that not all of the individuals in the marsupium are at the same stage of development. Photographs 2 and 3 illustrate the simultaneous occurrence of both *N. integer* eggs and embryos in the first stage of development.

Photo 2. Marsupium contents of a female *N. integer* (a – eggs, b – embryos)

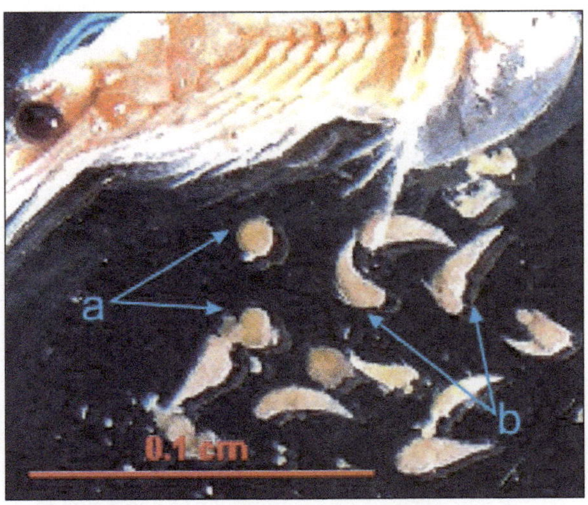

Photo 3. Marsupium contents of a female *N. integer* – close-up (a – eggs, b – embryos)

Proximate body composition of *n. integer*

The adaptations of *N. integer* to life in the estuarine zone consume a lot of energy [Verslycke and Janssen 2002]. These include this species's wide ranging tolerance of changes in salinity [Mauchline 1971], and its very efficient osmoregulation and respiration physiology [McLusky and Heard 1971, Roast et al. 1999]. This is why it is important to describe the proximate composition of *N. integer* populations. The proximate body composition of the common mysids examined in the current study is 9.34% protein, 1.22% fats, and 1.32% sugars (Table 5).

Table 5. Percentage share of the biochemical components of *N. integer*

Component	Percentage share
Water	86.28
Protein	9.34
Fat	1.22
Ash (mineral components)	1.84
Total carbohydrates	1.32

Verslycke and Janssen (2002) report a slightly different proximate composition of 7.39% protein, 3.99% fats, and 0.42% sugars. The differences in these results might have been the result of the material analysed originating from different regions and seasons, which means the specimens inhabited different environmental conditions, which included varied access to food, oxygenation, insolation, etc. While the differences in the protein fractions are not large, the fat content result from the current study is threefold lower. These results are similar to those of Ando and Nozaki [2007]; however, they assayed the fat contents in *Neomysis intermedia* [Czerniavsky, 1882]; this crustacean is a basic ingredient in the popular Japanese dish tsukudani and a southeastern Asian sauce. These authors report a 1% fat content that includes 21% palmitic acid, 12% linoleic acid, 19% EPA (eicosapentaenoic acid), and 14% DHA (docosahexaenoic acid). The results presented in Table 6 of the fatty acid composition were also similar to the results of Ando and Nozaki [2007]; however, 47.49% of the fat in *Neomysis integer* comprises polyunsaturated fatty acids. It is significant that EPA and DHA, which are Omega 3 fatty acids that are important to human health, comprised as much as 41.35% of the material originating from the Pomeranian Bay. This warrants further observation and study.

One additional question, which was not included in the current study but which occurs frequently in the literature about *N. integer*, should

Table 6. Fatty acid composition of *N. integer*

Fatty acids	Content [%]	
	Mean	SD
C14:0 (myristic)	0.98	0.05
C15:0 (pentadecanoic)	0.56	0.03
C16:0 (palmitic)	27.10	0.56
C16:1 (palmitoleic)	1.60	0.02
C17:0 (margaric)	1.41	0.04
C18:0 (stearic)	4.58	0.01
C18:1 (linoleic)	15.10	0.04
C18:2 (linoleic)	1.62	0.06
C20:3 (homo-gamma-linolenic)	1.13	0.02
C20:1 (eicosenoic)	1.20	0.01
C20:2 (eicosadienoic)	0.92	0.02
C20:4 (arachidonic)	2.46	0.00
C20:5 EPA (eicosapentaenoic)	17.13	0.13
C22:6 DHA (docosahexaenoic)	24.22	0.48
Total		
SFA (saturated fatty acids)	34.62	0.68
MUFA (monosaturated fatty acids)	17.89	0.01
PUFA (polyunsaturated fatty acids)	47.49	0.69

be posed at this juncture: what is the significance of this species as a bioindicator? This topic is omitted from studies of the biology of *N. integer* from the southern Baltic Sea, but the literature indicates that it is crucially important and common [Brandt et al. 1993, Emson and Crane 1994, Gaudy et al. 1991, Hunt et al. 1997, Jacobs and Grant 1974, Khan et al. 1992, Langdon et al. 1996, Lussier et al. 1985; 1999, Martin et al. 1989, McKenney 1998, Roast et al. 1999; 2000, Wildgust and Jones 1998]. With such a collection of wide-ranging, current data on the physiology and morphometry of this common mysid from the Pomeranian Bay, the logical undertaking would be to conduct further studies of its behavior under varied environmental conditions and its tolerance thresholds for harmful factors.

CONCLUSIONS

1. No dependencies between specimen size and collection location were noted in the material collected in 2006 and 2007 from the Pomeranian Bay.
2. In 2006 and 2007, the specimen length range of the population in the Pomeranian Bay was 2.64 to 18.84 mm. The mean length was 8.49 mm, with females at 9.9 mm, and males – 9.2 mm. The length of the smallest juvenile specimen caught was 2.64 mm, which indicates that *N. integer* exit the marsupium at this size to begin active life in the environment.
3. The size at which females achieve sexual maturity changes seasonally and is largely dependent on water temperature.
4. The mean wet weight of *N. integer* specimens from the Pomeranian Bay in 2006 and 2007 was 0.011 g, with females at 0.012 g, and males – 0.0099 g.
5. Three generations of *N. integer* were confirmed at most of the sampling sites in the coastal zone of the Pomeranian Bay in 2006 and 2007.
6. The sex ratio in the samples analyzed was fairly variable and depended on the season of the year (the ratio of females to males was 1.03:1 at the eastern Dźwirzyno site in spring 2006, while it was 45.5:1 at the eastern Kołobrzeg site in summer 2006). However, in most instances the number of females was greater than that of males. The 2006 and 2007 spring seasons were periods of sex ratio equilibrium, and as such the optimum time for reproduction.
7. The marsupia of *N. integer* contained from 14 to a maximum of 44 embryos, at an average of 34 embryos per marsupium.
8. The fat content of *N. integer* from the Pomeranian Bay was lower than that in specimens from other regions, while the fatty acids from the Omega 3 family comprised as much as 41.35% of all the fat content of the crustaceans assayed.

LITERATURE

1. Aaser H.J., Jeppesen E., Søndergaard M. 1995. Seasonal dynamics of the mysid *Neomysis integer* and its predation on the copepod *Eurytemora affinis* in a shallow hypertrophic brackish lake. Mar. Ecol. Prog. Ser.127: 47-56.
2. Ando Y., Nozaki M. 2007. Lipid Profile of Mysid *Neomysis intermedia* and Its Changes by Feeding DHA-Rich Marine Lipids. Journal of Oleo Science 56, 9: 471-478.
3. AOAC 1990: Official Methods of Analysis, 15th Ed. Association of Official Analytical Chemists, Washington DC.
4. AOAC Method 991.39. Fatty acids in encapsulated fish oils and fish oil methyl and ethyl esters.

5. Astthorsson O.S., Ralph R. 1984. Growth and moulting of *Neomysis integer* (Crustacea: Mysidacea). Mar. Biol. 79: 56-61.

6. Borghouts C.H. 1978. Population structure and life-cycle of *Neomysis integer* (Leach) (Crustacea, Mysidacea) in two types of inland waters. Verh. int. Verein Llmnol. 20: 2561-2565.

7. Brandt O.M., Fujimura R.W., Finlayson B.J. 1993. Use of *Neomysis mercedis* (Crustacea: Mysidacea) for estuarine toxicity tests. Trans. Am. Fish. Soc. 122: 279-288.

8. Bremer P., Vijverberg J. 1982. Production, population biology and diet of *Neomysis integer* (Leach) in a shallow Frisian lake (The Netherlands). Hydrobiologia 93: 41-51.

9. Chojnacki J.C. 1991. Population studies of Neomysis integer (Leach, 1815) from the marine coast of Wolin Island. Zesz. Nauk. AR w Szczecinie 143: 49-58. [in Polish].

10. Emson S., Crane M. 1994. A comparison of the toxicity of cadmium to the mysid shrimps *Neomysis integer* (Leach) and *Mysidopsis bahia* (Molenock). Water Res 28: 1711-1713.

11. Fockedey N., Mees J. 1999. Feeding of the hyperbenthic mysid *Neomysis integer* in the maximum turbidity zone of the Elbe, Westerschelde and Gironde estuaries. J. Mar. Sys. 22: 207-228.

12. Fockedey N., Ghekiere A., Bruwiere S., Janssen C.R., Vincx M. 2006. Effect of salinity and temperature on the intra-marsupial development of the brackish water mysid *Neomysis integer* (Crustacea: Mysidacea). Mar. Biol. 148: 1339–1356.

13. Gaudy R., Guérin J.P., Kerambrun P. 1991. Sublethal effects of cadmium on respiratory metabolism, nutrition, excretion and hydrolyse activity in *Leptomysis lingvura* (Crustacea: Mysidacea). Mar Biol. 109: 493-501.

14. Grieze I.I. 1972. Some peculiarities of Amphipoda from the Black Sea. Sea Biology 3: 1-32. [in Russian].

15. Hunt J.W., Anderson B.S., Turpen S.L., Englund M.A., Piekarski W. 1997. Precision and sensitivity of a seven-day growth and survival toxicity test using the west coast marine mysid crustacean *Holmesimysis costata*. Environ Toxicol Chem 16: 824-834.

16. Hunt J.W., Anderson B.S., Phillips B.M., Tjeerdema R.S., Puckett H.M., Stephenson M., Tucker D.W., Watson D. 2002. Acute and chronic toxicity of nickel to marine organisms: Implications for water quality criteria. Environ Toxicol Chem 21: 2423-2430.

17. Irvine K., Snook D., Moss B. 1995. Life histories of *Neomysis integer*, and its copepod prey, *Eurytemora affinis*, in a eutrophic and brackish shallow lake. Hydrobiologia 304: 59-76.

18. Jacobs F., Grant G.C. 1974. Acute toxicity of unbleached kraft mill effluent (UKME) to the opossum shrimp *Neomysis americana*, Smith. Water Res 8: 439-445.

19. Jansen. W. 1985. Basic biometric information on *Neomysis integer* Leach from the Darss-Zingst Bodden Chain (southern Baltic Sea)).Fischer. Forsch. 23, 4: 37-40. [in German].

20. Jażdzewski K., Konopacka A., Grabowski M. 2005. Native and alien malacostracen crustacean along the polish Baltic Sea coast in the twentieth century. Oceanological and Hydrobiological Studies 34; 175-193.

21. Jerling HL, Wooldridge TH. 1995. Feeding of two mysid species on plankton in a temperate South African estuary. J Exp Mar Biol Ecol 188: 243-259.

22. Kinne O. 1955. *Neomysis vulg*aris Thompson, autekological – biological study. Studie. Biol. 74: 160-202. [in German].

23. Khan A., Barbieri J., Khan S., Sweeney F. 1992. A new short-term mysid toxicity test using sexual maturity as an endpoint. Aquat. Toxicol. 23: 97-105.

24. Kolding S., Fenchel T.M. 1981. Patterns of reproduction in different populations of five species of the amphipod genus Gammarus. Oikos 37: 167-172.

25. Köhn J., Jones M.B., Moffat A. 1992. Taxonomy, biology and Ecology of (Baltic) Mysids (Mysidacea: Crustacea). International Expert Conf. Hiddensee, Germany.

26. Kuhlmann D. 1984. Effects of temperature, salinity, oxygen and ammonia on the mortality and growth of *Neomysis integer* Leach. Limnologica 15: 479-485.

27. Langdon C.J., Harmon V.L., Vance M.M., Kreeger K.E., Kreeger D.A., Chapman G.A. 1996. A 7-D toxicity test for marine pollutants using the Pacific mysid *Mysidopsis intii*. 1. Culture and protocol development. Environ. Toxicol. Chem. 15: 1815-1823.

28. Linde´n E., Lehtiniemi M. Viitasalo M. 2003. Predator avoidance behaviour of Baltic littoral mysids *Neomysis integer* and *Praunus flexuosus*. Marine Biology 143: 845-850.

29. Lussier S.M., Gentile J.H., Walker J. 1985. Acute and chronic effects of heavy metals and cyanide on *Mysidopsis bahia* (Crustacea: Mysidacea). Aquat Toxicol 7: 25-35.

30. Lussier S.M., Kuhn A., Comeleo R. 1999. An evaluation of the seven-day toxicity test with *Americamysis bahia* (formerly Mysidopsis bahia). Environ. Toxicol. Chem. 18: 2888-2893.

31. Mańkowski W. 1955. Zooplankton Atlas. MIR, Gdynia: 1-51. [in Polish].

32. Martin M., Hunt J.W., Anderson B.S., Turpen S.L., Palmer F.H. 1989. Experimental evaluation of the mysid *Holmesimysis costata* as a test organism for

effluent toxicity testing. Environ. Toxicol. Chem. 8: 1003-1012.

33. Mauchline, J. 1971. The biology of *Neomysis integer*. J. mar. biol. Ass. U.K. 51: 347-354.

34. Mauchline J. 1973. The broods of British Mysidacea (Crustacea). J Mar Biol Ass UK 53: 801-817.

35. Mauchline J. 1980. The biology of mysids and euphausiids. [In:] Blaxter J.H.S., F.S. Russel & M. Yonge (eds), Advances in Marine Biology (18). Academic press, London, s. 681.

36. Mauchline J. 1985. Growth in mysids and euphausiids. [In:] Wenner, A.M. (Ed.), Factors in Adult Growth, vol. 3. A.A. Balkema, Rotterdam, The Netherlands, pp. 337-354.

37. Mc Even G.E., Johnson M.M., Folsom R.R. 1954. A statistical analysis of the Folsom plankton sample splitter, based upon test obserwations. Arch. Meteorol. Geophys. Bioklim. S.A.Meteorol. Gophys., 7: 502-527.

38. McKenney C.L. Jr. 1998. Physiological Dysfunction in Estuarine Mysids and Larval Decapods with Chronic Pesticide Exposure. [In:] Wells PG, Lee K, Blaise C (eds) Microscale Testing in Aquatic Toxicology: Advances, Techniques, and Practice. CRC Press, Boca Raton, FL, USA, 465-476.

39. McLusky D.S., Heard V.E.J. 1971. Some effects of salinity on the mysid *Praunus flexuosus*. J. Mar. Biol. Assoc. UK 51: 709-715.

40. Mees J., Abdulkerim Z., Hamerlynck O. 1994. Life history, growth and production of Neomysis integer in the Westerschelde estuary (SW Netherlands). Mar Ecol Prog Ser 109: 43-57.

41. Morton M.G., Mayer F.L., Dickson K.L., Waller W.T., Moore J.C. 1997. Acute and chronic toxicity of azinphos-methyl to two estuarine species, *Mysidopsis bahia* and *Cyprinodon variegatus*. Arch Environ Contam Toxicol 32: 436-441.

42. Orav-Kotta H., Kotta J., Kotta I. 2009. Food competition between the benthic polychaete Hediste diversicolor and the semipelagic mysid *Neomysis integer* in the northern Baltic Sea. Estonian J. Ecol., 58: 324-331.

43. Parker M., West B. 1979. The natural history of *Neomysis integer* (Leach) in Lough Furnace, Co. Mayo, a brackishlough in the West of Ireland. Estuar. coast. mar. Sci. 8: 157-167.

44. Pawlicka M. 1978. Selected issues in the biology and ecology of Neomysis vulgaris (Thompson) in the Dead Vistula. Inst. Biol. Zakład Ekologii Zwierząt, Gdynia. pp. 4-37. [in Polish].

45. Remerie T., Bourgois T., Venreusel A. 2005. Morphological differentiation between geographically separated populations of *Neomysis integer* and *Mesopodopsis slabberi* (Crusstacea, Mysida). Hydrobiologia 549: 239-250.

46. Roast S.D., Thompson R.S., Widdows J., Jones M.B. 1998. Mysids and environmental monitoring: a case for their use in estuaries. Mar. Freshw. Res. 49: 827-832.

47. Roast S.D., Thompson R.S., Donkin P., Widdows J., Jones M.B. 1999. Toxicity of the organophosphate pesticides chlorpyrifos and dimethoate to *Neomysis integer* (Crustacea, Mysidacea). Water. Res. 33: 319-326.

48. Roast S.D., Widdows J., Jones M.B. 2000. Disruption of swimming in the hyperbenthic mysid *Neomysis integer* (Peracarida: Mysidacea) by the organophosphate pesticide chlorpyrifos. Aquat. Toxicol. 47: 227-241.

49. Roast S.D., Widdows J., Jones M.B. 2001a. Effects of salinity and chemical speciation on cadmium accumulation and toxicity to two mysid species. Environ. Toxicol. Chem. 20: 1078-1084.

50. Roast S.D., Widdows J., Jones M.B. 2001b. Impairment of mysid (*Neomysis integer*) swimming ability: an environmentally realistic assessment of the impact of cadmium exposure. Aquat. Toxicol. 52: 217-227.

51. Rudstam L.G., Hansson S., Larsson U. 1986. Abundance, species composition andproduction of mysid shrimps in a coastal area of the Northern Baltic Proper Ophelia 4: 225-238.

52. Rudstam L.G., Hansson S. 1990. On the ecology of *Mysis mixta* (Crustacea, Myidacea) in a coastal area of the northern Baltic proper. Ann. Zool. Fenn. 27: 259 -263.

53. Shvetsova G., Shvetsov F., Hoziosky S. 1992. Distribution, abudance and annual production of *Mysis mixta* Lilljeborg in Eastern and Southeastern Baltic. ICES C.M. 1992/L: 24-29.

54. Szypuła J., Ostrowski J., Margoński P., Krajewska–Sołtys A. 1997. Food of Baltic herring and spratin the years 1995–1996 in light of the availability of components. Bull. Sea Fish. Inst., Gdynia, 2 (141): 61-72.

55. Tattersall W.M., Tattersall O.S. 1951. The British Mysidacea. The Ray Society, London, s. 399-409.

56. Toda H., Takahashi M., Ichimura S. 1984. The effect of temperature on the post-embryonic growth of *Neomysis intermedia* Czerniawsky (Crustacea, Mysidacea) under laboratory conditions. J Plankton Res 6: 647–661.

57. Verslycke T., Janssen C.R. 2002. Effects of a changing abiotic environment on the energy metabolismin the estuarine mysid shrimp *Neomysis integer* (Crustacea; Mysidacea). J. Exp. Mar. Biol. Ecol. 279: 61-72.

58. Verslycke T., Vangheluwe M., Heijerick D., De Schamphelaere K., Van Sprang P., Janssen C.R. 2003. The toxicity of metal mixtures to the estua-

rine mysid *Neomysis integer* under changing salinity. Aquat. Toxicol. 64: 307-315.

59. Verslycke T., Fockedey N., McKenney C.L., Roast S.D., Malcom B.J., Mess J., Janssen C.R. 2004. Mysid Crustaceans as potential test organisms for the environmental endocrine distruption: areview. Environ. Toxicol. Chem. 23: 1219-1234.
60. Vorstman A.G. 1951. A year's lnvestlgatlon on the life cycleof *Neomysis vulgaris* Thornpson. Verh. Int. Verein. Limnol. 1: 437-445.
61. Wehrtmann I.S., Lopez G.A. 2003. Effects of temperature on the embryonic development and hatchling size of *Betaeus emarginatus* (Decapoda Caridea Alpheidae). J. Nat. Hist. 37: 2165-2178.
62. Wiktor, K. 1961. Observations of the biology of *Neomysis vulgaris* (Thompson 1928) in the Szczecin Lagoon and the Pomeranian Bay. Przegl. Zool. 5: pp. 36-42. [in Polish].
63. Wildgust M.A., Jones M.B. 1998. Salinity change and the toxicity of the free cadmium ion (Cd(aq)2+) to *Neomysis integer* (Crustacea: Mysidacea). Aquat. Toxicol. 41: 187-192.
64. Winkler G., Greve W. 2002. Laboratory studies of the effect of temperature on growth, moulting and reproduction in the co-occurring mysids *Neomysis integer* and *Praunus flexuosus*. Mar. Ecol. Prog. Ser. 235, 177-188.
65. Wittmann K.J. 1984. Ecophysiology of marsupial development and reproduction in Mysidacea (Crustacea). Oceanogr Mar. Biol. Ann. Rev. 22: 393-428.
66. Węsławski J. M. 1981. Observations of *Neomysis integer* (Thompson) 1928 aggregation formation under natural conditions. Zesz. Nauk. Wydz. Biol. and Nauk o Ziem Uniw. Gdańskiego 8; pp. 109-126. [in Polish].
67. Żmudziński L. 1974. Baltic pollution and its role in the biology and ecology of organisms. [in]: Baltic Sea Ecosystem Productivity. Ossolineum, Wrocław, pp. 275-284. [in Polish].
68. Żmudziński L. 1990. The Baltic Animal World. WSiP, Warszawa: pp. 1-195. [inPolish].

DYNAMICS OF WATER CONSUMPTION CHANGES IN A TOURIST RESORT

Izabela Bartkowska[1]

[1] Chair of Environment Engineering Systems, Faculty of Civil and Environmental Engineering, Bialystok University of Technology, Wiejska 45 A, 15-351 Białystok, Poland, e-mail: izabela.bartkowska@neostrada.pl

ABSTRACT

This work presents problems related to the water consumption in the city of Olecko situated in the north-east part of Poland in the Warmińsko-Mazurskie Province of the country, at the eastern border of the Mazurian Lake District along the trail connecting the land of the Large Mazurian Lakes with the Suwalsko-Augustowskie Lake District. Over 2011–2012 water extraction to the municipal water supply network was studied. The volume of water extracted every day was analyzed and the gathered volumes were analyzed statistically. The varying water extraction was also studied. The obtained results were presented in a graphic form. Basing on the descriptive stats and prepared diagrams certain general conclusions were drawn and the collected study figures and facts were summed up. This allowed to determine days of the highest and lowest water consumption. Also months of extreme water extraction and consumption were determined. The water extraction ranged from 1641 m^3/24h to 2607 m^3/24h, at an average value of 2077.4 m^3/24h. Over the period under study the day of the largest water extraction and consumption was in July and the day of the lowest water extraction and consumption in December. During a week inhabitants used the highest water amount on Saturdays and the lowest on Sundays and other feast-days. Basing on the conducted measurements also the coefficient of water consumption per capita was determined. The fluctuation of this coefficient was identical as that for the water consumption. Within the period of study it ranged from 73.3 l/M 24h to 116.5 l/M 24h. The average value of the specific water consumption was 92.8 l/M 24h. For the sake of discussion the obtained results were compared with observations across the country.

Keywords: indicators, tap water, variation trends, water consumption, water requirement index, water supply system.

INTRODUCTION

Water as a cultural symbol is one of the four elements. Nowadays as a useful substance it has many uses. It is used daily for household purposes, in industry and agriculture. One can hardly imagine life without it. Although our planet is up to 70% covered with water, it does not mean it is generally abundant. Water resources in individual regions are differentiated and the water deficit is pretty often. Our country with its water resource coefficient of 1460 m^3/inhabitant per year is perceived as one, where water deficit occurs and can be compared e.g. to Egypt [Hotloś et al 2012]. For this reason in the near future a more efficient and economic use of water will be preferred. A permanent monitoring both of water resources and its distribution contributes to a better efficiency of its use and distribution.

Ever-accelerating progress of the technology, ever-higher standard of living and people's awareness are reasons that we witness changes to the water supply systems. Water consumption is a very important issue both from a cognitive and practical point of view. These issues are essential particularly for engineering nowadays when the deficit of water resources is growing.

Many scientists point to a new economic situation in Poland since 1989; we witness a drop in water consumption by all groups of water users. Therefore, one can assess that with ubiquitous water resources deficit implementing of the market economy contributed seemingly to savings in this sector. Water consumption coefficients for the economy and population decreased significantly [Biedugnis et al. 2007, Budziło et al. 2005, Hotloś 2010, Kuczyński, Żuchowicki 2010, Podwójci et al. 2011, Żuchowicki, Gawin 2010, Żuchowicki, Kuczyński 2009, Żuchowicki, Telega 2010]. A more rational water management by users caused market-dependent water prices, a general metering of water consumption or the use of more-and-more modern and saving draw-off valves and taps. Important is also a better awareness of a consumer of advantages brought about by a rational water use or using showers, automatic washing machines and dishwashers. A falling trend of water consumption did not always comport with revamping and reconstruction of water supply systems. Water systems, which were designed much earlier, turned out to be significantly over-designed, with this overdesigning causing the flow speed to decrease. This decreased flow speed leads in turn to extensive scale, which causes secondary water contamination, increases the pipeline roughness and decreases the pipline throughput. This, in turn, very often causes an increase in the unit energe-consumption. In addition, water dwells in the system longer, therefore, the risk of its secondary contamination grows.

The analysis of this situation points to the need of a re-verification of earlier forecasts of the demand for water. A proper determination of the variation of the water consumption depending on time and place is essential for determination and calculation of a water supply system. Determination of real hydraulic parameters of the water supply system operation is important both for operational and economic reasons. It is noticed after years of observations that the trend of water consumption drop has weakened [Szopińska, Pasela 2009, Żuchowicki, Gawin 2013, Żuchowicki, Telega 2010]. Therefore, we have to take into account the need to verify the indicators adopted for determination of hydraulic parameters of previously designed and already existing and operated water systems. The determination of these factors shall be based on the volume of water distribution and consumption which after years of a falling trend remain at the same level. The purpose of this paper is to present the water draw-off amount on the example of the town of Olecko, and basing on the analysis of collected data determination of the water draw-off season dependency and observation of changes within the period of the conducted study.

AUTHOR'S STUDY

The study was carried out in Olecko, a town in the north-east part of Poland in the Warmian and Masurian province. The town is situated on the east end of the Masurian Lake District (Pojezierze Mazurskie) on the trail joining the Masurian Great Lake Land (Kraina Wielkich Jezior Mazurskich) with the Suwalsko-Augustowskie Lake District (Pojezierze Suwalsko-Augustowskie). It is a significant recreation and vacation region. The town is both the seat and the capital of the district.

CHARACTERISTIC OF THE STUDY SUBJECT AND THE METHODOLOGY

Between 2007 and 2011 a project including a rational reorganisation of the water & waste water management was realised. Under the project the waste water treatment plant and the water supply network were extended, the waste water treatment plant was modernized and the sewage system was repaired.

The water treatment plant is suplied from 6 deep wells tapping quaternary water form a depth of 45–74 m under the ground level. After extension works the capacity of the water treatment plant is 3200 m^3/24 h. The water treatment process bases on aeration and filtration. The total length of the water supply network is 211.6 km.

The research was carried out in 2011–2012. The 24h water take-up was analyzed. The results were recorded with a MAGFLO – Danfos electromagnetic flowmeter. For the analysis a 24h water supply to the network was assumed. Water is used mainly for household needs of the population. Tapping water from the network by few industrial facilities is low.

RESULTS AND DISCUSSION

Collected study results were subjected to a statistical anlysis. The author was using the program STATISTICA and the calcualtion sheet Excel. Values of selected desription stats are presented in Table 1.

Within the research study the average water supply to the municipal water supply network in Olecko was 20774 m^3/24h. The recorded minimum water supply was down to 26070 m^3/24h on Sunday, December 11. The maximum water

Table 1. Selected 24h and monthly description stats of the water take-up volumes

Distribution units	Water tapping volume	
	24h [m^3/24h]	monthly [m^3/month]
Average value	2077.4	63 376.6
Minimum value	1641.0	54 911.0
Maximum value	2607.0	72 068.0
Median	2068.5	62 542.0
Mode	1967.0	multiple
Variance	32231	25 403 398
Standard deviation	179.5	5040.2
Skewness factor	0.246	0.005
Kurtosis	-0.2	-0.9
Percentile 10%	1850.0	56 485.0
Percentile 90%	2310.0	69 070.0

supply was up to 2607.0 m^3/24h on Saturday, July 21. The median from the test as a consistent and asymptotically unloaded estimator of an expected value amounts to 20685 m^3/24h and it is as well a value below and above which exactly the half of observations is situated. A value of the water supply occuring most often is 1967.0 m^3/24h. The analysis of the 24h water supplies is characterized by a small right-hand asymmetry being confirmed by the skewness factor of 0.246. The distribution is approximately a normal one and this is confirmed by the Shapiro-Wilk test, for which $W = 0.9935$, and $p = 0.0643$. The 24h distrubution of the water supply is presented in a histogram in Figure 1.

Basing on the diagram one can see that most of the recorded values fall within 1800.0 m^3/24h and 2300.0 m^3/24h. Also the average value, the median and the mode occurs inside this range. In order to determine the diversity of the analyzed values the variance and the standard deviation were computed (Table 1). Obtained results allowed to recognize the studied set of values as uniform. The calculated percentiles allow stating that the probability of the occurrence of a water supply lower than 850.0 m^3/24h and higher than 2310.0 m^3/24h is 10%. The character of the water supply variation during the period under study is shown in Figure 2.

The diagram shows clearly the periods of increased water supplies, which result from the recreation and vacation character of the town. This is particularly clear when one considers the monthly water supply. The largest supply volumes occurred in summer months when the weather is warmer and more tourists visit the region. The water consumption in this season is often larger by above 20% than that in the autumn-winter time. It is well illustrated in Figure 3 which presents monthly water consumptions recorded in 2011.

In the period under study the largest water consumption was recorded in June. The day–24h–with the largest water consumption occurred on Saturday in July. Practically a gradual increase in water consumption is noted from January on. The water consumption starts again to decrease after the summer season. The day–24h–when the smallest water consumption occurred was a De-

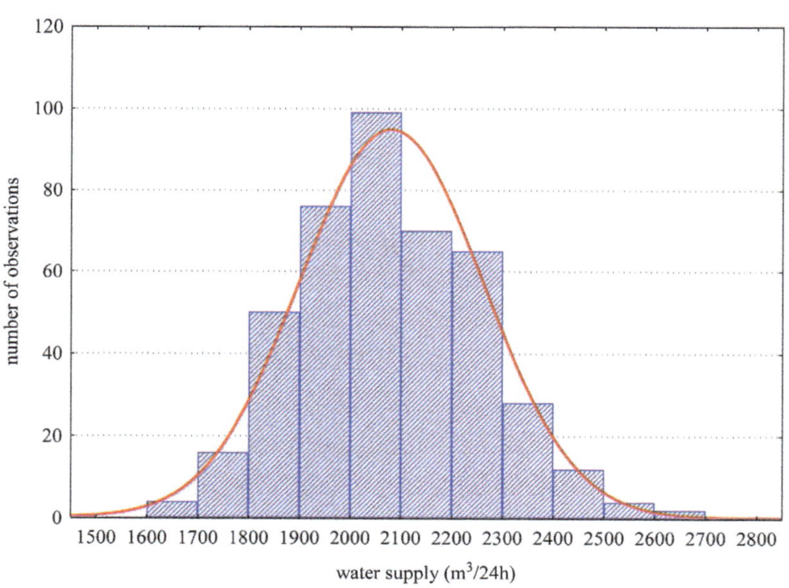

Figure 1. Histogram of 24h water supplies

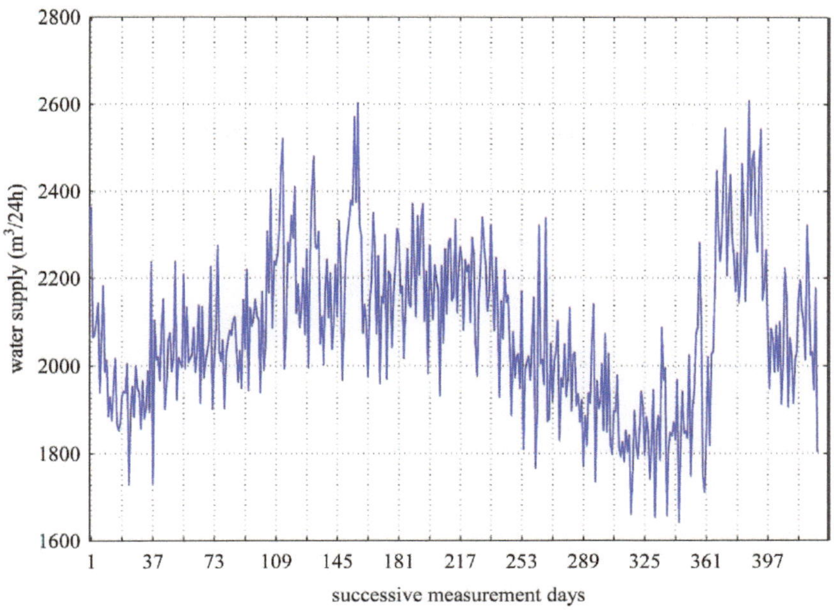

Figure 2. Variation of 24h water supplies

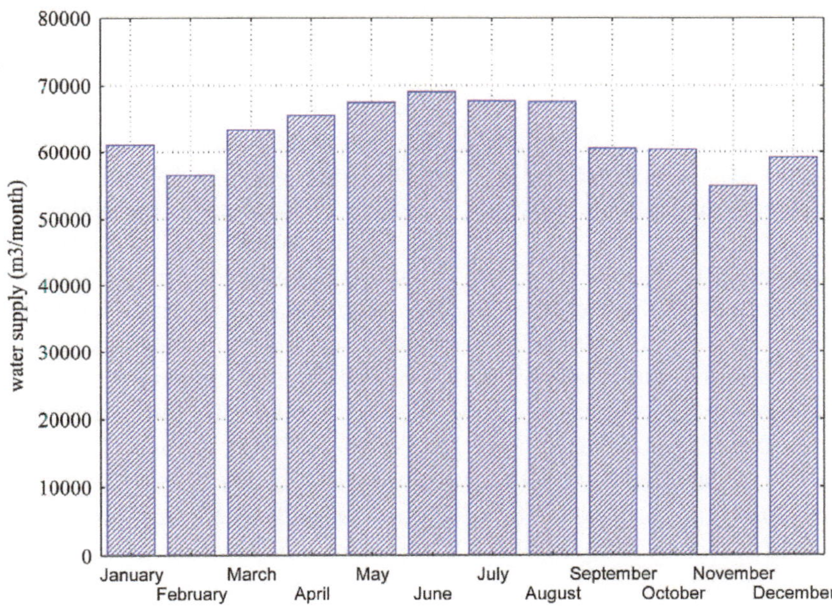

Figure 3. Variation of monthly water consumptions

cember Sunday. The occurrence of this phenomenon was observed also in other tourist and health resorts, such as Krynica [Budziło et al. 2005], Kołobrzeg [Usidus, Filon 2011] or Ustka [Usidus, Litewka 2013]. The season dependence of water consumption are reported as well by other authors and concern not only resorts attractive for tourists [Hotloś et al. 2012, Kuczyński, Żuchowicki 2010, Sawicka-Siarkiewicz, Gmitrzuk 2010, Tchórzewska-Cieślak 2005, Usidus, Drozdowicz 2010, Żuchowicki, Kuczyński 2009]. Apart from the said season of the year dependence there are also days in a week when the water consumption is larger. Figure 4 shows well these trends.

Ranges of water consumption levels in the diagram clearly indicate Sunday as the day of the lowest water consumption. Somewhat less water is used on Mondays, definitely more on Saturdays. On other days of a week (Tuesday to Friday) the water consumption differs a lot and it would be very difficult to find a scheme of these differences. For a summary, conclusions and some generalizations the achieved results were subjected to a statistical analysis taking into account water

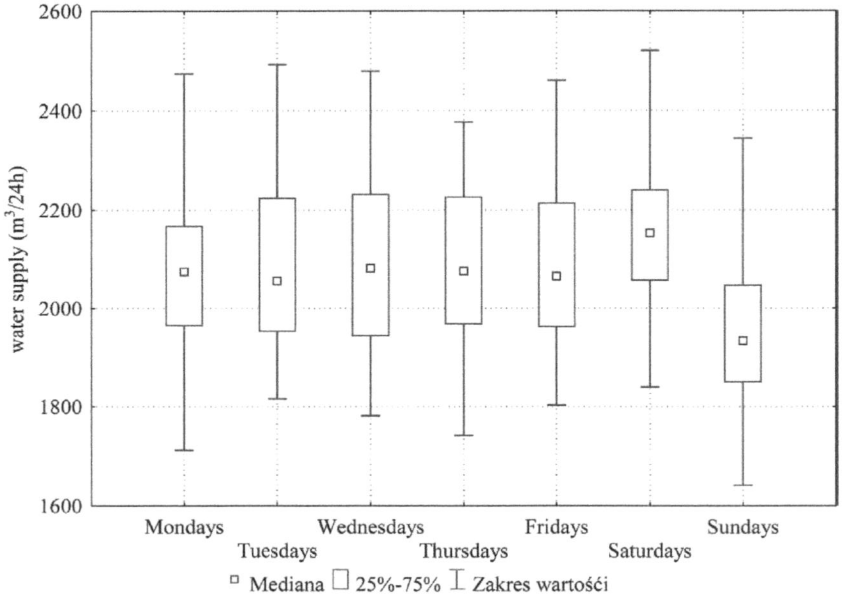

Figure 4. Range of water consumption on individual days of a week

consumption on individual week days. Calculation of the distribution unit values, in particular position measures (arithmetic mean, median and mode), diversity (standard deviation, variance and range), asymmetry (skewness coefficient) and concentration (curtosis) as well as percentiles allows to describe subject statistical data. Results are shown in Table 2.

Analyzing the calculated statistical values we can see that the lowest values of the measure of position occur for the water supply on Sunday. The diversity measures are also the smallest for Sunday recordings. Also the percentile values, which separate the put towards increasingly values of the variable, allow to state that the smallest water supplies fall on Sundays. Similar situation is with the results recorded on Sundays; when in this case the position measures are the largest like the diversity measures and percentiles. Within the period under study the Sunday water supply volumes were always by ten percent lower than those occurring on a Saturday before a given Sunday. For 80% of the analyzed values this difference was 8 to 18%. Many authors point to similar differences in the water consumption on individual week days. According to some publications the water consumption is larger on Saturdays by 5 to 10% [Kuczyński, Żuchowicki 2010]. Other authors report that the water consumption on weekends is higher than on working days on the average by 12% but sometimes even by 24% [Podwójci 2011]. Different water consumptions during a week are stressed by other authors. They notice also that the water consumption is higher on days preceding holidays but obviously lower on holidays [Sawicka-Siarkiewicz, Gmitrzuk 2010,

Table 2. Selected values of descriptive statistics of water consumption on days of a week

Distribution units	Water supply volume 24h [m³/24h]						
	Mondays	Tuesdays	Wednsdays	Thursdays	Fridays	Saturdays	Sundays
Average value	2072.8	2091.4	2088.4	2083.4	2090.8	2170.7	1945.5
Minimum value	1711.0	1816.0	1781.0	1741.0	1803.0	1840.0	1641.0
Maximum value	2602.0	2492.0	2479.0	2377.0	2460.0	2607.0	2374.0
Median	2075.0	2056.0	2082.0	2076.0	2066.0	2152.5	1934.0
Variance	32869	31217	28984	26695	28156	27946	25418
Standard deviation	181.3	176.7	170.2	163.4	167.8	167.2	159.4
Skewness factor	0.42	0.36	0.04	0.04	0.34	0.71	0.24
Kurtosis	0.4	-0.7	-0.8	-0.9	-0.5	0.5	0.1
Percentile 10%	1850.0	1879.0	1864.5	1872.0	1861.5	1967.0	1735.5
Percentile 90%	2273.0	2321.0	2307.0	2310.0	2292.0	2403.0	2141.0

Tchórzewska-Cieślak 2005, Usidus, Filon 2011, Usidus, Litewka 2013]. Schemes we watch do not depend on the size of a city or its character; they are first of all the result of the behaviour and manners of their inhabitants.

The 24h water supply volumes under study are characterized by a non-accidental order; successive values in the set of analyzed values represent successive measurements made at equal time intervals. In order to examine the phenomenon represented by the observation sequence we can use the analysis of time series [Ciężak et al 2011]. Thanks to this analysis the dynamics of the water supply within the study period will be better recognized.

First of all, the water supply time series with regard to the trend occurence was verified. Linear, logrythmical, second-order polynomial, power and exponential trends were analyzed. The model matching quality coefficient of the R^2 model was successively 0.080, 0.005, 0.416, 0.006 and 0,087. That means that the best of the analyzed models a second-order trend explains the variation Y only up to 41.6%. None of the models matched the actual data of the water take-up amount to the expected at least 60%. Figure 5 presents a diagram of a best-matched trend.

The subject of the analysis was also the search for the seasonality of an analyzed series. A dynamic econometric model with 0 et 1 variables – i.e. assuming for variables 1 or 0 for respective days of a week – was used for the study. The matching of the model turned out to be weak, because it explains the water take-up only to 21.7%, thus disqualifying it as a forecasting tool. The discussed seasonality diagram is shown in Figure 6.

In another analysis the transformed time series was verified with a Holt linear model. The Holt method (MH) is useful to smooth time series where accidental fluctuations and a development

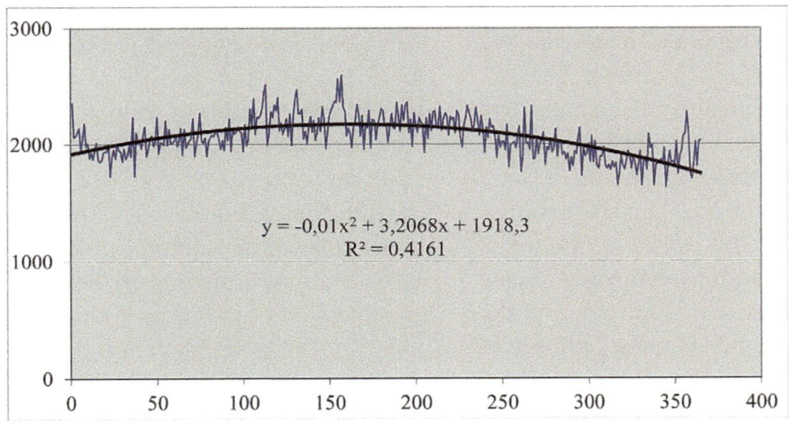

Figure 5. 24h- water supply volume along with the trend line

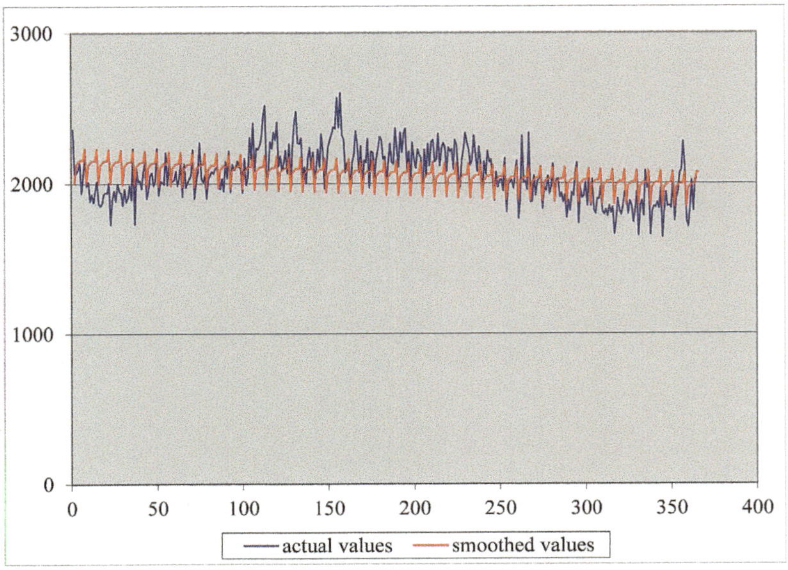

Figure 6. Results of the seasonality of studied time series

trend occurs. For construction of the model initial values of F_1 i S_1 were adopted. As F_1 the value of the first variable of the y_1 series and as S_1 the difference $y_2 - y_1$ were adopted. The values of the α and β parameters were selected minimazing the values of the mean squared error of extinct forecasts (α = 0.367956, β = 0.247053). A simulation of parameter computations describing the measure of matching the model to empirical data was carried out. The results are shown in Table 3.

Table 3. Results of correctness of matching model to actual series

Parameter name (unit)	Calculated value
Mean value of explained variable	2058.225
Mean value of forecast relative errors – MAPE (%)	5.328
Mean square of extinct forecasts	21 166.721
Mean root of mean forecast error (m³/24h)	145.490
Mean relative error of extinct forecasts (%)	7.070

The mean value of a variable estimated basing on the variable model is 2058.225 m³/24h. The values of the explained variable differ from empirical values by ± 145.49 m³/24h. The average value of forecast errors for the analyzed period is 5.328% of the actual values of variables. The standard deviation of the rest of a model, variable or a set is 7.07% of the explained variable mean value. Since the mean value of forecast relative errors and the mean relative error of extinct forecasts are lower than 10% we can allow using the Holt method for forecasting an analyzed time series. Figure 7 presents an initial and smoothed time series.

Basing on the analyzed data and the number of population the indicator of water consumption per one resident was calculated. The unit water consumption in Olecko ranged between 73.3 l/inhabitant per 24h to 116.5 l/inhabitant per 24h. Fluctuations of this factor were of course similar to the water supply variation during a year or week. The mean value was 92.8 l/inhabitant per 24h. During the period of study 80% of the values fell within a range between 82.7 to 103.2 l/inhabitant per 24h. These levels are much lower than those prescribed in the Ordinance of the Ministry of Infrastructure of January 14, 2002 on the average water consumption amount of 166 l/inhabitant per 24h. Lower water consumption levels were also achieved in a touristic community of a similar size where studies have shown a decrease in the unit water take-up in 2010 to a mean level of 94 l/inhabitant per 24h [Usidus, Litewka 2013]. According to other studies water consumption by one resident of a housing association in 2003 was within a range of 56.6–110.2 l/inhabitant per 24h [Mikołajczyk 2012, Tchórzewska-Cieślak 2005]. Also after a 6-year study of the multi-family housebuilding industry a stabilization of the unit water consumption was found out and in 2005 this unit was at a level of 98.02 l/inhabitant per 24h [Pasela, Gorączko 2013]. The analysis of water distribution among residents of a housing association of multi-family houses resulted in the average water take-up in 2006 of 115 l/person per 24h, 112 l/ 24h in 2007, and in 2008 already 95 l/inhabitant 24h [Żuchowicki, Kuczyński 2009]. In the same town within other housing estate in the same years higher levels of the 24h water con-

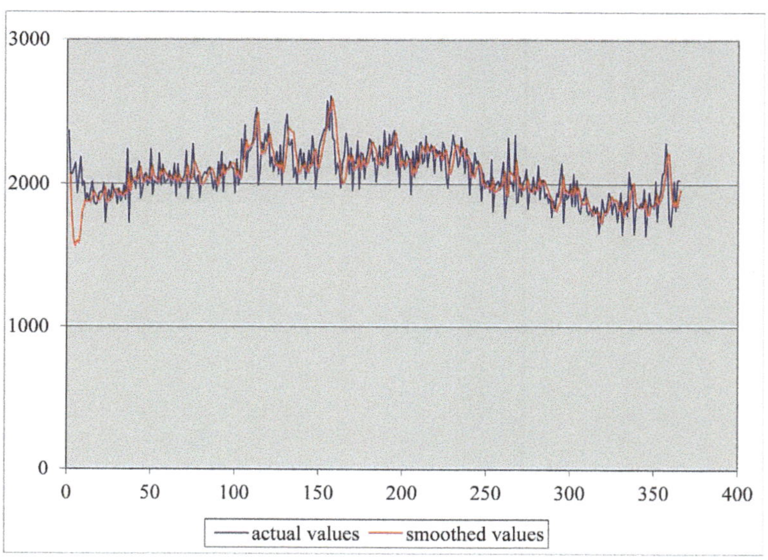

Figure 7. Diagram of a time series showing actual values and values matched with the MH line

sumption were achieved, respectively 137.6 l/inhabitant per 24h, 114.4 l/inhabitant per 24h and 115.6 l/inhabitant per 24h [Żuchowicki, Gawin 2010]. Observations conducted between 2000 and 2005 in Piła revealed water consumption in the first year of research 18 l/inhabitant per 24h and next it was down to 103 l/inhabitant per 24h in 2005 [Żuchowicki, Telga 2010]. Similarly, analyses carried out in larger cities also in housing associations between 2002 and 2006 resulted in determination of the unit water consumption respectively 98 l/inhabitant per 24h in Plock and 151 l/inhabitant per day in Warsaw [Biedugnis et al 2007]. Measurements within the area of single-family houses have shown the unit water consumption of 103 l/inhabitant per 24h in 2010 and 95 l/inhabitant per 24h in 2011 [Kępa et al 2013]. In another town in a settlement of a similar character the mean water consumption was between 2009 and 2010 at a level of 112 l/inhabitant per 24h [Żuchowicki, Gawin 2013]. Studies conducted between 1990 and 2008 within the Lower Silesia, Lubuskie and Opolskie Provinces revealed that after 2005 the falling trend of the unit water consumption stopped and stabilized at a level of 86÷112 l/inhabitant per 24h [Hotloś 2010]. This parameter was researched also for country areas, where it was determined at a level of 80–90 l/inhabitant per 24h, excepting commune administration seat, where it remained at 30–40 l/inhabitant per 24h [Mikołajczyk 2012, Usidus, Drozdowicz 2010]. Concluding, water consumption in Poland for cities is on the average 101 l/inhabitant per 24h, but it changes to a minimum level of 85 l/inhabitant per 24h in the podkarpackie and podlaskie province and to a maximum level of 120 l/inhabitant per 24h in the mazowieckie province [Kuczyński, Żuchowicki 2010, Sawicka-Siarkiewicz, Gmitrzuk 2010]. In rural areas it was on the average 67 l/inhabitant per 24h and ranges between 39 l/inhabitant per 24h in the Malopolska Region to 96 l/inhabitant per 24h in the Wielkopolska Region.

CONCLUSIONS

Due to contemporary engineering, technology and economy we can design, construct and operate reliable water supply networks, taking into account their size, local environmental, civilization and cultural conditions. Studies provide tools required for forecasting water demands and water distributions from water supply networks, planning best-structured water supply networks and devices, controlling the network operation, programming modernization and extension of these systems. Therefore, permanent monitoring of the water take-up and consumption by various users in the entire country is very important. Consumers' real water needs can be forecast basing on the analysis of water consumption. With no such data available determination of existing specific conditions and rationalization of new technical solutions is not feasible.

The studies we have carried out allowed to gather data on water consumption in the town of Olecko. These studies have conformed the seasonality resulting from the town character, season of the year or the atmospheric temperature. They proved also the occurrence of water consumption variation during a week resulting from habits of the population. The presented measurement results can be useful as a comparative material for a study of water take-up in weak industrialized towns, i.e. of a recreation and vacation character of a similar land development and population.

One should consider as well whether in the present situation the analysis of water consumption indicator values, which are recommended and used for designing, would not be required. Many authors indicate that after 2005 the water consumption decreasing trend stopped. Maybe it's high time to start a next change of the water consumption consumption factors, to give up indicated, obsolete standards of apartment furnishing as the basis of the selection of water consumption and basing on studies conducted at technical universities to suggest new, better solutions.

REFERENCES

1. Biedugnis S., Smolarkiewicz M., Podwójci P., Serejko G. 2007. Analiza zużycia wody w zabudowie wielorodzinnej na przykładzie wybranych spółdzielni Mazowsza. Zeszyty Naukowe Wydziału Budownictwa i Inżynierii Środowiska, nr 23, 245–255.
2. Budziło B., Bajer J., Kraus K. 2005. Analiza zaopatrzenia w wodę miasta Krynica. Monografie Komitetu Inżynierii Środowiska PAN, vol. 30, 35–50.
3. Ciężak W., Łomotowski J., Siwoń Z., Licznar P., Ciężak J. 2011. Modele neuronowe w analizie i prognozowaniu rozbiorów wody, Instal, nr 7–8, 61–64.

4. Hotloś H. 2010. Badania zmian poboru wody w wybranych miastach Polski w latach 1990–2008. Ochrona Środowiska, vol. 32, 3, 39–42.
5. Hotloś H., Głowacka J., Kołodziej A. 2012. Zmienność poboru wody w systemie wodociągowym Wrocławia, Ochrona Środowiska, 34(4), 23–28.
6. Kępa U., Stępniak L., Stańczyk-Mazanek E. 2013. Analiza zużycia wody i zmienność rozbiorów dla obszaru zasilania Kawie Góry na terenie miasta Częstochowy. Rocznik Ochrona Środowiska. Tom 15, 2546–2562.
7. Kuczyński W., Żuchowicki W. 2010. Ocena aktualnej sytuacji w zaopatrzeniu w wodę w Polsce na tle sytuacji w świecie. Rocznik Ochrona Środowiska. Tom 12, 419–465.
8. Mikołajczyk M. 2012. Analiza jednostkowych wskaźników zużycia wody w gospodarstwach domowych w warunkach wiejskich i miejskich, Instal, nr 2, 47–50.
9. Pasela R., Gorączko M. 2013. Analiza wybranych czynników kształtujących zużycie wody w budynkach wielorodzinnych. Rocznik Ochrona Środowiska. Tom 15, 1658–1672.
10. Podwójci P. 2011. Nierównomierność zużycia oraz rozbioru wody w budownictwie wielorodzinnym. Inżynieria Ekologiczna, Nr 26, 281–289.
11. Podwójci P., Kozłowski M., Krysiuk M. 2011. Zużycie wody w budownictwie wielorodzinnym – wybrane zagadnienia. Rocznik Ochrona Środowiska. Tom 13, 1653–1666.
12. Sawicka-Siarkiewicz H., Gmitrzuk N. 2010. Kształtowanie się jednostkowych wskaźników zużycia wody na terenach osiedli w grupach miast o liczbie mieszkańców od 50001 do 500000, Ochrona Środowiska i Zasobów Naturalnych, Nr 45, 63–82.
13. Szopińska K., Pasela R. 2009. Struktura zużycia wody w budownictwie wielorodzinnym, Instal, nr 4, 51–55.
14. Tchórzewska-Cieślak B. 2005. Analiza zużycia wody do spożycia przez mieszkańców miasta Jasła. Monografie Komitetu Inżynierii Środowiska PAN, vol. 30, 65–74.
15. Usidus D., Drozdowicz A. 2010. Analiza rozbiorów wody na wybranym terenie Pomorza Środkowego – w gminie Sianów. Rocznik Ochrona Środowiska. Tom 12, 543–558.
16. Usidus D., Filon A. 2011. Analiza zmienności rozbioru wody w miejscowości wypoczynkowej pasa nadmorskiego. Rocznik Ochrona Środowiska. Tom 13, 903–920.
17. Usidus D., Litewka A. 2013. Rozbiory wody w miejscowości Ustka. Rocznik Ochrona Środowiska. Tom 15, 1070–1085.
18. Żuchowicki A., Gawin R. 2010. Pobory wody przez mieszkańców budynku wielorodzinnego. Rocznik Ochrona Środowiska. Tom 12, 479–488.
19. Żuchowicki A., Gawin R. 2013. Struktura zużycia wody w budynkach jednorodzinnych. Rocznik Ochrona Środowiska. Tom 15, 924–929.
20. Żuchowicki A., Kuczyński W. 2009. Analiza porównawcza zmian w rozbiorach wody z uwzględnieniem sposobu jej dostarczania do odbiorców. Rocznik Ochrona Środowiska. Tom 11, 781–786.
21. Żuchowicki A., Telega M. 2010. Analiza produkcji i zużycia wody w Pile. Rocznik Ochrona Środowiska. Tom 12, 313–324.

Permissions

All chapters in this book were first published in JEE, by Polish Society of Ecological Engineering (PTIE); hereby published with permission under the Creative Commons Attribution License or equivalent. Every chapter published in this book has been scrutinized by our experts. Their significance has been extensively debated. The topics covered herein carry significant findings which will fuel the growth of the discipline. They may even be implemented as practical applications or may be referred to as a beginning point for another development.

The contributors of this book come from diverse backgrounds, making this book a truly international effort. This book will bring forth new frontiers with its revolutionizing research information and detailed analysis of the nascent developments around the world.

We would like to thank all the contributing authors for lending their expertise to make the book truly unique. They have played a crucial role in the development of this book. Without their invaluable contributions this book wouldn't have been possible. They have made vital efforts to compile up to date information on the varied aspects of this subject to make this book a valuable addition to the collection of many professionals and students.

This book was conceptualized with the vision of imparting up-to-date information and advanced data in this field. To ensure the same, a matchless editorial board was set up. Every individual on the board went through rigorous rounds of assessment to prove their worth. After which they invested a large part of their time researching and compiling the most relevant data for our readers.

The editorial board has been involved in producing this book since its inception. They have spent rigorous hours researching and exploring the diverse topics which have resulted in the successful publishing of this book. They have passed on their knowledge of decades through this book. To expedite this challenging task, the publisher supported the team at every step. A small team of assistant editors was also appointed to further simplify the editing procedure and attain best results for the readers.

Apart from the editorial board, the designing team has also invested a significant amount of their time in understanding the subject and creating the most relevant covers. They scrutinized every image to scout for the most suitable representation of the subject and create an appropriate cover for the book.

The publishing team has been an ardent support to the editorial, designing and production team. Their endless efforts to recruit the best for this project, has resulted in the accomplishment of this book. They are a veteran in the field of academics and their pool of knowledge is as vast as their experience in printing. Their expertise and guidance has proved useful at every step. Their uncompromising quality standards have made this book an exceptional effort. Their encouragement from time to time has been an inspiration for everyone.

The publisher and the editorial board hope that this book will prove to be a valuable piece of knowledge for researchers, students, practitioners and scholars across the globe.

List of Contributors

Allahbakhh Javid
School of Public Health, Shahroud University of Medical Sciences, Shahroud, Iran

Kamyar Yaghmaeian
School of Public Health, Tehran University of Medical Sciences, Tehran, Iran

Ebrahim Abbasi
School of Public Health, Shahroud University of Medical Sciences, Shahroud, Iran

Aliakbar Roudbari
Department of Environmental Health Engineering, School of Public Health, Shahroud University of Medical Sciences, Shahroud, Iran
School of Public Health, Shahroud University of Medical Sciences, Shahroud, Iran

Edyta Kosterna
Department of Vegetables Crop, Siedlce University of Natural Sciences and Humanities, B. Prusa 14, 08-110 Siedlce, Poland

Agnieszka Montusiewicz
Faculty of Environmental Engineering, Lublin University of Technology, Nadbystrzycka 40B, 20-618 Lublin, Poland

Magdalena Vaverková
Department of Applied and Landscape Ecology, Faculty of Agronomy, Mendel University in Brno, Zemědělská 1, 613 00 Brno, Czech Republic

Dana Adamcová
Department of Applied and Landscape Ecology, Faculty of Agronomy, Mendel University in Brno, Zemědělská 1, 613 00 Brno, Czech Republic

Jan Zloch
Department of Applied and Landscape Ecology, Faculty of Agronomy, Mendel University in Brno, Zemědělská 1, 613 00 Brno, Czech Republic

Urszula Wydro
Division of Sanitary Biology and Biotechnology, Bialystok University of Technology, Wiejska 45E, 15-351 Białystok. Poland

Elżbieta Wołejko
Division of Sanitary Biology and Biotechnology, Bialystok University of Technology, Wiejska 45E, 15-351 Białystok. Poland

Tadeusz Łoboda
Division of Sanitary Biology and Biotechnology, Bialystok University of Technology, Wiejska 45E, 15-351 Białystok. Poland

Marzena Matejczyk
Division of Sanitary Biology and Biotechnology, Bialystok University of Technology, Wiejska 45E, 15-351 Białystok. Poland

Andrzej Butarewicz
Division of Sanitary Biology and Biotechnology, Bialystok University of Technology, Wiejska 45E, 15-351 Białystok. Poland

Mohammadreza Azizzadeh
Department of Geography, Payam Noor University, PO Box 3697-19395, Tehran, Iran

Khadije Javan
Department of Geography, Payam Noor University, PO Box 3697-19395, Tehran, Iran

Jacek Sosnowski
Department of Grassland and Green Areas Creation, Siedlce University of Natural Sciences and Humanities, B. Prusa 14. 08-110 Siedlce, Poland

Beata Wiśniewska-Kadżajan
Department of Grassland and Green Areas Creation, Siedlce University of Natural Sciences and Humanities, B. Prusa 14. 08-110 Siedlce, Poland

Beata Wiśniewska-Kadźajan
Department of Grassland and Green Areas Creation, Siedlce University of Natural Sciences and Humanities, B. Prusa 14, 08-110 Siedlce, Poland

Kazimierz Jankowski
Department of Grassland and Green Areas Creation, Siedlce University of Natural Sciences and Humanities, B. Prusa 14, 08-110 Siedlce, Poland

Elżbieta Malinowska
Department of Grassland and Green Areas Creation, Siedlce University of Natural Sciences and Humanities, B. Prusa 14, 08-110 Siedlce, Poland

Jacek Sosnowski
Department of Grassland and Green Areas Creation, Siedlce University of Natural Sciences and Humanities, B. Prusa 14, 08-110 Siedlce, Poland

Roman Kolczarek
Department of Grassland and Green Areas Creation, Siedlce University of Natural Sciences and Humanities, B. Prusa 14, 08-110 Siedlce, Poland

Elżbieta Radzka
Department of Agrometeorology and Land Reclamation, University of Natural Sciences and Humanities, B. Prusa 14, 08-110 Siedlce, Poland

Krystyna Zarzecka
Chair of Plant Cultivation, University of Natural Sciences and Humanities in Siedlce, B. Prusa 14, 08-110 Siedlce, Poland

Marek Gugała
Chair of Plant Cultivation, University of Natural Sciences and Humanities in Siedlce, B. Prusa 14, 08-110 Siedlce, Poland

Alicja Baranowska
Pope John Paul II State School of Higher Education in Biala Podlaska, Sidorska 95/97, 21-500 Biala Podlaska, Poland

Elżbiata Radzka
University of Natural Sciences and Humanities in Siedlce, Prusa 14, 08-110 Siedlce, Poland

Katarzyna Rymuza
University of Natural Sciences and Humanities in Siedlce, Prusa 14, 08-110 Siedlce, Poland

Tomasz Lenartowicz
The Research Centre for Cultivar Testing in Słupia Wielka, 63-022 Słupia Wielka, Poland

Agnieszka Parzych
Environmental Chemistry Department, Institute of Biology and Environmental Protection, Pomeranian University, Arciszewskiego 22b St., 76-200 Slupsk, Poland

Małgorzata Cymer
Department of Land Reclamation and Environmental Management, University of Warmia - Mazury, Plac Łódzki 2, 10-719 Olsztyn, Poland

Teresa Noga
Department of Biological Foundations of Agriculture and Environmental Education, Faculty of Biology and Agriculture, University of Rzeszów, Ćwiklińskiej 2, 35-601 Rzeszów, Poland

Natalia Kochman
Department of Biological Foundations of Agriculture and Environmental Education, Faculty of Biology and Agriculture, University of Rzeszów, Ćwiklińskiej 2, 35-601 Rzeszów, Poland

Łukasz Peszek
International Doctoral Studies in Natural Sciences, Faculty of Biology and Agriculture, University of Rzeszów, Ćwiklińskiej 2, 35-601 Rzeszów, Poland

Jadwiga Stanek-Tarkowska
Department of Soil Studies, Environmental Chemistry and Hydrology, Faculty of Biology and Agriculture, University of Rzeszów, Ćwiklińskiej 2, 35-601 Rzeszów, Poland

Anita Pajączek
Department of Soil Studies, Environmental Chemistry and Hydrology, Faculty of Biology and Agriculture, University of Rzeszów, Ćwiklińskiej 2, 35-601 Rzeszów, Poland

Małgorzata Śliwka
Department of Environmental Engineering and Mineral Processing, Faculty of Mining and Geoengineering, AGH University of Science and Technology in Krakow, 30 Al. Mickiewicza Ave., 30-059 Krakow, Poland

Małgorzata Raczyńska
Department of Marine Ecology and Environmental Protection, Faculty of Food Sciences and Fisheries, West Pomeranian Technological University in Szczecin, Kazimierza Królewicza 4, 71-550 Szczecin, Poland

Juliusz C. Chojnacki
Department of Marine Ecology and Environmental Protection, Faculty of Food Sciences and Fisheries, West Pomeranian Technological University in Szczecin, Kazimierza Królewicza 4, 71-550 Szczecin, Poland

Monika Hałupka
Department of Marine Ecology and Environmental Protection, Faculty of Food Sciences and Fisheries, West Pomeranian Technological University in Szczecin, Kazimierza Królewicza 4, 71-550 Szczecin, Poland

Elzbieta Radzka
Department of Agrometeorology and Land Reclamation, University of Natural Sciences and Humanities, B. Prusa 14, 08-110 Siedlce, Poland

Izabela Bartkowska
Chair of Environment Engineering Systems, Technical University in Białystok, Wiejska 45A, 15-351 Białystok, Poland

Beata Wiśniewska-Kadżajan
Department of Grassland and Green Areas Creation, Siedlce University of Natural Sciences and Humanities, B. Prusa 14, 08-110 Siedlce, Poland

Teresa Noga
Department of Biological Foundations of Agriculture and Environmental Education, Faculty of Biology and Agriculture, University of Rzeszów, Ćwiklińskiej 2, 35-601 Rzeszów, Poland

Jadwiga Stanek-Tarkowska
Department of Soil Studies, Environmental Chemistry and Hydrology, Faculty of Biology and Agriculture, University of Rzeszów, Ćwiklińskiej 2, 35-601 Rzeszów, Poland

Anita Pajączek
Department of Soil Studies, Environmental Chemistry and Hydrology, Faculty of Biology and Agriculture, University of Rzeszów, Ćwiklińskiej 2, 35-601 Rzeszów, Poland

Łukasz Peszek
International Doctoral Studies in Natural Sciences, Faculty of Biology and Agriculture, University of Rzeszów, Ćwiklińskiej 2, 35-601 Rzeszów, Poland

Natalia Kochman
International Doctoral Studies in Natural Sciences, Faculty of Biology and Agriculture, University of Rzeszów, Ćwiklińskiej 2, 35-601 Rzeszów, Poland

Katarzyna Rymuza
Department of Quantitative Methods and Spatial Management, University of Natural Sciences and Humanities in Siedlce, Prusa 14, 08-110 Siedlce, Poland

Elżbieta Wołejko
Division of Sanitary Biology and Biotechnology, Faculty of Civil and Environmental Engineering, Bialystok University of Technology, 45d Wiejska Str., 15-351 Białystok, Poland

Andrzej Butarewicz
Division of Sanitary Biology and Biotechnology, Faculty of Civil and Environmental Engineering, Bialystok University of Technology, 45d Wiejska Str., 15-351 Białystok, Poland

Urszula Wydro
Division of Sanitary Biology and Biotechnology, Faculty of Civil and Environmental Engineering, Bialystok University of Technology, 45d Wiejska Str., 15-351 Białystok, Poland

Tadeusz Łoboda
Division of Sanitary Biology and Biotechnology, Faculty of Civil and Environmental Engineering, Bialystok University of Technology, 45d Wiejska Str., 15-351 Białystok, Poland

Leszek Pływaczyk
Institute of Environmental Development and Protection, Wrocław University of Environmental and Life Sciences, pl. Grunwaldzki 24, 50-363 Wrocław, Poland

Beata Olszewska
Institute of Environmental Development and Protection, Wrocław University of Environmental and Life Sciences, pl. Grunwaldzki 24, 50-363 Wrocław, Poland

Agnieszka Zawadzińska
Department of Horticulture, West Pomeranian University of Technology in Szczecin, Papieża Pawła VI 3, 71-459 Szczecin, Poland

Piotr Salachna
Department of Horticulture, West Pomeranian University of Technology in Szczecin, Papieża Pawła VI 3, 71-459 Szczecin, Poland

Mirosław Skorbiłowicz
Technical University in Bialystok, Wiejska 45A, 15-351 Białystok, Poland

Piotr Ofman
Technical University in Bialystok, Wiejska 45A, 15-351 Białystok, Poland

Sławomir Stankowski
Department of Agronomy, West Pomeranian Technological University in Szczecin, Papieża Pawła VI 3, 71-459 Szczecin, Poland

Bożena Michalska
Department of Meteorology and Green Areas Management, West Pomeranian Technological University in Szczecin, Papieża Pawła VI 3, 71-459 Szczecin, Poland

Janusz Smagacz
Institute of Soil Science and Plant Cultivation, State Research Institute, Czartoryskich 8, 24-100 Puławy, Poland

Marzena Gibczyńska
Department of General and Ecological Chemistry, West Pomeranian Technological University in zczecin, 71-434 Szczecin, Słowackiego 17, Poland

Piotr Salachna
Department of Horticulture, Faculty of Environmental Management and Agriculture, West Pomeranian University of Technology in Szczecin, Papieża Pawła VI 3, 71-459 Szczecin, Poland

Agnieszka Zawadzińska
Department of Horticulture, Faculty of Environmental Management and Agriculture, West Pomeranian University of Technology in Szczecin, Papieża Pawła VI 3, 71-459 Szczecin, Poland

Bożena Grabińska
Department of Land Improvement and Environmental Management, University of Warmia and Mazury in Olsztyn, Pl. Łódzki 2, 10-719 Olsztyn, Poland

Józef Koc
Department of Land Improvement and Environmental Management, University of Warmia and Mazury in Olsztyn, Pl. Łódzki 2, 10-719 Olsztyn, Poland

Ireneusz Cymes
Department of Land Improvement and Environmental Management, University of Warmia and Mazury in Olsztyn, Pl. Łódzki 2, 10-719 Olsztyn, Poland

Sławomir Szymczyk
Department of Land Improvement and Environmental Management, University of Warmia and Mazury in Olsztyn, Pl. Łódzki 2, 10-719 Olsztyn, Poland

Beata Wiśniewska-Kadżajan
Department of Grassland and Landscape Architecture, Siedlce University of Natural Sciences and Humanities, B. Prusa 14. 08-110 Siedlce, Poland

Szymon Szewrański
Department of Spatial Economy, Wroclaw University of Environmental and Life Sciences, Grunwaldzka 53, 50-357 Wrocław, Poland

Jan Kazak
Department of Spatial Economy, Wroclaw University of Environmental and Life Sciences, Grunwaldzka 53, 50-357 Wrocław, Poland

Marta Szkaradkiewicz
Department of Spatial Economy, Wroclaw University of Environmental and Life Sciences, Grunwaldzka 53, 50-357 Wrocław, Poland

Józef Sasik
Department of Spatial Economy, Wroclaw University of Environmental and Life Sciences, Grunwaldzka 53, 50-357 Wrocław, Poland

Roman Kolczarek
Institute of Agronomy, Faculty of Natural Sciences, Siedlce University of Natural Sciences and Humanities, B. Prusa 14, 08-110 Siedlce, Poland

Kazimierz Jankowski
Institute of Agronomy, Faculty of Natural Sciences, Siedlce University of Natural Sciences and Humanities, B. Prusa 14, 08-110 Siedlce, Poland

Małgorzata Raczyńska
Department of Marine Ecology and Environmental Protection, Faculty of Food Sciences and Fisheries, West Pomeranian Technological University in Szczecin, Kazimierza Królewicza 4, 71-550 Szczecin, Poland

Juliusz C. Chojnacki
Department of Marine Ecology and Environmental Protection, Faculty of Food Sciences and Fisheries, West Pomeranian Technological University in Szczecin, Kazimierza Królewicza 4, 71-550 Szczecin, Poland

Monika Hałupka
Department of Marine Ecology and Environmental Protection, Faculty of Food Sciences and Fisheries, West Pomeranian Technological University in Szczecin, Kazimierza Królewicza 4, 71-550 Szczecin, Poland

Agnieszka Policht-Latawiec
Department of Land Reclamation and Environmental Development, University of Agriculture in Krakow, Mickiewicza 24-28, 30-059 Krakow, Poland

Włodzimierz Kanownik
Department of Land Reclamation and Environmental Development, University of Agriculture in Krakow, Mickiewicza 24-28, 30-059 Krakow, Poland

Ksymena Grubka
Graduate student of the University of Agriculture in Krakow, Mickiewicza 24-28, 30-059 Krakow, Poland

Teresa Bowszys
Department of Agricultural Chemistry and Environmental Protection, University of Warmia and Mazury in Olsztyn, Oczapowskiego 8, 10-744 Olsztyn, Poland

Jadwiga Wierzbowska
Department of Agricultural Chemistry and Environmental Protection, University of Warmia and Mazury in Olsztyn, Oczapowskiego 8, 10-744 Olsztyn, Poland

Paweł Sternik
Department of Agricultural Chemistry and Environmental Protection, University of Warmia and Mazury in Olsztyn, Oczapowskiego 8, 10-744 Olsztyn, Poland

Marta Karolina Busse
Department of Agricultural Chemistry and Environmental Protection, University of Warmia and Mazury in Olsztyn, Oczapowskiego 8, 10-744 Olsztyn, Poland

Krystyna Zarzecka
Chair of Plant Cultivation, University of Natural Sciences and Humanities in Siedlce, B. Prusa 14, 08-110 Siedlce, Poland

Marek Gugała
Chair of Plant Cultivation, University of Natural Sciences and Humanities in Siedlce, B. Prusa 14, 08-110 Siedlce, Poland

Iwona Mystkowska
Chair of Plant Cultivation, University of Natural Sciences and Humanities in Siedlce, B. Prusa 14, 08-110 Siedlce, Poland

Magdalena Zarzecka
Medical University of Białystok, J. Kilinskiego 1, 15-089 Białystok, Poland

Anna Grzeszczyk-Kowalska
Department of Marine Ecology and Environmental Protection, Faculty of Food Sciences and Fisheries West Pomeranian University of Technology Szczecin, Kazimierza Królewicza str. 4, 71-550 Szczecin, Poland

Juliusz C. Chojnacki
Department of Marine Ecology and Environmental Protection, Faculty of Food Sciences and Fisheries West Pomeranian University of Technology Szczecin, Kazimierza Królewicza str. 4, 71-550 Szczecin, Poland

Małgorzata Raczyńska
Department of Marine Ecology and Environmental Protection, Faculty of Food Sciences and Fisheries West Pomeranian University of Technology Szczecin, Kazimierza Królewicza str. 4, 71-550 Szczecin, Poland

Mariusz Raczyński
Department of Fisheries Management, Faculty of Food Sciences and Fisheries West Pomeranian University of Technology Szczecin, Kazimierza Królewicza str. 4, 71-550 Szczecin, Poland

Izabela Bartkowska
Chair of Environment Engineering Systems, Faculty of Civil and Environmental Engineering, Bialystok University of Technology, Wiejska 45 A, 15-351 Białystok, Poland